THE
HAND

THE
HAND

Lee Milford, M.D.

Clinical Professor of Orthopaedic Surgery,
University of Tennessee Center for the Health Sciences;
Active Staff, Campbell Clinic, Memphis, Tennessee;
Chief of Staff Emeritus, Campbell Clinic;
Past President, Baptist Memorial Hospital Medical Staff;
Sterling Bunnell Lectureship, 1974;
Past President, American Society for Surgery of the Hand

with a contribution by
Phillip E. Wright II, M.D.

From the seventh edition of
CAMPBELL'S
OPERATIVE ORTHOPAEDICS
Edited by
A.H. Crenshaw, M.D.

Third edition
with 1130 illustrations and 4 color plates

The C. V. Mosby Company
ST. LOUIS • WASHINGTON, D.C. • TORONTO   1988

MOSBY

A TRADITION OF PUBLISHING EXCELLENCE

Acquisition editor: Eugenia A. Klein
Developmental Editor: Kathryn H. Falk
Project editor: Teri Merchant
Editing and production: Robert A. Kelly, Mary G. Stueck,
Suzanne C. Glazer
Design: John Rokusek

Third edition

Previous editions copyrighted 1971, 1982

International Standard Book Number 0-8018-3345-1

From the seventh edition of CAMPBELL'S OPERATIVE ORTHOPAEDICS,
copyright © 1987 by The C. V. Mosby Company

Printed in the United States of America

The C. V. Mosby Company
11830 Westline Industrial Drive, St. Louis, Missouri 63146

C/MV/MV   9 8 7 6 5 4 3 2 1

Preface

This book was first published in 1971 as a treatise on general hand surgery; it was an exact reproduction of the chapter on hand surgery in *Campbell's Operative Orthopaedics,* fifth edition. Included were numerous techniques for skin coverage, as well as techniques for treating the bones, muscles, tendons, and nerves of the hand. This made it appealing to plastic and general surgeons, as well as to orthopaedic surgeons. The book was translated into several languages, including French, Spanish, and Japanese.

The second edition, published in 1982, added techniques of microsurgery and an expanded section on the treatment of carpal bone pathology. Over 400 new references were listed and a selected bibliography was added, which suggested to the reader certain review articles and articles of special clinical significance.

The third edition of THE HAND follows the tradition of the first edition. The hand surgery and microsurgery chapters are an exact reproduction of those chapters in the seventh edition of *Campbell's Operative Orthopaedics.* This third edition has been completely reorganized, dividing the material into nineteen distinct chapters, each with its own bibliography. At the beginning of each chapter, the subject material is listed in detail to avoid the need for frequent index referral. *The Hand* has continued to be a single author publication with the exception of the chapter on microsurgery. This section has been expanded and more fully illustrated under the editorship of my friend and partner, Phillip E. Wright II, M.D. The past editions have proven popular with not only hand surgeons but also plastic and orthopaedic surgeons, general surgeons, and trauma surgeons. I am gratified to find that physical therapists and occupational therapists have also found it especially useful to review the illustrative material that accompanies the descriptions of surgical procedures.

I hope that this book continues to be helpful to all physicians and therapists who treat the complex problems of the hand.

Lee Milford, M.D.

Contents

CHAPTER 1

Surgical technique and aftercare

ARRANGEMENT AND ROUTINE IN OPERATING ROOM

The results of surgery largely depend on the skill and judgment of the surgeon: fatigue or uncertainty lowers efficiency. The surgeon should therefore set up a *standard routine* (Fig. 1-1), to which he adheres and on which every assistant can depend. The surgeon should never disrupt the functions of assistants by unexpected, sporadic, or irregular demands; they in turn should know exactly what is expected of them at every moment and should perform without hesitation or wasted motion. Only a standard routine can make this possible.

When local anesthesia is used, the atmosphere of the room should be quiet and pleasant, without sudden loud remarks or gushes of conversation that might alarm the patient.

The stool on which the surgeon sits should be firm and comfortable and absolutely stable. It should allow him to sit with the knees almost level with the hips, the feet resting flat on the floor without strain. The working surface should be at elbow height. To avoid shadows, the light should be above the surgeon's left shoulder (if he is right-handed) and should shine directly on the operative field. The assistant, seated opposite, should view the operative field from 7.5 to 10 cm higher than the surgeon so that he can see clearly without bending forward and perhaps obstructing the surgeon's view. The primary function of the assistant is to hold the patient's hand firm and motionless, with the fingers retracted, so as to present to the surgeon the best possible access to the operative field (Fig. 1-2). Several hand-holders have been designed but none have replaced a well-trained assistant.

The operating table should be immobile and should allow room both for the patient's hand and for the resting elbows of the surgeon and assistant; thus muscle fatigue is kept to a minimum. The surgeon should always sit at the axillary side of the extremity on which he is working so that the anatomy of either hand is always presented to him in the same relative position. The tray holding the basic instruments should be placed on an extension of the operating table that is level with the working surface. The instruments should always be arranged in the same order. To save time the surgeon routinely reaches for and selects instruments from the basic tray; with practice this can be done without looking. He discards an instrument after use, and the nurse returns it to its proper place on the tray. This is called the "drop technique." The nurse does not, however, remove the momentarily discarded knife, tissue forceps, or dissecting scissors from the operative field, for these are in almost constant use by the surgeon. Any special instruments are quickly handed to the surgeon on request from a large table nearby. Any special sutures should already be threaded, and additional knife blades should be waiting.

Preparation and draping for elective surgery

To standardize routine and permit mobility without contamination, the method of preparation and draping of the forearm and hand are always the same, no matter what the operation. Preparations of additional fields, however, for obtaining skin, tendon, and bone grafts will vary. After the patient is anesthetized, the tourniquet is applied by the surgeon but is not inflated. After scrubbing for surgery and putting on sterile gloves, the surgeon or assistant now sponges the hand and forearm to just above the elbow with half-strength tincture of iodine, which when dry is removed with alcohol, being careful that neither solution runs under the tourniquet. Next a sterile pad and then a sheet are placed under the arm of the patient. After putting on a gown and changing gloves, the surgeon completes the draping with sterile sheets as illustrated in Fig. 1-1.

The surgeon now takes his seat and the operating lights are adjusted. If multiple incisions are anticipated, they may be outlined on the skin by a sterile skin pencil or by methylene blue applied with a toothpick or sharpened applicator. Only then is the extremity exsanguinated by a gum rubber Martin bandage 4 inches (10 cm) wide and the tourniquet inflated.

Tourniquet

A bloodless field is essential for accurate dissection without damaging vital small structures. Although a tourniquet is necessary, it is dangerous and should always be handled with respect. The pneumatic tourniquet is less

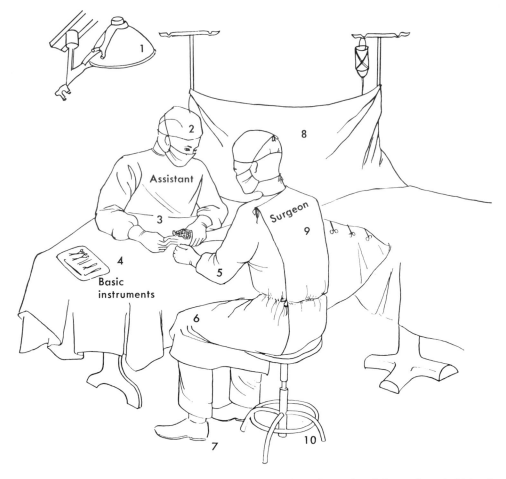

Fig. 1-1. Standard routine is used in operating room, regardless of procedure being performed. Light, *1*, passes over surgeon's left shoulder. Assistant's head, *2*, is 7.5 to 10 cm higher than surgeon's. Assistant holds patient's hand, *3*, firm and motionless. Basic instruments, *4*, are always arranged in same order. Surgeon's elbows, *5*, rest on sturdy table, knees, *6*, are almost level with hips, and feet, *7*, rest flat on floor. Vertical draping, *8*, prevents contamination of operative field by patient's face or by anesthesiologist. Surgeon holds back, *9*, comfortably erect and sits on a stool, *10*, which is firm and absolutely stable. See Fig. 2-1 for a description of hand table.

likely to cause permanent paralysis of the forearm and hand than an elastic or rubber bandage, but unless used judiciously, it may cause disproportionate or prolonged edema, stiffness, loss of acute sensation, and temporary weakness or paralysis. Pneumatic tourniquets are now available in several widths with Velcro fasteners, which are more efficient and less bulky than buckles (Fig. 1-3).

After several layers of cotton sheet wadding have been wrapped smoothly around the middle of the upper arm, an unwrinkled tourniquet is applied by the surgeon or an experienced assistant. The slightest wrinkle may pinch and blister the skin. The extremity is then either elevated for 2 minutes or is wrapped with a Martin bandage 4 inches (10 cm) wide from the fingertips to just distal to the tourniquet. The tourniquet is now inflated rapidly, thus preventing blood from being trapped in the arm after the first rise in pressure has obstructed the venous return.

The tourniquet pressure should not exceed 300 mm Hg for adults and 250 mm Hg or less for children; smaller cuffs are now available for the latter.

A recently designed tourniquet attached to an air pump and pressure gauge permits setting the pressure at 100 mm Hg above the systolic blood pressure and for a specific amount of time before an alarm is sounded. There is no rule as to how long a tourniquet may safely remain inflated on the arm. The usual limit is considered to be 1 hour or perhaps 1½ hours; this limit has sometimes been exceeded, but the risk of temporary or permanent paralysis is increased. If the operation lasts longer than 1½ hours, the tourniquet is released for 15 or more minutes while the arm is elevated about 60 degrees, and pressure is applied to the incisions with sterile dressings. Then, after the arm has been wrapped again with a Martin bandage, the tourniquet is reinflated.

Once the tourniquet is released, both it and the underlying cotton wrapping (sheet wadding) pad should be immediately removed to avoid venous congestion.

An excellent report by Wilgis seems to indicate that complete cellular anoxia for 2 hours requires marked extension of recovery time. Flatt has emphasized the need to

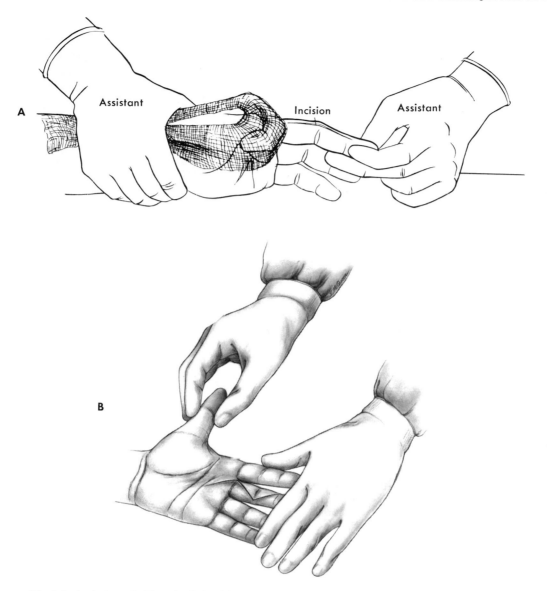

Fig. 1-2. A, Assistant holds patient's hand firm and motionless and exposes operative field for a midlateral digital incision. **B,** Ideal position for assistant to stabilize patient's hand as surgeon makes zigzag incision.

check the calibration of the pressure indicator gauge of tourniquets with a mercury manometer. He also reports that extreme pressures caused by a faulty gauge even over a short period of time may produce nerve damage that requires weeks for recovery.

For operations with the patient under local anesthetic and lasting less than 30 minutes, a Martin bandage alone may be used. Beginning at the fingertips and proceeding proximally, the bandage is applied in layers that overlap less than 6 mm. At the midforearm four or five layers are completely overlapped without wrinkles. Remember that each layer increases the pressure and that only moderate stretching is necessary. Beginning distally the bandage is then unwrapped up to the midforearm; here the layers are not disturbed until the operation is finished. When the patient is properly sedated, a pneumatic tourniquet may be used above the elbow for 30 minutes.

Neimkin and Smith recently reported using a double tourniquet on the upper arm to alternate the sites of pressure at 1 hour intervals. This ingenious method applies compression alternately to two segments of a nerve, and allows each segment to recover alternately. In 1000 consecutive patients, the proximal tourniquet was inflated about 100 mm Hg above the systolic blood pressure, but never over 280 mm Hg. After 1 hour, the proximal tourniquet was released and the distal cuff was inflated. The process was reversed at hourly intervals. They used the alternating cuff technique for up to 3½ hours in some patients. The constant pressure caused constant ischemia of the hand, but resulted in no apparent permanent clinical sequelae. They also devised a monitoring device for each tourniquet consisting of a mercury manometer attached by a T-tube to a freon gas tourniquet box and to the tourniquet cuff. It is placed so that the surgeon can see the manome-

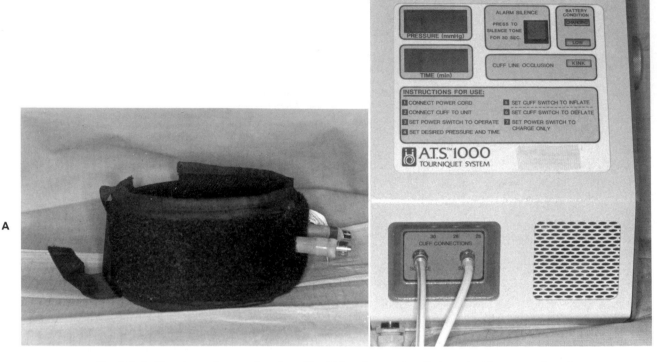

Fig. 1-3. A, Tourniquet of appropriate size with a Velcro fastener that eliminates buckle seems to be safer and more efficient. **B,** Air pump that permits continuous reading of pressure.

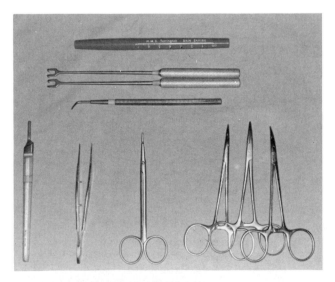

Fig. 1-4. Basic instruments for any surgical procedure on hand. Octagonal-shaped knife handle is preferable to flat handle because knife is more commonly held by precision pinch in hand surgery. Instruments are knife handle, small rat-tooth forceps, dissecting scissors, small hemostats, ruler-marking pencil, double-hook Lovejoy retractors, and probe.

ter at all times and thus can monitor the pressure dial on the tourniquet box at any time.

Instruments

For the accurate work required in hand surgery, instruments with small points are necessary; the handles, however, should be large enough to allow a firm, secure grip.

The four basic instruments are the knife, the small forceps, the dissecting scissors, and the mosquito hemostat (Fig. 1-4). The knife blade, which should be firmly attached to the handle, is changed often. The knife should be used for most dissection, rather than tearing through the tissues with a blunt instrument. The forceps should be carefully checked before surgery for cleanliness and precision of closure, since this instrument will touch the tissues most often. The scissors should have sharp double points, preferably curved, to dissect neurovascular bundles. Instruments used for fine surgery on soft tissues are shown in Figs. 1-5 to 1-10.

A mosquito hemostat is preferred for clamping vessels. Vessels should be clamped as seen, even when a tourniquet is used. An electric cautery is helpful; only minimal tissue is damaged if the vessels are carefully grasped by a small forceps or hemostat. Retractors should be of the

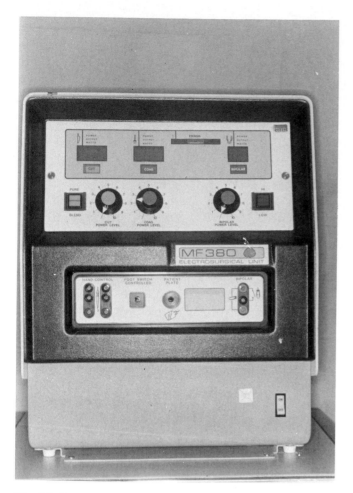

Fig. 1-5. Example of electric cautery with unipolar and bipolar power that can be controlled by hand or foot.

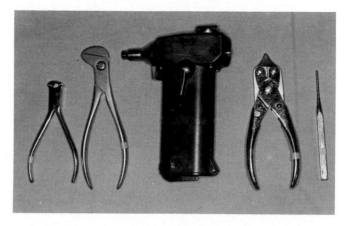

Fig. 1-6. Instruments useful for inserting Kirschner wires include end cutters, side cutters, lightweight battery-driven driver, pliers, and nail set.

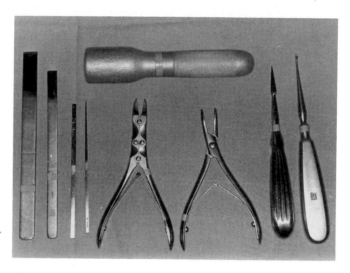

Fig. 1-7. Instruments for small bone surgery include ostetomes, bone cutter, rongeur, awl, small curet, and small hammer.

small single- or double-hook type and should have handles long enough to keep the assistant's hands away from the surgeon's working area.

For drilling holes in bone, small No. 60 carbon steel twist drill points with a mechanic's pin vice are satisfactory; small sharp-pointed Kirschner wires and a Bunnell or a battery driven hand drill may be used. If tying sutures with a needle holder is preferred, the Webster model, having the proper shape with smooth jaws for holding the finest materials, is helpful. Wire suture as well as nylon is now commercially packaged with curved or straight swaged needles of appropriate size.

Choice of anesthetic

Whatever the choice of anesthetic, if it fails to completely anesthetize the part, the results of surgery will be compromised. In hand surgery the operative procedure must be completely painless, since for accurate work the part must be held motionless. Because all anesthetics, whether general, regional, or local, carry some danger, the choice of agent and method of administration are determined by several factors. Some of these will be considered here; the techniques of their administration are not discussed.

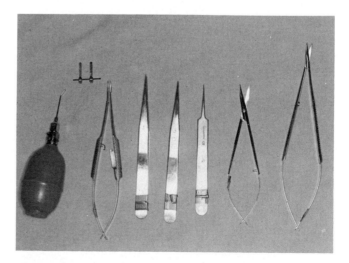

Fig. 1-8. Instruments useful in microvascular and digital nerve surgery include small irrigation bulb, microvascular clamp, microneedle holder, pickups, and scissors of assorted lengths.

Fig. 1-9. **A,** Certain dental instruments are often useful for dissection of ligaments and bone. **B,** Retractors of numerous designs have been used in hand surgery, but modified tonsil prong *(left)* has proved to be the most useful.

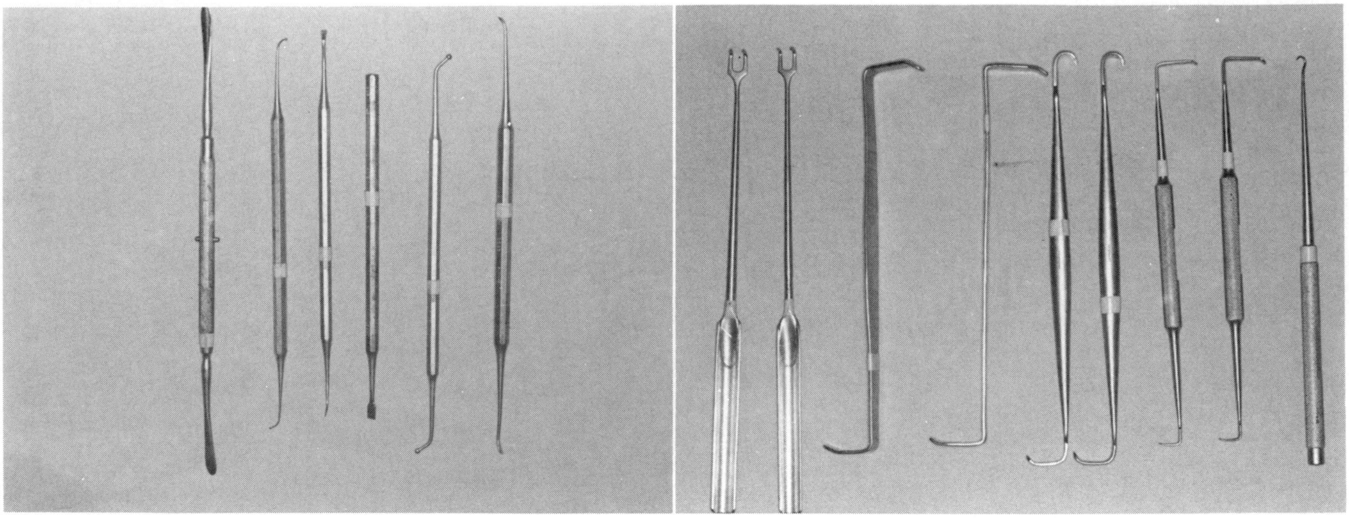

A general anesthetic is usually preferred for extensive hand surgery or when at the same time surgery is required elsewhere on the body. A general anesthetic or a proximal regional block may be demanded by the presence of infection, which a local injection may spread. The temperature and age of the patient are other considerations.

Regional anesthetics such as brachial, axillary, or peripheral nerve blocks are particularly desirable at times, especially so in acute injuries occurring after the patient has eaten. However, the surgeon must be patient when using such an anesthetic; after the injection he must wait at least 5 minutes by the clock, or longer in some instances, before making any incision. Satisfactory sedation before surgery is as important here as when general anesthetics are used. When peripheral nerve blocks are done, direct penetration of the nerves is neither necessary nor desirable (Fig. 1-11).

Pneumothorax has been reported after 1% to 5% of supraclavicular brachial blocks, depending on the skill of the operator, but this complication can be avoided by using axillary blocks. Brachialgia, although usually temporary, is rarely reported as another complication.

Axillary blocks are satisfactory for any operation distal to the elbow; in these instances encircling of the subcutaneous tissues of the proximal arm with a local anesthetic agent helps prevent pain from pressure of the tourniquet by blocking the intercostobrachial nerve.

A regional block at the wrist or at a more distal level is sometimes useful because then the patient can move his fingers during surgery; thus in certain operations, such as tenolyses and capsulotomies, the surgeon can observe any improvement in function and proceed accordingly.

A tourniquet is used and the patient is kept comfortable but semiconscious by adequate sedation before surgery and by an intravenous drip of thiopental (Pentothal) or by administration of fentanyl-droperidol (Innovar) during sur-

gery; under these circumstances the tourniquet is tolerated well for 30 minutes or longer.

The common digital nerves proximal to the finger webs may be more safely injected than the digital nerves at the base of the fingers. Although its cause is not completely understood, gangrene occasionally occurs after a local anesthetic is used to encircle the base of the finger; certainly this possibility is greater when epinephrine is added and a tight, narrow tourniquet is used at the base of the finger.

Intravenous anesthesia with double tourniquets at times makes a useful regional anesthetic. A solution of 1% lidocaine (Xylocaine) is introduced intravenously (40 ml in a 70 kg patient) into the arm after it is exsanguinated and the more proximal tourniquet is elevated. Only a few minutes are required to obtain a workable block. The more distal tourniquet is elevated over the anesthetized area once the more proximal tourniquet becomes painful. Leakage of solution into the wound sometimes is a worrisome problem. Surgery must be completed and the wound closed before release of the tourniquet because anesthesia is lost rapidly. The tourniquet should remain elevated for a minimum of 30 minutes after injection to avoid flooding of the anesthetic solution from the arm into the general circulation. After 30 minutes cellular fixation of the drug in the arm is sufficient.

Drugs used for local and regional anesthesia should become effective within a few minutes after injection, should cause no local irritation, and should have low systemic toxicity. Lidocaine seems to fulfill these requirements. For regional blocks, a total of up to 50 ml of 1% solution is the recommended safe dosage in a 70 kg adult. Mepivacaine (Carbocaine) acts longer but may be slower in onset and has been found to have about the same toxicity. The recommended dosage is up to 40 ml of 1% solution in a 70 kg adult. Bupivacaine (Marcaine) is preferred by many in replantation surgery, because it is effective for 8 hours

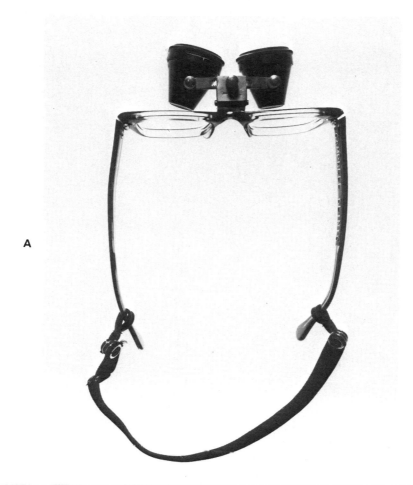

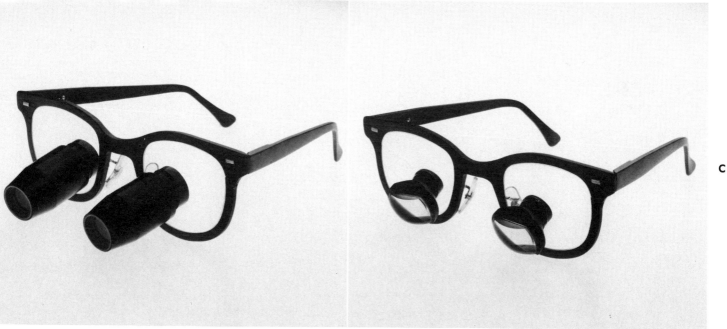

Fig. 1-10. A, Magnifying glasses for fine surgery on soft tissues. Either two or three magnifications may be chosen. Working distance is 20 to 25 cm. Glasses are stable, may be used over ordinary corrective glasses, and may be raised out of visual field at will. **B,** It is possible to achieve magnification up to 6× with magnification lens on glasses frame. However, magnification lens becomes too heavy for mounting if more than 6× magnification is needed. **C,** Magnification lens set within corrective lens is easy to use with magnification and pupillary distance set for individual surgeon.

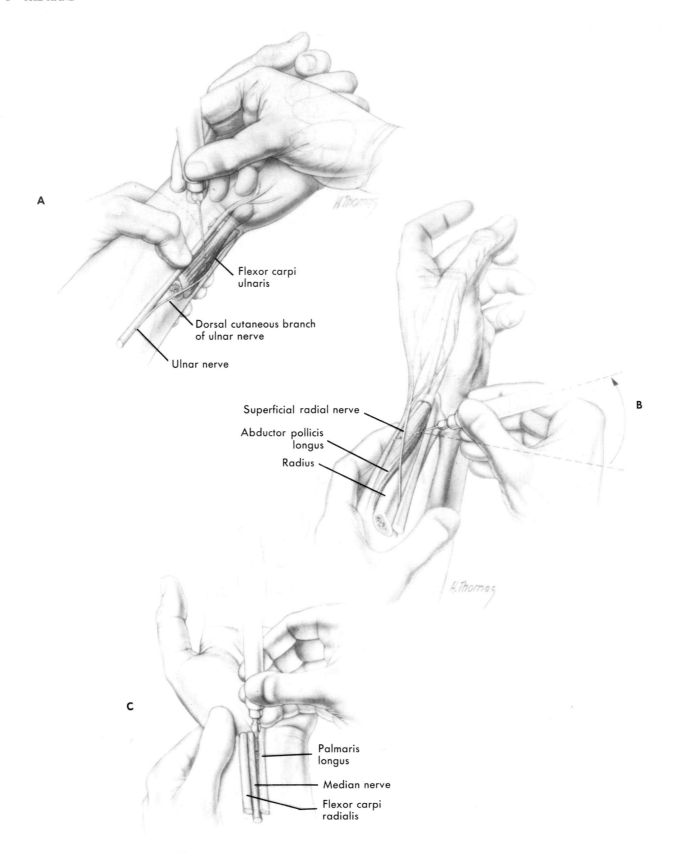

A

Flexor carpi
ulnaris

Dorsal cutaneous branch
of ulnar nerve

Ulnar nerve

Superficial radial nerve

Abductor pollicis
longus

Radius

B

C

Palmaris
longus

Median nerve

Flexor carpi
radialis

Fig. 1-11. Technique of peripheral nerve blocks. **A,** Ulnar nerve, superficial branch. **B,** Superficial radial nerve. **C,** Median nerve. (From Abadir, A.R.: Diagnostic nerve blocks. In Omer, G.E., Jr., and Spinner, M.: Management of peripheral nerve problems, Philadelphia, 1980, W.B. Saunders Co.)

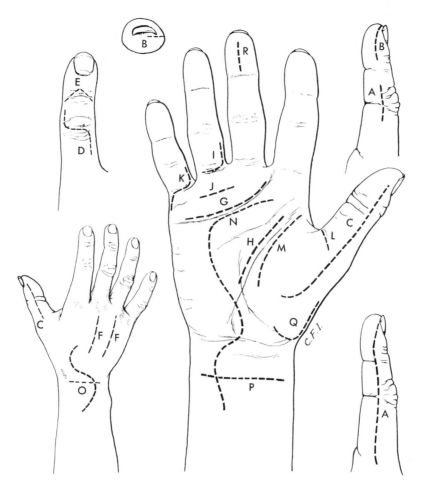

Fig. 1-12. Correct skin incisions in hand. *A,* Midlateral incision in finger. *B,* Incision for draining felon. *C,* Midlateral incision in thumb. *D,* Incision to expose central slip of extensor tendon. *E,* Inverted V incision for arthrodesis of distal interphalangeal joint. *F,* Incision to expose metacarpal shaft. *G,* Incision to expose palmar fascia distally. *H,* Incision to expose structures in middle of palm. *I,* L incision in base of finger. *J,* Short transverse incision to expose flexor tendon sheath. *K,* S incision in base of finger. *L,* Incision to expose proximal end of flexor tendon sheath of thumb. *M,* Incision to expose structures in thenar eminence. *N,* Extensive palmar and wrist incision. *O,* Incisions in dorsum of wrist. *P,* Transverse incision in volar surface of wrist. *Q,* Incision in base of thumb. *R,* Alternate incision to drain a felon. (Modified from Bunnell, S.: J. Bone Joint Surg. **14:**27, 1932; and Bruner, J.M.: Br. J. Plast. Surg. **4:**48, 1951.)

or more. It can be used as an axillary block to avoid the use of a general anesthesic.

BASIC SKIN TECHNIQUES
Incisions

As long as certain principles are observed, skin incisions can be made anywhere on the hand and not just in or near major skin creases (Figs. 1-12 and 1-13). In fact, they should not be placed within deep creases; here subcutaneous fat is scarce, and moisture tends to accumulate. An incision should be long enough to expose the deep structures without excessive stretching of the skin edges; greater exposure is possible if the skin and subcutaneous fat are dissected from the underlying fascia. The incision is always converted into a mobile oval or elliptic opening. Generally, shorter incisions are possible on the dorsum of the hand because here the skin is more mobile. For example, through a 7.5 cm lazy-S incision on the middorsum of the wrist, structures can be exposed from the extreme

radial side of the wrist to the extreme ulnar side or from the tendon of the extensor pollicis brevis to that of the extensor carpi ulnaris.

Rarely should an incision be made in a straight line. If gently curved, the scar is more graceful and less noticeable and conforms better to natural lines. A curved incision can also later be extended with freer choice of direction. Exposure is always better on the concave side of a semicircular incision; an S-shaped incision gives even more latitude.

The placement of an incision applies only to the skin; entries into deeper structures are made according to their anatomy and may be opposite in direction to those made in the skin. For example, the skin incision over the radial surface of the wrist in de Quervain's disease is transverse, but the underlying incision in the stenosed sheath is longitudinal.

Parallel or nearly parallel incisions that are too close together or too long should be avoided, for healing may

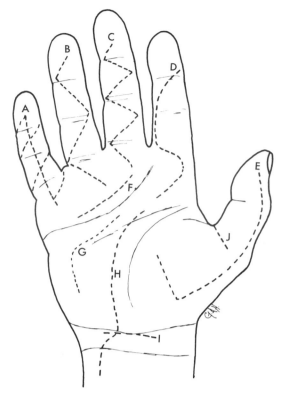

Fig. 1-13. Additional correct skin incisions in hand. *A,* Z-plasty incision often used in Dupuytren's contracture (McGregor). *B* and *C,* Zigzag incisions for Dupuytren's contracture or exposure of flexor tendon sheath. *D,* Volar flap incision. *E,* Incision to expose structures in volar side of the thumb and thenar area. *F,* Incision in distal palm for trigger finger or other affections of proximal tendon sheath. *G,* Incision to form flap over hypothenar area. *H,* Incision to expose structures in middle of palm; it may be extended proximally into wrist. *I,* Short transverse incision in volar surface of wrist. *J,* Short transverse incision to release trigger thumb.

be slow or a slough may even develop because of impairment of the blood supply. Scars that adhere to the underlying structures, especially bone, should be avoided if possible. The offset incision is helpful: the first incision is carried through the skin and subcutaneous fat, and after a flap is undermined on one side, the deep approach is made through the fascia and muscle parallel with but offset from the skin incision.

The plane of motion of a part is approximately perpendicular to the long axis of skin creases. Therefore an incision should not cross a crease at or near a right angle, since the resulting scar, being in the line of tension created by motion, will hypertrophy; indeed, it may limit motion, since a mature scar will not stretch like skin. Although true elsewhere in the body, this principle is more important when dealing with the hand.

At times incisions may be outlined on the skin with a sterile skin pencil especially if multiple incisions are needed. They may then be made without hesitation, thus saving time after the tourniquet is inflated.

FINGER INCISIONS

The first basic finger incision, the midlateral, has sometimes been misunderstood because of poor drawings and illustrations. With this incision the neurovascular bundle may be carried volarward with the volar lip of the incision, or it may be left in place by carrying the dissection superficial to it.

To carry the neurovascular bundle volarward, begin the incision on the midlateral aspect of the finger at the level of the proximal finger crease and carry it distally to the proximal interphalangeal joint just dorsal to the flexor skin crease; continue it distally along the middle phalanx, again dorsal to the distal flexor skin crease, and proceed toward the lateral edge of the fingernail (Fig. 1-14). Since flexor skin creases extend slightly over halfway around the finger, the incision is in fact slightly posterolateral. Develop the dorsal flap a little to aid in closure of the incision. On the radial sides of the index and middle fingers and the ulnar side of the little finger is a dorsal branch of the digital nerve that should be preserved when possible (Fig. 1-15). Develop the volar flap by continuing into the subcutaneous fat over the proximal and middle phalanges, but since fat is scanty over the proximal interphalangeal joint, be careful not to enter it by mistake. Immediately after incising the fat, carry the dissection volarward deep to the neurovascular bundle and expose the tendon sheath. The sheath can then be incised, or the neurovascular bundle can be exposed by further dissection (Fig. 1-16). The opposite neurovascular bundle can also be exposed because of its anterolateral position.

The second basic midlateral incision is developed superficial to the neurovascular bundle. Make the same midlateral skin incision, but just distal to the distal skin crease carry the incision obliquely into the pulp of the finger. As the volar skin flap is developed through the subcutaneous fat, carefully isolate the neurovascular bundle; it can best be found at the middle of the middle phalanx. Then expose the bundle by dissecting the fat from its volar surface and expose the flexor tendon sheath by carrying the dissection toward the bone. If necessary, the skin flap can be developed further by dissecting into the depths of the pulp distally, being careful not to disturb the nerves and arteries, and by extending the incision into the palm proximally.

Using the principles just outlined and illustrated, many less extensive exposures of the finger are possible.

Surprisingly, new incisions and approaches are still being described that allow more direct access to deep structures. The recently popularized zigzag finger incision (Fig. 1-13, *B* and *C*) does not require mobilizing either neurovascular bundle and directly exposes the volar surface of the flexor tendon sheath. However, when used on a contracted skin surface it tends to straighten out and result in a more linear scar than is desirable; here multiple Z-plasty incisions are more satisfactory. In either type of incision care must be taken to protect the neurovascular bundles.

THUMB INCISIONS

Midlateral incisions described for the fingers are also suitable for the thumb; the radial side is more accessible, and an incision here can be extended by curving its proximal end at the midmetacarpal area and creating a flap on the palmar surface of the thumb (Fig. 1-12, *C*). Care should be taken to avoid the dorsal branch of the superfi-

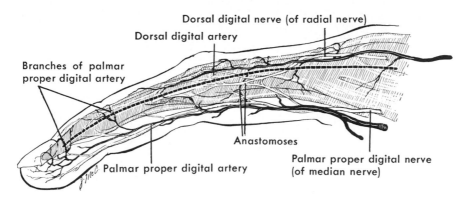

Dorsal digital nerve (of radial nerve)

Dorsal digital artery

Branches of palmar
proper digital artery

Anastomoses

Palmar proper digital artery

Palmar proper digital nerve
(of median nerve)

Fig. 1-14. Midlateral skin incision in finger extending from metacarpophalangeal joint to lateral edge of nail. To avoid flexor skin creases, it is placed slightly posterolateral. (Modified from Anson, J.B., and Maddock, W.G.: Callander's surgical anatomy, ed. 3, Philadelphia, 1952, W.B. Saunders Co.)

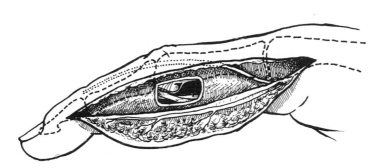

Fig. 1-15. Midlateral approach especially to expose flexor tendon sheath. On radial sides of index and middle fingers and on ulnar side of little finger is dorsal branch of digital nerve that should be preserved if possible. Volar flap containing neurovascular bundle has been developed and reflected. Window has been cut in sheath to show relations of flexor tendons.

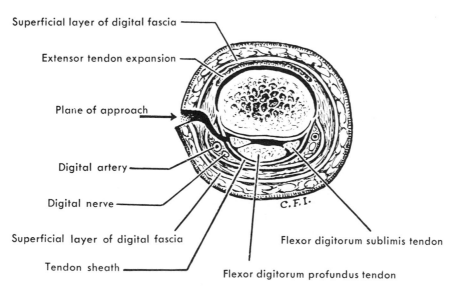

Superficial layer of digital fascia

Extensor tendon expansion

Plane of approach

Digital artery

Digital nerve

Superficial layer of digital fascia

Tendon sheath

Flexor digitorum sublimis tendon

Flexor digitorum profundus tendon

C.F.I.

Fig. 1-16. Cross section of finger to show midlateral approach when used to expose flexor tendons.

cial radial nerve to the radial side of the thumb. This incision may be used for tendon grafts without an additional palmar incision, since the flap can be developed sufficiently to expose most of the flexor surface of the thumb. Fat is scanty on the lateral aspects of the distal joint of the thumb, and the volar plate of the capsule may be opened by mistake when seeking the flexor tendon sheath.

When a transverse incision for trigger thumb is made at the level of the metacarpophalangeal joint, the two digital nerves of the thumb, located anterolaterally as in the fingers, must be carefully avoided (Fig. 1-12, *L*).

PALMAR INCISIONS

As a rule, distal palmar incisions are transverse; in the proximal palm they tend to be more longitudinal, with the distal end curving radially and paralleling the closest major skin crease, but at any desired distance from it. An incision of any desired length may be made across the palm, provided that the underlying digital nerves and other vital structures are protected. After the skin and underlying fat are incised, the latter is dissected from the palmar fascia and is carried with the skin flaps. It may be desirable, although tedious, to preserve small vessels perforating the palmar fascia if wide undermining of the skin flaps is necessary; otherwise most of the vital structures are deep to the palmar fascia. In the distal palm, structures lying between the metacarpal heads are not protected by the palmar fascia. After the skin flaps are retracted, the fascia can be incised in any direction necessary for ample exposure; excision of the fascia may be desirable. The tendons and their paralleling neurovascular bundles are then seen. The superficial volar arch can be ligated and cut at one end if deeper exposure is required. Incisions in the more proximal palm should parallel the thenar crease; however, when extended proximal to the wrist, they should not cross the flexor wrist creases at a right angle. The most important structure in the thenar area is the recurrent branch of the median nerve, which should be exposed and protected if its exact location is in doubt.

Basic skin closure techniques

Also see considerations for skin closure, p. 31.

Early closure but not necessarily immediate closure of all hand wounds lessens the chances of infection and excessive scarring, which destroy the gliding mechanism essential to hand movements. Immediate coverage is imperative when bone, cartilage, and tendon are laid bare, for without it these structures will not survive. Whenever possible, direct suture of the skin without tension is the best method of closure. On the dorsum of the hand or wrist this is sometimes possible even after considerable loss of the mobile skin by extending the wrist to relieve tension; care should be taken, though, not to hyperextend the metacarpophalangeal joints (Fig. 1-17). When a large defect here is closed in this manner, flexion of the wrist and fingers will be limited, and replacement of skin by grafting may be necessary later. The advantages of primary closure by direct suture are jeopardized unless each suture is accurately and patiently placed, for not just the epidermis, but each plane of tissue should meet its corresponding plane. Placing sutures too few in number and too close to the skin edges in attempting a ''plastic closure'' are errors of the

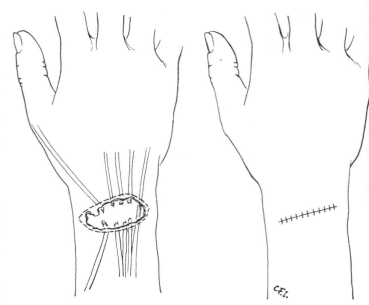

Fig. 1-17. Small defects in skin and subcutaneous tissue on dorsum of hand or wrist can be closed after wrist has been extended to relieve tension. This closure may require a graft later to permit wrist flexion while making a fist.

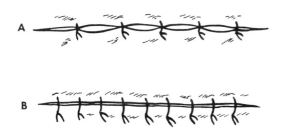

Fig. 1-18. A, Skin has been closed by an insufficient number of sutures placed too superficially and too close to skin edges. **B,** Skin has been closed by sufficient number of sutures placed more deeply and well away from skin edges.

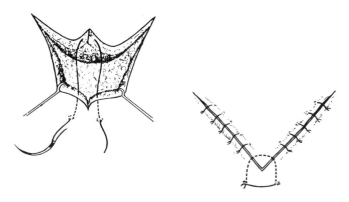

Fig. 1-19. Apical stitch is useful for suturing sharp angle in laceration or in elective flap.

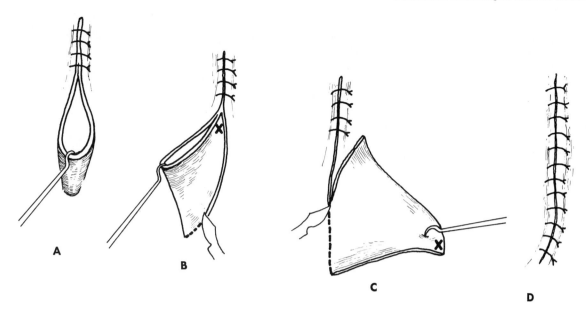

Fig. 1-20. Method of excising dog-ear. **A,** Fold of skin has been caught at its apex by a hook. **B,** Fold has been retracted to one side and skin is being incised along base of fold on opposite side; point *X* will form apex of flap. **C,** Skin has been unfolded and resulting flap is being excised. **D,** Skin closure has been completed.

Table 1-1. Types of skin grafts and their purposes*

Type			Purpose
Free grafts			
Divisions	*Synonyms*	*Thickness (inches)*	
Epidermal	Reverdin	0.008-0.010	Primary coverage for wound healing
Thin split	Thiersch	0.012	Wound healing; occasionally for permanent coverage
Medium split	Thiersch	0.015	Wound healing; occasionally for permanent coverage
Thick split	Ollier	0.025	Permanent coverage
Full-thickness	Wolf	0.030	Permanent coverage
Flap grafts			
Local			
Advancement			Permanent coverage of exposed vulnerable structures (nerves, joints, bones, and tendons)
Z-plasty			Reduction of tension lines that cause undesirable scars
Fillet			Introduction of new vascularity to recipient area
Distant			
Direct			
Nontubed open and closed			Same uses as local flaps; may be used to cover larger areas and therefore
Tubes			require greater number of operations
Indirect			

*After W.H. Frackelton.

inexperienced: the underlying tissues heal poorly, the skin edges tend to separate between the sutures, and necrosis occurs around the sutures (Fig. 1-18). The apical stitch is extremely useful for suturing a sharp angle in a laceration or in an elective flap, since it holds effectively without embarrassing the circulation at the apex (Fig. 1-19). A dog-ear may be excised one side at a time after splitting it down the middle to create two triangles; each triangle is then excised at its base. The line of excision of one side is used to mark the line of excision of the other. Another method of excising a dog-ear is shown in Fig. 1-20.

When closure without excessive tension by direct suture is impossible, some type of skin graft must be chosen without prolonged delay, usually within 5 days. The types of skin grafts most frequently used are outlined in Table 1-1.

FREE GRAFTS

When free skin grafts are to be obtained, it is well to remember that "the thinner the graft the better the take" and yet when the graft is expected to be permanent, "the thicker the graft the better the function." A thick graft is better able to withstand friction and constant use than a thin one and will contract only about 10%; a thin graft may contract up to 50%, or even 75%. For the graft to survive, it must reestablish its nutrition before death of its entire thickness occurs; great care is therefore needed, both in operative technique and in aftercare, to provide that it remains undisturbed and in direct contact with the recipient area during healing. This takes wise planning, especially in children. The graft will not survive if a hematoma separates it from the underlying vascular bed; rarely will it survive a gross infection. For primary coverage of acute wounds, free skin grafts are usually of thin or medium thickness. They will not easily survive on bare cortical bone, bare tendon, or bare cartilage. Full-thickness free skin grafts are not often used on the hand; but such a graft or a thick split graft is desirable for the palmar surface because it contains more elastic tissue and in growing children will contract less and will tend to stretch with growth. Since the survival of a full-thickness graft is so uncertain, it should be used only in elective surgery for skin coverage in the palm; it should never be used in acute injuries, with the possible exception of the fingertips. "Pinch grafts" should never be used.

SPLIT-THICKNESS SKIN GRAFTS

Frequently only a small or postage stamp graft is needed, and it may be obtained within the same operative field from the forearm; however, taking a graft from this area is undesirable in children and women because it will leave a slight scar. More suitable donor areas for these and larger grafts are the anterior and lateral aspects of the thigh and the medial aspect of the arm just inferior to the axilla. In some older women skin is available inferior to a pendulous breast without leaving a readily visible scar.

TECHNIQUE FOR OBTAINING GRAFTS WITH RAZOR BLADE. A small skin graft may be easily cut with an ordinary new razor blade held in a hemostat. Lubricate the surface of the skin with a drop of mineral oil and keep it taut with a tongue blade. Hold the side of the razor blade almost parallel to the skin with the edge of the blade in contact with it. Cut the split graft with to-and-fro motions of the blade; take care not to consciously force the blade forward, since it will automatically advance with the to-and-fro motions. A Weck knife (Fig. 1-21) is an instrument that refines this technique.

TECHNIQUE FOR OBTAINING GRAFTS WITH DERMATOME. An electrically powered dermatome such as the Stryker or Brown is not hard to assemble and use; even an inexperienced operator can cut consistently good grafts up to 7.5 cm wide. Skin glue is not required, but light lubrication of the skin with mineral oil or petroleum jelly is helpful. Bony prominences are not satisfactory donor sites with these dermatomes. The Reese dermatome does require skin glue and must be operated with precision but it is excellent for cutting grafts more than 7.5 cm wide; furthermore, it more accurately controls the thickness of the grafts. Three suggestions are offered in the use of this dermatome: (1)

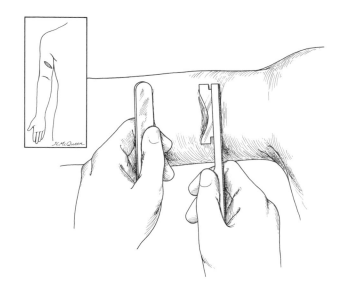

Fig. 1-21. Technique of removing split-thickness skin graft from flexor surface above elbow with Weck knife. This permits hiding scar at donor site much more effectively than if extensor surfaces distal to the elbow were used.

stretch the rubber tape tightly on the drum, (2) wait at least 3 minutes for the glue to dry before applying the dermatome to the skin, and (3) rotate the drum slowly and lift up gently while cutting the graft.

When using either type of dermatome, cut the graft somewhat larger than the recipient area.

TECHNIQUE FOR APPLYING SPLIT-THICKNESS GRAFT. The recipient area must have a vascular bed and be free of bleeding and gross infection. Place the graft on the recipient area without trimming or excessive handling. Suture it around its border with fine nylon to secure it in its new position and then trim the redundant edges (Fig. 1-22). When suturing the graft in place, it is much easier to insert a small curved needle first through the graft and then through the skin around the recipient area than to do the reverse.

A stent dressing may be applied, or finely meshed gauze impregnated with bismuth tribromophenate (Xeroform) may be placed over the graft and held in place with a bulky dressing secured by circumferential Kerlix or Kling conforming gauze and when necessary may be covered by a thin layer of plaster for splinting. The dressing is usually changed after about a week; any slough is then removed and a fresh dressing is applied. When an area of slough is large, regrafting is necessary. When a slough is anticipated or when another procedure is planned in which a split-thickness graft will be used, a graft larger than needed initially may be cut and refrigerated at between 0° and 5° C in a Ringer's solution or in a saline solution to which penicillin has been added; it can then be used at any later time up to 21 days.

The donor area is dressed with one layer of finely woven nylon or silk gauze. Otherwise the part is left uncovered, and drying of the area is encouraged. When the dressing prevents drying, the donor area tends to become macerated and secondarily infected, sloughing may occur, and this area may itself require skin grafting later. Bed sheets should be kept off the donor site with a bed cradle support.

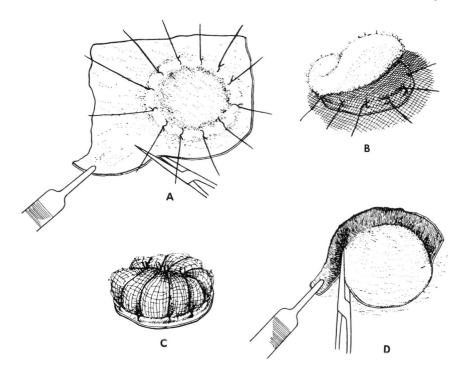

Fig. 1-22. Technique of applying split-thickness graft. **A,** Graft has been sutured over defect, and redundant edges of graft are being trimmed. **B,** Sheet of finely meshed gauze and pack of moist cotton or gauze have been placed over graft. **C,** Sutures have been tied over pack. **D,** Necrotic edges of graft are being trimmed away after graft has healed.

FREE FULL-THICKNESS GRAFTS

When a full-thickness graft is used, the recipient area must be free of infection and hemostasis must be complete. Preferably the graft is obtained from the groin or the medial aspect of the arm where the skin is thin, which is desirable, and where the defect created by removing the graft may be closed by undermining and suturing the skin edges (Fig. 1-23).

Sometimes an associated injury makes a detached piece of skin and underlying fat available; in this instance the skin may be stabilized on a dermatome drum and a full-thickness graft is excised from the fat.

TECHNIQUE. Make a pattern on sterile tape or gauze of the area to be covered. Using this pattern, outline the anticipated graft on the donor area with methylene blue; it should be slightly larger than the pattern to allow for the necessary margin in suturing and for shrinkage. Remove the graft with a sharp knife by dissection between the fat and the skin; do not take any fat with the graft. Suture it in place and excise the redundant edges. Apply a stent dressing and support the hand with a plaster splint for at least 7 to 10 days before redressing. At that time dark blisters of the superficial layer of the graft may be seen, but this must not be mistaken for a deep slough.

FLAP GRAFTS

A flap graft may be used in the primary closure of a hand wound or in a secondary procedure to replace scars, skin of poor quality, or sloughed skin. It may be obtained locally or from a distant part. When the area to be covered is small, a local flap may be indicated, like one of those shown in Fig. 1-24, or a Z-plasty, as shown in Fig. 1-27.

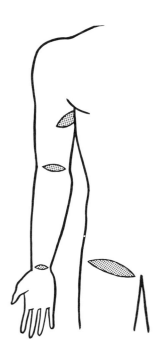

Fig. 1-23. Sites from which to obtain full-thickness skin grafts. Groin or medial aspect of arm is preferable (see text).

When the defect is on the volar surface of a finger and subcutaneous fat is needed, a cross finger flap (p. 32) may be used.

The advantages of a local flap over one from a distant part are that the involved hand is not tied to the distant donor and that in many instances finger motions may con-

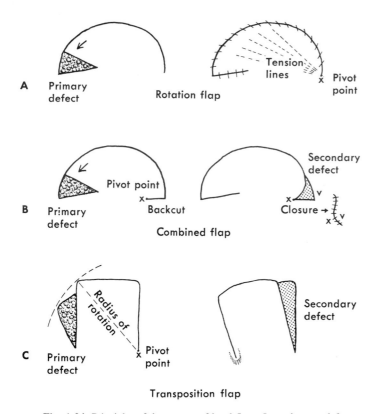

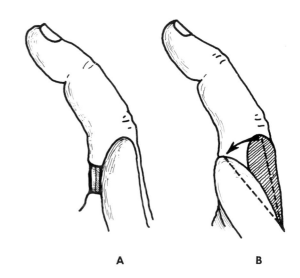

tinue. When the defect is too large to be covered with a local flap, a distant flap from the abdomen (p. 34) is indicated.

LOCAL FLAPS

Local flaps should usually be avoided when treating acute hand injuries; the skin that would be used as the flap is often already contused, and this insult combined with the amount of undermining necessary for closure often causes necrosis of the flap. Fig. 1-24 illustrates the principles of three types of local flaps.

TECHNIQUE FOR LOCAL FLAPS. Local flaps are usually of the simple transposition type. This type covers vital structures but leaves a defect that must in turn be covered with a split-thickness graft (Fig. 1-25). A common error in de-

Fig. 1-24. Principles of three types of local flaps. In each type, defect to be covered is converted into triangular one. Flap may be rotated, **A,** transposed, **C,** or both. **B. B,** Backcut in combined flap decreases tension on flap but also decreases blood supply to flap; defect created by this backcut is closed as shown. **C,** Defect created by transposing a flap must be covered with a split-thickness graft. (Redrawn from Rank, B.K., and Wakefield, A.R.: Surgery of repair as applied to hand injuries, Baltimore, 1960, The Williams & Wilkins Co.; and McGregor, I.A.: Fundamental techniques of plastic surgery and their surgical applications, Edinburgh, 1960, E. & S. Livingstone, Ltd.)

Fig. 1-25. Simple transposition type of local flap. **A,** Deep structures on anterolateral aspect of finger are exposed and must be covered by a local flap; flap has been outlined. **B,** Flap has been transposed, and defect thus created on posterolateral aspect of finger has been covered by a split-thickness graft. Note radius, *broken lines,* of arc, *arrow,* of transposition.

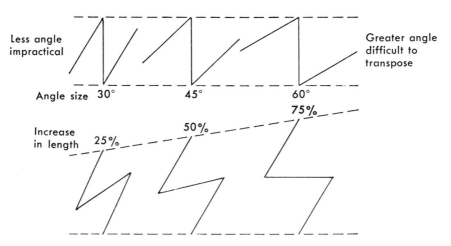

Fig. 1-26. Angles permissible in performing Z-plasties. Angles that central limb of Z make with each of other two limbs should be between 45 and 60 degrees. When angles are less than 45 degrees, blood supply to flap is impaired; when more than 60 degrees, flaps cannot be transposed without severe tension.

signing a local flap is to make it too short; it must be remembered that the fixed point of pivot from which the advancement is made is at that border of the base that is opposite the defect. If the corresponding border of the flap is not long enough, tension occurs when the flap is sutured in its new bed.

• • •

The Z-plasty is an application of the advancement type of local flap; suitably constructed skin flaps are brought from adjacent areas to release a contracture. Its primary use is in the release of a long, narrow contracture surrounded by tissue mobile enough to allow some shifting and manipulation without the danger of necrosis from impaired blood supply. The Z-plasty should not be used in attempting to close a wide fusiform defect. Nor should the

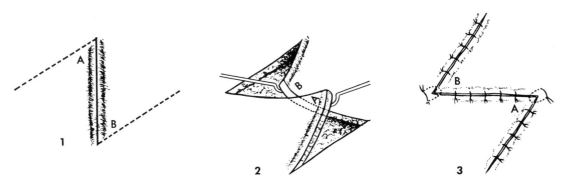

Fig. 1-27. Simple Z-plasty to release a long narrow contracture. **1,** Central limb of Z, *solid line,* is to be made among line of contracture, and other two limbs, *broken lines,* are to be made where shown. **2,** Incisions have been made and flaps are being shifted. **3,** Flaps have been sutured in their new positions. Note apical stitches at *A* and *B*.

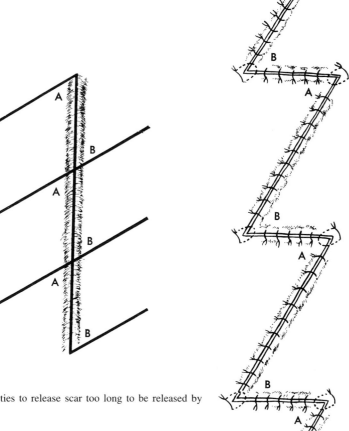

Fig. 1-28. Multiple Z-plasties to release scar too long to be released by single Z-plasty.

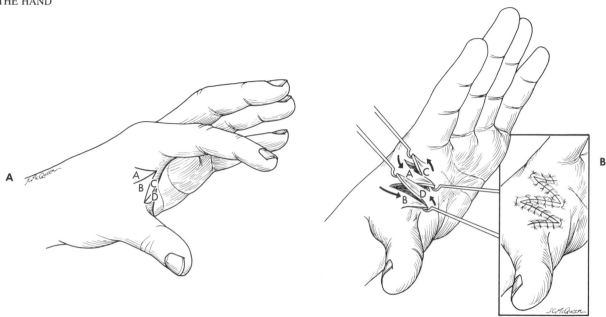

Fig. 1-29. Four-flap Z-plasties are useful in reducing first web contractures secondary to narrow linear scar and with normal elastic surrounding tissue. **A,** Outline of flaps. **B,** Flaps are rotated. *Inset,* Flaps are sutured in place.

Z-plasty be used in the primary closure of a wound unless the wound consists only of a laceration similar to a surgical incision.

TECHNIQUE OF Z-PLASTY (FIG. 1-26). Make the central limb of the Z along the line of the contracture to be released (Fig. 1-27). Now make the other two limbs of the Z equal in length to that of the central limb; the angle between each limb and the central limb must be equal one to the other and is usually about 60 degrees. An increase in this angle will not allow transposition of the flap without severe tension; a decrease makes the Z less effective in releasing tension and impairs the blood supply to each flap. Handle the points of the flaps with care, since they are most likely to undergo necrosis; suture each point with an apical stitch. Multiple Z-plasties (Figs. 1-28 to 1-30) may be used when a scar is too long to allow correction with one Z-plasty and when the scars resulting from the rotation of the flaps will be in a more desirable position.

McGregor has modified the standard multiple Z-plasty for use in the rather fixed palmar skin of the hand and fingers (Fig. 1-13, *A*). The length of its limbs may be varied, making adjoining flaps larger or smaller as desired; however, the length of the limbs of each individual Z must be equal. The oblique limbs are curved to broaden the tips of the flaps, thus increasing their blood supply. On the finger the oblique limbs end in the flexor skin creases; then when the flaps are shifted, the oblique limbs become transverse and fall within the creases.

CARE AFTER SURGERY

Care after surgery must be managed intelligently so that tissues are allowed to heal and functions of the affected part are restored as rapidly as possible. Such care begins with the application of the dressing; the routine dressing is applied as follows. A closely woven patch of gauze impregnated with Xeroform is placed over each incision.

Granulation tissue cannot grow through this material and cause it to adhere; the gauze also prevents the wound from becoming macerated. Then after the hand has been positioned properly, cotton sponges that have been soaked in saline solution and squeezed are carefully placed around it. Moist sponges conform to the contours of the hand more accurately and distribute pressure more evenly than do dry ones; furthermore, they promote absorption of blood because capillary tension is lowered and they prevent blood from collecting at the wound. Next a role of Kling bandage is applied to hold the wet gauze in place, then sheet wadding is wrapped around the hand and forearm. Finally an appropriate plaster splint is applied and is held in position with a roll of 2-inch (5 cm) gauze bandage. Splints and bandages on children tend to slip distally but can be controlled effectively by a tube of stockinette that encloses the entire extremity (Fig. 1-31). Immediately before the tourniquet is removed, keep the hand constantly elevated to prevent edema and hemorrhage after surgery. Elevation should be maintained for at least 48 hours; this can be done by positioning the hand on a pillow resting on the chest, by slight traction that elevates the hand and forearm while the elbow rests on the bed (Fig. 1-32, *A*), or by using a preformed rubber sponge block (Fig. 1-32, *B*).

Bed rest for 3 days or more is strongly recommended after major surgery on the hand. Body activity increases edema of the hand, and merely supporting it in a sling while the patient is ambulatory is not effective. Fingers not splinted should be exercised. The shoulder is likely to become stiff, especially in older patients, and should be abducted and extended toward the head several times daily; this is easily remembered if done at each mealtime.

Sutures of nylon or steel do not require removal until the splint is discarded, usually at 3 or 4 weeks; therefore complete redressing is unnecessary unless hematoma or infection is suspected, and in these instances the dressing

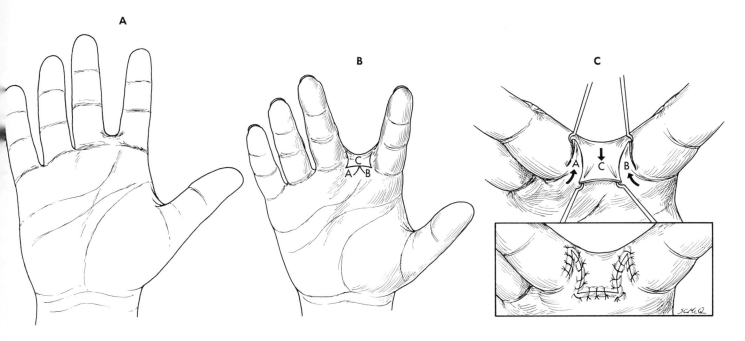

Fig. 1-30. To correct linear contracture of second, third, or fourth web caused by only a narrow scar, a dorsal flap may be fashioned using technique shown. **A,** Web contracture. **B,** Flaps are outlined. **C,** Flaps are rotated in place. *Inset,* Flaps are sutured.

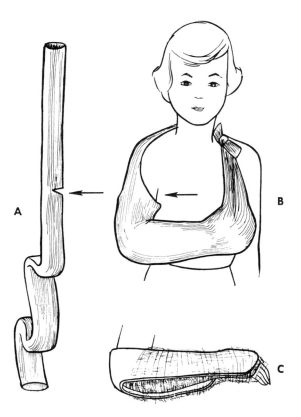

Fig. 1-31. Tube of stockinette may be used to enclose entire extremity of child after surgery. It prevents contamination of dressing and wound. **A and B,** Tube is opened, as indicated by arrows, is slipped on extremity, and is tied around neck. **C,** Plaster splint to immobilize forearm and wrist is prevented from slipping distally by including the elbow.

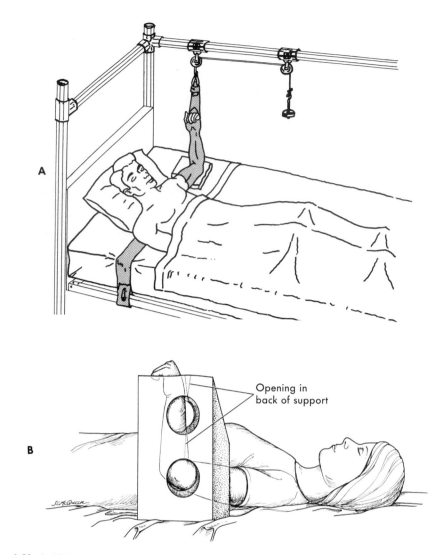

Fig. 1-32. A, Tube of stockinette and traction may be used to elevate hand and forearm after surgery. Tube is opened, is slipped on extremity, and is carried across bed beneath shoulders. At level of palm it is opened again to allow motion of digits. One end of stockinette is attached to mattress frame and the other to traction apparatus. **B,** Preformed rubber sponge block is also useful for elevating hand postoperatively while in bed. It was originally designed for postoperative foot protection. Although hand, forearm, and elbow are contained within block, they are not constricted. It is a mobile elevator so patient may take it home for nighttime use.

should be changed at 5 to 7 days, and the splint should be reapplied. Even when no complications are suspected, the wound should be inspected at about 7 days.

When extensor tendons have been sutured, immobilization is necessary for 4 weeks, and then gentle active exercises are begun, but the hand is supported between exercise periods until the fifth week. When flexor tendons have been sutured, immobilization is necessary for 3 weeks, and then guarded active exercises are begun; exercises against moderate resistance are begun during the fifth week. When tendons have been transferred, immobilization is necessary for 3 to 4 weeks; active motion is then begun, but some type of wrist support is continued until the fifth or sixth week.

Active use of the hand is the most effective way to reestablish motion after surgery. The use of splints, hand blocks, and putty are valuable, but occupational therapy is even better because it results in purposeful movement, a sense of accomplishment, and prevents the boredom that is frequent when a patient is requested to repeat a single motion dozens of times. Often the best occupational therapy is the patient's usual work, and if possible he should be offered the opportunity to return to it as part of the treatment, even if on a limited basis; the return also has a beneficial psychologic effect.

Applying excessive heat to the hand while it is held dependent or passive manipulation of joints by the physical therapist or surgeon is always contraindicated. Two points

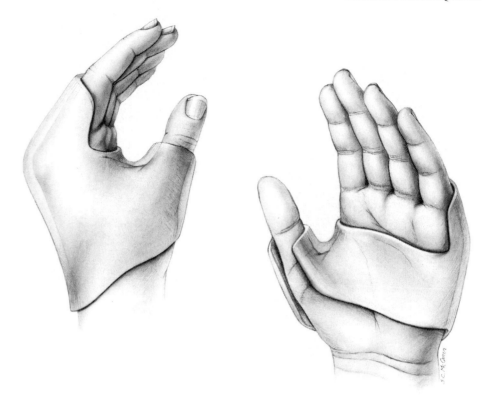

Fig. 1-33. Splint is comfortable and has very low profile, which encourages wearing while recovery is progressing in median-ulnar nerve palsies. It permits pinch and some grasping while maintaining metacarpophalangeal joints in slight flexion. It blocks metacarpophalangeal joint extension and thus prevents clawing.

are especially important in the care after surgery. First, the patient should never be sent to a physical therapist without specific written orders describing the exact treatment requested; the therapist cannot be expected to prescribe treatment, and when orders are not specific, the hand may be treated in a hot whirlpool bath in the dependent position. This feels good at the time, but edema and inflammation are likely to develop later. Second, the patient should not be required to carry out movements of the hand that are markedly painful. The cause of pain should be sought, and if necessary the part should be splinted. Often a neuroma is the cause of pain and should be excised if it impairs function.

These general principles of care after surgery are as important as the surgery itself; neglect of them will disrupt the results of the finest surgical skill.

SPLINTING

Splinting may have one of three purposes: (1) to immobilize all or part of the hand in a position that will promote healing and prevent deformity, (2) to correct an existing deformity and promote function in that part, and (3) to supply power to compensate for weakness, especially in muscles affected by peripheral nerve palsy.

The splint should permit unaffected parts to function as normally as possible. In the past some splints have restricted function too much. Thus a hand only partially disabled became completely so during use of the splint; therefore in some instances the splint was simply not used by the patient.

Immobilizing splints are most frequently used after an operation for a limited time only or intermittently to ensure correct position of joints and to relax muscles; they are also used to prevent further deformity, as in the arthritic hand. They should be comfortable and light. A splint maker should be available for making technical adjustments requiring special skills, but the patient himself should be able to apply the splint, remove it, and make minor adjustments. He should thoroughly understand the reason for wearing it and should be convinced of its value. As treatment progresses, whether he has faithfully used the splint can be determined by observing his skill in applying and removing it.

Some of the more useful splints are illustrated in Figs. 1-33 to 1-41. In Fig. 1-42 an important principle of hand splinting is illustrated.

The Joint Jack, a useful splint that gradually extends the contracted proximal interphalangeal joint, has been designed by Kirk Watson (Fig. 1-41). The tension is gradually increased by a screw in conjunction with a strap. It is used only intermittently. The patient should be cautioned about possible skin pressure problems that may occur over the proximal interphalangeal joint (see Fig. 3-25 for mallet-finger splint).

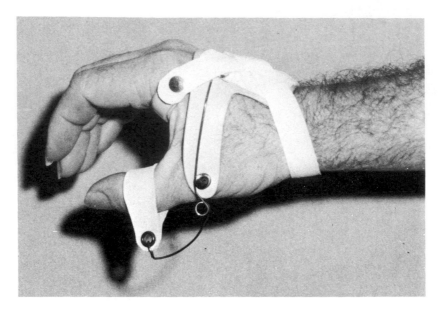

Fig. 1-34. Splint for low median nerve palsy dynamically holds thumb in abduction, extension, and opposition, thus preventing an adduction contracture of thumb. It is light and compact. (Courtesy Dr. George E. Omer, Jr.)

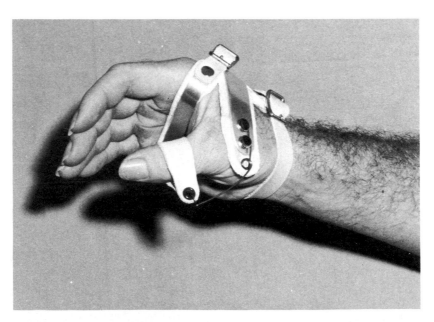

Fig. 1-35. Splint for low median nerve palsy functions like splint shown in Fig. 1-34. However, it is made of metal instead of plastic. (Courtesy Dr. George E. Omer, Jr.)

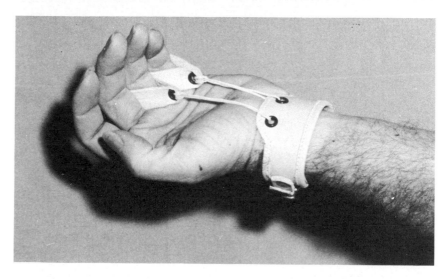

Fig. 1-36. Splint for ulnar nerve palsy dynamically forces metacarpophalangeal joints of ring and little fingers into flexion. Part of palm is covered by rubber bands, which is a disadvantage. (Courtesy Dr. George E. Omer, Jr.)

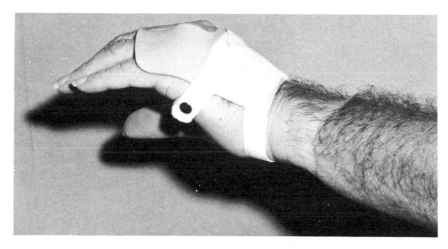

Fig. 1-37. Splint for ulnar nerve palsy prevents hyperextension deformity of metacarpophalangeal joints of ring and little fingers. Furthermore, it conforms to shape of transverse metacarpal arch and has no attachments that hinder function of hand. (Courtesy Dr. George E. Omer, Jr.)

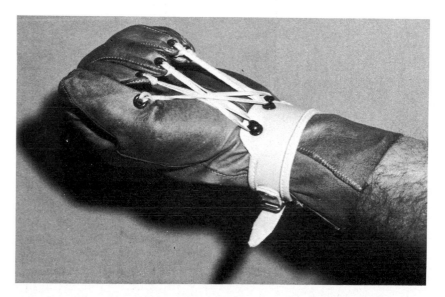

Fig. 1-38. Flexor glove dynamically forces fingers into flexion, exerting continuous force on proximal interphalangeal and metacarpophalangeal joints. Sometimes it also flexes wrist when proximal eyelets are too far proximal, and when desirable, it may be applied over volar wrist splint. (Courtesy Dr. George E. Omer, Jr.)

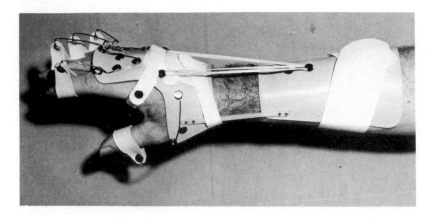

Fig. 1-39. Splint for high radial nerve palsy dynamically splints wrist and digits in extension. Furthermore, it is light and pliable and has none of large outriggers usually employed to extend digits. (Courtesy Dr. George E. Omer, Jr.)

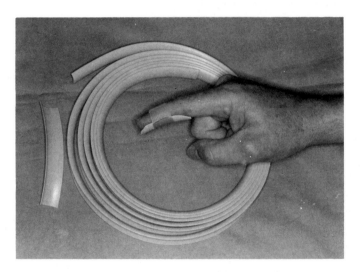

Fig. 1-40. Preformed plastic gutter splints for digits are easily adjustable in length and support soft tissue or fracture healing.

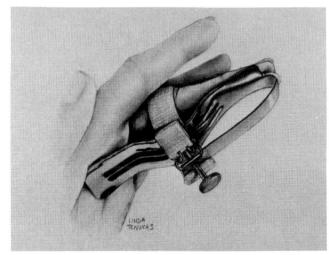

Fig. 1-41. Joint Jack finger splint. (Courtesy Joint-Jack Co., 198 Millstone Road, Glastonbury, Conn. 06033.)

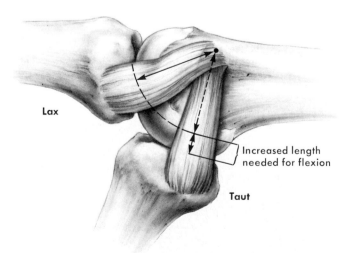

Lax

Taut

Increased length needed for flexion

Fig. 1-42. Because origins of collateral ligaments of metacarpophalangeal joints are eccentric, maximum length of ligaments is required to attain full flexion. Ligaments are relaxed in extension and taut in flexion; therefore joints should be splinted in flexion to prevent contracture of ligaments and loss of flexion.

REFERENCES
Surgical technique

Abadir, A.: Anesthesia for hand surgery, Orthop. Clin. North Am. 2:205, 1970.

Adams, J.P., Dealy, E.J., and Kenmore, P.I.: Intravenous regional anesthesia in hand surgery, J. Bone Joint Surg. 46-A:811, 1964.

Ashworth, C.R., et al.: Silicone-rubber interposition arthroplasty of the carpometacarpal joint of the thumb, J. Hand Surgery 2:345, 1977.

Barron, J.N.: Instruments for hand surgery, Hand 6:211, 1974.

Beasley, R.W.: Principles and techniques of resurfacing operations for hand surgery, Surg. Clin. North Am. 47:389, 1967.

Bolton, C.F., and McFarlane, R.M.: Human pneumatic tourniquet paralysis, Neurology 28:787, 1978.

Boyes, J.H.: Operative technique in surgery of the hand. In American Academy of Orthopaedic Surgeons: Instructional course lectures, vol. 9, Ann Arbor, 1952, J.W. Edwards.

Bruner, J.M.: Safety factors in the use of the pneumatic tourniquet for hemostasis in surgery of the hand, J. Bone Joint Surg. 33-A:221, 1951.

Bruner, J.M.: Problems of postoperative position and motion in surgery of the hand, J. Bone Joint Surg. 35-A:355, 1953.

Burnham, P.J.: Regional block at the wrist of the great nerves of the hand, JAMA 167:847, 1958.

Burnham, P.J.: A new incision for amputation of the index finger and its metacarpal, Am. J. Surg. 97:331, 1959.

Cannon, B.: Open grafting of raw surfaces of the hand, J. Bone Joint Surg. 40-A:79, 1958.

Curtis, R.M.: Cross-finger pedicle flap in hand surgery, Ann. Surg. 145:650, 1957.

Dellon, A.L., Curtis, R.M., and Chen, C.: Prevention of femoral vein occlusion by local injection of Thrombolysin in the rat, J. Hand Surg. 4:121, 1979.

Denny-Brown, D., and Brenner, C.: Paralysis of nerve induced by direct pressure and by tourniquet, Arch. Neurol. Psychiatr. 51:1, 1944.

Dery, R., Pelletier, J., and Jacques, A.: Metabolic changes induced in the limb during tourniquet ischemia, Can. Anaesth. Soc. J. 12:367, 1965.

Dupertuis, S.M.: An elevation of skin grafts for hand coverage, J. Bone Joint Surg. 34-A:811, 1952.

Eckhoff, N.L.: Tourniquet paralysis: a plea for the extended use of the pneumatic tourniquet, Lancet 2:343, 1931.

Flatt, A.E.: Tourniquet time in hand surgery, Arch. Surg. 104:190, 1972.

Frackleton, W.H., and Docktor, J.P.: Plastic surgery of the extremities. In American Academy of Orthopaedic Surgeons: Instructional course lectures, vol. 12, Ann Arbor, 1955, J.W. Edwards.

Gordon, L., and Buncke, H.J.: Universal microsurgical operating table, J. Hand Surg. 3:101, 1978.

Haas, L.M., and Landeen, F.H.: Improved intravenous regional anesthesia for surgery of the hand, wrist, and forearm: the second trap technique, J. Hand Surg. 3:194, 1978.

Hardy, S.B.: Principles in covering surface defects of the hand, Clin. Orthop. 13:63, 1959.

Hastings, D.E., and Evans, J.A.: The lupus hand: a new surgical approach, J. Hand Surg. 3:179, 1978.

Hunter, J.M., Schneider, L.H., Dumont, J., and Erickson, J.C., III: A dynamic approach to problems of hand function using local anesthesia supplemented by intravenous fentanyl-droperidol, Clin. Orthop. 104:112, 1974.

Iselin, F., Levame, J., and Godoy, J.: A simplified technique for treating mallet fingers: tendodermodesis, J. Hand Surg. 2:118, 1977.

Jones, K.G., Marmor, L., and Lankford, L.L.: An overview on new procedures in surgery of the hand, Clin. Orthop. 99:154, 1974.

Kasdan, M.L., et al.: Axillary block anesthesia for surgery of the hand, Plast. Reconstr. Surg. 46:256, 1970.

Kelikian, H., and Doumanian, A.: Skin grafts in hand surgery, Clin. Orthop. 9:205, 1957.

Kilgore, E.S., and Newmeyer, W.L.: In favor of standing to do hand surgery, J. Hand Surg. 2:326, 1977.

Klenerman, L.: The tourniquet in surgery, J. Bone Joint Surg. 44-B:937, 1962.

Klenerman, L., Biswas, M., and Hulands, G.H.: Systemic and local effects of the application of a tourniquet, J. Bone Joint Surg. 62-B:385, 1980.

LeWorthy, G.W.: Sole skin as a donor site to replace palmar skin, Plast. Reconstr. Surg. 32:30, 1963.

Linscheid, R.L.: Injuries to radial nerve at wrist, Arch. Surg. 91:942, 1965.

Lister, G.: Intraosseous wiring of the digital skeleton, J. Hand Surg. 3:427, 1978.

Lunseth, P.A., Burton, R.I., and Braun, R.M.: Continuous suction drainage in hand surgery, J. Hand Surg. 4:193, 1979.

Mandel, M.A., and Dauchot, P.J.: Radial artery cannulation in 1,000 patients: precautions and complications, J. Hand Surg. 2:482, 1977.

McGregor, I.A.: Fundamental techniques of plastic surgery and their surgical applications, Edinburgh, 1960, E. & S. Livingstone, Ltd.

McGregor, I.A.: The Z-plasty in hand surgery, J. Bone Joint Surg. 49-B:448, 1967.

Menon, J., and Gelberman, R.H.: Interphalangeal joint destruction: a complication of cryotherapy, J. Hand Surg. 5:600, 1980.

Micks, J.E., and Wilson, J.N.: Full-thickness sole-skin grafts for resurfacing the hand, J. Bone Surg. 49-A:1128, 1967.

Middleton, R.W.D., and Varian, J.P.: Tourniquet paralysis, Aust. N.Z. J. Surg. 44:124, 1974.

Miller, S.H., et al.: The acute effects of tourniquet ischemia on tissue and blood gas tensions in the primate limb, J. Hand Surg. 3:11, 1978.

Miller, S.H.: Effects of tourniquet ischemia and postischemic edema on muscle metabolism, J. Hand Surg. 4:547, 1979.

Neimkin, R.J., and Smith, R.J.: Double tourniquet with linked mercury manometers for hand surgery, J. Hand Surg. 8:938, 1983.

Ochoa, J., Fowler, T.J., and Gilliatt, R.W.: Anatomical changes in peripheral nerves compressed by a pneumatic tourniquet, J. Anat. 113:433, 1972.

Paletta, F.X., Willman, V., and Ship, A.G.: Prolonged tourniquet ischemia of extremities, J. Bone Joint Surg. 42-A:945, 1960.

Patterson, S., and Klenerman, L.: The effect of pneumatic tourniquets on the ultra structure of skeletal muscle, J. Bone Joint Surg. 61-B:178, 1979.

Peters, C.R., and Kleinert, H.E.: Office and emergency room care of the injured hand, South Med. J. 69:53, 1976.

Porter, R.W.: Functional assessment of transplanted skin in volar defects of the digits: a comparison between free grafts and flaps, J. Bone Joint Surg. 50-A:955, 1968.

Price, A.J., Jones, N.A.G., and Webb, P.J.: Do tourniquets prevent deep vein thrombosis? J. Bone Joint Surg. 62-B:529, 1980.

Pulvertaft, R.G.: Suture materials and tendon junctures, Am. J. Surg. 109:346, 1965.

Pulvertaft, R.G.: Twenty-five years of hand surgery: personal reflections, J. Bone Joint Surg. 55-B:32, 1973.

Rank, B.K., and Wakefield, A.R.: Surgery of repair as applied to hand injuries, Baltimore, 1960, Williams & Wilkins.

Rintala, A.: Palmar skin grafts, Acta Chir. Scand. 126:474, 1963.

Rob, C., and Smith, R., editors: Operative surgery: orthopaedics and plastic surgery, vol. 1, part X, Hand, Philadelphia, 1959, F.A. Davis Co.

Rorabeck, C.H.: Tourniquet induced nerve ischemia: an experimental investigation, J. Trauma 20:280, 1980.

Rudge, P.: Tourniquet paralysis with prolonged conduction block, J. Bone Joint Surg. 56-B:716, 1974.

Snedecor, S.T.: Bone surgery of the hand, Am. J. Surg. 72:363, 1946.

Solonen, K.A., Tarkkanen, L., and Narvanen, S.: Metabolic changes in the upper limb during tourniquet ischemia, Acta Orthop. Scand. 39:20, 1968.

Tanzer, R.C.: Prevention of postoperative hematoma in surgery of the hand: the use of the "compression suture," J. Bone Joint Surg. 34-A:797, 1952.

Tegtmeier, R.E.: Self-retaining retractors for hand surgery, Plast. Reconstr. Surg. 53:495, 1974.

Thomassen, E.H.: An improved method for application of the pneumatic tourniquet on extremities, Clin Orthop. 103:99, 1974.

Tountas, C.P., and Bergman, R.A.: Tourniquet ischemia: ultrastructural and histochemical observations of ischemic human muscle and of monkey muscle and nerve, J. Hand Surg. 2:31, 1977.

Vatashsky, E., et al.: Anesthesia in a hand surgery unit, J. Hand Surg. 5:495, 1980.

Verdan, C.: Basic principles in surgery of the hand, Surg. Clin. North Am. 47:355, 1967.

Wilgis, E.F.S.: Observations on the effects of tourniquet ischemia, J. Bone Joint Surg. 53-A:1343, 1971.

Splinting

Bunnell, S.: Spring splint to supinate or pronate the hand. J. Bone Joint Surg. **31-A**:664, 1949.

Bunnell, S., and Howard, L.D., Jr.: Additional elastic hand splints, J. Bone Joint Surg. **32-A**:226, 1950.

Fess, R.E., and Philips, C.: Hand splinting: principles and methods, St. Louis, 1986, The C.V. Mosby Co.

Kent, H.: Functional brace for the paralyzed hand: a preliminary report, J. Bone Joint Surg. **36-A**:1082, 1954.

Ketchum, L.D., Hibbard, A., and Hassanein, K.M.: Follow-up report on the electrically driven hand splint, J. Hand Surg. **4**:474, 1979.

Littler, J.W., and Tobin, W.J.: Thumb abduction splint, J. Bone Joint Surg. **30-A**:240, 1948.

Nickel, V., Perry, J., and Garrett, A.L.: Development of useful function in the severely paralyzed hand, J. Bone Joint Surg. **45-A**:933, 1963.

Peacock, E.E., Jr.: Dynamic splinting for the prevention and correction of hand deformities: a simple and inexpensive method, J. Bone Joint Surg. **34-A**:789, 1952.

Schottstaedt, E.R., and Robinson, G.B.: Functional bracing of the arm, Part II, J. Bone Joint Surg. **38-A**:841, 1956.

Stewart, J.E.: A plastic opponens splint for the thumb, J. Bone Joint Surg. **30-A**:783, 1948.

Strong, M.L.: A new method of extension-block splinting for the proximal interphalangeal joint: preliminary report, J. Hand Surg. **5**:606, 1980.

Thomas, F.B.: An improved splint for radial (musculospiral) nerve paralysis, J. Bone Joint Surg. **33-B**:272, 1951.

Weber, E.R., and Davis, J.: Rehabilitation following hand surgery, Orthop. Clin. North Am. **9**:529, 1978.

Zide, B.M., Bevin, A.G., and Hollis, L.I.: Examples of simply fabricated custom-made splints for the hand, J. Hand Surg. **6**:35, 1981.

Anatomy and miscellaneous

Albright, J.A., and Linburg, R.M.: Common variations of the radial wrist extensors, J. Hand Surg. **3**:134, 1978.

Anson, J.B. and Maddock, W.G.: Callander's surgical anatomy, ed. 3, Philadelphia, 1952, W.B. Saunders Co.

Armenta, E., and Lehrman, A.: The vincula to the flexor tendons of the hand, J. Hand Surg. **5**:127, 1980.

Ashby, B.S.: Hypertrophy of the palmaris longus muscle: report of a case, J. Bone Joint Surg. **46-B**:230, 1964.

Benedict, K.T., Jr., Chang, W., and McCready, F.J.: The hypothenar hammer syndrome, Radiology **111**:57, 1974.

Bora, F.W., Jr., Richardson, S., and Black, J.: The biomechanical responses to tension in a peripheral nerve, J. Hand Surg. **5**:21, 1980.

Bowers, W.H., et al.: The proximal interphalangeal joint volar plate: I, an anatomical and biomechanical study, J. Hand Surg. **5**:79, 1980.

Brand, P.W., Beach, R.B., and Thompson, D.E.: Relative tension and potential excursion of muscles in the forearm and hand, J. Hand Surg. **6**:209, 1981.

Brockis, J.G.: The blood supply of the flexor and extensor tendons of the fingers in man, J. Bone Joint Surg. **35-B**:131, 1953.

Brooks, D.: The place of nerve-grafting in orthopaedic surgery, J. Bone Joint Surg. **37-A**:299, 1955.

Bunnell, S.: Surgery of the hand, eds. 2 and 3, Philadelphia, 1948, 1956, J.B. Lippincott Co.

Burman, M.: Tendinitis of the insertion of the common extensor tendon of the fingers, J. Bone Joint Surg. **35-A**:177, 1953.

Burton, R.L., and Eaton, R.G.: Common hand injuries in the athlete, Orthop. Clin. North Am. **3**:809, 1973.

Campbell, J.B., et al.: Frozen-irradiated homografts shielded with microfilter sheaths in peripheral nerve surgery, J. Trauma **3**:303, 1963.

Cannon, B., and Peacock, E.E., Jr.: Plastic surgery: the hand, N. Engl. J. Med. **263**:184, 238, 1960.

Carlson, M.J., Linscheid, R.L., and Lucas, A.R.: Recognition of factitial hand injuries, Clin. Orthop. **122**:222, 1977.

Chase, R.A.: Surgical anatomy of the hand, Surg. Clin. North Am. **44**:1349, 1964.

Chase, R.A.: Surgery of the hand: part I, N. Engl. J. Med. **287**:1174, 1972.

Chase, R.A.: Surgery of the hand: part II, N. Engl. J. Med. **287**:1227, 1972.

Conway, H., and Stark, R.B.: Arterial vascularization of the soft tissues of the hand, J. Bone Joint Surg. **36-A**:1238, 1954.

Cooney, W.P., III, and Chao, E.Y.S.: Biomechanical analysis of static forces in the thumb during hand function, J. Bone Joint Surg. **59-A**:27, 1977.

Culver, J.E., Jr.: Extensor pollicis and indicis communis tendon: a rare anatomic variation revisited, J. Hand Surg. **5**:548, 1980.

Cuono, C., and Finseth, F.: Epidermolysis bullosa: current concepts and management of the advanced hand deformity, Plast. Reconstr. Surg. **62**:280, 1978.

Cuono, C.B., and Watson, H.K.: The carpal boss: surgical treatment and etiological considerations, Plast. Reconstr. Surg. **63**:88, 1979.

DeLee, J.C., Smith, M.T., and Green, D.P.: The reaction of nerve tissue to various suture materials: a study in rabbits, J. Hand Surg. **2**:38, 1977.

Dellon, A.L., and Seif, S.S.: Anatomic dissections relating the posterior interosseous nerve to the carpus, and the etiology of dorsal wrist ganglion pain, J. Hand Surg. **3**:326, 1978.

Doyle, J.R., and Blythe, W.F.: Anatomy of the flexor tendon sheath and pulleys of the thumb, J. Hand Surg. **2**:149, 1977.

Durksen, F.: Anomalous lumbrical muscles in the hand: a case report, J. Hand Surg. **3**:550, 1978.

Engber, W.D., and Gmeiner, J.G.: Palmar cutaneous branch of the ulnar nerve, J. Hand Surg. **5**:26, 1980.

Eyler, D.L., and Markee, J.E.: The anatomy and function of the intrinsic musculature of the fingers, J. Bone Joint Surg. **36-A**:1, 1954.

Felländer, M.: Tuberculous tenosynovitis of the hand treated by combined surgery and chemotherapy, Acta Chir. Scand. **111**:142, 1956.

Flatt, A.F.: Kinesiology of the hand. In American Academy of Orthopaedic Surgeons: Instructional course lectures, vol. 18, St. Louis, 1961, The C.V. Mosby Co.

Furnas, D.W.: Muscle-tendon variations in the flexor compartment of the wrist, Plast. Reconstr. Surg. **36**:320, 1965.

Gelberman, R.H., and Menon, J.: The vascularity of the scaphoid bone, J. Hand Surg. **5**:508, 1980.

Gillespie, T.E., Flatt, A.E., Youm, Y., and Sprague, B.L.: Biomechanical evaluation of metacarpophalangeal joint prosthesis designs, J. Hand Surg. **4**:508, 1979.

Hakstian, R.W., and Tubiana, R.: Ulnar deviation of the fingers: the role of joint structure and function, J. Bone Joint Surg. **49-A**:299, 1967.

Harris, H., and Joseph, J.: Variation in extension of the metacarpophalangeal and interphalangeal joints of the thumb, J. Bone Joint Surg. **31-B**:547, 1949.

Howard, L.D., Jr., Pratt, D.R., and Bunnell, S.: The use of compound F (Hydrocortone) in operative and non-operative conditions of the hand, J. Bone Joint Surg. **35-A**:994, 1953.

Huang, T.T., Blackwell, S.J., and Lewis, S.R.: Hand deformities in patients with snakebite, Plast. Reconstr. Surg. **62**:32, 1978.

Jabaley, M.E., Wallace, W.H., and Heckler, F.R.: Internal topography of major nerves of the forearm and hand: a current review, J. Hand Surg. **5**:1, 1980.

Johnson, R.K., and Shrewsbury, M.M.: The pronator quadratus in motions and in stabilization of the radius and ulna at the distal radioulnar joint, J. Hand Surg. **1**:205, 1976.

Kaplan, E.B.: Embryological development of the tendinous apparatus of the fingers: relation to function, J. Bone Joint Surg. **32-A**:820, 1950.

Kaplan, E.B.: Functional and surgical anatomy of the hand, Philadelphia, 1953, J.B. Lippincott Co.

Kelikian, H.: The crippled hand. In American Academy of Orthopaedic Surgeons: Instructional course lectures, vol. 14, Ann Arbor, 1957, J.W. Edwards.

Ketchum, L.D., et al.: A clinical study of forces generated by the intrinsic muscles of the index finger and the extrinsic flexor and extensor muscles of the hand, J. Hand Surg. **3**:571, 1978.

Kisner, W.H.: Double sublimis tendon to fifth finger with absence of profundus, Plast. Reconstr. Surg. **65**:229, 1980.

Kopell, H.P., and Thompson, W.A.L.: Pronator syndrome: a confirmed case and its diagnosis, N. Engl. J. Med. **259**:713, 1958.

Kulowski, J.: Segmental arterial spasm of the brachial artery: report of case treated by procaine into median nerve, Surgery **38**:1087, 1955.

Landsmeer, J.M.F.: The coordination of finger-joint motions, J. Bone Joint Surgery. **45-A**:1654, 1963.

Lassa, R., and Shrewsbury, M.M.: A variation in the path of the deep motor branch of the ulnar nerve at the wrist, J. Bone Joint Surg. **57-A**:990, 1975.

Last, R.J.: Specimens from the Hunterian collection: the ligaments of the tarsus: the interosseous ligaments of the wrist, J. Bone Joint Surg. **33-B**:114, 1951.

Lazar, G., and Schulter-Ellis, F.P.: Intramedullary structure of human metacarpals, J. Hand Surg. **5:**477, 1980.

Leaming, D.B., Walder, D.N., and Braithwaite, F.: A survey of 10,668 patients treated in the hand clinic of the Royal Victoria Infirmary, Newcastle upon Tyne, Br. J. Surg. **48:**247, 1960.

Lewis, R.A., et al.: The hand in mixed connective tissue disease, J. Hand Surg. **3:**217, 1978.

Linburg, R.M., and Comstock, B.E.: Anomalous tendon slips from the flexor pollicis longus to the flexor digitorum profundus, J. Hand Surg. **4:**79, 1979.

Littler, J.W.: The physiology and dynamic function of the hand, Surg. Clin. North Am. **40:**259, 1960.

Littler, J.W.: The finger extensor mechanism, Surg. Clin. North Am. **47:**415, 1967.

Lucas, G.L.: Volar plate advancement. Orthop. Rev. **4:**13, July 1975.

Lundborg, G.: The intrinsic vascularizaton of human peripheral nerves: structural and functional aspects, J. Hand Surg. **4:**34, 1978.

Lundborg, G., and Rydevik, B.: The vascularization of human flexor tendons within the digital synovial sheath region: structural and functional aspects, J. Hand Surg. **2:**417, 1977.

Mason, M.L.: Fifty years' progress in surgery of the hand, Int. Abstr. Surg. **101:**541, 1955.

McFarlane, R.M.: Observations on the functional anatomy of the intrinsic muscles of the thumb, J. Bone Joint Surg. **44-A:**1073, 1962.

Metcalf, W., and Whalen, W.: The surgical, social, and economic aspects of unit hand injury, J. Bone Joint surg. **39-A:**317, 1957.

Micks, J.E., Reswick, J.B., and Hager, D.L.: The mechanism of the intrinsic minus finger: a biomechanical study, J. Hand Surg. **3:**333, 1978.

Millender, L.H., and Nalebuff, E.A.: Preventive surgery: tenosynovectomy and synovectomy, Orthop. Clin. North Am. **6:**765, 1975.

Moes, R.J.: Andreas Vasalius and the anatomy of the upper extremity, J. Hand Surg. **1:**23, 1976.

Moller, J.T.: Lesions of the volar fibrocartilage in finger joints: a 2-year material, Acta Orthop. Scand. **45:**673, 1974.

Nissen, K.I.: The hand and forearm (editorial), J. Bone Joint Surg. **31-B:**498, 1949.

Ochiai, N., et al.: Vascular anatomy of flexor tendons: I, vincular system and blood supply of the profundus tendon in the digital sheath, J. Hand Surg. **4:**321, 1979.

Ogilvie-Harris, D.J., and Fornasier, V.L.: Pathologic fractures of the hand in Paget's disease, Clin. Orthop. **143:**168, 1979.

Paletta, F.X.: Surgical problems of the hand, South. Med. J. **48:**445, 1955.

Palmer, A.K., and Werner, F.W.: The triangular fibrocartilage complex of the wrist: anatomy and function, J. Hand Surg. **6:**153, 1981.

Parks, B.J., and Horner, R.L.: Medical and surgical importance of the arterial blood supply of the thumb, J. Hand Surg. **3:**383, 1978.

Peacock, E.E., Jr.: A study of the circulation in normal tendons and healing grafts, Ann. Surg. **149:**415, 1959.

Pimm, L.H., and Waugh, W.: Tuberculous tenosynovitis, J. Bone Joint Surg. **39-B:**91, 1957.

Pitzler, K.: Tuberculous tenosynovitis of the hand (Ueber die tuberkuloese Tendovaginitis an der Hand und ihre Behandlung), Zbl. Chir. **85:**529, 1960. (Abstracted by Joseph C. Mulier, Int. Abstr. Surg. **111:**489, 1960.)

Posner, M.A.: Injuries to the hand and wrist in athletes, Orthop. Clin. North Am. **8:**593, 1977.

Pulvertaft, R.G.: Surgery of the hand (editorial), J. Bone Joint Surg. **36-B:**3, 1954.

Pulvertaft, R.G.: Contractures of the hand. In Platt, H., editor: Modern trends in orthopedics (second series), New York, 1956, Paul B. Hoeber, Inc., Medical Book Department, Harper & Row, Publishers.

Reading, G.: Secretan's syndrome: hard edema of the dorsum of the hand, Plast. Reconstr. Surg. **65:**182, 1980.

Resnick, D.: Roentgenographic anatomy of the tendon sheaths of the hand and wrist: tenography, Am. J. Roentgenol. Radium Ther. Nucl. Med. **124:**44, 1975.

Robins, R.H.C.: Injuries and infections of the hand, Baltimore, 1961, The Williams & Wilkins Co.

Robins, R.H.C.: Tuberculosis of the wrist and hand, Br. J. Surg. **54:**211, 1967.

Schenck, R.R.: Variations of the extensor tendons of the fingers: surgical significance. J. Bone Joint Surg. **46-A:**103, 1964.

Shrewsbury, M.M., and Johnson, R.K.: A systemic study of the oblique retinacular ligament of the human finger: its structure and function, J. Hand Surg. **2:**194, 1977.

Shrewsbury, M.M., and Johnson, R.K.: Ligaments of the distal interphalangeal joint and the mallet position, J. Hand Surg. **5:**214, 1980.

Simmons, B.P., and Vasile, R.G.: The clenched fist syndrome, J. Hand Surg. **5:**420, 1980.

Snedecor, S.T.: Bone surgery of the hand, Am. J. Surg. **72:**363, 1946.

St. Onge, R., et al.: A preliminary assessment of Na-Hyaluronate injection into "no man's land" for primary flexor tendon repair, Clin. Orthop. **146:**269, 1980.

Stelling, F.H.: Surgery of the hand in the child. In American Academy of Orthopaedic Surgeons: Instructional Course lectures, vol. 15, Ann Arbor, 1958, J.W. Edwards.

Stelling, F.H.: Surgery of the hand in the child, J. Bone Joint Surg. **45-A:**623, 1963.

Taleisnik, J.: The ligaments of the wrist, J. Hand Surg. **1:**110, 1976.

Thompson, J.S., Littler, J.W., and Upton, J.: The spiral oblique retinacular ligament (SORL), J. Hand Surg. **3:**482, 1978.

Tubiana, R., and Valentin, P.: The physiology of the extension of the fingers, Surg. Clin. North Am. **44:**897, 1964.

Tubiana, R., and Valentin, P.: The physiology of the extension of the fingers, Surg. Clin. North Am. **44:**907, 1964.

Van Demark, R.E., Koucky, J.D., and Fischer, F.J.: Peritendinous fibrosis of the dorsum of the hand, J. Bone Joint Surg. **30-A:**284, 1948.

Vichare, N.A.: Anomalous muscle belly of the flexor digitorum superficialis: report of a case, J. Bone Joint Surg. **52-B:**757, 1970.

Walker, P.S., Davidson, W., and Erkman, M.J.: An apparatus to assess function of the hand, J. Hand Surg. **3:**189, 1978.

Wallace, W.A., and Coupland, R.E.: Variations in the nerves of the thumb and index finger, J. Bone Joint Surg. **57-B:**491, 1975.

Wenger, D.R.: Avulsion of the profundus tendon insertion in football players, Arch. Surg. **106:**145, 1973.

Wilson, J.N.: Dystrophy of the fifth finger: report of four cases, J. Bone Joint Surg. **34-B:**236, 1952.

Winkleman, N.Z.: Aberrant sensory branch of the median nerve to the third web space: a case report, J. Hand Surg. **5:**566, 1980.

Zook, E.G., et al.: Anatomy and physiology of the perionychium: a review of the literature and anatomic study, J. Hand Surg. **5:**528, 1980.

CHAPTER 2

Acute injuries

When the hand is acutely injured, the purpose of treatment is to restore its function; thus measures are necessary (1) to prevent infection. (2) to promote primary healing, and (3) to salvage injured parts. Realigning fractures and suturing nerves and tendons may be considered and sometimes may even be done during this primary treatment, but these repairs, except for fracture reduction, are secondary in importance to cleansing the wound of foreign matter and devitalized tissue and to providing complete skin coverage even if this requires skin grafting. The surgeon must therefore, by history and examination, personally appraise the injury to decide what procedures can be safely done primarily and what secondary procedures may be necessary later.

HISTORY

A statement such as "the patient injured his hand in a machine" is not informative enough. The history should provide the following information: (1) the exact time of injury, to determine the interval before treatment, (2) what first aid measures were given and by whom and where, (3) the nature, amount, and time of receiving of any medication, (4) the exact mechanism of injury, to determine the amount of crushing, contamination, and blood loss, (5) the nature and amount of food and liquid taken by the patient, and when it was taken—information necessary for selection of the anesthetic, and (6) the patient's age, occupation, place of employment, handedness, and general health status.

FIRST AID

The hand should be immediately covered with a sterile dressing to prevent further contamination. When the wound is severe and bleeding, the hand should be elevated with the patient lying supine; if bleeding is not controlled by elevation alone, pressure must be applied to the wound through the dressing. Frequently bleeding is quickly controlled by removing an improperly applied tourniquet; rarely is a tourniquet necessary.

Do not attempt to clamp arteries and veins at this time. Microsurgical techniques may be indicated to restore these vessels, and crushed vessels complicate this technique.

FIRST EXAMINATION

The wound is examined in two stages. The first examination is made before surgery, but under sterile conditions; masks should cover the faces of the examiner and the patient, and sterile instruments and gloves are used. The purpose of this examination is to estimate the size of the wound and to determine the extent of skin loss, if any, and of injury, if any, to the deep structures; these are not examined by probing the depth of the wound. The viability of the skin and any gross positional deformity are first noted. The wound is then covered, and tests are attempted to help in determining what deep structures are functioning; *each one of these structures must be considered damaged until proved otherwise*. To lessen the chance of error the assessment of each tissue should be orderly; attention is first directed to the skin and then to bones, tendons, and nerves.

The condition of the skin is noted as just suggested: the problem of primary closure by skin suture or by appropriate grafts is constantly kept in mind.

Roentgenograms of a severely injured hand are made routinely to reveal fractures or metallic foreign bodies.

Next perform an initial assessment of possible severed tendons and nerves. The technique for this evaluation is found in detail in Chapters 3 and 5, respectively.

From the first examination, as just described, the surgeon should obtain some idea as to what procedures will be necessary; however, final decision is withheld until the second examination is done during surgery. The patient is also examined for other injuries, and his general condition is evaluated. Antibiotics, sedation, blood transfusions, tetanus antitoxin, and other measures are started as indicated.

The patient should be advised as to the extent of his injuries and especially with regard to any possible amputation; he should also be forewarned when the use of skin grafts, especially of a distant flap graft, is anticipated.

Anesthesia

A general anesthetic or a regional block may be used, depending on the severity of the injury, the interval since the last ingestion of food or drink, and whether a distant flap will be necessary (see p. 5 for more details).

Tourniquet

A tourniquet is necessary while the wound is being cleansed and inspected and while the deep structures are being repaired, but when the viability of an area of skin is questionable because of a crushing or avulsing injury, a tourniquet should be used as briefly as possible. In these instances the ultimate viability of the skin may be jeopardized if the tourniquet is used for too long a time. When there is a large wound with fractures, elevation of the hand for 2 minutes is better than wrapping it with a Martin bandage before inflation of the tourniquet; thus further crushing of a severely injured hand and possible further displacement of any fractures are prevented (for more information on the tourniquet see p. 1).

Cleansing and draping of hand

After the patient or the part is anesthetized and the tourniquet is applied, the first aid dressing is removed and a sterile pad is placed over the major wound. The hand is held over a drain basin, and the surrounding uninvolved skin is shaved and then scrubbed with antiseptic soap and water to above the elbow. The nails and nail beds are cleansed, and the nails are trimmed. Next the wound is exposed and irrigated with normal saline solution (Fig. 2-1); antiseptics are not used in the wound, for they would irritate the tissues. Normal saline solution is poured into the wound from a large flask to provide a stream with enough force to loosen small foreign particles and to remove large hematomas. A gloved finger may be placed in

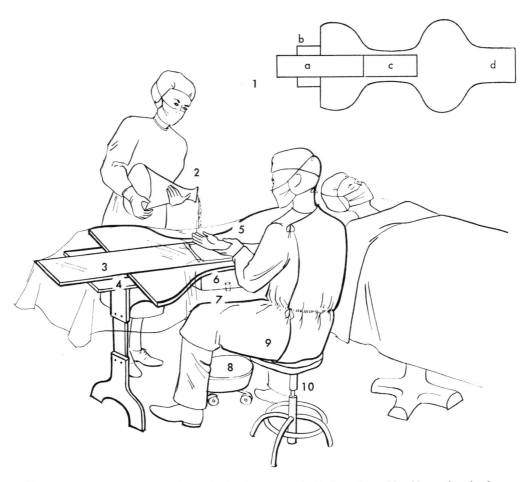

Fig. 2-1. Hand table and routine of cleansing hand. *1,* View of table from above. Movable panel, *a,* has been pushed to left so that pan may be inserted at *c.* When panel, *a,* has been pushed to right and table has been draped, basic instruments are placed on shelf, *b.* End of table, *d,* is placed beneath patient. *2,* Assistant irrigates open wound with normal saline solution. *3,* Movable panel has been pushed to left. *4,* Shelf on which basic instruments will be placed. *5,* Plastic sheet in which hole has been made to receive pan has been spread over table. *6,* Pan has been inserted to receive irrigating solution. *7,* Drainpipe in bottom of pan. *8,* Bucket into which irrigation solution flows from pan. *9,* Surgeon is seated comfortably on stool, *10,* which is firm and stable.

the wound to loosen the hematoma or to palpate the bones, but the depths of the wound should not be rubbed with a sponge or brush. Small bleeding vessels, which are sometimes more easily seen under water, are clamped with mosquito hemostats and later cauterized. Small villi of devitalized tissue seen floating in the solution are removed at their bases; these usually consist of fascia or fatty tissue. Nerve ends are not debrided. Ragged skin edges may be trimmed, but complete excision of the edges of the wound is not done routinely in the hand.

As the deeper parts of the wound are cleaned, they are carefully searched for foreign materials, especially if there is suspicion that they contain broken glass, wood, or pieces of glove, or when the wound has been caused by a gunshot. Cleaning should not be hurried and often may take up to one half of the total operating time; it must be thorough to prevent infection. Primary closure without infection is necessary to limit the scar and to allow additional early reconstruction, if necessary. When the cleaning is complete, all instruments, gloves, and drapes used during this process are discarded, and the hand is redraped (see other sources for the details of draping and of the routine in the operating room).

SECOND EXAMINATION

After a diligent effort has been made to convert the contaminated wound into a clean one in the operating room, the wound is examined again. The tissues in the depths of the wound, including exposed bones, tendons, and nerves, are assessed in an orderly anatomic manner to avoid error; the skin is also carefully examined. Only after an accurate assessment of the damage can correct decisions be made as to which structures may be repaired primarily. Bones and joints are inspected to see if segments are missing and to determine the extent of periosteal stripping, the expected time required for union, and the possibility of allowing early joint motion after Kirschner wire fixation of any fractures. Suspected damage to tendons and nerves must now be confirmed by direct vision since the conclusions drawn from the first examination are often wrong; usually the damage has been underestimated. Severed tendon ends are often brought into view by passive finger motion. When small hematomas are seen within synovial sheaths, further tendon injury should be suspected. Evaluating the skin damage is most important because primary closure may depend on this evaluation. Frequently some skin appears to be lost when actually it has only retracted; this is especially true of L-shaped wounds on the dorsum of the hand. When skin is crushed or flaps of skin are avulsed, the possiblity of necrosis must be weighed carefully; releasing the tourniquet may be necessary at this time for accurate evaluation. A valuable sign that skin is viable is a blush immediately after release of the tourniquet. The extent of bleeding from the skin edges, the color of the skin immediately after compression, and the amount of undermining of the skin edges must all be observed. Necrosis, infection, and scarring may all occur when flaps of doubtful viability are retained. The extent of skin loss either from the injury itself or from surgical excision of nonviable flaps must be evaluated, and plans must be made for complete coverage.

CONSIDERATIONS FOR AMPUTATION

Considerations for amputation are discussed on p. 181.

CONSIDERATIONS FOR SKIN CLOSURE

Primary skin closure is desirable and can usually be done in all incised, tidy wounds. The purpose of primary skin closure is to obtain early healing and to avoid infection, granulation tissue, and excessive scar production. Misjudgment leads to delayed healing resulting from hematoma, excessive swelling, and possible infection, any of which may require reopening the wound for drainage and an even greater delay in healing. Certain wounds should never be closed primarily; these include the severely contaminated or crushed wounds caused by farm machinery, human bites, tornado missiles, and augers. War wounds, such as high-velocity missile wounds, and wounds contaminated with human excreta or fertilizer also should not be closed primarily.

When in doubt, the wound should be left open after careful debridement under an anesthetic. Within 24 to 48 hours the wound should be reinspected, and if it is sufficiently clean, it can be closed by direct suture or by skin graft. A wound should not be left open to granulate and heal by scar unless it cannot be rendered clean, but should be closed if possible within 5 days after injury.

METHODS OF SKIN CLOSURE AND INDICATIONS FOR EACH

Unless severely contaminated or crushed, every wound of the hand (except those mentioned in the preceding paragraph) should be closed primarily, because healing by primary intention is the desired result. As stated previously, most *incised* wounds can be closed by simple direct suture of the skin; the subcutaneous tissue is not sutured separately. Careful hemostasis is necessary. Closure is easier when all viable skin edges have been preserved during the initial cleansing.

Wounds with distally attached flaps may have enough skin for primary closure but not enough venous drainage for the skin to survive. This deficient drainage causes engorgement and venous distention and finally thrombosis and necrosis; the color of the flap changes from a deep blue to purple and then to black. The retrograde flap is often the result of a crushing or tearing injury, and the very nature of this injury even further jeopardizes the survival of the flap. Such a flap on the dorsal surface of the hand or forearm is less likely to survive than one on the palm (Fig. 2-2). When there is doubt, the skin should be excised and replaced with a split graft (p. 14).

When skin is lost without exposing deep structures such as nerves, tendons, joints, or cortical bone, it should be immediately replaced with either a split-thickness graft or occasionally a full-thickness one. Sometimes a skin defect on the dorsal surface of the hand may be converted to a transverse elliptic one and closed in a transverse line; since the skin is mobile, this type of closure is possible here when the wrist is dorsiflexed.

When a skin defect exposes deep structures, a split-thickness or full-thickness free graft is insufficient. These structures require nutrition to survive; also they will not readily support a free graft. A flap graft is necessary to provide subcutaneous tissue for coverage and for sufficient nutrition. This flap may be a local one, but usually it is obtained from a distance.

Coverage of specific areas with flaps

A large skin defect on the dorsum of a finger that exposes tendons not covered with paratenon should be covered with a flap. Frequently a double local flap can be constructed by rotating a proximally based local flap on one side and a distally based local flap on the other side of the defect to cover the exposed tendon. The donor defects, of course, are covered with split grafts. When multiple fingers are involved or when there is a need for larger area coverage, a subpectoral flap may be in order. Thick subcutaneous fat, as is found on the lower abdomen, is not preferred on a flap.

Defects of the volar surface of a finger that expose tendons may be covered with a cross finger flap. The flap is

raised from the dorsal surface of an adjacent finger and extends from the middle of one of its lateral surfaces to that of the other; the flap is a little wider than the defect it is to cover. Although such a flap from the dorsal surface of one finger may be used to cover a defect on the volar surface of another, the reverse is never indicated. The use of flaps for amputations of the fingertip is discussed on p. 185 and of the thumb on p. 191.

Skin defects on the palm or dorsum of the hand that expose vital structures may be covered with a flap from an adjacent unsalvageable finger, from the opposite forearm or upper arm, or from the abdomen, depending on the size of the defect and the presence and location of any associated injuries. An abdominal flap from the same side allows the most comfortable position of the arm. To ensure survival of the flap (since it must be applied immediately), its base should be as wide as its length. The donor area and the raw part of the flap that will not make contact with the defect should be covered with split-thickness skin grafts. A small defect may be covered by rotating a local flap; this flap is unlikely to survive, however, if undermining of the skin must be extensive and if the skin is already crushed or contused from the injury. A filleted finger makes an excellent pedicle graft when this technique is applicable (p. 35).

CROSS FINGER FLAPS

Cross finger flaps (Fig. 2-3) are useful for covering a defect of the skin and other soft tissues on the volar surface of the finger when avascular structures are exposed. They are also useful for some amputations of the thumb (p. 191). These grafts should be avoided, however, in patients over 50 years of age, in hands with arthritic changes or a tendency to finger stiffness for some other reason, or when there is a local infection.

TECHNIQUE. Excise the edges of the defect so that it is rectangular, with its longer sides parallel to the long axis of the finger but not crossing skin creases. Then measure

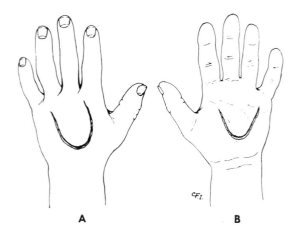

A   **B**

Fig. 2-2. Flap attached distally on dorsum of hand, **A,** is less likely to survive than is similar one on palm, **B.**

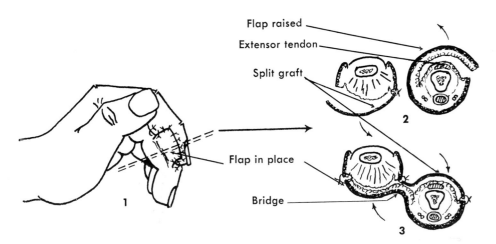

Fig. 2-3. Cross finger flap. *1,* Laterally based pedicle flap has been raised from middle finger and has been applied to distal pad of index. *2* and *3,* Cross sections of two fingers showing how cross finger flap has been applied and how raw surfaces of donor finger and of bridge between two fingers have been covered with split-thickness skin graft (see text). (From Hoskins, H.D.: J. Bone Joint Surg. **42-A:**261, 1960.)

its dimensions. Place the injured finger against the donor finger and determine where to locate the base of the proposed flap. Cut the flap from the donor finger through the skin and subcutaneous tissues, leaving its base attached to the side adjacent to the recipient finger (Fig. 2-4). Make the flap 6 mm wider than the defect and long enough both to cover the defect (allowing for normal skin contraction) and to provide a bridge between the fingers; if necessary the flap may be raised from one midlateral line of the donor finger to the other, but be careful to avoid incising the volar surface of the finger.

When raising the flap, make the incisions through the subcutaneous tissue but not through the peritenon of the extensor expansion (Fig. 2-5). When possible, avoid using skin distal to the distal interphalangeal joint so as not to injure the nail bed; the skin over the dorsal surface of the proximal interphalangeal joint should also be avoided unless necessary for width. When necessary, the base of the flap may be further freed by cutting the oblique fibers of the deep tissue that attaches the skin to the extensor tendon and periosteum along the side of the finger. Handle the flap with small hooks to prevent crushing and necrosis.

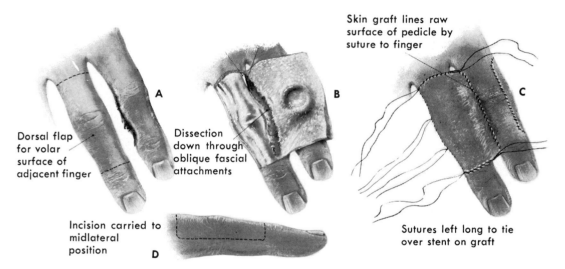

Fig. 2-4. Technique of applying a cross finger flap, using skin from dorsum of two phalanges (see text). (From Curtis, R.M.: Ann. Surg. **145:**650, 1957.)

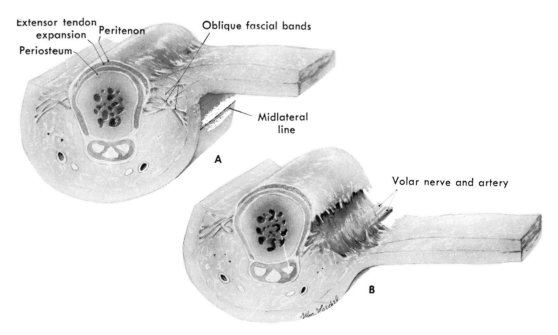

Fig. 2-5. Details of raising cross finger flap. **A,** Incision has been carried to but not through peritenon of extensor expansion; note oblique fascial bands. **B,** Incision has been continued in a way to sever oblique fascial bands but to avoid damaging volar digital artery and nerve. (Modified from Curtis, R.M.: Ann. Surg. **145:**650, 1957.)

Release the tourniquet and obtain absolute hemostasis; reinflate the tourniquet. Cut a thick split graft (up to 0.045 cm) from the forearm or thigh and suture it to the donor area and to the undersurface of the bridge. Now apply the flap to the recipient area and suture it in place with the finest suture (No. 5-0 or 6-0 nylon); the entire recipient area should be in contact with the flap. Leave the sutures long at the edges of the free split graft, and fashion a stent dressing. Cover the suture line with gauze impregnated with bismuth tribromophenate (Xeroform); then place moist cotton pledgets about the graft and apply gauze wrapping. To ensure immobility of the recipient finger, an oblique Kirschner wire through the interphalangeal joint may sometimes be used.

AFTERTREATMENT. The flap may be detached after 12 to 14 days. The skin margins of the recipient finger should be trimmed so that the junction of the normal skin with the graft is at a midlateral position on the finger. Motion of both fingers can be started the day after the flap is detached.

This technique may be varied so that the base of the flap is proximal rather than lateral; such a flap is useful for covering a defect near the tip of an adjacent finger or thumb (Fig. 2-6). Rotation of the flap is necessary, and care should be taken to prevent strangulation at the base and necrosis. Rotated flaps that are based proximally may be used to cover defects on the same finger (see Fig. 1-25).

ABDOMINAL FLAPS

An abdominal flap to be applied to the hand has its base either distal, toward the superficial epigastric vessels and on the same side as the hand to be covered, or proximal, above the umbilicus toward the thoracoepigastric vessels and on the opposite side (Fig. 2-7); the latter flap should not be used in barrel-chested patients with emphysema. Abdominal flaps generally should be obtained from above the umbilicus, thus avoiding the fat "storage area." Recipient areas usually gain weight as the storage area below the umbilicus adds weight.

TECHNIQUE. On sterile paper make a pattern of the defect and outline it on the abdomen; then outline the flap but make it enough larger than the pattern to allow for normal skin contraction and for the bridge between the abdomen and the defect. As a rule the flap should be rectangular to avoid a circular outline when the flap is attached to the hand. The most frequent mistake is making the flap too thick. When possible, follow the principles of placing hand incisions (p. 9) to avoid tension lines and excessive scarring. Using sharp dissection, raise the skin flap of the desired size and thickness (Figs. 2-8 and 2-9); maintain hemostasis and handle the fat carefully to prevent necrosis. Close the defect at the donor site by widely undermining its edges and suturing them together or by applying a split-thickness skin graft, or do both. With a split skin graft, cover that part of the undersurface of the flap that will not

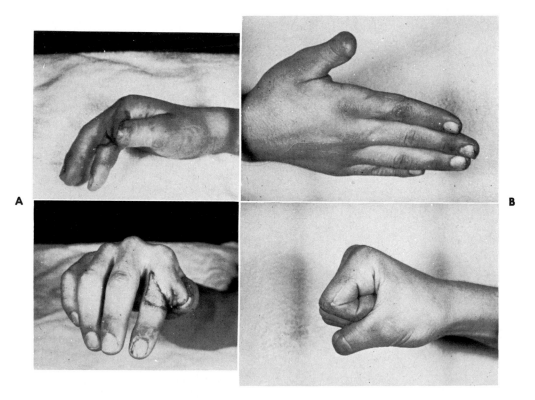

Fig. 2-6. Cross finger flap that is based proximally. **A,** Twenty days after surgery. Only radial part of the flap covers defect on thumb. Triangular defect on donor finger has been covered with split-thickness graft. **B,** Six months after surgery. Flap has been detached and has healed without redundancy. Donor finger has been disfigured little. (From Hoskins, H.D.: J. Bone Joint Surg. **42-A:**216, 1960.)

cover the hand defect. Slightly undermine the edges of the defect on the hand and apply the flap over the entire defect; suture the edges of the flap to those of the defect and suture the free edge of the split graft to that edge of the defect nearest to the base of the pedicle, thus covering all raw surfaces. Place strips of Xeroform gauze over the suture line and a dry dressing on the flap; be careful to prevent kinking, tension, and rotation at its base. Using flannel cloth reinforced with plaster, apply a bandage around the trunk and shoulder supporting the hand. The flap should be easily accessible for inspection through the dressing. When marked pronation or supination of the forearm is necessary to prevent tension on the flap, a heavy transverse pin through the radius and ulna just proximal to the wrist is helpful in maintaining this position.

AFTERTREATMENT. The flap should be inspected almost hourly during the first 48 hours for circulatory embarrassment produced by tension or torsion, or for the development of a hematoma.

If an area becomes necrotic, it should be excised and covered with a split graft. Gross infection from necrosis or hematoma will result in certain failure. The area should be redressed frequently to avoid offensive odor and decrease the chance of infection. Usually the flap may be safely detached after 3 weeks. In children this can be reduced to 2 weeks.

FILLETED GRAFTS

A filleted graft is a flap of tissue fashioned from a nearby part, usually a finger or toe, from which the bone has been removed but in which one or more neurovascular bundles have been retained. In the hand such a graft is indicated only when deep tissues such as tendons, nerves, and joints are exposed and when a nearby damaged finger is to be sacrificed because it is not salvageable; it is *never* used at the expense of a useful part.

A filleted graft is especially convenient when other injuries more proximal in the same extremity would interfere with positioning the hand to receive a flap from a distant part. The advantages of this graft are (1) it can be applied in a one-stage procedure at the time of injury and is obtained from within the same surgical field as the injured part, (2) its survival is almost assured because one or more of its neurovascular bundles are preserved, (3) its skin is similar to that which is to be replaced, (4) it is not attached to a distant part, and consequently after surgery the hand may be splinted in the position of function and elevated,

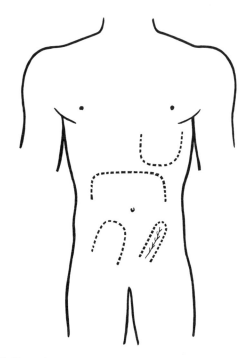

Fig. 2-7. Examples of abdominal flaps (see text for details as to length and width of flaps). Lower abdominal flap may be made narrower in relation to its length if it contains superficial circumflex iliac artery and vein *(lower right)* or superficial epigastric artery and vein.

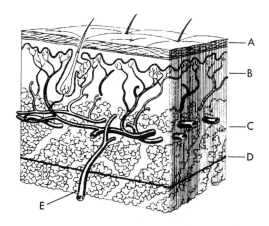

Fig. 2-8. Dissection of skin and subcutaneous fat. *A,* Epidermis; *B,* dermis; *C,* subdermal plexus of vessels; *D,* superficial fascia; *E,* arteries perforating the muscularis and deep fascia to join subdermal plexus of vessels. (From Kelleher, J.C., et al.: J. Bone Joint Surg. **52-A:**1552, 1970.)

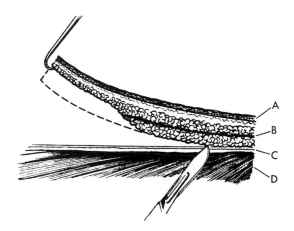

Fig. 2-9. Cross section of an abdominal pedicle flap being raised. *A,* Epidermis and dermis; *B,* superficial fascia of abdomen; *C,* deep fascia of abdomen; *D,* muscularis. *Dotted line* indicates extent of defatting of portion of pedicle to be applied to hand. Base or stem should retain sufficient fat to retain its shape sufficiently to prevent kinking. (From Kelleher, J.C., et al.: J. Bone Joint Surg. **52-A:**1552, 1970.)

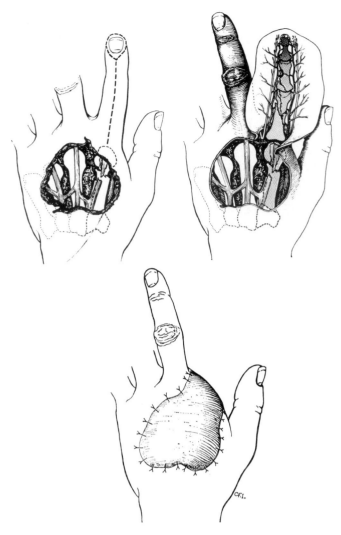

Fig. 2-10. Filleted graft fashioned from finger (see text).

and (5) it provides an adequate thumb web when the index finger is the donor.

TECHNIQUE. Because the main vessels course anterolaterally through the digit, it is easier to fashion a flap with its base anterior and cover a defect on the dorsum of the hand than vice versa (Fig. 2-10). Make a midline dorsal incision along the full length of the finger and skirt it around the nail distally. Deepen the dissection to the extensor tendon. Then remove this tendon, the underlying bone, and the flexor tendons and their sheath, but preserve the fat in which the neurovascular bundles are located; take great care to avoid damaging the bundles. Spread the flap thus created and place it on the donor area. If it is too wide, trim its edges, or if it is too long, excise its end; in the latter instance ligate the digital vessels and resect the digital nerves far enough proximally to prevent their being caught in the scar. Suture the flap in place so that it lies flat; avoid strangulating its base and trim only slightly any dog-ears that may be produced at the margins of the base so as to preserve the blood supply of the flap.

Order of tissue repair

To set priorities for repair of each tissue is important. After the wound is cleaned, the bony architecture must be reestablished immediately, or at least within a few days after the wound becomes clean; otherwise, the soft tissues will shorten and contract, thus making them impossible to repair without grafting. The wound should be closed within the first 5 days. Even if the wound is not closed, the bony architecture still should be reestablished. Tendons and nerves should be repaired at the time of skin closure or after secondary closure has been accomplished. While awaiting repair, nerves as well as tendons will contract, especially in the fingers and palm. Therefore, consideration should be given to tagging the nerve ends with a small suture, not necessarily together, but to the soft tissues of the palm. After closure, the nerves and then the tendons may be repaired, in that order. It is easier to suture the tendons and then repair the nerves than to attempt to suture the nerves first and maintain their continuity while manipulating the hand to suture the tendons.

Arterial injuries

When either the radial or ulnar artery alone is lacerated at the wrist level or more distally, the circulation in the hand remains sufficient. Even when both arteries are lacerated, the prognosis for survival of the hand is excellent in young people, good in those in early middle age, and fair in those who are older; in these circumstances circulation through the collaterals or an uninjured patent median artery or both is usually adequate. Even in older patients, repairing the radial or ulnar artery at the wrist is optional because arteriograms have demonstrated that these vessels may remain patent after surgery for only a few days at best. It may be argued, however, that circulation obtained in this manner, even though temporary, may be sufficient to sustain the hand while the collateral circulation is developing. Repair of arteries and veins distal to the wrist is resulting in increasingly successful survival of severely impaired or amputated digits. This requires thorough knowledge and skill in microsurgical technique (Chapter 19).

The diagnosis and treatment of closed arterial injuries of the hand and wrist are discussed on p. 149. Also see ulnar tunnel syndrome (Chapter 16).

REFERENCES
Open hand injuries (general)

Alpert, B.S., and Buncke, H.J.: Mutilating multidigital injuries: use of a free microvascular flap from a nonreplantable part, J. Hand Surg. **3:**196, 1978.

Aste, J.M.: Care of the injured hand, South. Med. J. **50:**600, 1957.

Bevin, A.G., and Chase, R.A.: The management of ring avulsion injuries and associated conditions in the hand, Plast. Reconstr. Surg. **32:**391, 1963.

Bilos, Z.J., and Eskestrand, T.: External fixator use in comminuted gunshot fractures of the proximal phalanx, J. Hand Surg. **4:**357, 1979.

Blair, W.F., and Marcus, N.A.: Extrusion of the proximal interphalangeal joint: case report, J. Hand Surg. **6:**146, 1981.

Burkhalter, W.E., et al.: Experiences with delayed primary closure of war wounds of the hand in Viet Nam. J. Bone Joint Surg. **50-A:**945, 1968.

Earle, A.S., and Vlastou. C.: Crossed fingers and other tests of ulnar nerve motor function, J. Hand Surg. **5**:560, 1980.

Flatt, A.E.: Minor hand injuries, J. Bone Joint Surg. **37-B**:117, 1955.

Flynn, J.E.: Problems with trauma to the hand, J. Bone Joint Surg. **35-A**:132, 1953.

Flynn, J.E.: Acute trauma to the hand, Clin. Orthop. **13**:124, 1959.

Furlong, R.: Injuries of the hand, Boston, 1957, Little, Brown & Co.

Holevich, J.: Early skin-grafting in the treatment of traumatic avulsion injuries of the hand and fingers, J. Bone Joint Surg. **47-A**:944, 1965.

Hoskins, H.D.: The versatile cross-finger pedicle flap: a report of twenty-six cases, J. Bone Joint Surg. **42-A**:261, 1960.

Hunt, J.C., Watts, H.B., and Glasgow, J.D.: Dorsal dislocation of the metacarpophalangeal joint of the index finger with particular reference to open dislocation, J. Bone Joint Surg. **49-A**:1572, 1967.

Imbriglia, J.E., and Sciulli, R.: Open complex metacarpophalangeal joint dislocation. Two cases: index finger and long finger, J. Hand Surg. **4**:72, 1979.

Kelleher, J.C., et al.: The distant pedicle flap in surgery of the hand, Orthop. Clin. North Am. **2**:227, 1970.

Kelleher, J.C., et al.: Use of a tailored abdominal pedicle flap for surgical reconstruction of the hand. J. Bone Surg. **52-A**:1552, 1970.

Kelly, A.P., Jr.: Primary tendon repairs: a study of 789 consecutive tendon severances, J. Bone Joint Surg. **41-A**:581, 1959.

Kilbourne, B.C., and Paul, E.G.: Do's and don'ts in the treatment of hand injuries, Surg. Clin. North Am. **38**:139, 1958.

Kleinert, H.E., et al.: Primary repair of flexor tendons, Orthop. Clin. North Am. **4**:865, 1973.

Lindsay, W.K., and McDougall, E.P.: Direct digital flexor tendon repair, Plast. Reconstr. Surg. **26**:613, 1960.

Littler, J.W.: The severed flexor tendon, Surg. Clin. North Am. **39**:435, 1959.

London, P.S.: Simplicity of approach to treatment of the injured hand, J. Bone Joint Surg. **43-B**:454, 1961.

Mason, M.L.: Primary tendon repair (Editorial), J. Bone Joint Surg. **41-A**:575, 1959.

Mason, M.L., and Bell, J.L.: The treatment of open injuries to the hand, Surg. Clin. North Am. **36**:1337, 1956.

Mason, M.L., and Bell, J.L.: The crushed hand, Clin. Orthop. **13**:84, 1959.

Maxim, E.S., Webster, F.S., and Willander, D.A.: The cornpicker hand, J. Bone Joint Surg. **36-A**:21, 1954.

McCormack, R.M.: Reconstructive surgery and the immediate care of the severely injured hand, Clin. Orthop. **13**:75, 1959.

McCormack, R.M.: Primary reconstruction in acute hand injuries, Surg. Clin. North Am. **40**:337, 1960.

Milford, L.: Shotgun wounds of the hand and wrist: with a report of four cases, South. Med. J. **52**:403, 1959.

Milford, L.: Resurfacing hand defects by using deboned useless fingers, Am. Surg. **32**:196, 1966.

Miller, H.: Acute open flexor tendon injuries of the hand, Clin. Orthop. **13**:135, 1959.

Moberg, E.: The treatment of mutilating injuries of the upper limb, Surg. Clin. North Am. **44**:1107, 1964.

Nemethi, C.E.: The primary repair of traumatic digital skeletal losses by phalangeal recession, J. Bone Joint Surg. **37-A**:78, 1955.

Phelps, D.B., Buchler, U., and Boswick, J.A., Jr.: The diagnosis of factitious ulcer of the hand: a case report, J. Hand Surg. **2**:105, 1977.

Posch, J.L.: Injuries to the hand in children, Am. J. Surg. **89**:784, 1955.

Rank, B.K., and Wakefield, A.R.: Surgery of repair as applied to hand injuries, Baltimore, 1960, The Williams & Wilkins Co.

Riordan, D.C.: Emergency treatment of compound injury of the hand, Orthopedics **1**:30, 1958.

Siler, V.E.: Primary tenorrhaphy of the flexor tendons in the hand, J. Bone Joint Surg. **32-A**:218, 1950.

Siler, V.E.: Combined nerve and tendon injuries in the hand and forearm, Am. Surg. **22**:764, 1956.

Sponsel, K.H.: Urgent surgery for finger flexor tendon and nerve lacerations: with emphasis on advancement of the divided profundus tendon distal to the level of laceration. JAMA **166**:1567, 1958.

Stromberg, W.B., Jr., Mason, M.L., and Bell, J.L.: The management of hand injuries, Surg. Clin. North Am. **38**:1501, 1958.

Sullivan, J.G., et al.: The primary application of an island pedicle flap in thumb and index finger injuries, Plast. Reconstr. Surg. **39**:488, 1967.

Van't Hof, A., and Heiple, K.G.: Flexor-tendon injuries of the fingers and thumb: a comparative study, J. Bone Joint Surg. **40-A**:256, 1958.

Verdan, C.E.: Primary repair of flexor tendons, J. Bone Joint Surg. **42-A**:647, 1960.

Verdan, C.E.: Practical considerations for primary and secondary repair in flexor tendon injuries, Surg. Clin. North Am. **44**:951, 1964.

Verdan, C.E.: Primary and secondary repair of flexor and extensor tendon injuries. In Flynn, J.E., editor: Hand surgery, Baltimore, 1966, The Williams & Wilkins Co.

Wakefield, A.R.: The management of flexor tendon injuries, Surg. Clin. North Am. **40**:267, 1960.

CHAPTER 3

Tendon injuries

EXAMINATION

Errors are always possible when making an examination for tendon injuries, because painful movements of the injured hand by the patient or by the examiner are necessary; this is equally true in examination for nerve injuries. Even when gross deformity is absent, the posture of the hand often provides clues as to which tendons are severed (Fig. 3-1).

Flexor tendons

When both flexor tendons of a finger are severed, the finger lies in an unnatural position of hyperextension. This may be further confirmed by passively extending the wrist, for in these instances this maneuver does not produce flexion of the fingers; by flexing the wrist, even greater unopposed extension of the affected finger is produced. Loss of normal tension in the involved finger may be found by gently pressing the fingertip and comparing the tension of it with that of each adjoining member. Tendon function is most frequently tested by active movements of the finger as directed by the examiner; this test is probably the least dependable and is almost worthless when the patient is an excited child or an adult under the influence of alcohol. When this test is used, the examiner must demonstrate with his own hand the maneuvers that are requested. When a wound is distal to the wrist, the examiner may need to stabilize the finger to obtain specific finger motions. The profundus flexor tendon is presumed severed when the distal interphalangeal joint cannot be flexed while the proximal interphalangeal joint is stabilized (Fig. 3-2). When neither the proximal nor distal interphalangeal joint can be actively flexed while the metacarpophalangeal joint is sta-

39

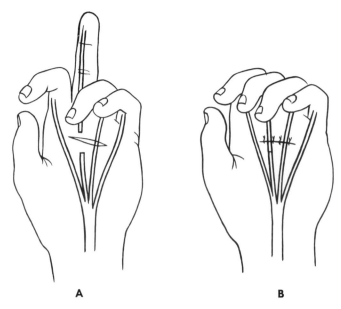

A   B

Fig. 3-1. A, If middle finger remains extended when hand is at rest, its flexor tendons have been severed. **B,** This finger becomes normally flexed after its profundus tendon or both this tendon and sublimis have been repaired.

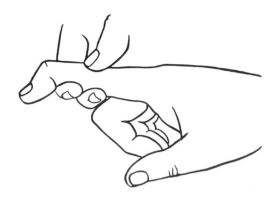

Fig. 3-2. If distal interphalangeal joint can be actively flexed while proximal interphalangeal joint is stabilized, profundus tendon has not been severed.

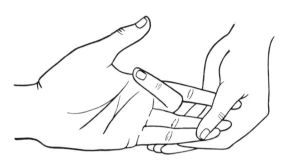

Fig. 3-3. If proximal interphalangeal joint can be actively flexed while adjacent fingers are held completely extended, sublimis tendon has not been severed (see text).

bilized, both flexor tendons are severed. To demonstrate the division of a sublimis tendon without that of the profundus, the two adjacent fingers are held in complete extension; this maneuver anchors the profundus tendon in the extended positon and prevents its flexing the proximal interphalangeal joint. Thus when a sublimis tendon is severed and the two adjacent fingers are held extended, flexion of the proximal interphalangeal joint is impossible (Fig. 3-3). To check the integrity of the flexor pollicis longus tendon, the metacarpophalangeal joint of the thumb should be stabilized so that flexion occurs at the interphalangeal joint only when this tendon is intact; flexion here is absent when the tendon is divided.

When a wound is at the wrist level, sometimes the joints of a finger may be actively flexed even when the tendons to that finger are severed because of the intercommunication of the profundus tendons at the wrist; this is particularly true of the little and ring fingers.

These tests for tendon injuries will not detect a partially divided tendon, for in this instance the tendon still functions; however, under these circumstances finger motion may be limited by pain, and the tests would then seem to indicate complete division of the tendon.

Extensor tendons

An extensor tendon (Fig. 3-4) is presumed to be divided between the proximal and distal interphalangeal joints when active extension of the latter joint is lost; initially a gross mallet finger deformity may be absent because the surrounding capsule and other soft tissues have not yet been stretched by the strong flexor digitorum profundus. The division of the central slip of an extensor tendon between the metacarpophalangeal and proximal interphalan-

geal joints results in loss of extension of the latter joint only after the lateral bands prolapse anteriorly; since the metacarpophalangeal and distal interphalangeal joints may both be actively extended, this lesion is easily overlooked during the initial examination. When the entire extensor expansion, including the lateral bands, is divided at this level, extension of the joints distal to the wound is lost; such a lesion is unlikely, however, since the expansion covers a convex surface of bone that usually blocks the injuring object before the division is complete. When the extensor tendon is divided just proximal to the metacarpophalangeal joint, the two distal finger joints can be extended by the lateral bands and their connecting transverse fibers, but extension of the metacarpophalangeal joint is incomplete. Partial or complete extension of the finger may be possible when a single extensor tendon is divided at the wrist because of the presence of accessory communicating tendons (vincular accessorium), as shown in Fig. 3-5.

When checking the long extensor tendon of the thumb, the examiner must stabilize the metacarpophalangeal joint and must carefully test for active extension of the interphalangeal joint. Division of this tendon is often overlooked because an intact short thumb extensor can actively extend the thumb as a unit. However, the short thumb extensor will not extend the interphalangeal joint alone.

Fig. 3-4. Lateral view of entire extensor mechanism.

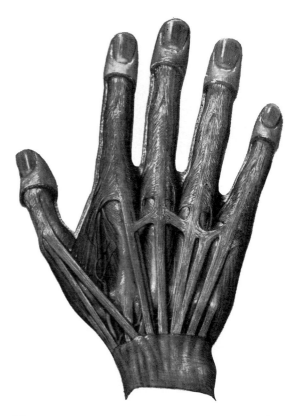

Fig. 3-5. Dorsal view of hand showing extensor tendons, accessory communicating tendons (vincular accessorium), and extensor expansions.

BASIC TENDON TECHNIQUES

The purpose of tendon suture is to approximate the ends of a tendon or to fasten one end of a tendon to adjoining tendons or to bone and to hold this position during healing. When tendons are being sutured, management should be delicate, causing as little reaction and scarring as possible. No blemish should be created on the gliding surface of the tendon; all crushed parts should be removed after the sutures are secured.

Multiple materials are now available for tendon suture that are less reactive than silk. However, monofilament No. 4-0 wire is still the strongest; it loses less tensile strength by square knotting than do the other materials. Wire is more difficult to use than the more pliable syn-

thetic materials, but it has more holding power at the knot than synthetic fibers.

The choice of technique in suture placement also affects the tensile strength of a tendon anastomosis. Those techniques that afford purchase of the tendon by the suture material in such a way as to avoid cutting through and out of the tendon are the most dependable (Figs. 3-6 and 3-8).

Tendons become increasingly soft, especially between the fifth and tenth day after suturing; therefore the purchase of the suture is diminished and separation is more likely during this period. After this, tendon strength increases rapidly for the next 2 weeks.

No suture material or technique can be relied on to maintain tendon anastomoses during unlimited active movement.

End-to-end suture

The most classic technique of end-to-end suture is the Bunnell crisscross stitch. It is not commonly used now because the rippling created by tension tends to make the ends of the tendon avascular. For historical reasons, however, it is described here (Fig. 3-6).

Ample exposure must be available to use this technique effectively. Some tendon shortening is also necessary. Use a wire suture (No. 34 monofilament) about 30 cm long, with a straight tendon needle attached to each end. Grasp the tendon transversely at one end with a hemostat; while applying slight tension and supporting the tendon with one fingertip, pass one needle obliquely through the center of the tendon so that it emerges on the opposite side about 6 mm further distally, but still 6 mm proximal to the tendon junction. Draw about one half of the suture through the tendon. Now reinsert the needle about 1 mm from the point of previous exit and pass it diagonally again through the tendon; usually this half of the suture pierces the tendon four times. Using the other needle and beginning 2 mm from the point of original entrance of the first needle, repeat the procedure but make one less pass through the tendon and emerge opposite the first. Make the final stitch with each needle emerging from the tendon through a notch made on each side next to its clamped end. Excise the end of the tendon with a knife, alternating from one side to the other so as not to leave a tail. With a hemostat, now grasp the other end of the tendon and in it make two small notches opposite each other for entrance of the needles. Pass one diagonally through this section of the tendon three times and the other four times, thus bringing them out finally on the same side of the tendon. Excise the

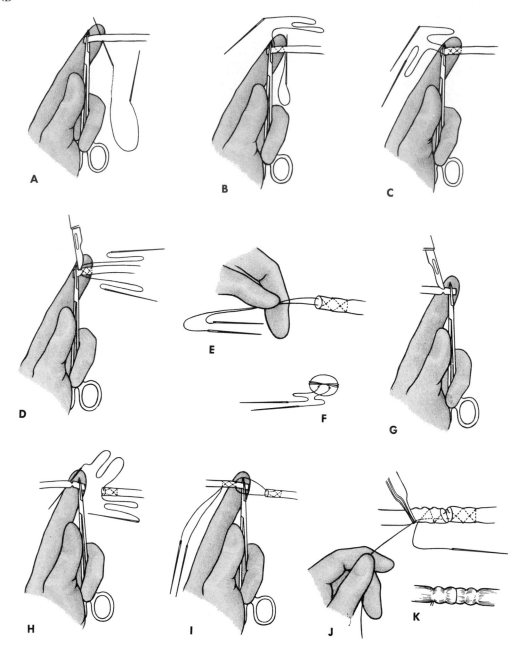

Fig. 3-6. End-to-end suture of a tendon using Bunnell crisscross stitch (see text).

traumatized end of this segment and tighten the suture by pulling on each wire, creating an accordian or pleated effect. Now tie the suture in a knot on the outside of the tendon. The accordian effect prevents separation of the junction after absorption takes place in the center of the tendon.

End-to-side anastomosis

End-to-side anastomosis is frequently used in tendon transfers when one motor must activate several tendons. Pierce the recipient tendon through the center with a No. 11 Bard-Parker knife blade and grasp the blade on the opposite side with a hemostat (Fig. 3-7). Then withdraw the blade, carrying the hemostat with it; with the latter, gently grasp the end of the tendon to be transferred and bring it through the slit. Repeat this technique with any adjacent tendons, placing the slits so that the transferred tendon approaches the recipient tendon at an acute angle to its line of pull. Bury the end of the transferred tendon in the last tendon pierced.

Double right-angled suture (modified Kessler)

To suture the severed ends of a tendon together without shortening, a double right-angled stitch (Fig. 3-8) can be used. Although the apposition of the tendon ends is not as neat as after the end-to-end suture just described, the method is easier and is more frequently used when multiple tendons are severed.

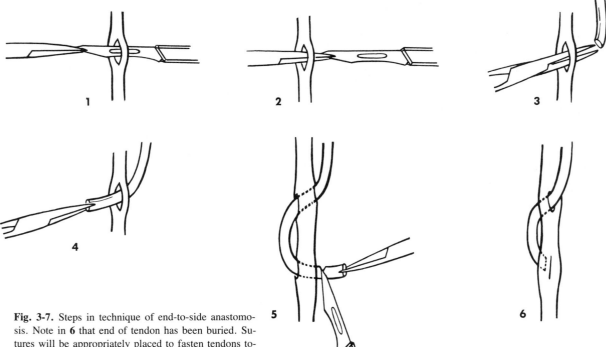

Fig. 3-7. Steps in technique of end-to-side anastomosis. Note in **6** that end of tendon has been buried. Sutures will be appropriately placed to fasten tendons together.

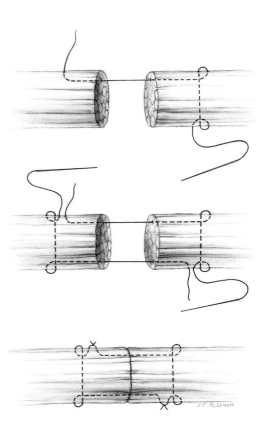

Fig. 3-8. "Modified Kessler stitch."

Fishmouth end-to-end suture (Pulvertaft)

A tendon of small diameter may be sutured to one of large diameter by the method shown in Fig. 3-9. This method is commonly used to suture tendons of unequal size.

Roll stitch

The roll stitch is especially useful for suturing extensor tendons over or near the metacarpophalangeal joints. Use a No. 4-0 monofilament wire or No. 4-0 monofilament nylon threaded on a small curved needle (Fig. 3-10). Pass the suture through the skin just medial or lateral to the divided tendon and through the proximal segment of the tendon near its margin from superficial to deep and then through the deep surface of the distal segment to emerge on its superficial surface. Next pass it proximally and through the opposite margin of the proximal segment and bring it out through the skin on the opposite side of the tendon from which it was introduced. At about 4 weeks the suture can be removed by pulling on one of its ends.

Attachment of tendon to bone

To attach a tendon to bone, use a small osteotome or dental chisel to roughen the site of insertion or raise a small area of cortex to accept the tendon (Fig. 3-11). If several tendon ends are to be fixed to bone, they are best inserted into a large hole drilled in the bone. After an area of cortex has been elevated or a large hole made, perforate the bone with a small Kirschner wire in a Bunnell drill. Then using the first needle as just described for the end-to-end suture, run the suture diagonally two or three times

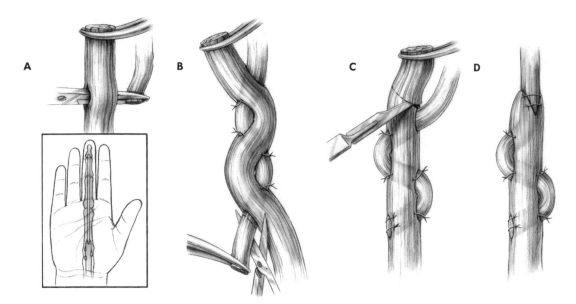

Fig. 3-9. Pulvertaft technique of suturing tendon of small diameter to one of larger diameter. **A,** Smaller tendon is brought through the larger and anchored with one or two sutures after the tension is adjusted. **B,** Tendon is then brought through a more proximal hole and is once again anchored with one or two sutures after the tension is adjusted. **C,** After the excess is cut flush with the larger tendon the exit hole can be closed with one or two sutures. **D,** The excess of the larger tendon is then trimmed as shown to permit a central location of the smaller tendon. This so-called fishmouth is then closed with sutures.

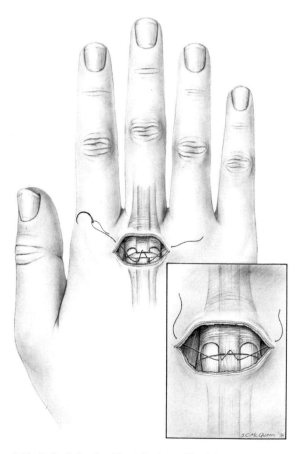

Fig. 3-10. Roll stitch using No. 4-0 wire or No. 4-0 monofilament nylon is especially useful in suturing a lacerated extensor tendon over or near head of metacarpal.

through the end of the tendon. Loop a pull-out wire over the second needle and complete the crisscross diagonal suture. Using the needles or a loop, pass the two ends of the suture through the bone and snug the tendon against it.

INDICATIONS FOR PRIMARY TENDON SUTURE

When a wound is caused by a sharp object such as a knife and is reasonably clean, some tendons of the hand should be repaired at the time of primary wound closure. Some factors determining which tendons may be so repaired are as follows: (1) the time since injury, (2) the area of the hand affected, (3) the number of tendons divided, (4) the severity of wound contamination, (5) the presence or absence of unstable fractures, and (6) the general condition of the surrounding tissues. In the following instances divided tendons are usually not sutured primarily but should be left for secondary repair: (1) when the injury is produced by severe crushing, in which the tendons will be more edematous and will heal with more adhesions than after simple division, (2) when the site of tendon division is near one or more displaced fractures, for the sutured tendon is more likely to adhere to the surrounding tissues, and (3) when the skin is avulsed or lost and skin grafting is necessary over the divided tendon or tendons.

A lacerated *flexor tendon* is no longer considered an absolute surgical emergency. The definitive repair can be delayed several days provided the wound is clean or rendered clean, the hand is splinted, and prophylactic antibiotics are started. This is especially true when the patient must be transported to a center where a surgeon skilled in the techniques of tendon repair is available. However, prolonged delay in repair may permit unacceptable retraction of ten-

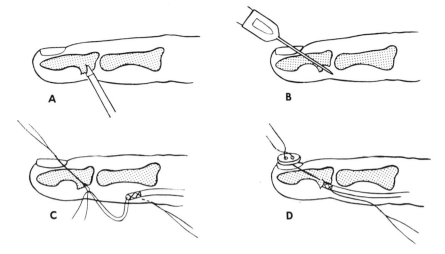

Fig. 3-11. One method of attaching a tendon to bone. **A,** Small area of cortex is being raised with osteotome. **B,** Hole is being drilled through bone with Kirschner wire in Bunnell drill. **C,** Bunnell crisscross stitch has been placed in end of tendon, and wire suture is being drawn through hole in bone. **D,** End of tendon has been drawn against bone, and suture is being tied over button.

dons and especially of nerves and retraction of the latter is not easily overcome after 2 weeks. As a general rule, all flexor tendons should be repaired at whatever level they are severed. Because of the vincular system of the profundus tendon, when both have been severed some surgeons believe the results are better when both are repaired than when the profundus tendon alone is repaired. It is better to stabilize fractures and suture digital nerves and tendons at the time of initial repair than to delay and do a secondary procedure for tendon repair. Indeed, later it may be necessary to do a tendon graft. The principle of maintaining maximum coverage of tendons with the flexor sheath has been well established; in many instances the entire sheath can be repaired after a tendon repair. Furthermore, it is essential that certain areas of the flexor sheath be preserved, such as the A2 and A4 pulley systems; otherwise a flexion deformity of the finger will develop and excursion of the tendon will be lost.

The *extensor tendons* of the fingers can usually be repaired primarily at almost any level in clean, incised wounds. When there are multiple fractures, severe skin loss, or severe contamination, repair may have to be delayed, but even without suturing, extensor tendons often eventually unite with the scar tissue and, because they are surrounded by loose areolar tissue and fat, are able to glide and function satisfactorily. Extension is often incomplete after this type of healing, but it may be complete when the wrist is splinted in extension during healing.

FLEXOR TENDON REPAIR

Because of certain anatomic differences in the flexor surface of the hand, it will be divided into five zones to describe the appropriate repair procedures for flexor tendons lacerated in each of these zones (Fig. 3-12). *Zone I* extends from just distal to the insertion of the sublimis tendon to the site of insertion of the profundus tendon. *Zone II* is in the critical area of pulleys (Bunnell's "no

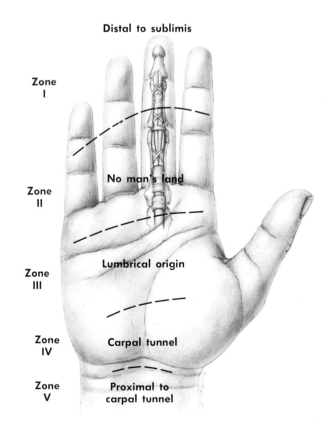

Fig. 3-12. Flexor zones of hand. Designated zones on flexor surface of hand are helpful since treatment of tendon injuries may vary according to level of severance.

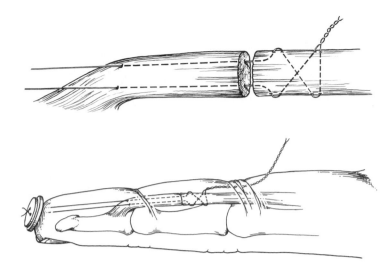

Fig. 3-13. This pull-out suture technique is useful when distal segment of tendon is too long for tendon advancement to bone.

man's land'') between the distal palmar crease and the insertion of the sublimis tendon. *Zone III* comprises the area of the lumbrical origin between the distal margin of the transverse carpal ligament and the beginning of the critical area of pulleys or first annulus. *Zone IV* is the zone covered by the transverse carpal ligament. *Zone V* is the zone proximal to the transverse carpal ligament and includes the forearm.

Zone I

The flexor digitorum profundus tendon may be repaired primarily by direct suture to its distal stump or by advancement and direct insertion into the distal phalanx when the distance is 1 cm or less. A pull-out wire technique may be used to attach the proximal tendon end to its distal stump (Fig. 3-13) or directly to the bone following advancement. When there is a delay in the diagnosis of interruption of this tendon and it has retracted into the palm, then its vinculum has been torn loose and repair probably should not be attempted because it would require threading the tendon through the bifurcation of the sublimis. This may be difficult or even impossible when the profundus tendon has become edematous after several days of lying in the palm, and, in addition, normal sublimis function may be impaired even if the profundus is repaired primarily by rethreading it through the bifurcation because it will act much like a free graft since the vinculum is avulsed. This is especially true when an avulsed flexor digitorum profundus in an athlete has so retracted. The indication for a free tendon graft for patients diagnosed late depends considerably on the patient's age, occupation, and specific need for distal interphalangeal flexion, as in playing a stringed instrument, for this graft may jeopardize flexor digitorum sublimis function. Occasionally when a profundus repair or graft is not done, the tip of the index finger may have to be stabilized later by tenodesis of the distal interphalangeal joint. Rarely is tenodesis or arthrodesis needed in the same situation in the other three fingers.

Zone II

The critical area of pulleys, or Bunnell's ''no man's land,'' has been a subject of considerable discussion for years as to which, if any, tendon divided here should be repaired primarily. Here especially, the primary surgeon has the greatest influence on the final result. To be qualified to make the decision and perform a primary repair, a surgeon should be sufficiently skilled to perform a tendon graft or tenolysis later, should there be a failure of the primary repair.

Primary repairs at this level frequently fail as a result of adhesions in the area of the pulleys. Primary repairs have a better chance of success in children than in adults. Exacting wound care is critical. When there is doubt as to the timing of tendon repair, the wound should be cleaned and the repair done later by an experienced surgeon.

Most surgeons now recommend that both the sublimis and profundus tendons be repaired at this level. During the repair procedure, it is essential that the A2 and A4 pulleys be preserved (Figs. 3-14 and 3-15). Ideally the entire pulley system is repaired following necessary exposure of the underlying tendons. Various techniques of tendon suture are recommended by different surgeons. The suture material used is also a controversial point in technique. Some surgeons prefer 4-0 monofilament wire, others a synthetic woven fiber, and still others a nylon suture. At this level pull-out sutures are unnecessary. In this zone a tenolysis is indicated when flexor function does not improve during a period of 6 months and the wound is stable. A tenolysis is as technically critical as a tendon graft and is complicated by postoperative tendon rupture in about 10% of patients. The incidence of required tenolysis following flexor tendon repair varies from 18% to 25%, depending on the author. The operation may improve function by as much as 50%. In patients with impaired sensibility and in those over 45 to 50 years old the results will be much less satisfactory than in younger patients with normal sensibility. The Strickland technique of tendon repair in zone II is detailed in Figs. 3-16 to 3-18.

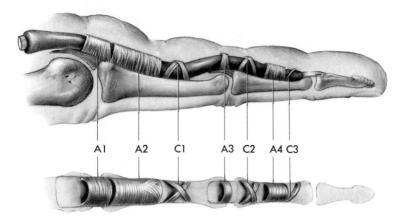

A1 A2 C1 A3 C2 A4 C3

Fig. 3-14. This anatomic diagram of various parts of flexor sheath is helpful in understanding gliding of tendon. Maintenance of second annulus *(A2)* and fourth annulus *A4)* is essential to retain appropriate angle of approach and prevent ''bowstringing'' of flexor tendons or tendon graft. (From Doyle, J.R., and Blythe, W.: In American Academy of Orthopaedic Surgeons: Symposium on tendon surgery in the hand, St. Louis, 1975, The C.V. Mosby C .)

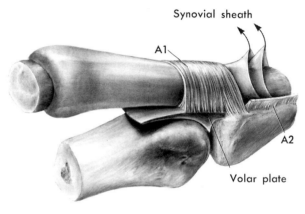

Synovial sheath

A1

A2

Volar plate

Fig. 3-15. Diagram of relationship to synovial layers (there are two) and annulus. (From Doyle, J.R., and Blythe, W.: In American Academy of Orthopaedic Surgeons: Symposium on tendon surgery in the hand, St. Louis, 1975, The C.V. Mosby Co.)

Zone III

At the zone III level the muscle bellies of the lumbricals as well as the tendons are frequently interrupted. Additional incisions are often needed to further expose this area. All tendons may be repaired primarily or delayed only a few days. Primary anastomoses of severed nerves are extremely important since delaying the repair even a few weeks results in the development of significant gaps between the nerve ends. Lumbrical muscle bellies should not be sutured because this may increase the tension of these muscles and result in a ''lumbrical plus'' finger.

Zone IV

All tendons and nerves in zone IV may be repaired primarily; however, for exposure it may be necessary to completely or nearly completely excise the transverse carpal ligament. Should complete excision be necessary, the wrist should not be placed in flexion past neutral position but the fingers should be brought into slightly more flexion than usual to permit relaxation of the musculotendinous units. Flexion of the wrist beyond neutral may permit subluxation of the repaired tendons out of their normal bed and bowstring them just under the sutured skin. When it is technically possible to accomplish tendon repair and retain part of the transverse carpal ligament, this problem is eliminated. Alternatively a portion of the transverse carpal ligament can be repaired to create a pulley for these tendons. Remember that the flexor digitorum profundus tendons at this level may not be well defined and there may be frequent interdigitations.

Zone V

Since zone V is proximal to the transverse carpal ligament, tendon gliding after repair is better than in more distal zones. All tendons and nerves lacerated in this area should be repaired primarily when wound conditions are satisfactory, as advised earlier in this section. The chief difficulty of repair here usually is one of exposure, which requires a proximal extension and possibly a distal extension of the transverse laceration usually present. Severed tendons can usually be located by the blood clots within the sheaths. At this level the profundus tendons are not completely differentiated into individual tendon units. The sublimis tendons are better differentiated and the severed ends are usually more easily matched. When the necessary expertise is not available, primary repair may be delayed and the wound cleaned. Results probably are not compromised by a delay of a day or even longer when desired or indicated. At this level excision of some of the synovial covering is necessary to identify and remove the hematoma; however, a generalized synovectomy is not indicated. An isolated laceration of the palmaris longus tendon does not require repair.

Delayed repair

Delayed repair in any zone may be necessary in the face of severe wound contamination, soft tissue loss, or lack of surgical skill. Delayed repairs of tendons are reasonable also when other injuries that require immediate surgery are

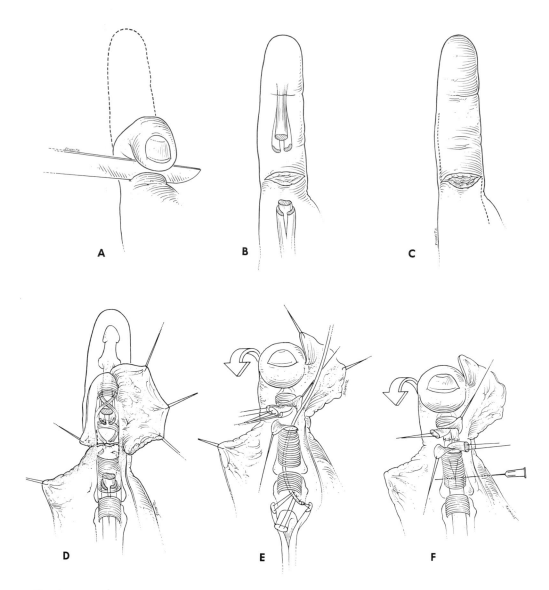

Fig. 3-16. Strickland's technique of flexor tendon repair in zone II is illustrated. **A,** Knife laceration through zone II with digit in full flexion. **B,** Level of flexor tendon retraction of same finger following digital extension. **C,** *Dotted lines* depict radial and ulnar incisions to allow wide exposure of flexor tendon system. **D,** Flexor tendon system of involved finger after reflection of skin flaps. In this case, laceration has occurred through C1 cruciate pulley area. Note proximal and distal position of severed flexor tendon stumps resulting from flexed attitude of finger at time of injury. *Dotted lines* indicate lateral incisions in cruciate-synovial portions of sheath, which will be used to provide exposure for tendon repair. **E,** Reflection of small triangular flaps at cruciate-synovial sheath allows distal flexor tendon stumps to be delivered into wound by passive flexion of distal interphalangeal joint. Profundus and sublimis stumps are retrieved proximal to A1 pulley and, following placement of tendon sutures, are delivered back through proximal portion of tendon sheaths by using small catheter or infant feeding gastrostomy tube. **F,** Proximal flexor tendon stumps are maintained at repair site by means of transversely placed small-gauge hypodermic needle, followed by repair of flexor digitorum sublimis slips.

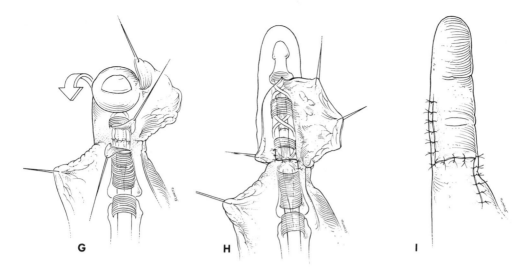

G   H   I

Fig. 3-16, cont'd. G, Completed repair of both tendons with the distal interphalangeal joint in full flexion. **H,** Extension of distal interphalangeal joint delivers repairs under intact distal flexor tendon sheath. Repair of cruciate *(C1)* synovial pulley has been completed. **I,** Wound repair at conclusion of procedure. (From Strickland, J.W.: Hand Clin. **1:**55, 1985.)

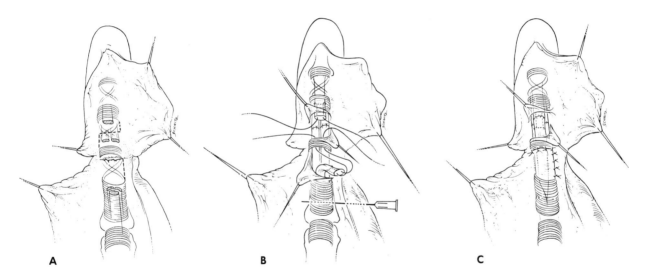

A   B   C

Fig. 3-17. More distal severance of profundus and sublimis tendons resulting in inability to deliver 1 cm of distal tendon stumps into C1 cruciate synovial area by distal interphalangeal joint flexion. Cruciate-synovial incisions on both sides of A3 pulley will be necessary to provide adequate exposure for suture placement and repair. **A,** Appearance of flexor system following tendon interruption at level just proximal to proximal interphalangeal joint. Position of distal flexor tendon stumps can be seen. Full passive distal interphalangeal joint flexion will not provide adequate exposure in C1 cruciate-synovial area for suture placement and repair. *Dotted lines* indicate cruciate-synovial incisions in C1 and C2 areas on both sides of A3 pulley. **B,** Sutures placed in distal stumps in C2 cruciate-synovial area and in proximal stumps through C1 opening. Proximal stumps are delivered distally beneath A3 pulley. Sublimis tendon slips have already been repaired. **C,** Digit following repair of both tendons with repair of C1 cruciate-synovial area completed. Repair of C2 cruciate-synovial sheath should now be possible to complete surgery. (From Strickland, J.W.: Hand Clin. **1:**55, 1985.)

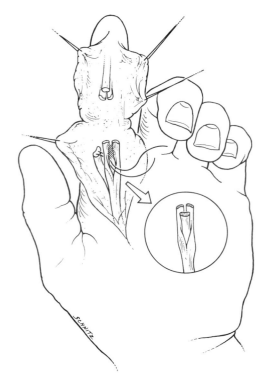

Fig. 3-18. Separated position of two tendon ends in distal palm following flexor tendon interruption and proximal retraction. Correctly position profundus in sublimis hiatus before passing tendons distally into digit. Reestablish anatomic relationship of profundus and sublimis tendon stumps so they may be correctly repaired to corresponding distal tendon stumps. In some cases, profundus will have to be passed back through the hiatus created by the sublimis slips to lie palmar to Camper's chiasma and recreate position of tendons that was present at the level of tendon laceration. (From Strickland J.W.: Hand Clin. **1:**55, 1985.)

present and there is not sufficient surgical time to repair the tendons also. There is no known reason to expect complications when the repair of tendons is delayed for 2 to 3 days, providing the wound has been cleaned. Prolonged delay may permit unacceptable retraction of tendons and nerves.

REPAIR OF FLEXOR TENDON OF THUMB

The thumb may be arbitrarily divided into zones as are the other digits. Zone I covers the area distal to the metacarpophalangeal joint and involves the proximal phalanx, zone II involves the pulley system at the metacarpophalangeal joint, and zones III, IV, and V correspond precisely to the zones described for the other digits (Fig. 3-12).

Zone I

When the long flexor tendon of the thumb is divided in zone I within 1 cm of its insertion, it can be sutured primarily either to the distal stump or advanced and sutured directly into the bone. Some of the flexor sheath may have to be divided. When this tendon is divided more proximally than 1 cm from its insertion, further advancement will be necessary and lengthening of the tendon by Z-plasty just proximal to the wrist should be carried out. This

tendon is unique in that it may be advanced without disturbing its blood supply since it does not have a vinculum. Urbaniak and Goldner recommend advancement in preference to tendon grafting since there are fewer paratendinous adhesions after advancement than after a free tendon graft.

Zone II

In zone II the critical pulley area at the metacarpophalangeal joint, a portion of the pulley may be excised to lessen the possiblity of adherence to the pulley of the site of the tendon suture. Primary repair, however, is unpredictable and a later graft might be the better choice unless the surgeon is experienced in tendon repair. Advancement of the tendon distally to be sutured to a stump that is shortened to lie distal to the metacarpophalangeal pulley has the advantage of not placing a suture line under the pulley (Figs. 3-19 and 3-20). Lengthening of the tendon at the wrist by Z-plasty is required for this procedure also. Urbaniak and Goldner recommend a reinforcing graft at the site of lengthening just proximal to the wrist.

Zone III

In zone III with a laceration of the flexor pollicis longus tendon, the proximal end frequently will retract to near the wrist level. Regardless, as a rule it is still easy to recover the proximal end, and an attempt to seek this out in the area of the thenar muscles is advisable. Primary repairs in this zone can be performed when the two ends are recoverable by flexing the wrist and the distal joint of the thumb and fashioning a simple connecting anastomosis. When the proximal tendon has to be recovered by making an additional incision at the wrist, the tendon should be carefully rethreaded through the original site. This may be done by inserting a wire loop or a tendon carrier through the sheath from the distal end, picking up the proximal end of the tendon, and threading it through, proximally to distally.

Zone IV

In zone IV the tendon is rarely cut since it is protected in part by a shelf of the radiocarpal bones. There is no contraindication to repair at this level as long as the repair technique is atraumatic and the two ends are recoverable. An effort should be made to avoid the creation of a mass of foreign material (suture) sufficient to secondarily compress the median nerve within the closed space of the carpal tunnel.

Zone V

In zone V primary repair of the flexor pollicis longus tendon, like all other tendons, is indicated.

TECHNIQUE FOR PRIMARY SUTURE OF FLEXOR TENDONS. Further exposure of the tendon to be sutured may be necessary. These additional incisions (Fig. 3-21) should be made without crossing flexion creases at a nearly right angle. Usually less exposure is needed distally than proximally, since the distal segment of the tendon may be delivered into the wound by flexing the distal joints; also the segment is not subject to the action of a muscle as is the proximal segment. Because of their common muscle origin, when sublimis tendons are divided, particularly at the wrist, they may be delivered into the wound as a group by

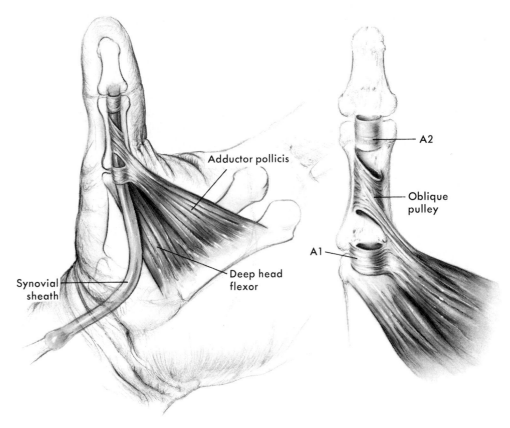

Fig. 3-19. Relationship of thumb intrinsic muscles and annular bands. (Courtesy Dr. James R. Doyle.)

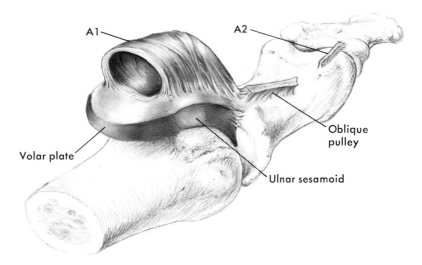

Fig. 3-20. Relationship of flexor sheath, volar plate, and metacarpophalangeal joint can be noted in this diagram of volar aspect of thumb. (Courtesy Dr. James R. Doyle.)

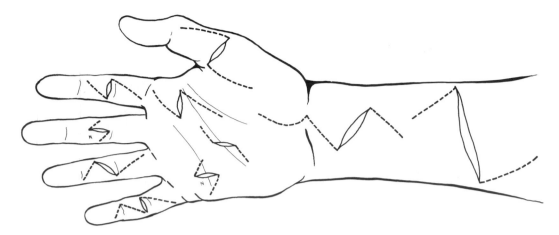

Fig. 3-21. Exposures for primary suture of tendons. Examples of skin lacerations are indicated by *solid lines* and direction in which they may be enlarged to obtain additional exposure by *broken lines* (see text).

finding and pulling one distally; this is also true of the profundus tendons. The tendon ends must be properly matched by careful attention to their levels in the wound, their diameters, the shape of their cross sections, their relation to neighboring structures, and the angle of the cuts through each tendon. It is no disgrace to open an anatomy book in the operating room to be certain of anatomic relations; it is a disgrace and a calamity to sew the median nerve to the flexor pollicis longus or palmaris longus tendon. The median nerve lies at a different level, it is slightly more yellow and usually has a midline vessel, and nerve filaments can easily be identified in its severed ends.

The selection of the suture material is not as important as the cleaning and handling of the tissues. The sutures should be small but strong enough to hold the tendon when only one well-placed suture is used in each. No. 4-0 monofilament wire for adults and No. 5-0 monofilament wire for children are still preferred for suturing flexor tendons. Other alternatives are Mersilene and nylon.

Each tendon selected for repair (see discussion of indications, p. 44) is picked up with a small Adson tissue forceps, and a through-and-through suture is inserted near its tip for use in retraction or handling; the end of the tendon should never be crushed by an Allis forceps or a Kocher clamp. The extreme tip of a tendon may be held with a small hemostat, but this crushed segment must be excised before the suture is tied; in clean, incised wounds this shortens the tendon unnecessarily. The suturing technique must be exact to hold the tendon ends together accurately and to prevent distraction or exposure of raw surfaces at the junction (the techniques of tendon suture are discussed on p. 41).

AFTERTREATMENT. Controlled postoperative mobilization of recently repaired tendons is now accepted by many surgeons as beneficial. This technique seems to reduce the formation of adhesions that prevent the tendons from gliding. It also improves the healing process of the tendon. There are two ways of accomplishing this. The first is a technique described by Duran and Houser in which the wrist is splinted in 20 degrees flexion, the metacar-

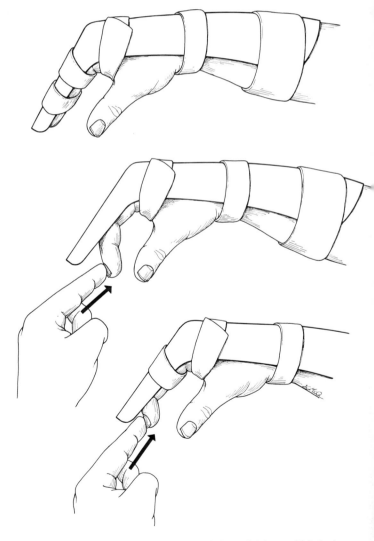

Fig. 3-22. Passive flexion of the interphalangeal joints, which is done several times a day for 4 to 5 weeks. Duran and Hauser popularized early passive motion following tendon repairs. (Redrawn from Strickland, J.W.: Orthop. Clin. North Am. **14**:844, 1983.)

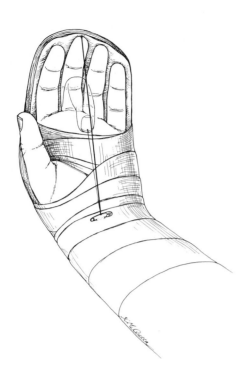

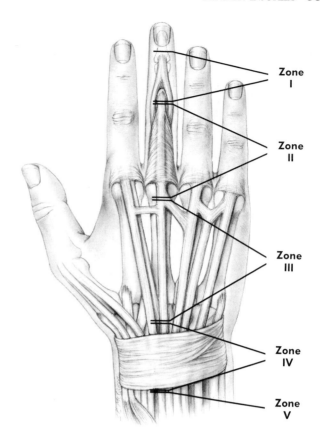

Fig. 3-23. After primary flexor tendon repair or flexor tendon graft, wrist and hand are held in a posterior plaster splint. Additionally, involved finger is held in flexion by elastic band attached at wrist level and at fingernail by wire through nail or glued-on garment hook. This permits active finger extension and protected passive flexion. (Redrawn from Kleinert, H.: Orthop. Clin. North Am. **4:**874, 1973.)

Fig. 3-24. Indications for surgery of extensor tendon lacerations vary according to level of pathology; therefore various zones have been designated.

pophalangeal joint in about 45 degrees flexion, and the interphalangeal joints in almost full extension (Fig. 3-22). Before closure it is determined how much passive movement of the finger tip is necessary to cause a 3 to 5 mm excursion of the tendon. This amount of movement is started at about 5 days after surgery, when a splint is applied that the patient can remove and passively flex the interphalangeal joints about 8 repetitions at a time, twice a day. This is continued for 4½ weeks when active flexion and extension are permitted. Passive extension is not permitted for 6 weeks. This technique is difficult if not impossible to carry out in children under 10 to 12 years old, and other patients must be reliable in this immediate postoperative period.

A second technique for carrying out early passive motion is to insert a suture through the tip of the fingernail or to glue a garment hook to the nail to which an elastic band is then attached (Fig. 3-23). The wrist is held in 20 degrees flexion, the metacarpophalangeal joint in 20 degrees, and the interphalangeal joints in 10 degrees. The elastic band is anchored at the wrist with a safety pin and pulls the fingers toward the base of the thumb over the scaphoid area. The elastic band or bands permit passive flexion of the finger or fingers from a posture of flexion while active extension can be initiated by the patient. However, extension is limited by a posterior plaster splint. The patient is encouraged to actively extend the fingers for 3 weeks.

Only the involved finger or fingers are carried through this technique. At 3 weeks the posterior splint is removed and the injured finger remains anchored to a wrap at the wrist level by the elastic band for an additional 2 weeks.

EXTENSOR TENDON REPAIR

The extensor surface of the hand may be divided into zones to comply with the different anatomic relationships of the extensor tendons and their attachments (Fig. 3-24).

Zone I

Zone I extends from the distal insertion of the extensor tendon to the attachment of the central slip at the proximal end of the middle phalanx. Mallet finger deformities result from an avulsion of the insertion of the tendon, sometimes including a small bone fragment, and may be treated by splinting alone (see mallet finger, p. 57). Lateral tendons lacerated proximal to the insertion may be sutured with a very small, single-stitch suture or a roll stitch, as is used more proximally.

Zone II

Zone II, extending from the metacarpal neck to the proximal interphalangeal joint, includes the extensor mechanism with its contoured surface encompassing the phalanx and the metacarpal head. Tendon lacerations at this level require different techniques of suture from those at other levels. Here they should be repaired with a roll stitch or some other suture technique that permits total removal of the suture later (Fig. 3-10). Suture material and other foreign material tend to cause more inflammatory reaction over joints than over nonmoving parts. No foreign material should be left permanently within this zone, especially over the metacarpophalangeal joint.

Zone III

Zone III extends proximally from the metacarpal neck to the distal border of the dorsal carpal ligament. The tendons are lying free in this area without ligamentous attachment and are covered only by paratenon and fascia. Tendons may be sutured individually with a mattress-type suture of monofilament wire or other suture material that does not necessarily have to be removed since there is much less reaction to foreign materials in this area.

Zone IV

Zone IV is the area of the wrist under the dorsal carpal ligament. At this level the tendons have mesotenon. They are held by the dorsal carpal ligament, which acts as a pulley, and are ensheathed in canals not unlike the theca on the flexor surface. Thus repaired tendons are likely to become stuck in their canals as they heal. Consequently primary repair of extensor tendons here may be done with a mattress or similar suture, and the sutured area should be released by excising the overlying carpal ligament. This may permit some bowstringing of the repaired tendon at the wrist when the wrist is in extension, but it does help to avoid the adherence of the sutured tendon at this site and the loss of normal excursion. Splinting the wrist after repair in a position of moderate extension instead of full extension helps limit the bowstring effect.

Zone V

Zone V is the zone proximal to the proximal margin of the dorsal carpal ligament. In this zone many extensor tendons are still contained within their respective muscles. The tendinous portion of the musculotendinous unit may be sutured with a carefully placed stitch since sutures tend to pull out of muscle tissue. The wrist is placed in full extension postoperatively to permit maximum relaxation of the musculotendinous unit since it is quite difficult to maintain muscle-to-muscle anastomosis by any suture technique.

AFTERTREATMENT. Aftertreatment consists of applying a volar splint extending from just distal to the elbow to the proximal interphalangeal joints. The wrist is held in appropriate extension and the metacarpophalangeal joints in about 30 degrees of flexion with the proximal interphalangeal joints left free. This splint or some similar protection is needed for 4 to 5 weeks.

Traumatic dislocation of extensor tendon at metacarpophalangeal joint of middle finger

Traumatic dislocation of the extensor tendon toward the ulnar aspect of the metacarpophalangel joint occurs most commonly in the middle finger. The mechanism of dislocation is apparently a tear in the proximal portion of the shroud ligament and the more proximal fascia as the middle finger is suddenly extended against force as in a flicking or thumping motion. When seen within the first few days, this dislocation can be treated effectively by splinting the metacarpophalangeal joint and wrist in extension for a 3-week period. When the condition is chronic, a repair using a section of the central fibers of the extensor mechanism at the metacarpophalangeal joint may be successful.

TECHNIQUE. Make a curved incision on the radial side of the metacarpophalangeal joint to expose the joint area and subluxating extensor tendon. Create a loop by removing a 5 cm lateral margin of the central tendon at this level, maintain the distal insertion of this segment, pass the transferred segment in and out of a window in the superficial portion of the joint capsule, and suture the proximal end to the extensor tendon. Adjustment of tension is essential to maintain the central alignment of the subluxating extensor tendon. Maintain the finger in a slightly flexed position for 3 weeks.

TENDON RUPTURES
Flexor tendon rupture

Although rupture of flexor tendons is not as common as that of extensor tendons, it does occur and is often not diagnosed. In the athlete the most common tendon to be avulsed is the flexor digitorum profundus at its insertion in the ring finger. Traumatic rupture usually occurs at the insertion of the tendon. Frequently the patient's initial complaint is that of a mass in the palm without awareness of any loss of finger function. The flexor tendons most frequently ruptured are the profundus tendons, and more rarely the sublimis tendons or the flexor pollicis longus. These ruptures are most frequent in men during the third and fourth decades and about 20% are associated with synovitis. (See tendon ruptures in arthritis, p. 281.)

TREATMENT. The treatment, if any, usually consists of excision of the segment of tendon from within the palm; rarely is reconstructive surgery necessary. When a profundus tendon is ruptured, it should be reinserted on the distal phalanx only if the rupture is discovered before the tendon has retracted into the palm. Otherwise the tendon should be excised; then if pinch is impaired, particularly of the index finger, an arthrodesis (p. 170) or tenodesis (p. 68) of the distal interphalangeal joint can be done.

When a rupture of the flexor pollicis longus tendon is seen early, the tendon may be reattached to the distal phalanx; when seen late, a tendon graft may be necessary because of muscle shortening and tendon degeneration (p. 59). Other than for the thumb a restoration by graft is usually not indicated.

Extensor tendon rupture (mallet finger)

When an extensor tendon is avulsed from its insertion into the distal phalanx, the treatment is usually nonsurgical. The distal interphalangeal joint should be held in hyperextension on a splint (Fig. 3-25) constantly for 6 to 8 weeks and during the night only for 1 additional week; this much time is necessary to prevent the tendon from stretching after the splint is removed. The last few degrees of flexion of the distal joint may be limited, but the flexion deformity of the joint is usually corrected. This

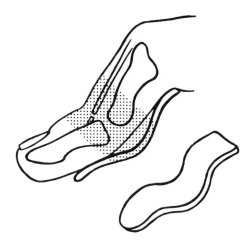

Fig. 3-25. Mallet finger uncomplicated by avulsion of large fragment of bone can be treated effectively on biconcave splint that extends distal joint alone. Plastic adhesive dressing that holds splint in place may be changed daily by patient.

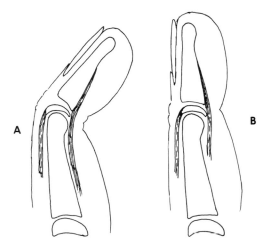

Fig. 3-26. A, Displacement of epiphysis of distal phalanx may cause digit to assume a mallet finger posture. **B,** Hyperextension of phalanx usually affords satisfactory reduction of displaced epiphysis.

treatment has been successful even as late as 3 months after injury. (The treatment of old mallet finger is discussed on p. 57.)

Mallet finger deformities in children may be caused by traumatic separation of epiphyses (Fig. 3-26). These should be recognized since it is only necessary to hyperextend the distal interphalangeal joint to obtain reduction. Healing is rapid when compared with injury of the extensor tendon. Although rare, growth disturbance is possible.

Tendon ruptures in the rheumatoid hand are discussed on p. 281.

Rupture of central slip of extensor expansion (buttonhole deformity)

Rupture (or laceration) of the central slip of the extensor expansion at or near its insertion results in loss of active extension of the proximal interphalangeal joint and there-

fore in persistent flexion of the joint. Eventually the collateral ligaments and volar plate of the joint become contracted. The lateral bands of the extensor expansion subluxate volarward and are held there by the transverse retinacular ligaments, which also become contracted. Thus the buttonhole deformity is established. The lateral bands, because they lie volar to the transverse axis of the proximal interphalangeal joint, act as flexors of the joint. The tight oblique retinacular ligaments and lateral bands force the distal interphalangeal joint into hyperextension that is increased by any attempt to passively extend the proximal interphalangeal joint.

When a buttonhole deformity is traumatic, treatment differs from that useful for the same deformity caused by arthritis (p. 274). When rupture (or laceration) of the central slip is acute, it should be exposed surgically and repaired. Then the proximal interphalangeal joint is placed in full extension and is held in this position by a Kirschner wire inserted across it. At 3 weeks the wire is removed and guarded flexion is gradually begun. Unfortunately such injuries are usually seen late, after the secondary deformities mentioned previously have already developed. In these instances the treatment described on p. 57 may be indicated. Should the rupture be incomplete with some active extension remaining at the proximal interphalangeal joint, then splinting may be used to treat this. The splint maintains the proximal interphalangeal joint in full extension while permitting full, active flexion of the distal interphalangeal joint.

OTHER CAUSES OF BUTTONHOLE DEFORMITY

Buttonhole deformity may also be caused by traumatic rotation of a digit at the proximal interphalangeal joint while in partial flexion. Rotation causes a lateral condyle of the proximal phalanx to protrude through the capsule in the area between the lateral band and central tendon. At this point the extensor expansion is thin. This condylar herniation causes a volar subluxation of the lateral band. A rupture in the extensor mechanism occurs at this point, but not necessarily a rupture of the central tendon. There may also be a partial rupture of the collateral ligament or a momentary anterior dislocation of the proximal interphalangeal joint. After hemorrhage and swelling occur, the proximal interphalangeal joint cannot be fully extended and the joint remains in a flexed position, the subluxated lateral band becomes shortened, and the herniation of the condyle is maintained. The transverse retinacular ligament eventually contracts, holding even more securely the subluxated lateral band. Spinner and Choi have shown experimentally that with anterior dislocation of the proximal interphalangeal joint, complete rupture of the middle slip and lateral ligament may occur as seen in Fig. 3-27.

To repair this injury, reposition the lateral bands after a release by cutting the transverse retinacular ligament. Then repair the collateral ligament if completely torn. Maintain the proximal interphalangeal joint in full extension for a period of 3 weeks with a small transfixing Kirschner wire. Aftertreatment consists of gradual exercise of the proximal interphalangeal joint and night splinting to maintain full extension should there be a lag in active extension.

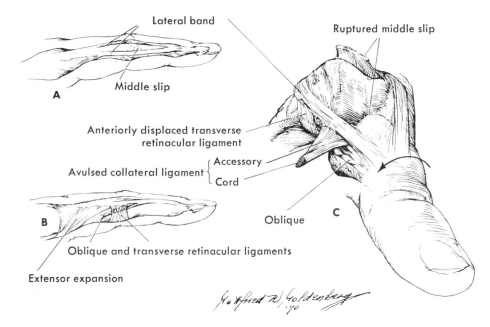

Fig. 3-27. **A,** Dorsolateral and, **B,** lateral view of extensor mechanism. **C,** Anterior dislocation of proximal interphalangeal joint with rupture of middle slip, avulsed collateral ligament, and partial tear of distal fibers of transverse retinacular ligament. Lateral band is displaced anterioraly. (From Spinner, M., and Choi, B.Y.: J. Bone Joint Surg. **52-A:**1329, 1970.)

SECONDARY REPAIR OF TENDONS

Before tendons are secondarily repaired, certain requirements must be met: (1) wound erythema and swelling suggest inflammation or potential infection and should not exist; (2) skin coverage must be adequate; (3) the tissues in which the tendon is expected to glide must be relatively free of scar; (4) the alignment of bones must be satisfactory, and any fractures must be healed or fixed securely; (5) joints must have a useful range of passive motion; and (6) sensation in the involved digit must be undamaged or restored, or it must be possible to suture damaged nerves at the time of tendon repair. When all these requirements are met, the tendon may be repaired as late as 2 weeks after injury; however, whether it should be repaired then mainly depends on the condition of the wound itself or the associated tissues. If the delay has been long, scar tissue may be abundant, and a two-stage procedure, such as insertion of a Hunter rod (see p. 61), may be indicated. Secondary repair of tendons may also have to wait, not only because of the factors listed above, but because reconstruction of the flexor pulleys followed later by tendon grafting is preferred to reconstruction of the pulleys at the time of tendon grafting. The critical pulleys that must be reconstructed are the A2 and A4. In these reconstructions a Hunter rod or similar implant is useful to maintain the lumen of the tendon sheath while the grafted pulleys are healing. This is followed later by the insertion of the flexor tendon graft. Generally, tendons may be repaired secondarily by direct suture at the site of division, by tendon graft, or even by tendon transfer.

Extensor tendons of fingers

An extensor tendon can usually be repaired secondarily by direct suture at the level of the metacarpophalangeal joint or on the dorsum of the hand. However, after approximately 5 weeks, when the time since injury has been so long that the proximal segment has retracted, or when a segment of tendon has been destroyed, an extensor proprius tendon may be transferred to the distal segment, or the distal segment may be sutured to an intact adjoining extensor tendon, or a segmental tendon graft may be inserted.

For a severe injury in which whole segments of tendons are lost, grafting may be necessary; when the muscle has also been damaged and has become fibrotic, transferring the flexor carpi ulnaris or other appropriate tendons may be necessary to provide a motor (see section on technique of tendon suture, p. 41).

CENTRAL SLIP OF EXTENSOR EXPANSION (BUTTONHOLE DEFORMITY)

Not all buttonhole deformities result from lacerations of the central tendon of the extensor mechanism. This deformity can be the result of avulsion or stretching of the insertion of the central tendon caused by direct contusion or rotational stretch. The deformity may take a month or longer after the trauma to develop from the imbalance created by loosening the central tendon.

Buttonhole deformities that are diagnosed early in closed wounds before fixed contractures occur may be treated conservatively. The conservative treatment consists of splinting the proximal interphalangeal joint in full extension while permitting the distal interphalangeal joint to be actively flexed. Avoid excessive pressure and resultant skin necrosis over the proximal interphalangeal joint area. Extension should be maintained constantly for 4 to 6 weeks and continued at night for several more weeks.

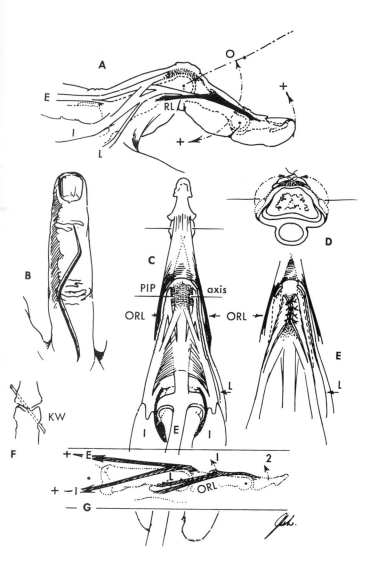

Fig. 3-28. Littler technique for repair of old buttonhole deformity. **A,** Typical deformity with flexion of proximal interphalangeal joint and extension of distal interphalangeal joint. Lateral bands have subluxated volarward. **B,** Dorsal curved longitudinal incision is made. **C,** Insertions of lateral bands are completely freed except for radialmost fibers of radial lateral band. **D** and **E,** Lateral bands are shifted dorsally and proximally and are sutured together and to soft tissues over proximal third of middle phalanx and to central tendon. **F,** Proximal interphalangeal joint is fixed in full extension by Kirschner wire. **G,** After repair proximal interphalangeal joint is extended by extensor hood and distal interphalangeal joint by preserved lumbrical muscle and oblique retinacular ligament. (From Littler, J.W., and Eaton, R.G.: J. Bone Joint Surg. **49-A:**1267, 1967.)

on the radial side separate by sharp dissection and leave intact the radial fibers of the lateral band that are continuations of the insertions of the lumbrical muscle and the oblique retinacular ligament; thus active extension of the distal interphalangeal joint is preserved. Now the insertions of the lateral bands are completely free except for the radialmost fibers of the radial lateral band. Shift the bands dorsally and proximally and suture them to the scar tissue and periosteum over the proximal third of the middle phalanx (Fig. 3-28, *E*). Also suture them to the attenuated central tendon while the proximal interphalangeal joint is held in full extension. With this joint still in full extension, insert a Kirschner wire obliquely across it. Now close the wound, leaving the divided transverse retinacular ligaments unsutured.

AFTERTREATMENT. At 3 weeks the Kirschner wire is removed and guarded motion is begun.

MALLET FINGER

As late as 12 weeks after injury a mallet finger caused by avulsion of the extensor tendon from the distal phalanx should be treated by splinting as described for an acute injury (p. 54). After 12 weeks if the distal phalanx droops severely but passive motion in the distal interphalangeal joint is still satisfactory, surgery may be indicated.

TECHNIQUE. Make a V-shaped incision with the point of the V 6 mm proximal to the nail base on the dorsum of the finger. Develop the incision to expose the extensor tendon and its intervening scar. Sever the tendon transversely proximal to the joint, leaving the insertion of the tendon into the bone. Resect enough of the scar or the tendon to take up the slack and maintain extension of the distal joint when the tendon ends are sutured end to end. Use No. 4-0 monofilament nylon or No. 4-0 monofilament wire as a pull-out roll stitch. This suture is all that is used on the tendon (Fig. 3-10). Close the skin with interrupted sutures and maintain the finger in extension with a compressed wet cotton dressing.

AFTERTREATMENT. At 2 weeks, the sutures are removed and only the distal joint is maintained in extension with a small metal splint for a total of 8 weeks, as in conservative treatment.

TECHNIQUE (FOWLER). Make a midlateral finger incision (p. 10) from just distal to the proximal interphalangeal joint to a point level with the middle of the proximal phalanx. Open the deep tissues until the edge of the lateral band of

Buttonhole deformities occurring in rheumatoid disease are described on p. 274.

In buttonhole deformity the central slip of the extensor expansion has retracted and the lateral bands have become loose and subluxated volarward after their dorsal transverse fibers have ruptured farther. This subluxation causes the proximal interphalangeal joint to flex; the lateral bands become contracted and in time cause a fixed flexion contracture of this joint and hyperextension of the distal interphalangeal joint.

Reconstruction after rupture or laceration of the central slip of the extensor expansion is difficult. A precise and extensive procedure is necessary not only to restore the function of the damaged central slip but to release the associated contracture as well.

TECHNIQUE (LITTLER, MODIFIED). Make a dorsal curved incision centered over the proximal interphalangeal joint (Fig. 3-28, *B*) and expose the lateral bands. Then with the point of a probe dissect deep to each transverse retinacular ligament from its origin near the volar plate to its insertion on the border of the lateral band; with small scissors divide each ligament near its middle. Next free the insertions of the lateral bands so that they can be replaced dorsally. But

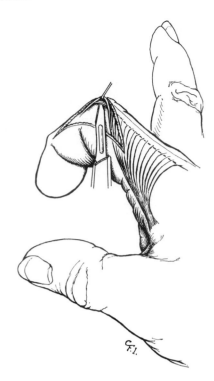

Fig. 3-29. Fowler operation to correct mallet finger (see text).

the extensor hood is located; then elevate this edge with a small hook (Fig. 3-29) and, with the finger in extension, continue elevating the expansion until the deep surface of the central slip is exposed at the proximal interphalangeal joint. Now elevate the entire extensor hood from the proximal phalanx; with the point of a No. 11 Bard-Parker knife blade and beginning on the deep surface of the central slip, free the insertion of the central slip from the proximal edge of the middle phalanx. Releasing this central slip allows the entire extensor mechanism to displace proximally; thus the tension increases on its distal end and shortens the avulsed tendon where it has become too long after healing to the distal phalanx by scar.

AFTERTREATMENT. A splint is applied with the proximal interphalangeal joint in no more than 30 degrees flexion and the distal joint in extension; prevention of acute flexion of the proximal interphalangeal joint prevents the capsule of this joint from being torn after release of the central slip. At 3 weeks this splint is removed and another is applied that immobilizes only the distal interphalangeal joint; it is held in extension for 4 more weeks on a small hammered metal splint that allows full motion of the more proximal joints.

CORRECTION OF OLD MALLET FINGER BY TENDON TRANSFER

The technique used for correcting hyperextension locking deformity of the proximal interphalangeal joint by transfer of a lateral band of the extensor mechanism may also be used for correction of old mallet finger deformity when there is satisfactory passive motion and no more than moderate arthritic changes in the distal joint (Fig. 8-13).

TECHNIQUE. Make a lateral incision on the less scarred side of the digit, and expose the extensor mechanism and

the flexor sheath. Detach one lateral band just beyond the metacarpophalangeal joint of the finger, and strip it loose entirely to its insertion distally. Now make a small pulley within the flexor tendon sheath opposite the proximal interphalangeal joint. Pass the tendon slip through the pulley, and then bring the end back to be sutured to the extensor hood a bit dorsal to its original position on the lateral side of the extensor mechanism. Correct tension on this transfer is essential and will hold the proximal interphalangeal joint slightly flexed while the distal joint is fully extended. Maintain the proximal interphalangeal joint in this position of slight flexion for 3 weeks by transfixing the joint with a small Kirschner wire. Suture the wound and enclose the finger in a moist cotton dressing. At 3 weeks, remove the wire and begin gradual motion of the finger.

Long extensor of thumb

When an extensor pollicis longus tendon has been divided at the interphalangeal joint, its proximal segment does not retract appreciably because the adductor pollicis, abductor pollicis brevis, and extensor pollicis brevis insert into the extensor expansion; consequently the tendon can be repaired secondarily without grafting or tendon transfer. But when the tendon has been divided at the metacarpophalangeal joint or more proximally, its proximal segment retracts rapidly, and by 1 month after injury a fixed contracture of the muscle has usually developed. The contracture may often be overcome by rerouting the tendon from around Lister's tubercle and placing it in a straight line; when this maneuver does not provide enough length, the extensor indicis proprius tendon may be transferred, and in this instance only one suture line is necessary instead of the two a graft would require. When the tendon is divided at a level far enough proximal for the distal end of the palmaris longus tendon to reach the end of its distal segment, this tendon may be transferred instead of the extensor indicis proprius. A graft is necessary to bridge a long defect when a tendon transfer is either impossible or undesirable.

AFTERTREATMENT. A splint is applied with the wrist in near full extension and the thumb extended and abducted. The splint should begin distal to the elbow but extend to the thumb tip and to the distal palmar crease. This immobilizes the thumb but releases the movement of the fingers. The splint should be maintained for 4 weeks and then the thumb is gradually permitted to move as only the wrist is splinted in extension another week.

Flexor tendons of fingers
DISTAL HALF OF FINGER (ZONE I) (FIG. 3-12)

When the profundus tendon has been lacerated or avulsed, it should be reattached within few days, before it retracts into the palm and before avulsion of the vinculum occurs, or else it should not be repaired. After a few days it swells and is impossible to rethread through the bifurcation of the sublimis. Rethreading the swollen profundus may also jeopardize proximal interphalangeal joint movement. Profundus function may be restored by a tendon graft but only when indicated. When treated early, an avulsed or lacerated profundus tendon can be advanced 1 cm and reattached as discussed under primary repair (p. 46).

In string musicians the motion at the distal interphalangeal joint that is provided by an intact flexor digitorum profundus is critical to enable the finger pulp to accurately fret the strings. This must be considered in those persons who have lost flexor digitorum profundus tendon function if other criteria are met.

CRITICAL AREA OF PULLEYS (ZONE II)

When the sublimis tendon alone is divided in the critical area of pulleys, it should not be repaired secondarily because the profundus tendon provides satisfactory function, except rarely, as in the index finger of the dominant hand, for writing. Even then, there exists a risk of jeopardizing profundus function. Hyperextension deformities of the proximal interphalangeal joint occasionally occur following the laceration of a sublimis tendon in a very flexible hand. This can be dealt with by means other than sublimis tendon suture. When the profundus tendon alone is divided in this area, the sublimis tendon provides ample flexion of the proximal interphalangeal joint and the profundus tendon should not be repaired as a delayed or secondary procedure. In the rare instance in which the distal joint is in need of stabilization, this can be accomplished by tenodesis (see p. 68). Tenodesis is occasionally necessary in the index finger but rarely in the other digits.

FOREARM AND PALM (ZONES III, IV, AND V)

As late as 3 or 4 weeks after injury, flexor tendons in the forearm and palm may be repaired by direct suture since flexing the wrist provides ample slack to overcome contracture of the muscles. After 4 or 5 weeks the muscles become contracted and fixed, and a graft is necessary at times to bring together the tendon ends. This may be in the form of a "minigraft" or short segmental graft between the tendon ends (Fig. 3-30). When there has been destruction of tendons, profundus tendons take priority and a combination of distal profundus tendon and available sublimis tendon motors can be used. The attachment of tendons should be done with a simple buried stitch, usually of monofilament wire.

TECHNIQUE OF PROFUNDUS ADVANCEMENT (WAGNER). Make a midlateral incision (p. 10); incise the tendon sheath and retract it. Usually the proximal end of the profundus will have retracted into the palm; make a transverse incision near the distal palmar crease to recover it. Carefully thread it through the bifurcation of the sublimis and into the distal end of the finger; when this cannot be done accurately and with assurance that the relation of the sublimis to the profundus is normal, abandon the procedure because it will end in failure not only of profundus function, but also of sublimis function. Next, resect the distal segment of the profundus at a level just proximal to the distal interphalangeal joint and split its distal stump in a transverse plane (Fig. 3-31). With a Bunnell pull-out wire suture, fix the distal end of the proximal segment of tendon into the split profundus stump and tie the suture at the end of the finger through a button. Do not disturb the capsular attachment of the profundus stump because it protects the volar plate and helps to assure a gliding surface. Repair any divided digital nerves at this time. Close the wound.

AFTERTREATMENT. Immediately after surgery a splint is applied with the wrist in enough flexion to allow full ex-

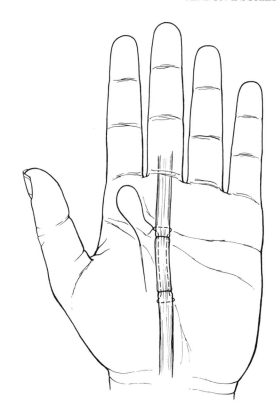

Fig. 3-30. Long-standing flexor tendon interruptions in palm may require a short segmental graft or "minigraft" to avoid too much tension.

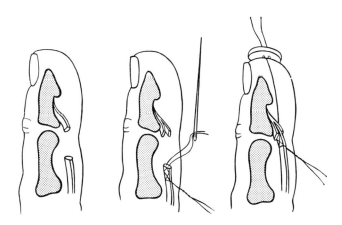

Fig. 3-31. Wagner technique of profundus advancement (see text).

tension of the two interphalangeal joints. At 3 weeks the pull-out suture is removed, and exercises are gradually begun. Full active extension of the finger with the wrist in dorsiflexion may not be possible for several more weeks.

REPAIR OF FINGER FLEXORS BY TENDON GRAFT

When the sublimis and profundus tendons have both been divided in the critical area of pulleys, restoring flexion of the finger by grafting is indicated when the skin is pliable, any wounds are well healed, and edema has subsided; the joints must allow a full passive range of motion, and sensation in the finger should be normal, or at least one digital nerve should be intact (one divided digital nerve may be sutured at the time of grafting if the other nerve is intact). The A2 and A4 pulley systems should also be intact; otherwise, these should be reconstructed in a separate staged procedure before tendon grafting (see technique below). Failure of tendon grafts in patients over 50 years of age is very common.

TECHNIQUE. Make a zigzag incision on the volar aspect of the finger to expose the underlying flexor sheath up to the proximal finger crease or make a midlateral incision (see Fig. 1-15); carefully avoid entering the proximal interphalangeal joint, which is covered laterally with little fat. The neurovascular bundles are also easily damaged in the zigzag incision. Expose the flexor sheath and preserve as much of the unscarred sheath as possible. Excise no more than absolutely necessary of the A2 and A4 pulley systems; complete excision of either will result in functional failure of the grafted tendon. Free the scarred sublimis and profundus tendons and carefully preserve all of the volar plate of the proximal and distal interphalangeal joints. Divide both tendons at their insertions and bring them out through a transverse palmar incision made over the midbelly of the lumbrical muscles. Then with a small chisel, raise a flap of bone just distal to the distal interphalangeal joint on the volar surface of the distal phalanx for later insertion of the profundus tendon graft (see Figs. 3-11 and 3-32). Under this bone flap, drill a small hole with a Kirschner wire large enough to accommodate two No. 4-0 monofilament wire sutures.

An alternate method for distal insertion is to split the remaining distal tendon stump (Fig. 3-31). The stump is

large enough to receive and hold a suture. This is the preferred method in children. Now through the proximal palmar incision, divide the sublimis tendon as far proximally as possible and discard it; retain the profundus tendon for attachment to the graft.

In approximately 15% of people the palmaris longus tendon is absent, but when it is present, it can be taken from the same forearm and used as the graft. Expose the tendon through a transverse incision just proximal to its insertion at the wrist and through another in the upper forearm. Divide the tendon at its musculotendinous junction proximally and detach it distally after dissecting out the various portions of its insertion; then draw the tendon out through the forearm incision. Place a monofilament No. 34 pull-out wire suture in the distal end of the palmaris longus tendon before dividing it proximally. This suture is much more easily placed when the proximal end of the tendon is stabilized. Thread the pull-out wire through the hole in the distal phalanx. Bring the tendon proximally through the intact tendon sheath and into the palm by wetting the tendon and pushing it through with a probe. To prevent bowstringing, determine if the pulley system is intact while bringing the tendon proximally.

A careful attempt must be made to anastomose the tendon to the graft under appropriate tension. Place the wrist in neutral and the finger in full extension. Now place tension on the musculotendinous junction proximally and at the point where the tendon and graft are to be joined; mark the junction with a methylene blue pen. Adjust the tension so that when the wrist is in extension, the finger will automatically be brought into about the same degree of flexion as the adjoining digits. Increase flexion a little in the more ulnar digits. Several methods may be used to suture the proximal junction. One is a direct suture by the crisscross Bunnell method using a monofilament wire (see Fig. 3-6); another is the fishmouth insertion as described by Pulvertaft (see Fig. 3-9). Do not suture the lumbrical muscle to the tendon junction lest the tension in the lumbrical muscle be increased. Attach the tendon ends and close the wound without subcutaneous sutures. Insert a drain in the proximal palmar wound and remove it after 24 hours. Place the wrist in gentle flexion with all the fingers in slight extension; the wrist should not be placed in forced flexion because this increases pain postoperatively and may cause pressure on the median nerve. Cover the wounds with a layer of nonadhesive material and then a moist molded dressing. Apply a posterior plaster splint to hold the wrist in flexion. Some prefer to keep the finger in flexion by gluing a garment hook to the fingernail and attaching it to an elastic band anchored at the wrist level by a pin (see Fig. 3-23).

AFTERTREATMENT. The wrist is elevated for the first 24 to 48 hours. In children, a long arm cast is applied to keep the dressing from shifting distally. It is extremely important to prevent a postoperative hematoma. Therefore, some surgeons release the tourniquet before wound closure and keep pressure on the wound manually for 5 minutes. We prefer to apply a moist, conforming dressing to the volar aspect of the finger with the wrist held in flexion by a posterior splint and to elevate the hand immediately and throughout recovery from the anesthetic. The care is then

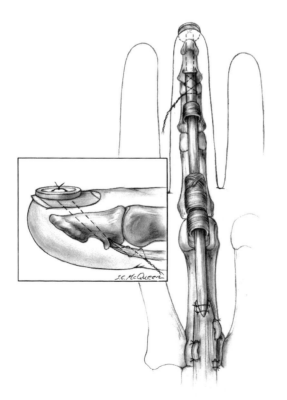

Fig. 3-32. Repair of finger flexor tendons by tendon graft Graft has been sutured in place. Note the proximal and distal pulleys have been narrowed.

similar to that after primary suture of a flexor tendon. Passive motion with a 5 mm excursion carried out 3 or 4 times a day under the direction of a therapist may help reduce adhesions. Also, the technique shown in Fig. 3-23 does not seem to put a disruptive pull on the tendon sutures. At 3 to 4 weeks, the pull-out wire suture is removed. Further active motion is started, but protected motion is continued until 4 weeks. Extending the wrist during finger exercises helps create a tenodesing effect on the graft, whereas flexing the wrist tends to relax it. Furthermore, having the therapist manually stabilize the metacarpophalangeal joint helps focus motion at the proximal interphalangeal joint, resulting in a greater arc of motion at this critical level than the combination of flexion at the metacarpophalangeal and proximal interphalangeal joints.

RECONSTRUCTION OF FLEXOR TENDON PULLEYS BEFORE TENDON GRAFTING

It is useless to perform a flexor tendon graft in a digit without intact portions of the A2 and A4 pulleys. These pulleys may have been destroyed by previous surgery during an attempt to repair the tendons primarily. When destroyed, the angle of approach of the tendon to its insertion is altered and a flexion contracture develops at the proximal interphalangeal joint or metacarpophalangeal joint, or both. Reconstruction of the A2 and A4 pulley systems should not be attempted unless there are strong indications for later reconstruction by a flexor tendon graft. Ideally, the digit should be free of fractures, neurovascular defects, and heavy scarring. The results are also better in a young adult or an adolescent because multiple procedures are more likely to result in joint stiffness, excessive scarring, and other complications in older patients. Generally, tendon pulleys cannot be reconstructed effectively at the time of flexor tendon grafting because they will be stretched by the pressures of early motion required by the gliding tendon graft even if motion is delayed for 3 weeks. A tendon sheath spacer should be inserted and left in place for several weeks after grafting of the pulley system. A Hunter rod or similar space filler is ideal. The donor graft for the pulley may be a split sublimis tendon or the palmaris longus. However, the palmaris longus tendon may be needed later for grafting of the flexor tendon itself. (See the discussion of donor sites for tendon grafts on p. 65.)

TECHNIQUE. Make a zigzag (see Fig. 1-13, *B* and *C*) or midlateral (see Fig. 1-15) incision exposing the area of the flexor tendons. Make the exposure wide enough to show all of the flexor pulley system. Excise the scarred tendons and surrounding scar tissue. However, retain any part of the sheath that is not scarred, especially in the area of the distal joint and the A1 pulley system in the palm. Bring the tendons out through an additional palmar incision and complete their excision. Insert a Hunter silastic rod of appropriate size and attach it distally either to the remaining profundus tendon stump or to the bone by a small screw (Fig. 3-33) as described for the two-stage Hunter rod technique. Obtain an appropriate donor tendon 6 cm long and 0.25 cm wide (about one half the width of the palmaris longus). Weave the strip over the rod beginning at the A2 pulley level. Weave it through slots made in the intact border of the sheath as it inserts into the bone; this margin of

the fibroosseous tunnel as it arises from the phalanx provides strong tissue in which to anchor the graft. Bring the graft back and forth over the rod and suture it at both the A2 and A4 levels; do not weave it over the proximal interphalangeal joint. Attach the rod proximally at the forearm or palmar level in a scar-free area away from the profundus tendon. Leave the profundus tendon attached to the lumbrical muscle to maintain its length. Close the wound loosely and support the hand with a dorsal splint.

AFTERTREATMENT. After the swelling has subsided, passive motion of the finger joints is carried out for 6 weeks or longer; then the rod may be removed and the tendon graft is inserted.

TWO-STAGE PROCEDURE FOR FLEXOR TENDON REPAIR (HUNTER ROD TECHNIQUE)

For patients with excessive scarring and joint stiffness and possibly with need of nerve anastomoses, a two-stage procedure for tendon repair may be indicated. The first stage consists of excising the tendon and scar from the flexor tendon bed and preserving the flexor pulley system. A Dacron-impregnated silicone rod is inserted to maintain the tunnel in the area of the excised tendons until passive motion and sensitivity have been restored to the digit. The rod is attached distally to bone or tendon stump. The degree of scarring within the palm determines whether it is attached proximally in the midpalm or above the wrist. It should not be attached to muscles or tendons proximally because this will create a scar at the site of future anastomosis. The second stage consists of removal of the rod and insertion of a tendon graft.

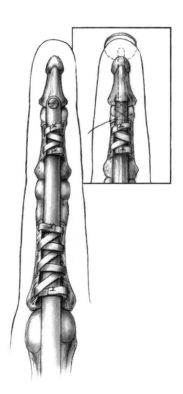

Fig. 3-33. Reconstruction of flexor tendon pulleys (see text). (Redrawn from Kleinert, H.E., and Bennett, J.B.: J. Hand Surg. 3:297, 1978.)

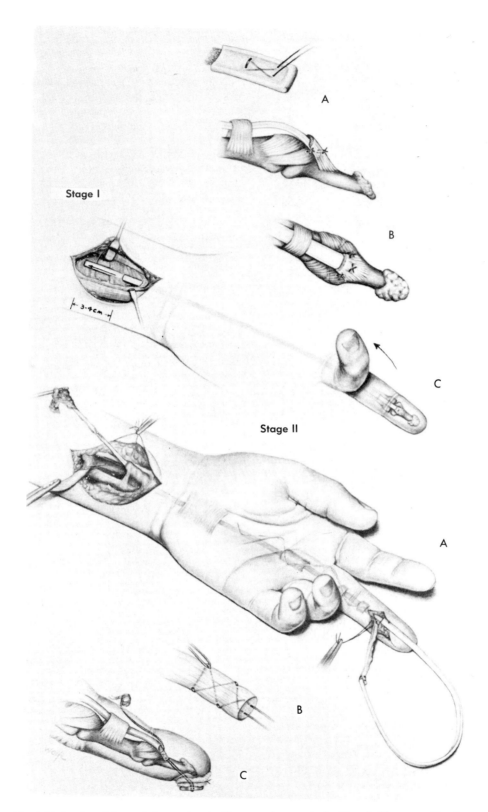

Stage I

Stage II

Fig. 3-34. Passive gliding technique, using the Hunter tendon prosthesis. **Stage I:** Placement of tendon prosthesis after excision of scar and formation of pulleys. *A,* Figure-eight suture in distal end of prosthesis. *B,* Distal end of prosthesis sutured to stump of profundus tendon and adjacent fibrous tissue on distal phalanx. *C,* Prosthesis in place showing free gliding and excursion of its proximal end during passive finger flexion. **Stage II:** Removal of prosthesis and insertion of tendon graft. *A,* Graft has been sutured to proximal end of prosthesis and then pulled distally through the new tendon bed. Note mesentery-like attachment of new sheath visible in the forearm. *B,* Distal anastomosis. Bunnell pull-out suture in distal end of tendon graft. *C,* Distal anastomosis. Complete Bunnell suture with button over fingernail. Reinforcing sutures are usually placed through stump of profundus tendon.

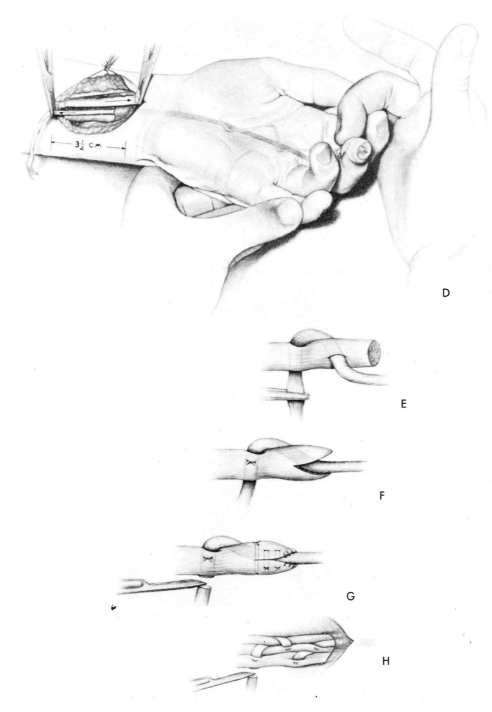

Fig. 3-34, cont'd. *D,* Proximal anastomosis. Measuring excursion of tendon graft and selecting motor. If procedure is done under local anesthesia (see text) the true amplitude of active muscle contraction can be measured. *E,* Proximal anastomosis. Graft is threaded through tendon motor muscle two or three times for added strength. *F,* Proximal anastomosis. Stump is fish-mouthed after the method of Pulvertaft, the tension is adjusted, and one suture is inserted as shown. Further adjustment of the tension can be accomplished simply by removing and shortening or lengthening as need be. *G,* Proximal anastomosis. After appropriate tension is selected, anastomosis is completed. *H,* Proximal anastomosis. Technique when graft is anastomosed to common profundus tendon (see text). (From Hunter, J.M. and Salisbury, R.E.: J. Bone Joint Surg. **53-A:**829, 1971).

TECHNIQUE

Stage one. Make a zigzag (see Fig. 1-13, *B* and *C*) or midlateral (see Fig. 1-15) incision to expose the entire flexor sheath area. Expose the palm either by a continuation of the zigzag incision or through an additional incision at the level of the A1 pulley. Excise the profundus and sublimis tendons, but retain a stump of the profundus tendon 1 cm long at the distal phalanx. Retain only the unscarred portions of the flexor pulley system, but some portions of the A2 and A4 pulleys are vital. Now extend the dissection into the palm; if the lumbrical muscle is scarred, excise it. Select a Dacron silicone rod of appropriate size and wash it free of all lint. Insert it into the palm, or if this area is scarred, continue blunt dissection proximally so that the rod extends to or above the wrist level (Fig. 3-34, *A*). Excising the entire sublimis tendon may be necessary to make room for the rod at the wrist. Attach the rod distally to the stump of the profundus or into the distal phalanx with a screw. Attach the proximal end to the fascia above the wrist or in the palm. After seating the prosthesis, passively flex the fingers to observe any tendency toward buckling. Traction on the prosthesis will determine the need for possible further modification of the pulley system, either by excising more scar tissue or reconstructing a defective area in the system especially at A2 and A4 levels.

AFTERTREATMENT. After closing the wound, the hand is supported by a splint and gentle passive motion of the finger joints is started during the second to the fourth week. The hand should be examined regularly for synovitis or buckling of the rod. If synovitis develops, prompt and complete immobilization is indicated and the second stage of the procedure should be carried out before chronic fibrosis develops. This usually can be done after 2 months.

Stage two. Using appropriate anesthesia, make a midlateral or zigzag incision at the distal phalanx and another incision proximally in either the palm or the forearm. Obtain a tendon graft of appropriate length and attach it to the freed prosthesis at its distal end. A plantaris tendon graft that is long may be necessary if the graft should extend above the wrist. Extract the prosthesis through the proximal incision and attach the graft at the distal phalanx as in a routine tendon graft; then attach the graft proximally to the profundus tendon (Fig. 3-34, *B*).

An alternate method is to attach the graft proximally to the profundus tendon and pull it distally. Appropriate tension on the graft is essential and can be adjusted according to the technique for standard tendon grafting (p. 60).

AFTERTREATMENT. The care after surgery is the same as that described for a flexor tendon graft (Fig. 3-23).

Long flexor of thumb

The flexor pollicis longus tendon may be repaired secondarily by direct suture at any level within the thumb if the two ends can be approximated without excessive tension. Within the first few weeks after injury, tension of the muscle from contracture may be overcome by flexing the wrist. The pulley mechanism should be avoided opposite the metacarpophalangeal joint when the tendon has been divided near this level because the suture line is likely to adhere to the pulley. This can be accomplished by tendon advancement as described under primary repair (p. 50). Grafting may be indicated when secondary repair has been delayed for a long time or when substance of the tendon has been lost. When repair by either method is impossible, the interphalangeal joint should be stabilized by arthrodesis (p. 170) or tenodesis (p. 68).

TECHNIQUE FOR FLEXOR TENDON GRAFT. Make an incision on the radial side of the thumb from a point near the base of the nail to near the middle of the metacarpal and then angle it toward the palm to end near the middle of the thenar eminence (Fig. 1-13, *E*). Elevate the skin and subcutaneous tissue as a flap with its base toward the palm; carefully dissect the branch of the radial nerve and its corresponding vessel, and retract them with this flap. Also dissect the digital neurovascular bundle, which lies well toward the anterior aspect of the thumb. Identify the pulley, and open the tendon sheath and the pulley sufficiently to insert the tendon graft, but leave a segment of the pulley at least 1 cm wide intact over the metacarpophalangeal joint to prevent "bowstringing" of the tendon. Also leave intact the oblique pulley at the proximal and middle thirds of the proximal phalanx. Free the flexor tendon, but take care not to enter the interphalangeal joint or to damage its volar plate. Now make a transverse incision 2.5 cm long proximal to the flexor crease of the wrist and identify the flexor pollicis longus tendon and withdraw it; if possible, tag the distal end of the tendon with a suture before withdrawing it and use this suture as a guide in threading the graft into the thumb.

Now obtain a tendon graft from an appropriate site (see below). Anchor the graft at the point of insertion of the original tendon as described for a flexor tendon graft of a finger (p. 60). Suture the proximal end of the graft to the distal end of the flexor pollicis longus tendon at a level proximal to the wrist so that the juncture will not enter the carpal tunnel or encroach on the median nerve when the thumb and wrist are extended. The graft should be under enough tension to slightly flex the interphalangeal joint of the thumb when the wrist is in the neutral position. As an alternative method, the proximal end of the graft may be sutured to the tendon opposite the middle of the thumb metacarpal; this requires only one incision and helps prevent the median nerve neuritis that sometimes develops after suture at the level of the wrist.

Close the wound with No. 5-0 nylon or a similar suture without subcutaneous sutures. Apply a splint to the wrist and hand with the wrist in 45 degrees flexion and the interphalangeal joint of the thumb in extension.

AFTERTREATMENT. At 3 weeks the pull-out suture is removed and active motion is begun. At 7 weeks, active flexion can be increased to approach full exertion by the muscle.

Tenolysis after tendon grafting

In addition to the complications that may occur after any surgical procedure, tendon grafting may be complicated by the inability of the tendon to glide and restore motion. In some instances this complication may be treated by tenolysis, but this operation should not be carried out until at least 3 months after the grafting and then only when there is no detectable active motion at the proximal interphalangeal joint. Further delay is indicated when accurately measured joint motion indicates improvement week by week, but if tenolysis is delayed too long, joints become stiff.

Tenolysis is just as delicate as the original grafting procedure and must be carried out through a full midlateral or volar zigzag approach with sharp dissection and careful hemostasis. A regional or local block anesthesia and proper sedation are useful here so that the patient can move his finger during surgery (see discussion of anesthetics, p. 5). The graft is freed in the palm and finger under direct vision; tendon strippers should not be used. Instilling hydrocortisone into the wound may prevent some immediate adhesions. Motion of the finger is begun immediately after surgery. The incidence of tendon rupture after tenolysis is greater than 10%.

Boyes and others perform a new tendon graft rather than a tenolysis after a graft failure.

Donor tendons for grafting

Donor tendons for grafting, in order of preference, are the palmaris longus, the plantaris, the long extensors of the toes, the extensor indicis comminus, and the flexor digitorum sublimis.

PALMARIS LONGUS

The palmaris longus is the tendon of choice because it fulfills the requirements of length, diameter, and availability without producing a deformity. The presence of this tendon should be determined before any grafting procedure; its presence can be demonstrated by having the patient oppose the tips of the thumb and little finger while flexing the wrist (Fig. 3-35). The tendon is reported to be present in one arm in 85% of people and in both arms in 70%. The tendon is flat, is surrounded by paratenon, and

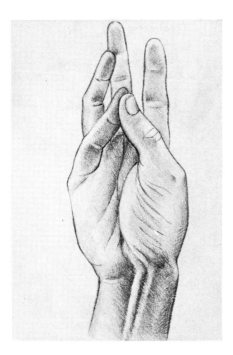

Fig. 3-35. Method of demonstrating presence of palmaris longus tendon (see text).

is long enough for a graft about 15 cm in length. Excise the tendon as follows. Make a short transverse incision directly over the tendon just proximal to the flexion crease of the wrist. Divide the tendon, grasp its end with a hemostat, and apply traction so that it can be palpated easily in a proximal direction. Then make a second transverse incision over the tendon at the junction of the middle and proximal thirds of the forearm. Identify the tendon, divide it, and withdraw the segment to be used as a graft. Should paratenon be desired, a long curved incision over the forearm is necessary; this method is much more disabling, and as an alternative use a tendon stripper similar to that devised by Brand for the plantaris tendon (Fig. 3-38). Occasionally there is a double palmaris longus tendon, or it has multiple premature insertions or an associated aberrant muscle. Any of these will make it difficult to withdraw from two transverse incisions.

PLANTARIS TENDON

The plantaris tendon is equally as satisfactory for a graft as the palmaris tendon, has the advantage of being almost twice as long (enough to provide two grafts), but is not as accessible (Figs. 3-36 and 3-37). It is present in 93% of people. The tendon lies anteromedial to the tendo calcaneus proximal to the heel and may be obtained for grafting as follows. Make a small medial longitudinal incision just anterior to the insertion of the tendo calcaneus. Identify the tendon as a slip distinctly separate from the tendo calcaneus (Fig. 3-38). Divide the tendon near its insertion and thread it through the loop of a tendon stripper made for this purpose. Keep the knee in full extension. Clamp the distal end of the tendon with a hemostat and hold it taut and pass the stripper up the leg until the resistance of the gastrosoleus fascia is encountered; overcome this resistance with additional force on the stripper. Advance the stripper proximally for a total of about 25 cm, where resistance is again met as the loop of the stripper meets the belly of the muscle. Now palpate the loop of the stripper through the skin and over it make a longitudinal incision 5 cm long. Then free from around the plantaris tendon the gastrocnemius muscle belly, divide the tendon under direct vision, and withdraw the tendon distally. If a tendon stripper is not available, remove the tendon through multiple short transverse incisions.

LONG EXTENSORS OF TOES

Tendons of the long extensors of the toes are not as desirable as the palmaris longus or the plantaris tendons because there are many communications between them, especially as the cruciate ligament is approached proximally. Every toe except the little one has an extensor brevis tendon to dorsiflex it after excision of the long extensor tendon. The extensor hallucis longus is much larger than the other extensors, and the extensor of the second toe is much more intimately related to the dorsalis pedis artery. The extensor of the third toe is probably easiest to remove and use. Make multiple short transverse incisions over the tendon and remove it by elevating the skin proximal to each incision and dissecting to a more proximal level; then make another incision at this point and repeat the procedure. Extract the divided end of the tendon

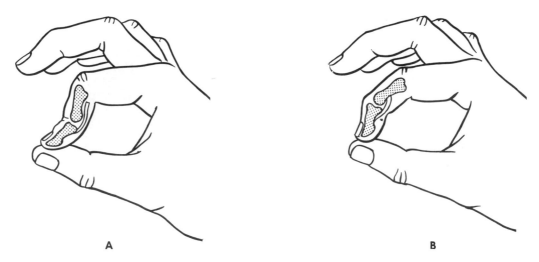

A   B

Fig. 3-40. Tenodesis for irreparable damage to profundus tendon. **A,** Before tenodesis distal interphalangeal joint is unstable and hyperextends during pinch. **B,** After tenodesis joint is stable and remains partially flexed during pinch.

through each successive incision and remove it through the proximal incision, as shown in Fig. 3-39, *A*.

A tendon may be removed much more easily through a long curved incision along the course of the tendon, as shown in Fig. 3-39, *B,* but such an incision temporarily prevents weight bearing and results in an unsightly scar.

EXTENSOR INDICIS PROPRIUS

The extensor indicis proprius tendon is usually long enough for a single flexor tendon graft but is rarely used. Divide it distally at its insertion into the extensor hood through a small transverse incision and proximally just distal to the dorsal carpal ligament through another incision. Suture the distal end of the proximal segment of tendon to the extensor digitorum communis to help maintain independent extension of the index finger.

FLEXOR DIGITORUM SUBLIMIS

A flexor digitorum sublimis tendon should not be excised simply as a graft, but at times one is available when removed in conjunction with an amputation or a flexor tendon grafting. The tendon is usually too thick so that its central part undergoes necrosis when it is used as a graft; thus a local reaction is produced that causes adhesions. The tendon may be split longitudinally to make it thinner, but this leaves a raw surface where adhesions are even more likely to develop.

Tenodesis

Tenodesis, as already mentioned, is useful when the profundus is damaged, flexor tendon grafting is impossible, and the fingertip is more useful functionally when partially flexed and stabilized than when extended; this is usually true of the index finger (Fig. 3-40), and for certain occupations of other fingers as well. The operation is possible only when the distal stump of the profundus tendon is long enough to anchor proximal to the distal interphalangeal joint.

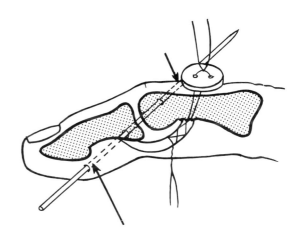

Fig. 3-41. Technique of tenodesis (see text). Kirschner wire is cut off beneath skin at points indicated by arrows.

TECHNIQUE. Make a midlateral incision (p. 10) and identify the stump of the tendon. Flex the distal interphalangeal joint 30 degrees and note the length of profundus tendon required for tenodesis. Insert a Bunnell pull-out wire suture in the tendon and excise any redundant tendon (Fig. 3-41). With the joint in the desired position, insert a Kirschner wire obliquely across it and cut off the wire beneath the skin. Then at the level of intended tenodesis roughen the bone of the middle phalanx with a dental chisel and drill two small holes through it from anterior to posterior. With straight needles, thread the ends of the wires through the holes to the dorsum of the finger and tie threm through a button over the middle phalanx; bring the pull-out wire through the volar surface of the finger. Close the wound with No. 5-0 nylon. No external splinting is necessary.

AFTERTREATMENT. The pull-out wire suture is removed at 3 weeks and the Kirschner wire at 5 or 6 weeks.

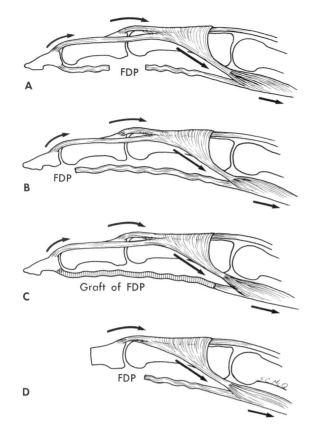

Fig. 3-42. Conditions that cause "lumbrical plus" finger. **A,** Severance of flexor digitorum profundus *(FDP)* (produces paradoxical extension). **B,** Avulsion of flexor digitorum profundus. **C,** Overly long flexor tendon graft. **D,** Amputation through middle phalanx. (Redrawn from Parkes, A.: J. Bone Joint Surg. **53-B:**236, 1971.)

"Lumbrical plus" finger

The "lumbrical plus" finger results from loss of tension of the flexor digitorum profundus tendon distal to the lumbrical muscle origin. This tension loss permits proximal retraction of the profundus tendon and the origin of the lumbrical muscle and thus produces unwanted secondary extension of the proximal and distal interphalangeal joints from "overpull" by the lumbrical muscle. This may occur with an amputation through the middle phalanx, with an avulsion of the insertion of the flexor digitorum profundus or division of this tendon, or with a loosely inserted tendon graft for the distal segment of the flexor digitorum profundus (Fig. 3-42). The middle finger is most commonly involved, apparently because of the origin of the lumbrical muscle on the radial aspect of the middle finger profundus and its ulnarward displacement by the adjoining lumbrical origin of the ring finger. The lumbrical tendon is pulled around the head of the metacarpal and its tension is increased.

This diagnosis should be considered when there is incomplete flexion of the proximal interphalangeal joint of the middle finger with no other apparent cause. Functionally, when finger flexion is attempted, there is initial flexion of the metacarpophalangeal joint that continues to nearly complete flexion before any proximal interphalangeal joint flexion is accomplished, instead of the normal smooth sweeping motion of the tip of the finger. This is called "paradoxical extension" of the interphalangeal joints. The test for "lumbrical plus" finger, according to Parkes, is to demonstrate first that the patient has full passive flexion of all joints of the finger; then, with strong gripping or flexing of all fingers with the profundus tendons, flexion becomes incomplete and active partial extension of the interphalangeal joints occurs.

Treatment of this abnormality consists of division or resection of the involved lumbrical while the patient is under local anesthesia.

REFERENCES

Araico, J., and Ortiz, J.M.: Subcutaneous flexor pollicis longus tendon graft technique, Plast. Reconstr. Surg. **45:**578, 1970.

Arons, M.S.: Purposeful delay of the primary repair of cut flexor tendons in "some-man's land" in children, Plast. Reconstr. Surg. **53:**638, 1974.

Bader, K., and Curtin, J.W.: Clinical survey of silicone underlays and pulleys in tendon surgery in hands, Plast. Reconstr. Surg. **47:**576, 1971.

Becker, H., Orak, F., and Deponselle, E.: Early active motion following a beveled technique of flexor tendon repair: report on fifty cases, J. Hand Surg. **4:**454, 1979.

Bell, J.L., Mason, M.L., et al.: Injuries to flexor tendons of the hand in children, J. Bone Joint Surg. **40-A:**1220, 1958.

Bennett, J.B.: Flexor tendon laceration in Ehler-Danlos syndrome: a case report, J. Bone Joint Surg. **59-A:**259, 1977.

Bluestone, L.: Pin fixation technique in tendon repair, Surgery **60:**506, 1966.

Bora, F.W., Jr.: Profundus tendon grafting with unimpaired sublimus function in children, Clin. Orthop. **71:**118, 1970.

Bowers, W.H., and Hurst, L.C.: Chronic mallet finger: the use of Fowler's central slip release, J. Hand Surg. **3:**373, 1978.

Boyes, J.H.: Flexor-tendon grafts in the fingers and thumb: an evaluation of end results, J. Bone Joint Surg. **32-A:**489, 1950.

Boyes, J.H.: Evaluation of results of digital flexor tendon grafts, Am. J. Surg. **89:**1116, 1955.

Boyes, J.H.: Why tendon repair? (editorial), J. Bone Joint Surg. **41-A:**577, 1959.

Boyes, J.H., and Stark, H.H.: Flexor-tendon grafts in the fingers and thumb: a study of factors influencing results in 1000 cases, J. Bone Joint Surg. **53-A:**1332, 1971.

Boyes, J.H., Wilson, J.N., and Smith, J.W.: Flexor-tendon ruptures in the forearm and hand, J. Bone Joint Surg. **42-A:**637, 1960.

Brand, P.W.: Evaluation of the hand and its function, Orthop. Clin. North Am. **4:**1127, 1973.

Brand, P.W., Beach, R.B., and Thompson, D.E.: Relative tension and potential excursion of muscles in the forearm and hand, J. Hand Surg. **6:**209, 1981.

Braun, R.M.: Palmaris longus tendon transfer for augmentation of the thenar musculature in low median palsy, J. Hand Surg. **3:**488, 1978.

Broder, H.: Rupture of flexor tendons, associated with a malunited Colles' fracture, J. Bone Joint Surg. **36-A:**404, 1954.

Browne, E.Z., Jr., Teague, M.A., and Synder, C.C.: Prevention of extensor lag after indicis proprius tendon transfer, J. Hand Surg. **4:**168, 1979.

Bruner, J.M.: Surgical exposure of the flexor pollicis longus tendon, Hand **7:**241, 1975.

Bunnell, S.: The early treatment of hand injuries, J. Bone Joint Surg. **33-A:**807, 1951.

Carter, S.J., and Mersheimer, W.L.: Deferred primary tendon repair: results in 27 cases, Ann. Surg. **164:**913, 1966.

Chacha, P.: Free autologous composite tendon grafts for division of both flexor tendons within the digital theca of the hand, J. Bone Joint Surg. **56-A:**960, 1974.

Casscells, S.W., and Strange, T.B.: Intramedullary wire fixation of mallet-finger, J. Bone Joint Surg. **39-A:**521, 1957.

Christophe, K.: Rupture of the extensor pollicis longus tendon following Colles' fracture, J. Bone Joint Surg. **35-A:**1003, 1953.

Chuinard, R.G., et al.: Tendon transfers for radial nerve palsy: use of superficialis tendon for digital extension, J. Hand Surg. **3:**560, 1978.

Chuinard, R.G., Dabezies, E.J., and Mathews, R.E.: Two-stage superficialis tendon reconstruction in severely damaged fingers, J. Hand Surg. **5:**135, 1980.

Clark, C.B.: A reevaluation of selective tendon repair, Orthop. Rev. **4:**31, July 1975.

Clark, C.B.: Primary flexor tendon repair in the fingers, Orthop. Rev. **5:**67, May 1976.

Crosby, E.B., and Linsheid, R.L.: Rupture of the flexor profundus tendon of the ring finger secondary to ancient fracture of the hook of the hamate: review of the literature and report of two cases, J. Bone Joint Surg. **56-A:**1076, 1974.

Dolphin, J.A.: Extensor tenotomy for chronic boutonnière deformity of the finger: report of two cases, J. Bone Joint Surg. **47-A:**161, 1965.

Doyle, J.R., and Blythe, W.: The finger flexor tendon sheath and pulleys: anatomy and reconstruction. In American Academy of Orthopaedic Surgeons: Symposium on tendon surgery in the hand, St. Louis, 1975, The C.V. Mosby Co.

Duran, R.J., Houser, R.G.: Controlled passive motion following flexor tendon repair in zones 2 and 3. In American Academy of Orthopaedic Surgeons: Symposium on tendon surgery in the hand, St. Louis, 1975, The C.V. Mosby Co.

Chirurgie des Tendons de la main, monographie du G.E.M., 1976, Expansion Scientifique.

Elliott, R.A., Jr.: Injuries to the extensor mechanism of the hand, Orthop. Clin. North Am. **2:**335, 1970.

Entin, M.A.: Repair of extensor mechanism of the hand, Surg. Clin. North Am. **40:**275, 1960.

Entin, M.A.: Flexor tendon repair and grafting in children, Am. J. Surg. **109:**287, 1965.

Entin, M.A.: Philosophy of tendon repair, Orthop. Clin. North Am. **4:**859, 1973.

Erdelyi, R.: Reconstruction of the flexor digitorum profundus with the aid of the flexor profundus split from an adjoining finger, Plast. Reconstr. Surg. **37:**13, 1966.

Farkas, L.G.: Use of interposed flap of tendon sheath to prevent adhesions after the repair of a cut flexor profundus tendon: experimental study in chickens, Plast. Reconstr. Surg. **62:**404, 1978.

Farkas, L., and Lindsay, W.K.: Functional return of tendon graft protected entirely by pseudosheath—experimental study, Plast. Reconstr. Surg. **65:**188, 1980.

Fetrow, K.O.: Tenolysis in the hand and wrist: a clinical evaluation of two hundred and twenty flexor and extensor tenolyses, J. Bone Joint Surg. **49-A:**667, 1967.

Ford, J.C., Smith, J.R., and Carter, J.E.: Primary tendon grafting in injuries of the thumb flexor, South. Med. J. **64:**78, 1971.

Fowler, S.B.: The management of tendon injuries (editorial), J. Bone Joint Surg. **41-A:**579, 1959.

Freehafer, A.A., Peckham, H., and Keith, M.W.: Determination of muscle-tendon unit properties during tendon transfer, J. Hand Surg. **4:**331, 1979.

Furlow, L.T., Jr.: The role of tendon tissues in tendon healing, Plast. Reconstr. Surg. **57:**39, 1976.

Gaisford, J.C., Hanna, D.C., and Richardson, G.S.: Tendon grafting: a suggested technique, Plast. Reconstr. Surg. **38:**302, 1966.

Gama, C.: Results of the Matev operation for correction of boutonnière deformity, Plast. Reconstr. Surg. **64:**319, 1979.

Gelberman, R.H., Amiff, D., Gonsalves, M., et al.: The influence of protected passive mobilization on the healing of flexor tendons: a biochemical and microangiographic study, Hand **13:**120, 1981.

Gelberman, R.H., Woo, S.L., Lothringer, K., et al.: Effects of early intermittent mobilization on healing canine flexor tendons, J. Hand Surg. **7:**170, 1982.

Goldner, J.L.: Deformities of the hand incidental to pathological changes of the extensor and intrinsic muscle mechanisms, J. Bone Joint Surg. **35-A:**115, 1953.

Green, W.L., and Niebauer, J.J.: Results of primary and secondary flexor-tendon repairs in no man's land, J. Bone Joint Surg. **56-A:**1216, 1974.

Hall, T.D., and Alves, A.B.: Treatment of mallet finger by complete metacarpophalangeal flexion, Surg. Gynecol. Obstet. **106:**233, 1958.

Hallberg, D., and Lindholm, A.: Subcutaneous rupture of the extensor tendon of the distal phalanx of the finger: "mallet finger": brief review of the literature and report on 127 cases treated conservatively, Acta Chir. Scand. **119:**260, 1960.

Hamas, R.S., Horrell, E.D., and Pierrett, G.P.: Treatment of mallet finger due to intra-articular fracture of the distal phalanx, J. Hand Surg. **3:**361, 1978.

Hamlin, C., and Littler, J.W.: Restoration of the extensor pollicis longus tendon by an intercalated graft, J. Bone Joint Surg. **59-A:**412, 1977.

Hamlin, C., and Littler, J.W.: Restoration of power pinch, J. Hand Surg. **5:**396-401, 1980.

Harris, C., Jr., and Rutledge, G.L., Jr.: The functional anatomy of the extensor mechanism of the finger, J. Bone Joint Surg. **54-A:**713, 1972.

Hauge, M.F.: The results of tendon suture of the hand: a review of 500 patients, Acta Orthop. Scand. **24:**258, 1954-1955.

Herzog, K.H.: Treatment of extensor tendons of the fingers (Zur Versorgung der Fingerstrecksehnenverletzungen). Arch. Klin. Chir. **293:**225, 1960. (Abstracted by Joseph C. Mulier, Int. Abstr. Surg. **111:**183, 1960.)

Hillman, F.E.: New technique for treatment of mallet fingers and fractures of distal phalanx. JAMA **161:**1135, 1956.

Hoffman, S., Simon B.E., and Nachamie, B.: Unusual flexor tendon ruptures in the hand, Arch. Surg. **96:**259, 1968.

Holm, C.L., and Embick, R.P.: Anatomical considerations in the primary treatment of tendon injuries of the hand, J. Bone Joint Surg. **41-A:**599, 1959.

Hunter, J.: Artificial tendons: early development and application, Am. J. Surg. **109:**325, 1965.

Hunter, J.M., and Salisbury, R.E.: Use of gliding artificial implants to produce tendon sheaths: techniques and results in children, Plast. Reconstr. Surg. **45:**564, 1970.

Hunter, J.M., and Salisbury, R.E.: Flexor-tendon reconstruction in severely damaged hands: a two-state procedure using a silicone-Dacron reinforced gliding prosthesis prior to tendon grafting, J. Bone Joint Surg. **53-A:**829, 1971.

Hunter, J.M., et al.: The pulley system, Orthop. Trans. **4:**4, 1980.

Jaffe, S., and Weckesser, E.: Profundus tendon grafting with the sublimis intact: an end-result study of thirty patients, J. Bone Joint Surg. **49-A:**1298, 1967.

Kahn, S.: A dynamic tenodesis of the distal interphalangeal joint for use after severance of the profundus alone, Plast. Reconstr. Surg. **51:**536, 1973.

Kaplan, E.B.: Anatomy injuries and treatment of the extensor apparatus of the hand and the digits, Clin. Orthop. **13:**24, 1959.

Kessler, I.: The "grasping" technique for tendon repair, Hand **5:**253, 1973.

Ketchum, L.D.: Primary tendon healing: a review, J. Hand Surg. **2:**428, 1977.

Ketchum, L.D., et al.: The determination of moments for extension of the wrist generated by muscles of the forearm, J. Hand Surg. **3:**205, 1978.

Ketchum, L.D., Martin, N.L., and Kappel, D.A.: Experimental evaluation of factors affecting the strength of tendon repairs, Plast. Reconstr. Surg. **59:**708, 1977.

Kettelkamp, D.B., Flatt, A.E., and Moulds, R.: Traumatic dislocation of the long-finger extensor tendon: a clinical, anatomical, and biomechanical study, J. Bone Joint Surg. **53-A:**229, 1971.

Kilgore, E.S., Jr., et al.: Atraumatic flexor tendon retrieval, Am. J. Surg. **122:**430, 1971.

Kilgore, E.S., Jr., et al.: Correction of ulnar subluxation of the extensor communis, Hand **7:**272, 1975.

Kilgore, E.S., Jr., et al.: The extensor plus finger, Hand **7:**159, 1975.

Kleinert, H.E., and Meares, A.: In quest of the solution to severed flexor tendons, Clin. Orthop. **104:**23, 1974.

Kleinert, H.E., et al.: Primary repair of flexor tendons, Orthop. Clin. North Am. **4:**865, 1973.

Kleinert, H.E., and Bennett, J.B.: Digital pulley reconstruction employing the always present rim of the previous pulley, J. Hand Surg. **3:**297, 1978.

Lane, C.S.: Reconstruction of the unstable proximal interphalangeal joint: the double superficialis tenodesis, J. Hand Surg. **3:**368, 1978.

Leddy, J.P.: Flexor tendons—acute injuries. In Green, D.P. (ed.): Operative hand surgery, New York, Churchill Livingstone, 1982, pp. 1347-1373.

Leddy, J.P., and Packer, J.W.: Avulsion of the profundus tendon insertion in athletes, J. Hand Surg. **2**:66, 1977.

Leffert, R.D., and Meister, M.: Patterns of neuromuscular activity following tendon transfer in the upper limb: a preliminary study, J. Hand Surg. **1**:181, 1976.

Lister, G.D.: Reconstruction of pulleys employing extensor retinaculum, J. Hand Surg. **4**:461, 1979.

Lister, G.D., et al.: Primary flexor tendon repair followed by immediate controlled mobilization, J. Hand Surg. **2**:441, 1977.

Littler, J.W.: Principles of reconstructive surgery of the hand, Am. J. Surg. **92**:88, 1956.

Littler, J.W.: Basic principles of reconstructive surgery of the hand, Surg. Clin. North Am. **40**:383, 1960.

Littler, J.W., and Eaton, R.G.: Redistribution of forces in the correction of the boutonnière deformity, J. Bone Joint Surg. **49-A**:1267, 1967.

Lundborg, G., and Rank, F.: Experimental intrinsic healing of flexor tendons based upon synovial fluid nutrition, J. Hand Surg. **3**:21, 1978.

Lundborg, G., et al.: Superficial repair of severed flexor tendons in synovial environment: an experimental, ultrastructural study on cellular mechanisms, J. Hand Surg. **5**:451, 1980.

Madsen, E.: Delayed primary suture of flexor tendons cut in the digital sheath, J. Bone Joint Surg. **52-B**:264, 1970.

Mahoney, J., Farkas, L.G., and Lindsay, W.K.: Silastic rod pseudosheaths and tendon graft healing, Plast. Reconstr. Surg. **66**:746, 1980.

Mangus, D.J., et al.: Tendon repairs with nylon and a modified pullout technique, Plast. Reconstr. Surg. **48**:32, 1971.

Manktelow, R.T., and McKee, N.H.: Free muscle transplantation to provide active finger flexion, J. Hand Surg. **3**:416, 1978.

Manske, P.R., Bridwell, K., and Lesker, P.A.: Nutrient pathways to flexor tendons of chickens using tritiated proline, J. Hand Surg. **3**:352, 1978.

Manske, P.R., Lesker, P.A., and Bridwell, K.: Experimental studies in chickens on the initial nutrition of tendon grafts, J. Hand Surg. **4**:565, 1979.

Manske, P.R., McCarroll, H.R., Jr., and Hale, R.: Biceps tendon rerouting and percutaneous osteoclasis in the treatment of supination deformity in obstetrical palsy, J. Hand Surg. **5**:153, 1980.

Manske, P.R., Whiteside, L.A., and Lesker, P.A.: Nutrient pathways to flexor tendons using hydrogen washout technique, J. Hand Surg. **3**:32, 1978.

Marshall, K.A., Wolfort, F.G., and Edlich, R.F.: Immediate insertion of silicone rubber rods in fingers with cut flexor tendons, Plast. Reconstr. Surg. **61**:77, 1978.

Mason, M.L.: Primary versus secondary tendon repair, Q. Bull. Northwestern Univ. Med. School **31**:120, 1957.

Matev, I., Karancheva, S., Trichkova, P., et al.: Delayed primary suture of flexor tendons cut in the digital theca, Hand **12**:158, 1980.

McCormack, R.M., Demuth, R.J., and Kindling, P.H.: Flexor-tendon grafts in the less-than-optimum situation, J. Bone Joint Surg. **44-A**:1360, 1962.

McDowell, C.L., and Synder, D.M.: Tendon healing: an experimental model in the dog, J. Hand Surg. **2**:122, 1977.

McKenzie, A.R.: Function after reconstruction of severed long flexor tendons of the hand: a review of 297 tendons, J. Bone Joint Surg. **49-B**:424, 1967.

Mendelaar, H.M.: Posttraumatic ruptures of the tendon of the musculus extensor pollicis longus. Arch. Chir. Neerl. **12**:146, 1960. (Abstracted by Preston J. Burnham, Int. Abstr. Surg. **111**:490, 1960).

Miller, R.C.: Flexor tendon repair over the proximal phalanx, Am. J. Surg. **122**:319, 1971.

Murray, G.: A method of tendon repair, Am. J. Surg. **99**:334, 1960.

Neviaser, R.J., and Wilson, J.N.: Interposition of the extensor tendon resulting in persistent subluxation of the proximal interphalangeal joint of the finger, Clin. Orthop. **83**:118, 1972.

Neviaser, R.J., Wilson, J.N., and Gardner, M.M.: Abductor pollicis longus transfer for replacement of first dorsal interosseous, J. Hand Surg. **5**:53, 1980.

Nichols, H.M.: Repair of extensor-tendon insertions in the fingers, J. Bone Joint Surg. **33-A**:836, 1951.

North, E.R., and Littler, J.W.: Transferring the flexor superificalis tendon: technical considerations in the prevention of proximal interphalangeal joint disability, J. Hand Surg. **5**:498, 1980.

Parkas, L.G., et al.: An experimental study of the changes following Silastic rod preparation of a new tendon sheath and subsequent tendon grafting, J. Bone Joint Surg. **55-A**:1149, 1973.

Parkes, A.: The "lumbrical plus" finger, J. Bone Joint Surg. **53-B**:236, 1971.

Peacock, E.E., Jr.: Some technical aspects and results of flexor tendon repair, Surgery, **58**:330, 1965.

Peacock, E.E., Jr., and Madden, J.W.: Human composite flexor tendon allografts, Ann. Surg. **166**:624, 1967.

Peacock, E.E., Jr., et al.: Postoperative recovery of flexor-tendon function, Am. J. Surg. **122**:686, 1971.

Pennington, D.G.: The locking loop tendon suture, Plast. Reconstr. Surg. **63**:648, 1979.

Posch, J.L.: Primary tenorrhaphies and tendon grafting procedures in hand injuries, Arch. Surg. **73**:609, 1956.

Posch, J.L., Walker, P.J., and Miller, H.: Treatment of ruptured tendons of the hand and wrist, Am. J. Surg. **91**:669, 1956.

Potenza, A.D.: Flexor tendon injuries, Orthop. Clin. North Am. **2**:355, 1970.

Potenza, A.D., and Melone, C.: Evaluation of freeze-dried flexor tendon grafts in the dog, J. Hand Surg. **3**:157, 1978.

Pratt, D.R.: Internal splint for closed and open treatment of injuries of the extensor tendon at the distal joint of the finger, J. Bone Joint Surg. **34-A**:785, 1952.

Pratt, D.R., Bunnell, S., and Howard, L.D., Jr.: Mallet finger: classification and methods of treatment, Am. J. Surg. **93**:573, 1957.

Pulvertaft, R.G.: Tendon grafts for flexor tendon injuries in the fingers and thumb: a study of technique and results, J. Bone Joint Surg. **38-B**:175, 1956.

Pulvertaft, R.G.: The treatment of profundus division by free tendon graft, J. Bone Joint Surg. **42-A**:1363, 1960.

Pulvertaft, R.G.: Problems of flexor-tendon surgery of the hand, J. Bone Joint Surg. **47-A**:123, 1965.

Reynolds, B., Wray, R.C., Jr., and Weeks, P.M.: Should an incompletely severed tendon be sutured? Plast. Reconstr. Surg. **57**:36, 1976.

Riddell, D.M.: Spontaneous rupture of the extensor pollicis longus: the results of tendon transfer, J. Bone Joint Surg. **45-B**:506, 1963.

Rix, R.R.: Combined nerve and tendon injury in the palm, JAMA **217**:480, 1971.

Robb, W.A.T.: The results of treatment of mallet finger, J. Bone Joint Surg. **41-B**:546, 1959.

Robertson, D.C.: The place of flexor tendon grafts in the repair of flexor tendon injuries to the hand, Clin. Orthop. **15**:16, 1959.

Schmitz, P.W., and Stromberg, W.B., Jr.: Two-stage flexor tendon reconstruction in the hand, Clin. Orthop. **131**:185, 1978.

Schneider, L.H., et al.: Delayed flexor tendon repair in no man's land, J. Hand Surg. **2**:452, 1977.

Schultz, R.J.: Traumatic entrapment of the extensor digiti minimi proprius resulting in progressive restriction of motion of the metacarpophalangeal joint of the little finger, J. Bone Joint Surg. **56-A**:428, 1974.

Smith, R.J.: Non-ischemic contractures of the instrinsic muscles of the hand, J. Bone Joint Surg. **53-A**:1313, 1971.

Smith, R.J., and Hastings, H.: Principles of tendon transfers to the hand. In American Academy of Orthopaedic Surgeons: Instructional course lectures, St. Louis, 1980, The C.V. Mosby Co.

Snow, J.W.: Use of a retrograde tendon flap in repairing a severed extensor in the PIP joint area, Plast. Reconstr. Surg. **51**:555, 1973.

Snow, J.W.: A method for reconstruction of the central slip of the extensor tendon of a finger. Plast. Reconstr. Surg. **57**:455, 1976.

Snow, J.W., and Littler, J.W.: A non-suture distal fixation technique for tendon grafts, Plast. Reconstr. Surg. **47**:91, 1971.

Souter, W.A.: The problem of boutonnière deformity, Clin. Orthop. **104**:116, 1974.

Southmayd, W.W., Millender, L.H., and Nalebuff, E.A.: Rupture of the flexor tendons of the index finger after Colles' fracture: case report, J. Bone Joint Surg. **57-A**:562, 1975.

Spinner, M., and Choi, B.Y.: Anterior dislocation of the proximal interphalangeal joints: a cause of rupture of the central slip of the extensor mechanism, J. Bone Joint Surg. **52-A**:1329, 1970.

Strickland, J.W.: Management of acute flexor tendon injuries, Orthop. Clin. North Am. **14**:841, 1983.

Strickland, J.W., and Glogovac, S.V.: Digital function following flexor tendon repair in zone II: a comparison of immobilization and controlled passive motion techniques, J. Hand Surg. **5**:537, 1980.

Street, D.M., and Stambaugh, H.D.: Finger flexor tenodesis, Clin. Orthop. **13**:155, 1959.

Suzuki, K.: Reconstruction of the post-traumatic boutonnière deformity, Hand **5:**145, 1973.

Suzuki, K., et al.: Free graft of fascial tube in flexor tendon repair in the digital sheath of the hand: an attempt at a composite tissue autograft, Acta Orthop. Scand. **47:**36, 1976.

Thompson, R.V.: An evaluation of flexor tendon grafting, Br. J. Plast. Surg. **20:**21, 1967.

Tsuge, K., Ikuta, Y., and Matsuishi, Y.: Intra-tendinous tendon suture in the hand: a new technique, Hand **7:**250, 1975.

Tubiana, R.: Incisions and technics in tendon grafting. Am. J. Surg. **109:**339, 1965.

Tubiana, R.: Results and complications of flexor tendon grafting, Orthop. Clin. North Am. **4:**877, 1973.

Urbaniak, J.R., and Goldner, J.L.: Laceration of the flexor pollicis longus tendon: delayed repair by advancement, free graft or direct suture: a clinical and experimental study, J. Bone Joint Surg. **55-A:**1123, 1973.

Urbaniak, J.R., et al.: Vascularization and the gliding mechanism of free flexor-tendon grafts inserted by the silicone-rod method, J. Bone Joint Surg. **56-A:**473, 1974.

Verdan, C.E.: Half a century of flexor-tendon surgery: current status and changing philosophies, J. Bone Joint Surg. **54-A:**472, 1972.

Versaci, A.D.: Secondary tendon grafting for isolated flexor digitorum profundus injury, Plast. Reconstr. Surg. **46:**57, 1970.

Wagner, C.J.: Delayed advancement in the repair of lacerated flexor profundus tendons, J. Bone Joint Surg. **40-A:**1241, 1958.

Wakefield, A.R.: Late flexor tendon grafts, Surg. Clin. North Am. **40:**399, 1960.

Watson, A.B.: Some remarks on the repair of flexor tendons in the hand, with particular reference to the technique of free grafting, Br. J. Surg. **43:**35, 1955.

Weckesser, E.C.: Tendolysis within the digit using hydrocortisone locally and early postoperative motion, Am. J. Surg. **91:**682, 1956.

Weinstein, S.L., Sprague, B.L., and Flatt, A.E.: Evaluation of the two-stage flexor-tendon reconstruction in severely damaged digits, J. Bone Joint Surg. **58-A:**786, 1976.

White, W.L.: Secondary restoration of finger flexion by digital tendon grafts: an evaluation of seventy-six cases, Am. J. Surg. **91:**662, 1956.

White, W.L.: Tendon grafts: a consideration of their source, procurement and suitability, Surg. Clin. North Am. **40:**403, 1960.

White, W.L.: The unique, accessible and useful plantaris tendon, Plast. Reconstr. Surg. **25:**133, 1960.

Williams, S.B.: New dynamic concepts in the grafting of flexor tendons, Plast. Reconstr. Surg. **36:**377, 1965.

Wilson, R.L., Carter, M.S., Holdeman, V.A., and Lovett, W.L.: Flexor profundus injuries treated with delayed two-staged tendon grafting, J. Hand Surg. **5:**74, 1980.

Winspur, I., Phelps, D.B., and Boswick, J.A., Jr.: Staged reconstruction of flexor tendons with a silicone rod and a "pedicled" sublimis transfer, Plast. Reconstr. Surg. **61:**756, 1978.

Wray, R.C., Jr., Holtman, B., and Weeks, P.M.: Clinical treatment of partial tendon lacerations without suturing and with early motion, Plast. Reconstr. Surg. **59:**231, 1977.

Wray, R.C., Jr., Moucharafieh, B., and Weeks, P.M.: Experimental study of the optimal time for tenolysis, Plast. Reconstr. Surg. **61:**184, 1978.

Young, R.E.S., and Harmon, J.M.: Repair of tendon injuries of the hand, Ann. Surg. **151:**562, 1960.

CHAPTER 4

Fractures

PRINCIPLES OF TREATMENT

The principles of fracture treatment applying elsewhere in the body also apply to the hand, except that, since the tolerance of this part is low, fractures here must be more accurately reduced to restore function.

Closed treatment by manipulation and splinting in the position of function is indicated for almost all stable fractures of the metacarpals and phalanges; this cannot be emphasized too much. There are exceptions, however, when treatment should consist of open or closed reduction and internal fixation, usually with a Kirschner wire. Internal fixation is indicated in the following instances: (1) when a displaced fracture with a fragment too small to manipulate involves a joint; exact reduction is necessary to restore smooth joint motion; (2) when a fracture is displaced so severely that interposition of soft tissue prevents realignment by manipulation; (3) when a fracture is so unstable that muscle contracture or joint motion will soon displace it; (4) when fractures are multiple and the hand cannot be held in the position of function without internal fixation; (5) when a fracture is open (compound); internal fixation allows dressing the wound after surgery without loss of reduction; (6) when a fracture traps a tendon between the fragments, making closed reduction impossible (Fig. 4-1). Severely comminuted closed fractures should not be opened, for the same reasons that apply to the long bones of the extremities.

Internal fixation usually consists of one or two oblique or intramedually Kirschner wires or two parallel or crossed wires (p. 76). Movable parts, such as tendons and ligaments about the fracture, should not be caught by the transfixing wire. Pinning of adjacent joints should be avoided unless specifically required. A circumferential or pull-out wire suture may occasionally be used, but small plates and screws are rarely if ever used.

When an attempt to restore position is made, angulation and lack of apposition are much more obvious immediately than an error of rotation. Rotation at the fracture may become obvious only after healing when a fist can again be made; then one finger may override another or may deviate to one side (Fig. 4-2). Observing the plane of the fingernails (Fig. 4-3) at the time of fixation helps to determine rotation; passively flexing all fingers at one time also helps to verify the position after internal fixation, and incorporating one adjacent finger in the dressing may help to prevent malrotation. (See also complications of fractures of long bones, p. 90.)

The little finger has a normal tendency to overlap the ring finger. This becomes most apparent when it can only be partially flexed while the ring finger is fully flexed. This overlap is permitted by the rotation allowable at the fifth carpometacarpal joint. In cases of fractures causing limited flexion of this finger, it is very worrisome to the patient and physician until it is understood that eventually full

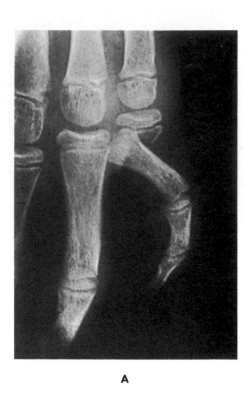

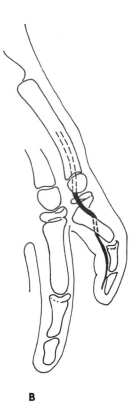

A   B

Fig. 4-1. Fracture of finger impossible to reduce by closed methods because of entrapment of tendons. **A,** Fracture of base of proximal phalanx in boy 7 years old. **B,** Flexor tendons lay on dorsum of phalangeal shaft and were trapped, making closed reduction impossible. (From von Raffler, W.: J. Bone Joint Surg. **46-B:**229, 1964.)

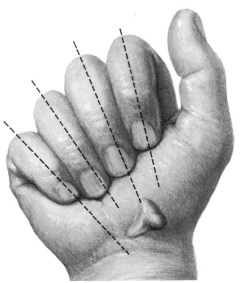

Normal flexion of fingers

Flexion of fingers with
malrotation of ring finger

A   B

Fig. 4-2. Any malrotation of metacarpal or phalangeal fracture must be corrected. **A,** Normally all fingers point toward region of scaphoid when fist is made. **B,** Malrotation at fracture causes affected finger to deviate.

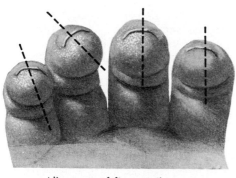

Alignment of fingernails
with malrotation of ring finger

Normal alignment of fingernails

Fig. 4-3. Observing plane of fingernails helps to detect any malrotation at fracture.

flexion of the little finger will align it normally with the ring finger. It is interesting to note that once full flexion of the little finger is accomplished, external rotation is not possible while further internal rotation is. Therefore at times apparent internal malrotation at the fracture site may not be real (Fig. 4-4).

When percutaneous or closed pinning is indicated, it is usually wise to do it before edema distorts and obscures the anatomic landmarks. If edema is already severe, the hand should first be elevated for 24 hours with the patient in bed.

Exact anteroposterior and lateral roentgenograms are necessary to determine position of the fragments before and after reduction. Even when the fracture is being reduced under direct vision, roentgenograms may prevent errors in alignment and reveal small fragments of bone not seen before reduction. When a joint cannot be completely extended, it is viewed more accurately by placing the bone segment distal to the joint parallel to the film. Cardboard cassettes are best for detail. Waiting for conclusive roentgenographic evidence of union before starting motion is unnecessary; most fractures are healed enough to permit motion at 3 weeks, which is long before union appears in the roentgenograms.

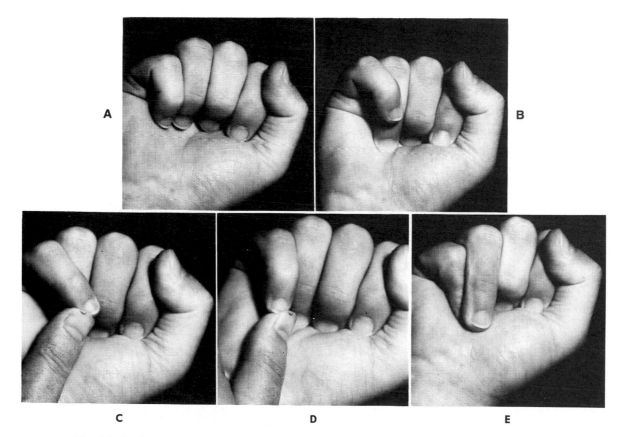

Fig. 4-4. A, Note normal alignment of normal little finger. **B,** Normal little finger can be made to overlap ring finger. With incomplete flexion this overlap may be perceived as rotational deformity. **C,** Rotation of little finger at carpometacarpal joint may be accentuated by passive help. **D,** From normal alignment of little finger external rotation is not possible. **E,** Incomplete flexion of ring finger may simulate rotational deformity as it tends to overlap little finger.

FRACTURES ASSOCIATED WITH OPEN WOUNDS

After cleaning and examination of the wound and after plans are made for its closure, fractures, if present, should be considered next. Usually fractures should be reduced, and if necessary, stabilized with small Kirschner wires to allow finger motion as soon as soft tissue healing permits; this type of fixation also permits the wound to be inspected or dressed without losing alignment of the fractures.

When the hand is severely traumatized with multiple tissues crushed and compound fractures, additional incisions are usually unnecessary to expose the fractures. The fractures should be fixed with Kirschner wires under direct vision or percutaneously to maintain the normal architectural position. In segmental defects of tubular bones, spacers made of wire bent in the shape of a U may help prevent collapse while the wound is healing. Judgment must be made to determine whether the wound is sufficiently clean to permit primary closure or whether the wound should be only debrided and cleaned; when in doubt, do not close. Furthermore, closure with tension on the traumatized and compromised skin edges may cause necrosis, since increase in edema within the first 48 hours will create even more tension. At 48 hours the wound may be reevaluated in the operating room, and plans made at that time for closure. The target is to close the wound within the first 4 to 5 days before granulating tissues form and contractures develop. Exposed tendons without their paratenon or sheath will soon necrose without appropriate coverage. (See p. 12 for methods of and indications for skin closure.) Cultures of wound tissue are indicated, and appropriate prophylatic antibiotics should be given initially.

BASIC BONE TECHNIQUES

Little equipment is usually needed for treating the bones of the hand. The same instruments used in handling the soft tissues are used to manipulate the bones; a straight Kocher clamp or towel clip is sufficient for a metacarpal shaft and a hemostat for smaller fragments. Because the bones of the hand are so small, dental chisels, a small rasp, and small bone cutters are useful.

Rarely is more fixation needed than that afforded by Kirschner wires and external splinting. The wires should be sharpened on both ends so that after being drilled in one direction they can, if necessary, be drilled retrograde, and each point should be centered, like a trocar point, to make insertion easier and more precise. (A point made by cutting the wire obliquely is not centered and makes accurate insertion difficult.) A small drill operated by hand is needed for accurate fixation with a wire. The wire is never allowed to project from the drill more than 5 cm to prevent bending it during insertion; when more than 5 cm is needed, the first 5 cm is inserted first, and the drill is then moved on the wire. After insertion the wires should be cut off flat, and the ends should be worked well beneath the skin; end-cutting wire nippers are useful here. Kirschner wires can usually be removed in the office with the patient under local anesthesia, using a pointed extractor with grooved, corrugated, and parallel jaws.

In treatment of fractures one oblique Kirschner wire is usually perferable to two wires crossed because one alone will allow some impaction at the fracture site, whereas two may tend to hold the fragments apart. But sometimes a second wire is needed to control rotation.

A lightweight battery-driven drill without a cumbersome cord attachment is extremely useful for insertion of Kirschner wires up to 0.065 inch (0.1625 cm) in size.

THUMB METACARPAL

The integrity of the carpometacarpal joint of the thumb is far more important than that of any other in the function of the thumb and thus of the whole hand (Fig. 4-5). Unless accurately reduced, metacarpal fractures involving this joint may cause limitation of motion, pain, and weakness of pinch and of grip.

Bennett's fracture

Edward H. Bennett, an Irish surgeon, described Bennett's fracture in 1881. It is an intraarticular fracture through the base of the first metacarpal. The shaft is laterally dislocated by the unopposed pull of the abductor pollicis longus (Fig. 4-6), but the medial projection or "hook" remains in place or slightly rotated because of its capsular attachment. Reduction by traction is easy but is difficult to maintain. Rubber band traction through a transverse pin in the proximal phalanx is not dependable: im-

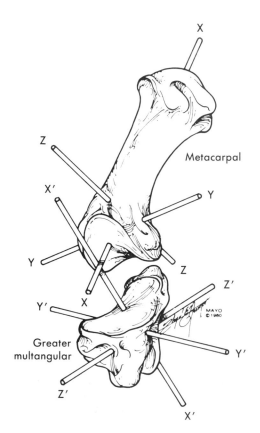

Fig. 4-5. Anatomic locations of X′, Y′, and Z′ axes of greater multangular and X, Y, and Z axes of thumb metacarpal, indicating multiple possibilities of motion between these two bones. Greater multangular is central reference for thumb metacarpal about which these movements take place. (From Cooney, W.P., III, et al.: J. Bone Joint Surg. **63-A:**1371, 1981.)

mobilization is incomplete, and verification of alignment by roentgenograms through the overlying cast is difficult. The use of a cast that maintains reduction by pressure on the base of the metacarpal is also unsatisfactory: too much pressure causes skin necrosis, and too little allows loss of reduction.

The technique of closed pinning described by Wagner (Figs. 4-7 to 4-9) is preferred, but should reduction be unsatisfactory, open reduction is indicated.

TECHNIQUE OF CLOSED PINNING (WAGNER). Maintaining reduction of the fracture by manual traction and pressure, drill a Kirschner wire into the base of the metacarpal across the joint and into the greater multangular. Check the reduction by roentgenograms; if it is accurate, cut the

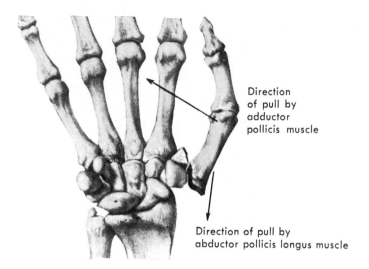

Direction of pull by adductor pollicis muscle

Direction of pull by abductor pollicis longus muscle

Fig. 4-6. In Bennett's fracture, first metacarpal shaft is displaced by divergent pull of muscles.

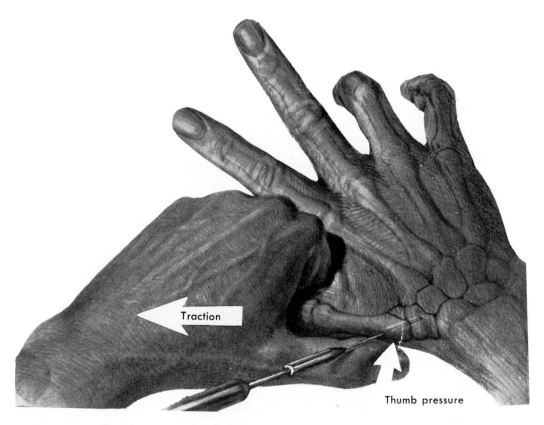

Traction

Thumb pressure

Fig. 4-7. Wagner technique of closed pinning of Bennett's fracture (see text).

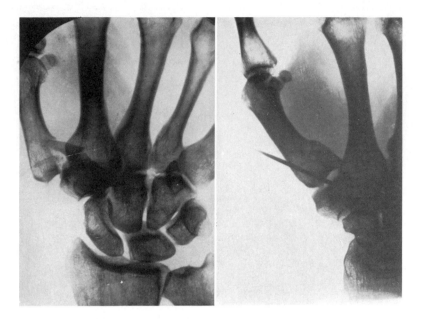

Fig. 4-8. Bennett's fracture treated by closed pinning with one Kirschner wire.

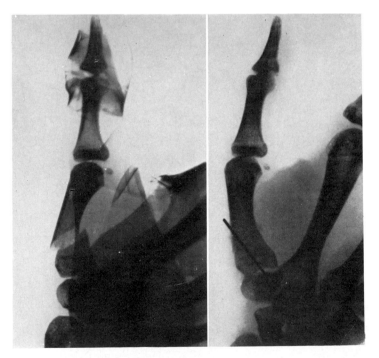

Fig. 4-9. Fracture of proximal shaft of first metacarpal treated by closed pinning as for Bennett's fracture.

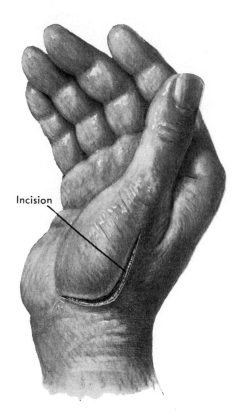

Fig. 4-10. Wagner skin incision used in approaching carpometacarpal joint of thumb.

wire near the skin. Then apply a forearm cast, holding the wrist in extension and the thumb in abduction; leave the distal thumb joint free.

TECHNIQUE OF OPEN REDUCTION (WAGNER). Begin a curved incision on the dorsoradial aspect of the first metacarpal and curve it volarward at the wrist crease (Fig. 4-10). To expose the fracture, partially strip the soft tissue from the proximal end of the metacarpal shaft and incise the carpometacarpal joint. Align the articular surface of the larger fragment with that of the smaller, and under direct vision drill a wire across the joint, maintaining the reduction. If fixation by a single wire is insecure, a second wire may be added (Fig. 4-11). After closing the wound, apply a forearm cast as described previously.

TECHNIQUE OF OPEN REDUCTION (MOBERG AND GEDDA). Expose the joint by an incision similar to that just described or by the incision shown in Figs. 4-12 and 4-13. Pass the transfixing wire through the skin of the palm into the smaller fragment until the tip of the wire is visible at the fracture. Place around the Kirschner wire tip a small loop of fine steel wire and use it to guide the fragment into accurate position. Then drill the Kirschner wire across the fracture. Remove the wire loop and complete the pinning. Close the wound and apply a cast as described previously.

AFTERTREATMENT. The cast is removed for wound inspection at 2 to 3 weeks but is replaced and worn until 4 weeks after surgery. The wire can then be removed, but immobilization may be necessary for 2 to 4 more weeks.

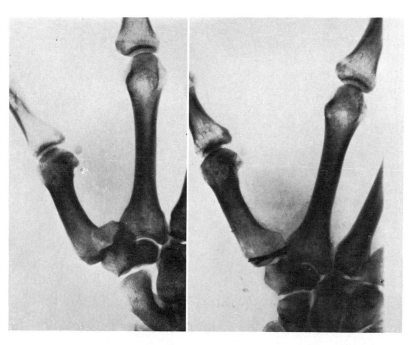

Fig. 4-11. Comminuted Bennett's fracture treated by open reduction. Two Kirschner wires were necessary to keep articular fragments reduced.

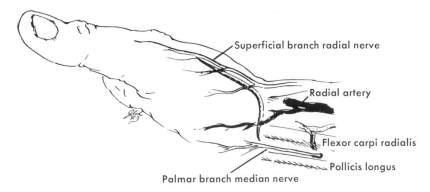

Superficial branch radial nerve

Radial artery

Flexor carpi radialis

Pollicis longus

Palmar branch median nerve

Fig. 4-12. Incision for exposure of carpometacarpal joint. Preserve branches of superficial radial nerve and radial artery and extend incision only to flexor carpi radialis tendon as shown. (From Eaton, R.G., and Littler, J.W.: J. Bone Joint Surg. **55-A:**1655, 1973.)

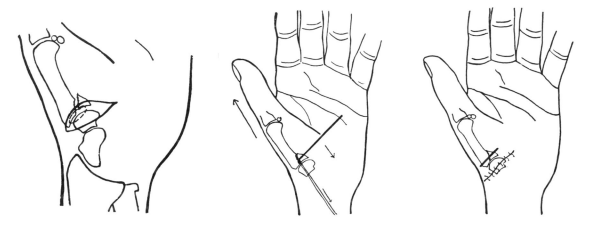

Fig. 4-13. Moberg and Gedda technique of open reduction of Bennett's fracture (see text). (From Gedda, K.O.: Acta Chir. Scand. [Suppl.] 193, 1954.)

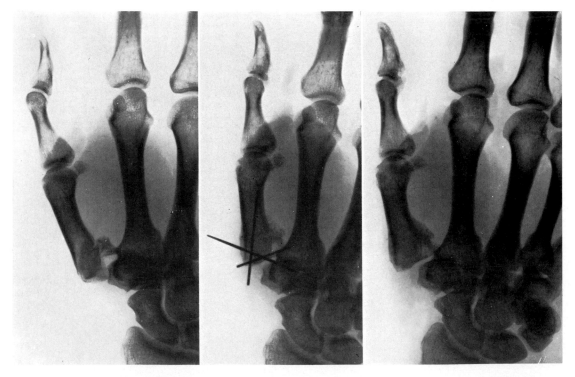

Fig. 4-14. Rolando's fracture treated by open reduction and fixation with three Kirschner wires. Early result has been excellent, but arthritic changes later may impair function.

Rolando's fracture (comminuted first metacarpal base)

Severe comminution of the first metacarpal involving the proximal articular surface may be reduced by traction and held by immediate closed pinning in satisfactory alignment, or open reduction may be attempted to reassemble the articular fragments (Fig. 4-14).

The immediate aftertreatment is the same as just given. However, after either treatment, traumatic arthritic changes usually occur and eventually require arthrodesis of the joint.

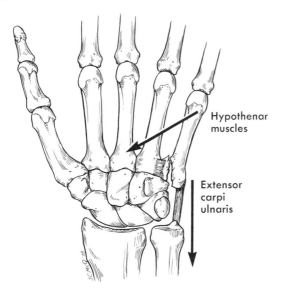

Fig. 4-15. Unstable fracture of base of fifth metacarpal may permit proximal displacement of shaft similar to Bennett's fracture (see text).

FOUR MEDIAL METACARPALS
Carpometacarpal fracture-dislocation of finger rays

Fracture-dislocation of the proximal ends of the metacarpals is often not recognized because of swelling. All four may be dislocated but usually only two or three. A true lateral roentgenogram is needed for accurate diagnosis. When the injury is seen early, manual reduction is easy, but Kirschner wire fixation is usually necessary to prevent redislocation. When seen late, the injury requires open reduction, and sometimes the proximal end of the metacarpal must be resected and the carpometacarpal joint must be fused.

Intraarticular fracture of base of fifth metacarpal

Bora and Didizian have called attention to the potentially disabling intraarticular fracture at the base of the fifth metacarpal (Fig. 4-15). When the injury is not reduced properly and malunion results, weakness of grip as well as a painful joint results. The joint consists of the base of the fifth metacarpal articulating with the hamate and the adjoining fourth metacarpal. The extensor carpi ulnaris tendon attaches to the dorsum of the proximal portion of the fifth metacarpal. The joint permits approximately 30 degrees of normal flexion and extension and the rotation necessary in grasp and in palmar cupping. This displaced intraarticular fracture might be compared with Bennett's fracture since there is a great tendency for the pull of the extensor carpi ulnaris to displace proximally the metacarpal shaft, similar to the thumb metacarpal displacement in Bennett's fracture. In addition to the routine anteroposterior and lateral views, a roentgenogram should be made with 30 degrees of pronation to give a better view of the

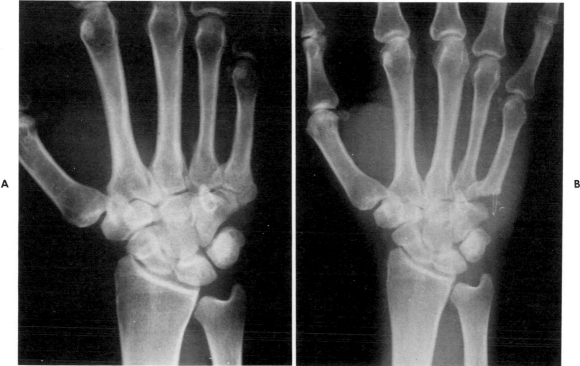

Fig. 4-16. A, This malunited fracture of base of fifth metacarpal was painful. **B,** Resection arthroplasty is preferred over osteotomy. Tendon of extensor carpi ulnaris must be reattached.

articular surface for accurate diagnosis. This fracture often may be reduced by traction and percutaneous pinning and is then protected by a cast. However, those fractures that are not recognized early and are healing in a displaced position should have either correction by osteotomy of the malunion or resection arthroplasty (Fig. 4-16).

Fracture of metacarpal shaft or neck

Fracture of a metacarpal shaft is usually best treated by closed methods, but when several metacarpals are fractured and there is open soft tissue trauma, internal fixation is indicated (Fig. 4-17). Correct rotational alignment is the most important factor in reduction, Introduce the Kirschner

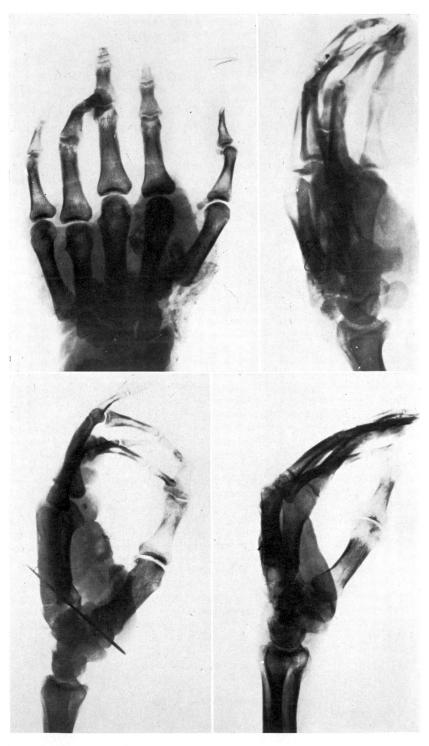

Fig. 4-17. Severely comminuted fractures of metacarpals complicated by severe injuries of soft structures require reduction and internal fixation to restore both longitudinal and transverse metacarpal arches.

wire at the fracture site and drill it out through the skin at the metacarpal base; while drilling, force a bow in the wire convex toward the palm and hold the wrist in flexion so that the wire emerges on the dorsum of the wrist. Then reduce the fracture and drill the wire in the opposite direction into the distal fragment, stopping just proximal to the metacarpophalangeal joint. Cut off the proximal end under the skin (Figs. 4-18 and 4-19). Apply a splint holding the wrist in extension. A fracture of the metacarpal neck can

be treated similarly if open reduction is necessary (Fig. 4-19, *B*).

An alternate method applicable in a few metacarpal shaft fractures is percutaneous pinning. With the metacarpophalangeal joint acutely flexed, introduce a 0.062-inch (0.1550 cm) Kirschner wire into the metacarpal head and drill it to the level of the fracture. By manual pressure and manipulation of the wire and with the aid of an image intensifier, reduce the fracture and drill the wire out the

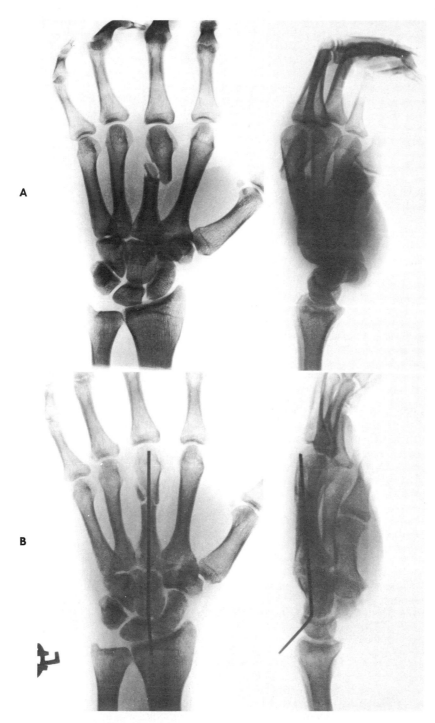

Fig. 4-18. Metacarpal shaft fracture treated by open reduction. **A,** Third metacarpal shaft is comminuted and is angulated radially and dorsally. **B,** After open reduction and medullary fixation.

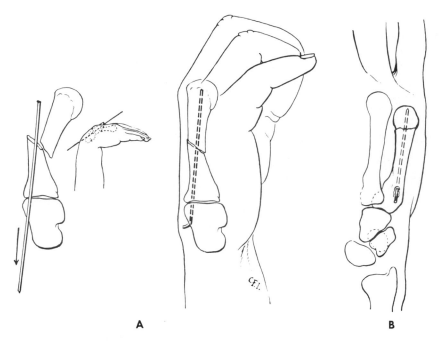

Fig. 4-19. Techniques of open reduction and medullary fixation of fracture of metacarpal shaft, **A,** and of metacarpal neck, **B** (see text).

dorsum of the wrist as just described. Withdraw the wire until the distal tip is just proximal to the metacarpophalangeal joint.

Metacarpal head fractures

Intraarticular metacarpal head fractures, especially of the fourth and fifth metacarpals, are often caused by the patient striking an opponent's teeth in a fist fight. Compound injuries here are almost always caused by human bites. (See section on human bite treatment on p. 391 for care of wound and antibiotics.) Many intraarticular head fractures require open reduction and internal fixation, particularly if the articular surface is displaced so that an incongruous joint would result. These should be fixed with Kirschner wires. Occasionally these fractures result in avascular necrosis of the displaced fragment (Figs. 4-20 and 4-21).

PHALANGES
Fracture of middle or proximal phalanx

A direct blow on the dorsum of the fingers is often the cause of fractures of the middle and proximal phalanges. Angulation is toward the palm, and the fingers may assume a claw position. When multiple or compound, these fractures should be treated with longitudinal or oblique Kirschner wires. They may be approached through a longitudinal dorsolateral incision or, for a proximal phalangeal fracture, through one placed dorsally over the phalanx (Fig. 4-22). The latter extends from the metacarpophalangeal joint to the proximal interphalangeal joint in an S curve.

Expose the extensor tendon and incise it longitudinally in its center; retract it to each side to expose the fracture site. Drill a Kirschner wire into the distal fragment under

direct vision, and then, after reducing the fracture, drill it retrograde. Care should be taken to correct any rotational deformity, although some shortening may be accepted. With a running suture of No. 34 monofilament wire, repair the extensor tendon. Support the finger in the position of function and the wrist in extension.

Sometimes an unstable oblique fracture of a middle or proximal phalanx can be treated by closed reduction and percutaneous pinning with a Kirschner wire inserted across the fracture. Then the finger is splinted for 2 to 3 weeks and the wire is removed at 3 to 4 weeks.

Some severely comminuted phalangeal fractures may be aligned and held by percutaneous Kirschner wires. The wires are then joined by a segment of polymethylmethacrylate. Final alignment of the bone is permitted while the plastic sets (Fig. 4-23).

Fracture-dislocation of proximal interphalangeal joint

Fracture-dislocations at the proximal interphalangeal joint as a rule result in an unstable volar displacement of the middle phalanx caused by disruption of the attachment of the volar fibrocartilaginous plate. When there is a large single volar fragment involving more than 50% of the joint surface, open reduction and internal fixation can be carried out with one or more Kirschner wires or a wire loop pullout. When the fragment or fragments include less than 50% of the articular surface, the technique described by McElfresh, Dobyns, and O'Brien, allowing active motion of the proximal interphalangeal joint while maintaining the finger in an extension block splint, gives satisfactory results especially in those without gross displacement. In proximal interphalangeal fracture-dislocations that have a comminuted surface of the proximal phalanx of 40% or less with displacement of fragments (as in persistent dorsal

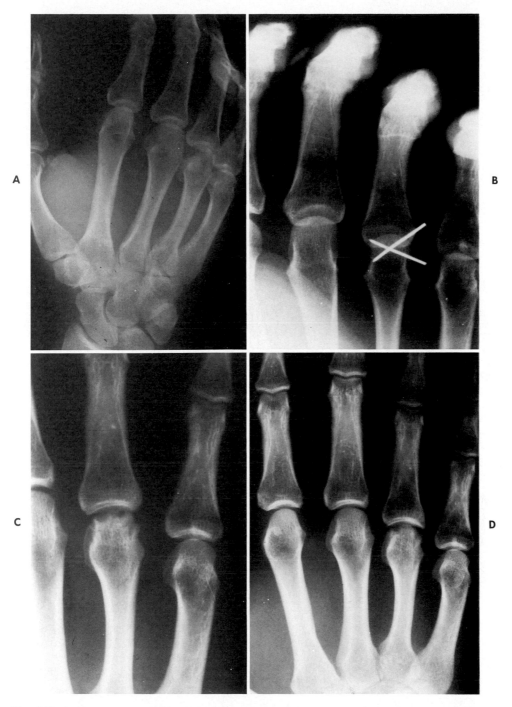

Fig. 4-20. A, Roentgenogram of hand of 20-year-old man who sustained horizontally directed fracture of fourth metacarpal head with palmar fragment displaced proximally. **B,** Fracture was reduced and held in place with 0.089 mm Kirschner wires. **C,** At 4 months roentgenograms showed early avascular necrosis of metacarpal head. **D,** At 2½ years roentgenograms showed some remodeling but definite incongruities of metacarpal head. (From McElfresh, E.C., and Dobyns, J.H.: J. Hand Surg. **8:**383, 1983.)

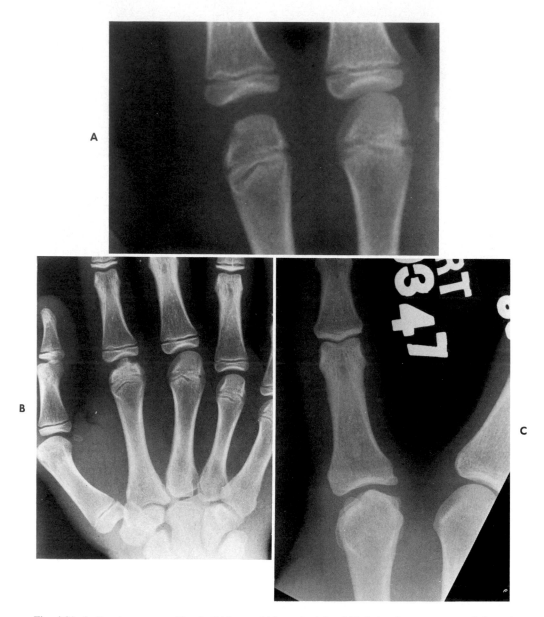

Fig. 4-21. A, Roentgenogram of hand of 14-year-old boy who injured his index finger metacarpophalangeal joint while playing baseball. **B,** At 9 months roentgenograms showed avascular necrosis of metacarpal head. **C,** At 8 years roentgenograms showed deformity of metacarpal head with early osteoarthritic changes. (From McElfresh, E.C., and Dobyns, J.H.: J. Hand Surg. **8:**383, 1983.)

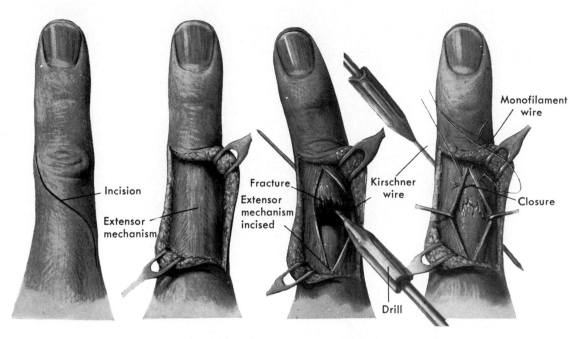

Fig. 4-22. For rare phalangeal shaft fractures that require open reduction, technique of Pratt is useful (see text).

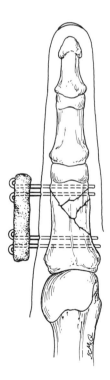

Fig. 4-23. Method for maintaining reduction of comminuted fractures of middle or proximal phalanx by using two or more percutaneous wires. These are externally stabilized by segment of polymethylmethacrylate.

fracture-dislocation with preserved condyles of the proximal phalanx), the method of Eaton and Malerich is recommended. They have used this technique in old healed displaced fractures up to 2 years after injury (Fig. 4-24, *A* to *D*).

TECHNIQUE (McELFRESH, DOBYNS, AND O'BRIEN). A malleable metal dorsal splint is incorporated in a forearm gauntlet plaster cast so that the involved finger is maintained in flexion at the proximal interphalangeal joint and the metacarpophalangeal joint (Fig. 4-25). Since instability occurs when the proximal interphalangeal joint is extended, the angle at which it occurs can be ascertained before application of the plaster. The proximal interphalangeal joint should be blocked in flexion 15 degrees short of this demonstrated position of instability. The proximal phalanx must be held securely against the dorsal splint to avoid extension at the proximal interphalangeal joint caused by further flexion of the metacarpophalangeal joint. Immediate flexion motion of the proximal interphalangeal joint is permitted. Full extension is not permitted for 6 to 12 weeks; however, an increased amount of extension may be permitted each week and encouragement is given to increase flexion.

TECHNIQUE (EATON AND MALERICH). Make a volar incision using an elongated V with the flap based radially. Excise the flexor tendon sheath from the proximal phalanx sufficiently to allow the tendons to be retracted to one side so as to view the entire joint. Hyperextend the joint to identify the fracture site in fresh cases. The volar plate will still be attached to the bone fragments of the middle phalanx. Detach the accessory collateral ligament from both sides, thus freeing the volar plate. Also detach the bone fragments by sharp dissection at the distal margin of the volar plate. In acute cases, the collateral ligaments and joint capsule need not be incised. Drill two small holes at

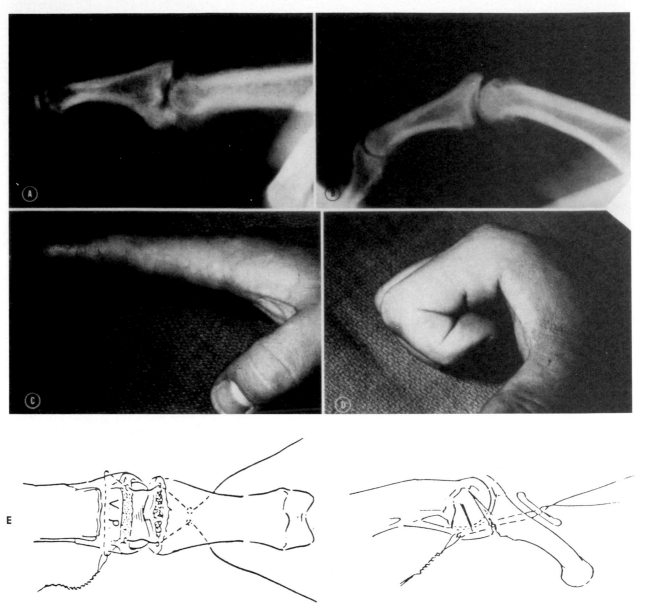

Fig. 4-24. A, Roentgenogram 1 year following fracture-dislocation. Patient had pain and only 20 degrees of motion. **B,** Roentgenogram 14 months following arthroplasty. Note smooth, congruous articular arc. **C,** Active extension 14 months after arthroplasty of proximal interphalangeal joint. **D,** Active flexion 14 months after arthroplasty of proximal interphalangeal joint. **E,** Schema of volar plate advancement. (From Eaton, R.G., and Malerich, M.M.: J. Hand Surg. **5:**260, 1980.)

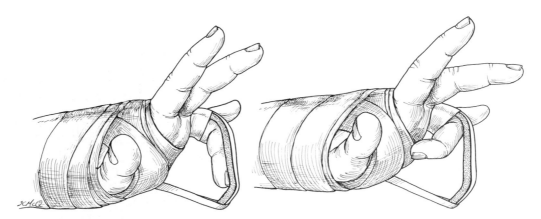

Fig. 4-25. Extension-block splinting (see text). (Redrawn from McElfresh, E.C., Dobyns, J.H., and O'Brien, E.T.: J. Bone Joint Surg. **54-A:**1705, 1972.)

the extreme margin of a trough created at the middle phalanx by the bone deficit, and possibly by some small carpentry. Place the pull-out wire through each corner of the volar plate and then through the drill holes to emerge dorsally. Place traction on these wires to snug the volar plate into the articular defect, thus effectively resurfacing the joint. Reduction can be maintained by flexing the joint no more than 35 degrees (see Fig. 4-24, *E*). Congruity of reduction should be checked by a roentgenogram. A Kirschner wire is inserted across the joint to maintain reduction. Place the hand and finger in a plaster of Paris cast for 2 weeks. After 2 weeks, the Kirschner wire is removed and active guarded flexion is started with a dorsal block splint. At 5 weeks, full extension should be accomplished, and if not, a dynamic splint should be used. The pull-out wires can be removed at 3 weeks.

In late cases in which the fractures have malunited, the volar plate is divided as far distally as possible. It may be necessary to excise both collateral ligaments. A transverse trough at the proximal edge of the middle phalanx must be created and extend completely across the bone to avoid an angular deformity when attaching the volar plate. The passive motion at the proximal interphalangeal joint should be 110 degrees, so as to easily touch the distal palmar crease with the fingertip. If the passive motion is not 110 degrees, perform a dorsal capsular release. Then attach the volar plate as described above.

• • •

The following technique is appropriate when the fracture is not comminuted and the fragment is large enough (50% of the articular surface) to be fixed in place by a Kirschner wire.

TECHNIQUE. Make a midlateral incision (p. 10) on the proximal interphalangeal joint and divide the transverse retinacular ligament, exposing the collateral ligament and joint capsule. Detach the accessory collateral ligament at its distal insertion and expose the fibrocartilaginous volar plate. Locate the avulsed osseous fragment but preserve its periosteal attachment. Replace the fragment anatomically and fix it in position with the smallest available Kirschner wire inserted in a dorsal direction; draw the wire dorsally until its volar end lies just beneath the articular surface of the fragment and will not interfere with flexion of the joint. Cut off the wire even with the dorsal surface of the phalanx. Then place the joint in functional position and fix it with an obliquely inserted Kirschner wire. Now suture the accessory collateral and transverse retinacular ligaments and close the wound.

When the fracture is a month old or older, osteotomy may be necessary to free the small fragment; then any resulting osseous defect is filled with bone from either the proximal ulna or the volar side of the proximal phalanx. If necessary for reduction of the dislocation, both collateral ligaments may be detached.

AFTERTREATMENT. A pressed-out wet cotton cast is applied and held in place with a Kling bandage. At 3 weeks the transarticular Kirschner wire is removed, and motion is begun slowly.

Fracture of distal phalanx

Fractures of the distal phalanx are usually caused by crushing injuries and thus are usually comminuted; they require only splinting. When a circular wound is present that nearly amputates the fingertip, a Kirschner wire is of

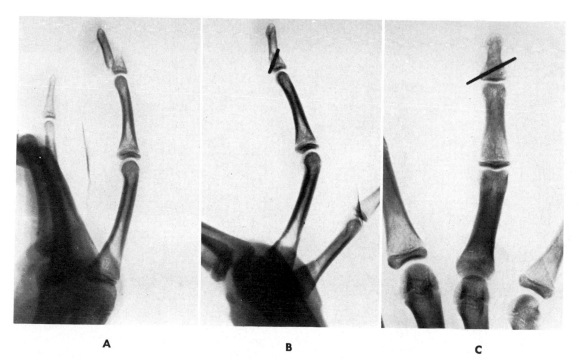

A   **B**   **C**

Fig. 4-26. A, Compound fracture of a distal phalanx in which fingertip has been almost amputated. **B** and **C,** Fracture has been fixed with a Kirschner wire.

value in supporting the bone while the soft tissues heal (Fig. 4-26). Prolonged tenderness and hypesthesia of the fingertip after the fracture is a result of the injury to the soft tissue, not to the bone.

INTRAARTICULAR FRACTURES

See also the discussion of complications of fractures (p. 90).

Intraarticular fractures with a single fragment involving one third or more of the joint surface are usually accompanied by subluxation and require reduction and fixation with a suture or a Kirschner wire. (See fracture-dislocation of the proximal interphalangeal joint, p. 84.) Closed reduction is sometimes accomplished by flexing the finger and thus apposing the larger fragment to the smaller; the joint is then transfixed with a Kirschner wire. Another closed method is three-point skeletal traction using a vertical traction ring. Open reduction, however, is usually preferred (Fig. 4-27). Drill a Kirschner wire into the smaller fragment, reduce the fracture, and bring the wire out through the larger fragment. Then attach the drill to the opposite end of the wire and extract it until its tip is just beneath the articular cartilage of the smaller fragment. Motion can usually be started at 2 weeks, and the wire can be removed at 4 weeks (Fig. 4-28).

Intraarticular fractures include avulsion fractures at the insertions of tendons and ligaments. The fragments are usually widely displaced by the pull of the tendon or ligament and should be reduced and fixed internally to restore tendon or ligament function as well as joint integrity (Fig. 4-29). When the fragment is small (less than one fourth of

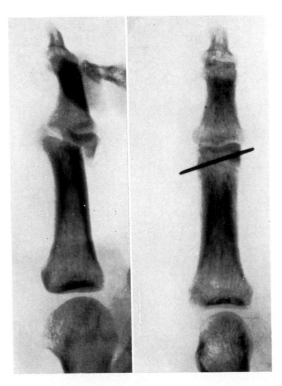

Fig. 4-27. Intraarticular fracture treated by open reduction and fixation with Kirschner wire (see text).

the joint surface), treatment is directed toward the soft tissue avulsion and may consist of open reduction and splinting or splinting alone in the position of function.

Hemicondylar fractures produced by lateral stress (usually at the proximal interphalangeal joint) require internal fixation if displaced. Open reduction often is necessary, but closed reduction and percutaneous pinning (Fig. 4-30) may be attempted.

COMPLICATIONS OF FRACTURES OF LONG BONES

Complications of fractures include malunion, nonunion, adhesions of tendons to the fracture site, infection (p. 391), and limitation of joint motion (see also reconstruction, p. 163).

When multiple tissues must be reconstructed, the repair of bones and joints is third in the order of priority. When good skin coverage is absent, repair will fail, and when the hand is insensitive, repair is futile. Therefore bone and joint reconstruction is indicated only after good skin coverage has been obtained and when at least protective sensation is present or is forthcoming.

Malunion

When fractures of one or more bones of the hand unite in poor position, the resulting disturbance of muscle balance causes weakness of grasp and pinch, especially when the metacarpals and proximal phalanges are involved. The kinesthetic sense also seems to be disturbed. Rotational deformity and angulation cause deviation of the digits that flexion usually increases.

Not every malunited fracture should be treated. It is the function of the fingers and the hand, not the roentgenographic appearance, that determines whether treatment is necessary. Ill-advised treatment usually fails to improve function and sometimes makes it worse. Unless a deformity is gross, it should usually be accepted when motion in the surrounding joints is satisfactory, for treatment by osteotomy may lead not only to nonunion but also to difficulty in reestablishing satisfactory joint motion. This is especially true in patients beyond middle age.

Most malunited fractures of the *metacarpal neck* should not be treated, particularly those of the neck of the fifth metacarpal. Flexion deformity of the neck of this bone of 40 degrees can easily be accepted with good function. When the fifth metacarpal head is displaced volarward, the carpometacarpal joint allows dorsal displacement of the distal end of the bone so that the palm can yield when a hard object is grasped; this is also true to a lesser extent of the ring finger. For the second and third metacarpals, however, there is little or no motion in the carpometacarpal joints, and when the head of one of these bones is displaced volarward, it remains as a hard unyielding mass in the palm and may be painful when a firm object is grasped; then treatment is usually indicated. When a metacarpal head is markedly displaced, hyperextension of the metacarpophalangeal joint and secondary contracture of the collateral ligaments often occur; a capsulotomy (p. 174) as well as an osteotomy may then be necessary.

TECHNIQUE FOR CORRECTING MALUNION OF METACARPAL NECK. Make a longitudinal dorsal incision just proximal and lateral to the metacarpal head; expose the extensor hood and

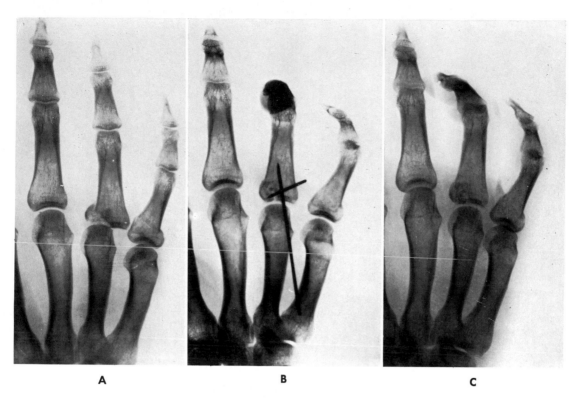

A	**B**	**C**

Fig. 4-28. A, Impacted intraarticular fracture. **B,** Fracture treated by open reduction and fixation with Kirschner wires. **C,** Result at 4 weeks when wires were removed. Eventually motion only 20 degrees less than normal was restored.

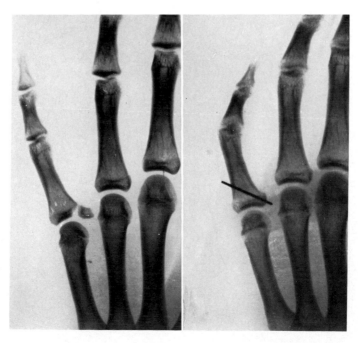

Fig. 4-29. Avulsion fracture of proximal phalanx treated by open reduction and fixation with Kirschner wire.

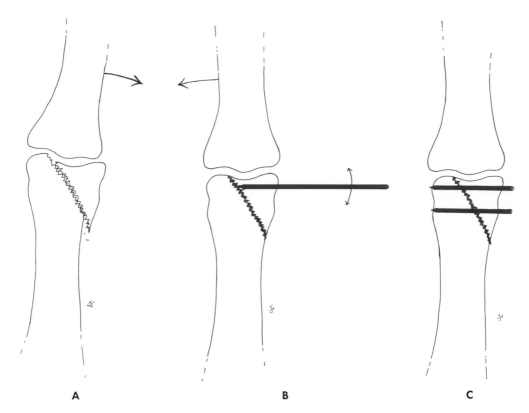

Fig. 4-30. A, Displaced unstable condylar fracture usually requires open reduction and fixation. **B,** Manipulation of fracture using intact collateral ligament may permit Kirschner wire insertion to hold reduction. **C,** Two wires may be necessary to avoid rotation of reduced fragment.

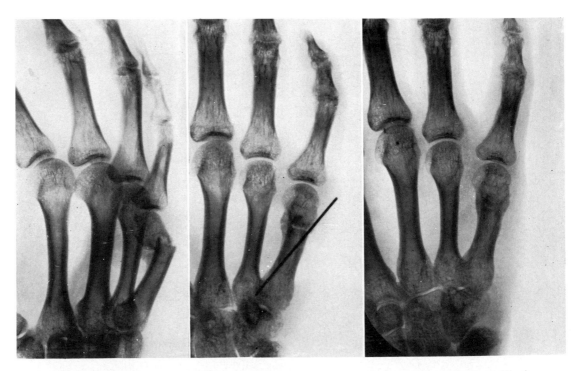

Fig. 4-31. Malunited fracture of fifth metacarpal neck treated by open reduction and fixation with one Kirschner wire inserted obliquely. This is rarely necessary because the normal motion of the fifth carpometacarpal joint permits tolerance of up to 30 degrees angulation at fracture site.

free it on one side of the metacarpal neck with a sharp knife. Dissect the interosseus muscle from the lateral side of the neck and the extensor tendon and expansion from its dorsum as necessary for sufficient exposure. If the callus is hard, drill across the old fracture site transversely; otherwise cut across it with an osteotome. Drill the medullary canal proximally and distally so that it will accept a medullary cortical bone peg a little larger than a matchstick. The peg may be obtained from the proximal ulna or proximal tibia. Insert it proximally into the medullary canal of the shaft; then cap it with the metacarpal head. Carefully check rotational alignment and then impact the fragments. Pack cancellous bone chips about their juncture as needed. When the osteotomy is unstable despite the bone peg, insert a Kirschner wire obliquely across it (Fig.

4-31). Next examine the metacarpophalangeal joint for passive flexion; when the collateral ligaments are contracted and allow little or no motion, capsulotomy (p. 174) may be indicated. Now suture the lateral expansion of the extensor hood in place with fine suture. Hold the finger in moderate flexion at all joints and apply a volar splint.

Malunion of a metacarpal shaft or of a phalanx may also be treated with a medullary cortical bone peg; the peg must be carefully shaped to fit snugly (Fig. 4-32). Figs. 4-33 and 4-34 illustrate malunited phalangeal fractures treated by osteotomy and fixation with Kirschner wires.

Malrotation of a proximal phalanx at any level should be treated by osteotomy at the base of the phalanx. The base of the phalanx heals quite well and is cut with less difficulty than the hard cortical bone in the middle third.

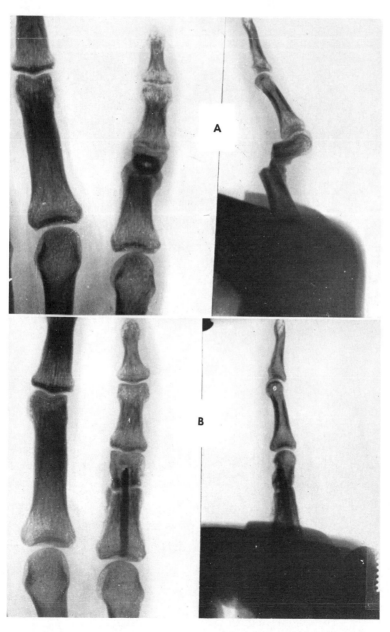

Fig. 4-32. A, Malunited phalangeal fracture. **B,** Result is satifactory after treatment by osteotomy and fixation with medullary bone peg.

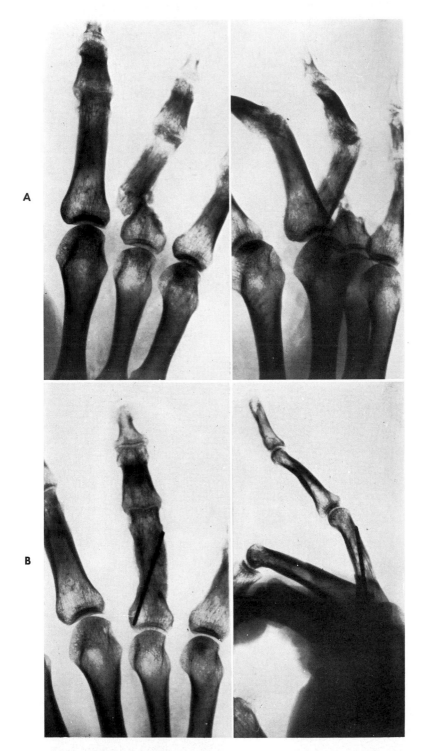

Fig. 4-33. A, Malunited phalangeal fracture in which fragments are severely displaced. **B,** After treatment by osteotomy and fixation with Kirschner wire.

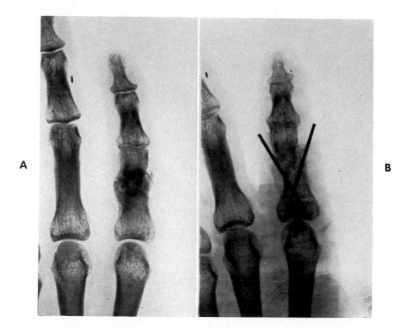

Fig. 4-34. **A,** Malunited phalangeal fracture in which there is rotational deformity (Fig. 4-3). **B,** After treatment by osteotomy through proximal end of bone and fixation with two Kirschner wires. Healing is usually more rapid after osteotomy at this level than after one at old fracture.

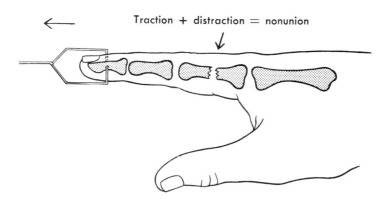

Traction + distraction = nonunion

Fig. 4-35. Nonunion of phalanx is most often caused by distraction of fragments by traction.

It is important to make an orientation mark on each side of the proposed osteotomy line so that these reference points can be used to determine when rotation is corrected.

Nonunion

Nonunion in the phalanges is most often caused by distraction of the fragments by traction (Fig. 4-35); other causes are infection, lack of fixation, and bone loss. When the nonunion is associated with nerve and tendon injuries that severely impair function, amputation must be considered; this is especially true when only one finger is involved. Nonunions of comminuted fractures of the tuft of the distal phalanx usually require no treatment; the fragments commonly unite or are finally absorbed. (These fractures are often the result of a crushing injury, and any local pain is evoked by the soft tissue injury, not by the presence of small bone fragments.)

Nonunions of transverse fractures of the distal phalanx, however, may require surgical treatment to obtain union when they are painful. The differentiation between pain from the nonunion and pain from scar tissue about nerve endings is obviously important. Lateral bending stress on the nonunion site should cause pain from a symptomatic nonunion. Simple tapping of the finger tuft when the nerve endings are tightly bound with dense scar tissue should cause pain similar to that of a neuroma.

Nonunion in the *metacarpals* is most often produced by bone loss. For a nonunion in which no bone substance is lost, the technique of repair is the same as that just described for malunion. For one in which bone substance is lost, Littler's method is recommended (Fig. 4-36).

TECHNIQUE (LITTLER). Success of bone grafting to replace a metacarpal defect and thus to restore normal architecture and function depends on two precautions. First, the dor-

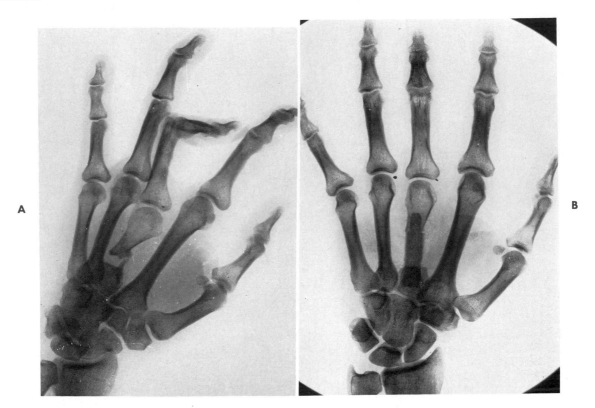

Fig. 4-36. A, Metacarpal nonunion in which bone substance has been lost. **B,** After grafting by Littler technique. (From Littler, J.W.: J. Bone Joint Surg. **29:**723, 1947.)

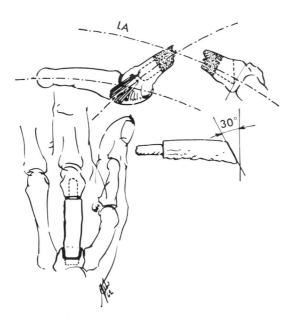

Fig. 4-37. Littler technique for grafting metacarpal nonunion in which bone substance has been lost (see text). (Courtesy Dr. J.W. Littler.)

sum of the hand must be well covered by skin and subcutaneous tissue, even if an abdominal pedicle flap is required (p. 34). Second, the fine details of what Bunnell called "bone carpentry" must be exact for the recipient bone is too small for ordinary plates and screws and fixation of the graft depends on exactitude of size and fit. Very small plates and screws such as the AO minifragment set are available. They may serve adequately for internal fixation. Expose the defective metacarpal with a longitudinal or curved dorsal incision, depending on location of existing scars. Dissect all scar tissue from the extensor tendons, but preserve the paratenon intact. Dissect the fibrous tissue *en bloc* from between the fragments so that traction can restore normal finger length. Usually the proximal fragment must be sacrificed as far as its base; resect it with an osteotome at an angle of 30 degrees (Fig. 4-37) that makes a recess in the bone. Cut the end of the distal fragment transversely with a circular saw or rongeur and open the medullary canal to receive the doweled end of the graft.

With traction on the finger, measure exactly the defect between the fragments and take from the tibia a graft at least 1.3 cm longer than the estimated defect. Fashion a dowel at one end of the graft and cut the other end obliquely at 30 degrees. Insert the doweled end into the medullary canal of the distal fragment and press the proximal end into the prepared metacarpal or carpal recess. Compression of the graft between the two fragments will hold it in place. If necessary, stabilize the graft by passing one or more Kirschner wires through it and into adjacent uninvolved metacarpals. Close the periosteal sheath, if present, and the soft tissues over the graft with fine sutures.

AFTERTREATMENT. With the hand in the position of function, a plaster cast is applied that extends to the proximal interphalangeal joints. This cast is then immediately split to allow for postoperative swelling. On about the twelfth day a new cast is applied that immobilizes only the grafted metacarpal and the proximal phalanx; it is left in place for 2 months. Administration of prophylactic antibiotics just before or during surgery should be considered, and an antibiotic should be given for several days after surgery since the injury producing the bone defect is always open. Thus the region is potentially infected even though the original wound has healed.

FRACTURES AND DISLOCATIONS OF CARPAL BONES

The diagnosis of fractures and dislocations of the carpal bones is difficult for several reasons. Because there are eight, small, tightly packed, cuboidlike bones in this region, overlap in all roentgenographic views is inevitable, except in the anteroposterior view, and even in this view at least one bone overlaps another. All views must be interpreted with a thorough understanding of the normal anatomy. Futhermore, the carpal bones normally shift in their relationship to one another during the various arcs of wrist motion.

Because of the difficulty in recognizing fractures in acute injuries, many fractures in this region are missed until late. Articular damage and ligamentous injuries are even more difficult to evaluate. The latter may permit abnormal rotations and subluxations of the various bones. Special roentgenographic techniques are helpful, but even with their use, a precise diagnosis may be difficult to make. Often prognosis is uncertain because of the peculiarities of the blood supply of these bones, especially of the scaphoid (Figs. 4-38 and 4-39).

Fractures of scaphoid

Fracture of the carpal scaphoid bone is the most common fracture of the carpus and the most commonly undiagnosed fracture of the upper extremity. A delay in diagnosis of this fracture alters the prognosis for union. A wrist sprain that is sufficiently severe to require roentgenographic examination initially should be treated as a possible fracture of the scaphoid, and the roentgenograms are repeated at 2 weeks even though the original films are negative.

For the purposes of understanding fractures of the scaphoid and selecting the indicated treatment, Cooney, Dobyns, and Linscheid classify them as either undisplaced and stable or displaced and unstable (Fig. 4-40). For an undisplaced acute stable fracture we use a forearm cast from just below the elbow proximally to the base of the thumbnail and the proximal palmar crease distally, that is, the so-called thumb spica with the wrist in slight radial deviation and in neutral flexion (see Fig. 4-42). The expected rate of union is 95% within 10 weeks. During this time, the fracture is observed roentgenographically for healing. Should collapse of a fragment occur, then treatment is altered accordingly.

For a displaced and unstable fracture in which the fragments are offset more than 1 mm in the anteroposterior or oblique view, or the lunocapitate angulation is greater than 15 degrees, or the scapholunate angulation is greater than 45 degrees in the lateral view, a different course of treatment is required. Initially reduction may be attempted by longitudinal traction, compression of the carpus, and application of a long-arm cast, or one may elect to proceed immediately with internal fixation. The rate of union is reported to be 54% in these fractures. The average time for union is 16½ weeks. Cooney et al. also found that the rate of union in fractures in which internal fixation was achieved by means of a Matti or Russe procedure using an inlay bone graft was better than the rate of union achieved in those in which a dorsal bone peg was used. Thus an inlay bone graft such as in the Matti or Russe procedure should be considered rather than the dorsal bone peg procedure. The Russe technique has a reported 86% rate of union. In these bone grafting procedures we have frequently used a staple for internal fixation.

When the scaphoid is fractured, the anatomy peculiar to this carpal bone predisposes the injury to delayed union or nonunion and also to result in disability of the wrist in several ways. First, it articulates with the distal radial cup as well as with four of the remaining seven carpal bones. The scaphoid moves in nearly any movement performed by the carpus and especially the very important movement of volar flexion; therefore any alteration of its articular surface through fracture, dislocation, or subluxation or any alteration of its stability by ligamentous rupture causes severe secondary changes throughout the entire carpus.

Second, its blood supply is precarious. Obletz and Halbstein have shown that only 67% of scaphoid bones have arterial foramina throughout their length, including the distal, middle, and proximal thirds. Of the remaining bones, 13% have blood supply predominantly in the distal third and 20% have most of the arterial foramina in the waist area of the bone with no more than a single foramen near the proximal third. This suggests that one third of the fractures occurring in the proximal third may be without adequate blood supply, and this seems to be borne out clinically; avascular necrosis occurs in 35% of fractures at this level. Taleisnik and Kelly sectioned injected and uninjected specimens to determine the blood supply of the carpal scaphoid. They demonstrated that vessels entered this bone from the radial artery, both laterovolarly and dorsally as well as distally. The laterovolar and dorsal systems share in the blood supply to the proximal two thirds of the scaphoid (Fig. 4-41).

Etiology. This fracture has been reported in people from 10 to 70 years of age although it is found most commonly in young adult men. It is caused by a fall on the outstretched palm, resulting in severe hyperextension and slight radial deviation of the wrist. It is associated with other fractures of the carpus and forearm in 17% of cases, including transscaphoid perilunar dislocations, fractures of the greater multangular, Bennett's fractures, fractures of the radial head, dislocations of the lunate, and fractures at the distal end of the radius.

Prognosis. Prognosis is excellent in an undisplaced fracture diagnosed early. Plaster immobilization with the wrist

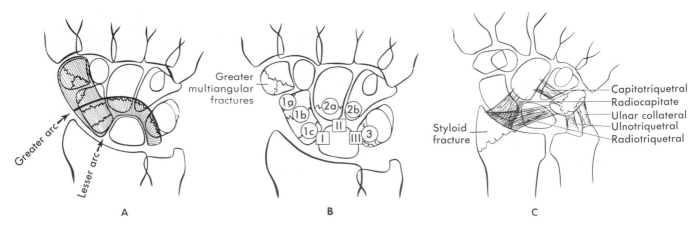

Fig. 4-38. A, Vulnerable zone of carpus. Most carpal injuries that have been produced experimentally and seen clinically fall within shaded area. Immediate evaluation of multiple views of wrist with stress should be obtained to carefully document any fractures or dislocations in this zone. Pure greater arc injury would be transscaphoid, transcapitate, transhamate, transtriquetral fracture-dislocation. Pure lesser arc injury would be perilunate or lunate dislocation. Many combinations of these two extremes are seen clinically. B, Three stages of fractures and perilunar instability. First stage represents fractures and dislocation of radial carpals. Most injuries and residual complaints involve radial carpals. They transmit force from thumb and second metacarpal to distal radius through the scaphoid, the "keystone of the wrist." The symbols are: 1a, scaphoid tuberosity compression fracture; 1b, scaphoid waist fracture; 1c, scaphoid proximal pole fracture; I, scapholunate dislocation, first stage of perilunar instability that involves scapholunate interosseous ligament disruption and scapholunate diastasis; 2a, capitate fracture; II, capitolunate dislocation, second stage of perilunar instability; 2b, hamate tail fracture; 3, triquetral fractures; III, triquetrolunate dislocation, third stage of perilunar instability or full perilunate or lunate dislocation. C, Avulsion fractures of wrist. Subtle fractures of margin of triquetrum may be only indication of full perilunar dislocation and are result of avulsion by ligaments that attach to triquetrum. The two strongest and largest ligaments are ovular: radiocapitate and radiotriquetral. If carpus is displaced and styloid fractured, these ligaments are usually intact and cause styloid to migrate with displaced carpus. Fixation of styloid will restore stability to the carpus. Symptomatic triquetral residual complaints directly related to these avulsion fractures are rare; hence open reduction and stabilization of these small avulsion fractures appear to be rarely indicated. Their primary importance appears to be diagnostic. (From Johnson, R.P.: Clin. Orthop. 149:33, 1980.)

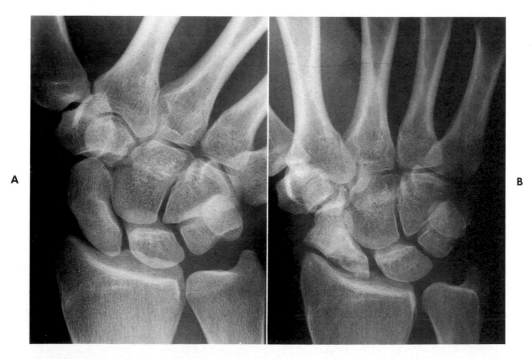

Fig. 4-39. A, Woman, 24 years old, with no history of any injury was initially seen by her general practitioner for painful wrist of two months duration. Roentgenogram taken at that time showed no abnormality. B, Roentgenogram taken 18 months later showed collapse of the proximal pole of scaphoid and secondary degenerative changes. (From Allen, P.R.: J. Bone Joint Surg. 65-B:333, 1983.)

Scaphoid nonunion associated with the location of fracture and amount of displacement		
Location	Number of fractures	Percentage of union
Distal third	2	100
Middle third	56	80
Proximal third	32	64
Displacement	Number of fractures	Percentage of union
Stable	48	85
Unstable	42	65

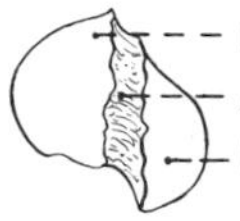

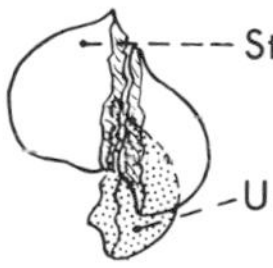

Fig. 4-40. Union of scaphoid after bone grafting is influenced significantly by both location of fracture and amount of displacement. (From Cooney, W.P., Dobyns, J.H., and Linscheid, R.L.: J. Hand Surg. **5:**343, 1980.)

at neutral dorsiflexion and slight radial deviation is the usual treatment. The cast extends from the uppermost forearm to the distal joint of the thumb (Fig. 4-42). The thumb is maintained in a functional position, and the fingers are free to move from the metacarpophalangeal joints distally. However, should the fractures be displaced, the diagnosis be delayed, or the fracture occur in the proximal third, the prognosis is less favorable for union. Contrary to early thinking, a bipartite scaphoid is so rare as to be of no clinical significance.

NONUNION

Nonunion in fracture of the scaphoid is influenced by delayed diagnosis and gross displacement as well as associated injuries of the carpus. Of these fractures, an estimated 40% are undiagnosed at the time of the original injury; thus a large number of ununited fractures is probable. The incidence of avascular necrosis is probably 30% to 40% as a result of the fact that the greatest percentage of fractures of the scaphoid occur at the waist, with about 30% in the proximal third. This correlates with the 30% incidence of impaired blood supply to the proximal one third in studies by Obletz and Halbstein. According to reports of Russe, 70% of the scaphoid fractures are located in the middle third, 20% in the proximal third, and 10% in the distal third.

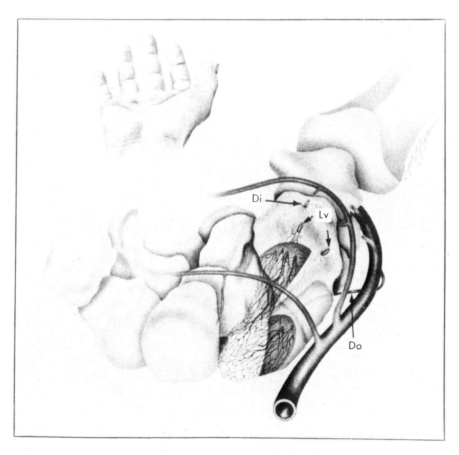

Fig. 4-41. Representation of extraosseous and intraosseous blood supply of scaphoid. *Lv,* Laterovolar vessels; *Do,* dorsal vessels; and *Di,* distal vessels. (From Taleisnik, J., and Kelly, P.J.: J. Bone Joint Surg. **48-A:**1125, 1966.)

Fig. 4-42. Cast for immobilizing fractured scaphoid. It extends from base of thumbnail and proximal palmar crease distally to 1 inch distal to elbow proximally. Note that full flexion of metacarpophalangeal joints of fingers is possible.

Cystic changes in the scaphoid and the adjoining bones may occur in untreated fractures and avascular necrosis may ensue, but according to Russe and Stewart, this is not an absolute indication for surgery. Mazet and Hohl, as well as Stewart, have reported several patients in whom fractures of the scaphoid healed after a delay in diagnosis of 5 months. Union followed cast immobilization of 8 to 12 months. Many nonunions of the scaphoid have minimal symptoms and can be tolerated well in sedentary occupations. These of course should have no treatment. Ancient fractures of the scaphoid with nonunion may result ultimately in degenerative arthritis, but this may take years to develop, depending on the amount of chronic stress applied and the activity of the wrist. In ancient fractures with arthritis, symptoms may be decreased by excision of the radial styloid just proximal to the fracture in middle third fractures; however, other reconstructive surgery, especially for severe arthritic degeneration, may be indicated. Resection of the proximal row or arthrodesis of the wrist joint is more dependable. When other associated injuries of carpal bones require open reduction, the fractured scaphoid should be accurately reduced also. Some moderately displaced fractures of the waist or middle third of the scaphoid may unite without open reduction with satisfactory results.

INDICATIONS FOR OPERATIVE TREATMENT

When diagnosis of an undisplaced fracture of the scaphoid has been delayed for several weeks, treatment should begin with cast immobilization. This treatment should be maintained for at least 16 to 20 weeks before surgery is considered. Surgery should be considered only when there is no indication of new healing activity and no indication of union after a trial of cast immobilization for 20 weeks. Fractures with delayed diagnosis and displacement of fragments and those occurring through the proximal third have less chance of uniting and should be operated on earlier. Regardless of the technique of bone grafting, there will almost always be some loss of motion even if the fracture unites, whereas with prolonged cast immobilization, once union is obtained, a near normal range of motion can be expected.

OPERATIONS FOR NONUNION

We have found the following operations to be useful for nonunions of the scaphoid: (1) styloidectomy, (2) excision

of the proximal fragment, (3) excision of the proximal row of carpal bones, (4) grafting, and (5) arthrodesis of the wrist.

STYLOIDECTOMY

Styloidectomy alone is probably of little value in treating nonunions of the scaphoid. But when arthritic changes involve only the radial margin of the radiocarpal joint, styloidectomy is indicated in conjunction with any grafting of the scaphoid or excision of its ulnar fragment for nonunion.

EXCISION OF PROXIMAL FRAGMENT

Excising *both* fragments of the scaphoid is unwise; although the immediate result may be satisfactory, eventual derangement of the wrist is likely. Soto-Hall reports that in his own cases and in those of others he has observed, the capitate has gradually migrated into the space previously occupied by the scaphoid; 5 to 7 years may pass before this shift is sufficient to cause a disability.

When indicated, excising the ulnar fragment is usually satisfactory: the loss of one fourth or less of the scaphoid impairs wrist motion less than any other operation for nonunion, and because immobilization after surgery is brief, function usually returns rapidly. Strength in the wrist is usually decreased to some extent, although this may be difficult to detect clinically.

The indications for excising the proximal fragment of a *nonunion* are as follows:

1. When the fragment consists of one fourth or less of the scaphoid. Regardless of its viability, grafting of such a small fragment will usually fail but not always.

2. When the fragment consists of one fourth or less of the scaphoid and is sclerotic, comminuted, or severely displaced. The comminuted fragments should usually be excised early to prevent arthritic changes; a severely displaced fragment should also be excised early when it cannot be accurately replaced by manipulation. In a few patients we have replaced the fragments with a Silastic prosthesis. It must be sculptured to fit the defect, and the capsule is closed securely.

3. When the fragment consists of one fourth or less of the scaphoid and grafting has failed.

When a nonviable proximal fragment consists of more than one fourth of the scaphoid, some other treatment is preferable to excision because the result of excising a major part of the scaphoid is similar to that of excising the entire bone.

When arthritic changes are present in the region of the radial styloid, styloidectomy is indicated in conjunction with excision of the proximal fragment.

TECHNIQUE. At the level of the styloid process of the radius, make a transverse skin incision 5 cm long on the dorsoradial aspect of the wrist. Retract the tendons of the thumb abductors volarward and the tendon of the extensor pollicis longus ulnarward. Incise the joint capsule and expose the scaphoid. To avoid excising a normal carpal bone, place a metal marker on the bone thought to be the ulnar fragment of the scaphoid and identify the fragment in an anteroposterior roentgenogram. Grasp the fragment to be excised with a towel clip, apply traction, and remove the fragment by dividing its soft tissue attachments. Close the wound.

AFTERTREATMENT. The wrist is immobilized in a cock-up splint for 2 weeks. Active exercises are then begun and are continued until function is restored.

EXCISION OF PROXIMAL ROW OF CARPAL BONES

The proximal row of carpal bones should usually be excised only after the more complicated carpal injuries such as those involving both the scaphoid and the lunate. As noted in the discussion on fresh fracture, many of these injuries can be treated satisfactorily without surgery; however, if the fragments have not been repositioned accurately, wrist function may be poor, and then surgery may be indicated.

Arthrodesis of the wrist is preferable for those who engage in heavy manual labor and excision of the proximal row of carpal bones for those in whom mobility of the wrist is important. Usually best results are obtained in patients under 40 years of age.

As a general rule, when any bones of the proximal row of the carpus are excised, the lunate may be excised by itself, but when either the triquetrum or the scaphoid must be excised, all three should be excised. Excising the pisiform is unnecessary, for it is a sesamoid bone in the tendon of the flexor carpi ulnaris and is not a part of the wrist joint.

TECHNIQUE. Make a transverse incision on the dorsum of the wrist 6 mm distal to the radiocarpal joint and extending from the dorsal aspect of the ulnar styloid to that of the radial one. Deepen the incision to the extensor retinaculum but preserve the sensory branches of the radial and ulnar nerves. Ligate and divide the superficial veins. Now divide the retinaculum longitudinally both on the radial and on the ulnar sides of the extensor digitorum communis tendons; avoid damaging the extensor pollicis longus tendon that crosses the wound diagonally. Expose the dorsum of the proximal row of carpal bones through two longitudinal incisions in the capsule—one in the interval between the extensor digitorum communis tendons and the extensor carpi ulnaris and the second between the extensor carpi radialis brevis tendon and the extensor digitorum communis. (Because the extensor pollicis longus tendon crosses this area diagonally, it can be retracted either medially or laterally as necessary.) Now expose the lunate by elevating

the capsule of the wrist beneath the extensor digitorum communis tendons; insert a threaded pin into the lunate, apply traction to the bone through the pin, and excise the bone by dividing its capsular attachments with sharp pointed scissors. Next insert the pin into the triquetrum and excise it in a similar manner. (The lunate and triquetrum are excised first to provide more space for the more difficult excision of the scaphoid.) Now through the more radial of the two incisions in the capsule, first excise the ulnar fragment of the scaphoid in the manner just described above and then the radial fragment, but dissect close to this fragment to avoid injuring the radial artery. Close the wound.

AFTERTREATMENT. The wrist is immobilized in slight extension and with the hand in the functional position in a plaster sugar-tong splint for 2 or 3 weeks. Active motion of the digits is encouraged soon after surgery and is continued throughout the convalescence. When the soft tissues have healed, active motion of the wrist is gradually increased. Active exercises to strengthen grip are of utmost importance.

• • •

Neviaser (1983) reported on 23 men and one woman who had proximal row carpectomies for posttraumatic disorders of the wrist. Diagnoses included 10 transscaphoperilunar dislocations with arthritis, 10 ununited scaphoid fractures with arthritis, three scapholunate disassociations with arthritis, and one acute carpal dislocation (Figs. 4-43 and 4-44). He followed these patients from 3 to 10 years and observed that the patients' grip on the operated side was nearly equal to that on the opposite side. They had approximately 70% of normal motion, and 95% were free of pain. The best results were in those in whom the head of the capitate and the lunate fossae were only minimally arthritic. Neviaser recommends the procedure, when indicated, as an alternative to arthrodesis.

TECHNIQUE (NEVIASER). Make a dorsal oblique or straight dorsal longitudinal incision. Preserve the extensor retinaculum by reflecting it laterally. Then make a T-shaped incision in the dorsal capsule and dissect it from the proximal carpal row, including the scaphoid, the lunate, and the triquetrum. Excise these bones piecemeal, but leave a thin shell of cortical bone adherent to the palmar capsule if necessary. Avoid injuring the proximal articular surface of the capitate and allow it to settle into the lunate fossa. If the greater multangular abuts the radial styloid, thus preventing radial deviation, perform a radial styloidectomy. Cut the styloid transversely to remove the radial edge of the lunate fossa. Repair the dorsal capsule. After closure of the skin immobilize the wrist in slight extension. Continue immobilization 3 weeks and then begin progressive exercises.

GRAFTING OPERATIONS

Many techniques in the past have been reported to produce satisfactory results for nonunion of the scaphoid without associated fractures or dislocations of surrounding bones and without severe arthritic changes. In the presence of severe arthritic changes or associated unreduced frac-

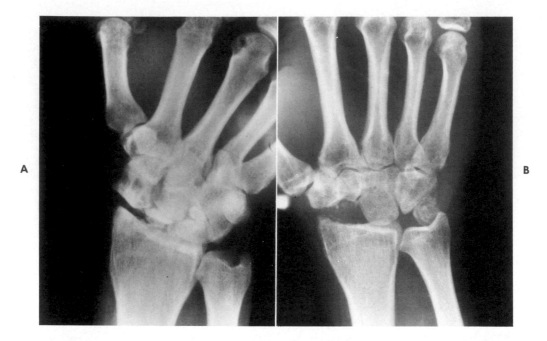

Fig. 4-43. A, Long-standing scaphoid nonunion with arthritis, avascular necrosis, collapse of proximal pole, and settling of capitate into proximal row. **B,** Postoperative roentgenogram of proximal row carpectomy with radial styloidectomy. (From Neviaser, R.J.: J. Hand Surg. **8:**301, 1983.)

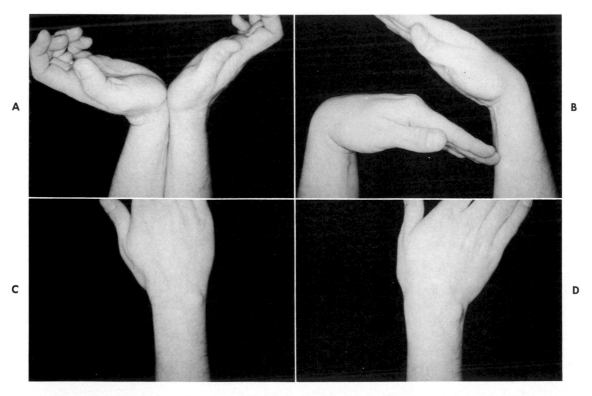

Fig. 4-44. Typical results at 6 years showing **A,** extension; **B,** flexion; **C,** radial deviation; and **D,** ulnar deviation in a construction worker. (From Neviaser, R.J.: J. Hand Surg. **8:**301, 1983.)

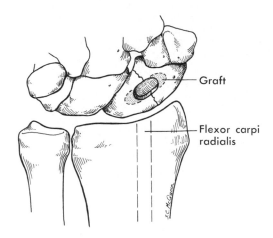

Fig. 4-45. Matti-Russe technique of bone grafting for nonunion of carpal scaphoid.

tures, resection of the proximal carpal row or arthrodesis of the wrist may be indicated. Here only one technique will be described for nonunion of the scaphoid: that reported by Russe with bone grafting and a volar approach. Mulder reported 97% bony union in 100 operations using the Matti-Russe technique. No patient with severe osteoarthritis was subjected to bone grafting. There were, however, 28 wrists with small marginal osteophytes before surgery. Following the surgery, 13 wrists of this series developed slight osteoarthritic changes. Russe advises against bone grafting when there is a small proximal fragment.

TECHNIQUE (MATTI-RUSSE). With general anesthesia and tourniquet control, make a longitudinal incision 3 to 4 cm in length on the volar aspect of the wrist, thus avoiding the dorsal blood supply to the scaphoid. Place the incision just radial to the flexor carpi radialis tendon. Retract the tendon ulnarward and continue the incision through the wrist capsule to the scaphoid bone and expose the nonunion. It may be seen more clearly by marked dorsiflexion of the wrist. Freshen the sclerotic bone ends with a small gouge and form a cavity that extends well into each adjacent fragment. From the opposite iliac crest, obtain a piece of cancellous bone and shape it into a large peg to fit into the preformed cavity and stabilize the two fragments (Fig. 4-45). Place multiple small bone chips around the peg. Make a roentgenogram at surgery to assure filling of the entire cavity. After removal of the tourniquet, suture the capsule and close the skin. Apply a cast from the elbow to the tip of the thumb and to the metacarpophalangeal joints of the other digits with the wrist in neutral position. Split the cast down the volar aspect immediately after application. Remove the sutures at 8 to 10 days and apply a new cast. For a total of 12 to 16 weeks, check the patient every week or two and replace the cast when necessary. As an alternative, apply a well-formed volar splint to include the thumb, wrap in cotton gauze, and change it at 10 days to a solid cast. Newer cast materials of lightweight and water-resistant plastic make prolonged cast immobilization more acceptable.

Naviculocapitate fracture syndrome

Fewer than 12 cases of naviculocapitate fracture syndrome have been reported; however, it is mentioned here

for diagnostic purposes and should be considered among those associated injuries that can occur with a fracture of the scaphoid. Axial compression of a dorsiflexed wrist forces further dorsiflexion, and after the scaphoid fractures, the dorsal lip of the radius forcefully impacts the head of the capitate causing it to fracture. As the wrist continues into further dorsiflexion, after both the scaphoid and capitate are fractured, the capitate head then rotates 90 degrees. The hand, when returned to neutral position, brings the proximal fragment of the capitate into 180 degrees rotation (Fig. 4-46). This injury can be associated with dorsoperilunate dislocation (Fig. 4-47, *B*) or fractures of the distal end of the radius. At times the scaphoid may not be fractured by subluxation, as with dorsal dislocation of the capitate with rotary luxation of the scaphoid. Open reduction is necessary to derotate the capitate fragment. Some surgeons have excised this fragment, but others have replaced it, reduced both the scaphoid and capitate fractures, and maintained them with external fixation alone.

Anterior dislocation of lunate

The most common carpal dislocation is anterior dislocation of the lunate. On a lateral roentgenographic view of the normal wrist the half-moon–shaped profile of the lunate articulates with the cup of the distal radius proximally and with the rounded proximal capitate distally. On an anteroposterior view the normal rectangular profile of the lunate when dislocated becomes triangular because of its tilt. An anteriorly dislocated lunate may cause acute compression of the median nerve (Fig. 4-47, *A*). This continuous medial nerve compression may result in a permanent palsy and should be relieved by reducing the lunate bone as an emergency procedure. When the injury is treated early, manipulative reduction is usually easy, and immobilization for 3 weeks with the wrist in slight flexion is required. When treated after 3 weeks, the injury may be difficult to reduce by manipulation and open reduction may be necessary. A dorsal approach to clean out the space to receive the lunate is suggested by Campbell and Thompson. However, Hill suggests a volar approach to decompress the median nerve as the lunate is reduced. When the lunate cannot be reduced by open reduction, a reconstructive procedure such as resection of the proximal row of carpal bones or an arthrodesis may be necessary.

Volar transscaphoid perilunar dislocations

Volar transscaphoid perilunar dislocations are extremely rare and are mentioned only for completeness. A report by Aitken and Nalebuff describes the mechanism of injury as a fall on the dorsum of the flexed wrist. This is directly opposite the mechanism producing a dorsal perilunar dislocation (Fig. 4-47, *B*). In their patient, reduction was early, and the fracture was found to be stable with the wrist in dorsiflexion but unstable in flexion. Avascular changes were noted later in the scaphoid, and even though they cleared eventually, nonunion of the scaphoid persisted.

Dorsal transscaphoid perilunar dislocations

Like the fractured scaphoid bone alone, this entity is frequently diagnosed late. It may be associated with other

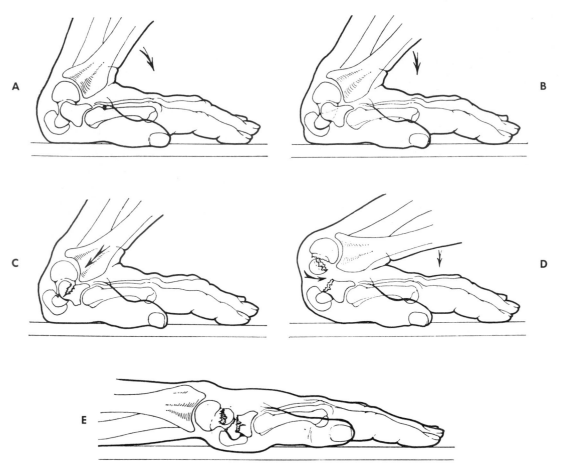

Fig. 4-46. Mechanism of carpal fractures from falls on outstretched hand with wrist going into marked dorsiflexion. **A,** Wrist in marked dorsiflexion. Note that capitate is at 90-degree angle to radius. **B,** Scaphoid fractures as result of increased dorsiflexion at midcarpal joint. **C,** Dorsal lip of radius strikes capitate, causing it to fracture. **D,** Proximal fragment of capitate is rotated 90 degrees. **E,** Return of wrist to neutral position. Note that proximal fragment of capitate is now rotated 180 degrees. (From Stein, F. and Siegel, M.W.: J. Bone Joint Surg. **51-A:**391, 1969.)

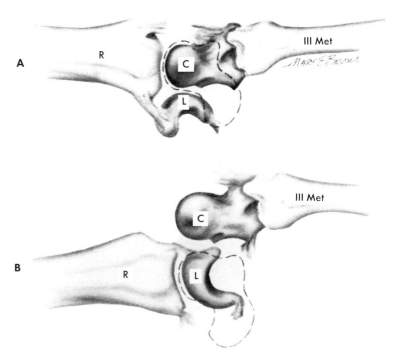

Fig. 4-47. Anterior dislocation of lunate and perilunar dislocation of carpus. **A,** Anterior dislocation of lunate. **B,** Dorsal perilunar dislocation of carpus. (From Hill, N.A.: Orthop. Clin. North Am. **1**(2):275, 1970.)

injuries of the upper extremity. Early reduction by closed manipulation is best. When accurate reduction of the scaphoid fracture is not obtained, open reduction with a bone graft or other internal fixation is usually indicated. Boyes has reported an open reduction as late as 6 weeks following injury. Bone grafting of the scaphoid and maintenance of reduction of the dislocation gave good results at evaluation 13 months later. A closed reduction can usually be carried out up to 3 weeks following injury. Later many of these injuries require open reduction. Internal fixation with Kirschner wires may be necessary for stability. After 2 months, open reduction may not be possible, and resection of the proximal carpal row or arthrodesis of the wrist is indicated.

Rotary subluxation of scaphoid

Rotary subluxation of the carpal scaphoid may be misdiagnosed as a wrist sprain following acute dorsiflexion of the wrist. It may be caused by and accompany fracture-dislocations of the wrist, especially transscaphoid perilunar fracture-dislocations. A tear of the scapholunate interosseous ligament permits this subluxation (Fig. 4-48). This ligament attaches to the proximal pole of the scaphoid and to the lunate bone and extends dorsally along with the dorsal radiocarpal ligament. Volarly it is continuous with the volar radiocarpal ligament. The scaphocapitate interosseous ligament attaches at the distal part of the scaphoid and traverses to the capitate. This is a strong ligament that usually withstands acute dorsiflexion injuries; however, the scapholunate ligament at times does not. This permits the proximal pole of the scaphoid to rotate dorsally and become more vertical in position and permits separation of the scaphoid from the lunate. The diagnosis is made on an anteroposterior roentgenographic view when a gap is noted between the scaphoid and the lunate bones of greater than 2 mm. This gap is particularly significant when the carpus is in slight radial deviation and supinated. The rotation of the scaphoid causes it to appear to be shortened and produces a so-called ring sign on the anteroposterior view. The lateral view of the wrist shows the more vertical orientation of the rotated scaphoid. Occasionally the capitate migrates proximally into the gap created by the separation of the scaphoid and lunate especially when an axial force is exerted on the capitate as when the patient is making a fist.

On examination, pain and tenderness along the radiocarpal articulation, with or without edema, are usually present with some mild limitation of motion particularly in volar flexion. Frequently the patient cannot recall the injury. Degenerative arthritic changes may eventually occur as a result of the abnormal position of these bones. A separation of 2 mm at the scapholunate articulation may not always be symptomatic. A comparative roentgenogram of the opposite wrist may be helpful. This separation may be accentuated by the patient making a fist while the anteroposterior roentgenographic view is exposed.

Closed treatment for the acute rotary subluxation of the scaphoid consists of attempting reduction by placing the wrist in neutral flexion and a few degrees of ulnar deviation. The preferred treatment is open reduction through a dorsal approach with closure of the scapholunate gap and internal fixation of the lunate and scaphoid with Kirschner wires. In addition, repair of the dorsal radiocarpal ligament is recommended, but this may be difficult. Management of an old rotary subluxation of the scaphoid may demand at times reconstruction of the scapholunate interosseous ligament with a segment of the extensor carpi radialis brevis tendon plus a Kirschner wire for fixation after the graft has been passed through the scaphoid into the adjoining lunate. Insufficient experience with this procedure has been reported in the literature to provide data for comparing results with nontreatment. For this technique, we suggest reviewing the work of Dobyns et al., Linscheid et al., and Howard et al.

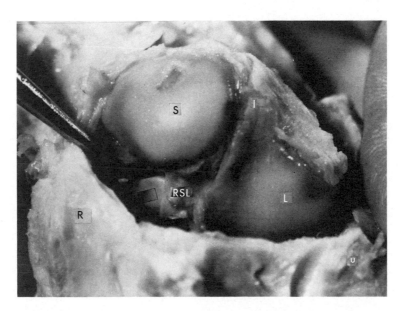

Fig. 4-48. The radioscaphoid ligament *(RSL)*. *S*, scaphoid; *I*, scapholunate interosseous ligament; *L*, lunate; *R*, radius; *RSL*, radioscaphoid ligament. (From Mayfield, J.: Clin. Orthop. **149**:45, 1980.)

Fracture of hook of hamate

A fracture of the hook of the hamate is sometimes difficult to demonstrate. Symptoms at the heel of the hand persist, consisting of pain on firm grasp and on pressure against the bony prominence. A carpal tunnel view may show the fracture but some are better demonstrated by computed tomography.

When using the latter technique, place the patient's hands together in the praying attitude. This makes the diagnosis easier, and showing both wrists eliminates the possibility of a congenital variation of the hamate, which is usually bilateral (Figs. 4-49 to 4-51).

Greater multangular ridge fractures

Palmer reported on three patients with greater multangular ridge fractures resulting from falls on the dorsiflexed

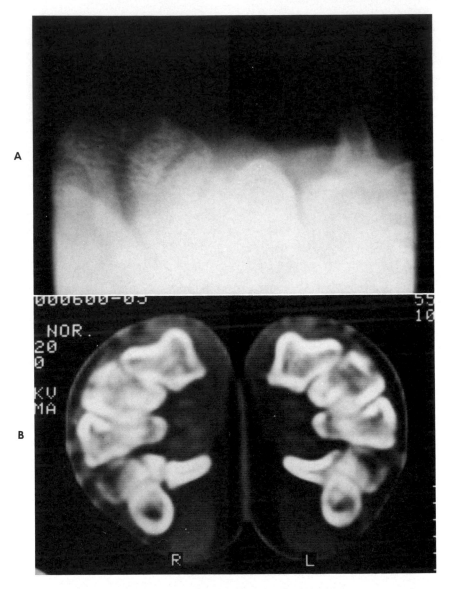

Fig. 4-49. A, Carpal tunnel view and **B,** CT image of patient who complained of severe pain in his right hand. He was injured by taking full swing at baseball resulting in foul tip. Fracture of hook of hamate is hardly seen in carpal tunnel view, but it is clearly demonstrated in CT image. (From Egawa, M., and Asai, T.: J. Hand Surg. **8:**393, 1983.)

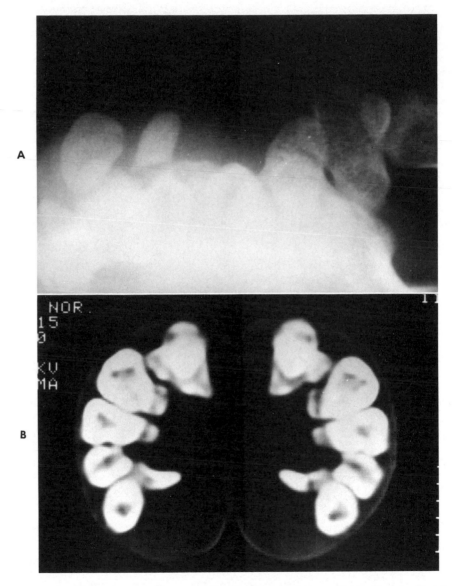

Fig. 4-50. A, Carpal tunnel view and **B,** CT image of patient with fracture of hook of hamate. He injured his left hand on full-swing foul ball. (From Egawa, M. and Asai, T.: J. Hand Surg. **8:**393, 1983.)

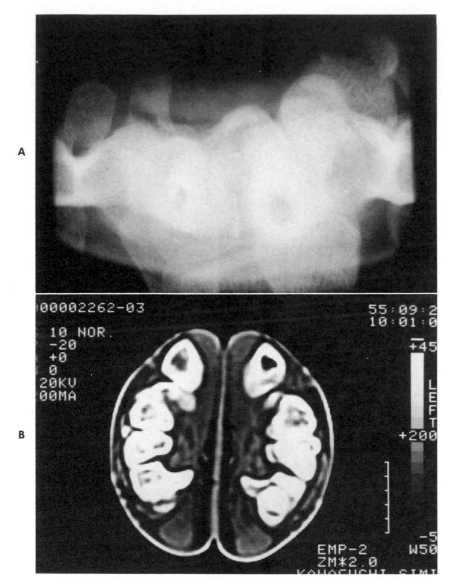

Fig. 4-51. A, Carpal tunnel view and **B,** CT image demonstrate fracture of hook of hamate in patient who landed on her outstretched left hand when she fell from her motorcycle. (From Egawa, M., and Asai, T.: J. Hand Surg. **8:**393, 1983.)

wrist. These fractures are demonstrated roentgenographically only on the carpal tunnel view of the wrist. He classified them into two types: type I is a fracture of the base of the ridge, and it may heal when treated by immobilization in plaster (Figs. 4-52 and 4-53); type II is an avulsion at the tip of the ridge, and it usually fails to heal when immobilized (Fig. 4-54).

KIENBÖCK'S DISEASE

Kienböck's disease is a painful disorder of the wrist of unknown cause in which roentgenograms show avascular necrosis of the carpal lunate. It occurs more frequently between the ages of 15 and 40 years and in the dominant wrist of adult males engaged in manual labor.

In 75% of the patients the disorder is preceded by severe

trauma, usually with the wrist in severe dorsiflexion. Recently Armistead et al. have demonstrated in some patients occult fractures of the lunate with computed tomography (Fig. 4-55, *A*). Untreated the disease usually results in fragmentation of the lunate, collapse and shortening of the carpus (Fig. 4-55, *B*), and secondary arthritic changes throughout the proximal carpal area. Symptoms may develop as early as 18 months before roentgenograms show evidence of the disease. Conservative management includes casting of the wrist for several weeks if warranted followed by repeated roentgenograms in search of occult fracture or tardy avascular changes of the lunate or other disorders that may be more evident later, including previously undiagnosed fractures of the carpal scaphoid.

The treatment of established Kienböck's disease is not

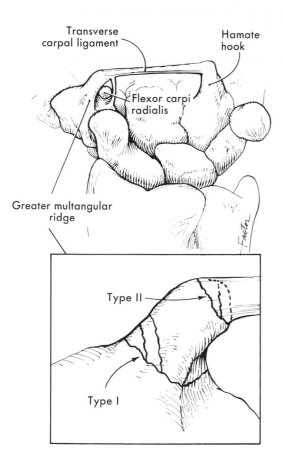

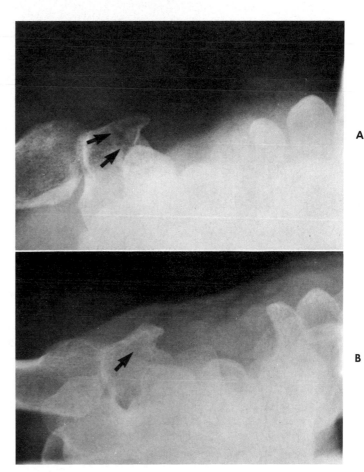

A

B

Fig. 4-52. Carpal tunnel view showing flexor carpi radialis cradled by palmar ridge of greater multangular. *Inset* shows type I fracture at base of volar ridge of greater multangular (direct loading) and type II fracture at tip (avulsion). (From Palmer, A.K.: J. Hand Surg. **6:**561, 1981.)

Fig. 4-53. A, Carpal tunnel view demonstrates fracture of base of palmar ridge of greater multangular *(arrows)*. This is designated as type I greater multangular ridge fracture. **B,** Fracture healed following 3 months of plaster immobilization. (From Palmer, A.K.: J. Hand Surg. **6:**561, 1981.)

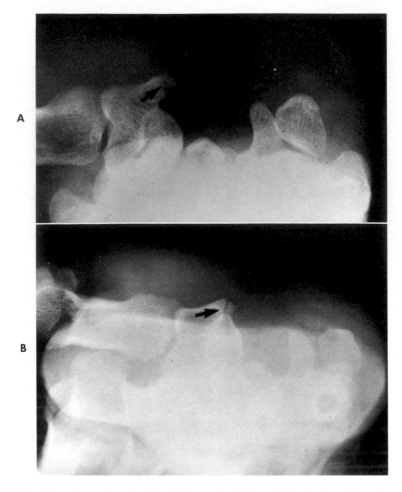

Fig. 4-54. Type II greater multangular ridge fractures. **A,** Fracture of tip of palmar ridge of greater multangular *(arrow)* caused by fall on dorsiflexed wrist. **B,** Carpal tunnel view demonstrates fracture of tip of palmar ridge of greater multangular *(arrow)* sustained when patient fell. (From Palmer, A.K.: J. Hand Surg. **6:**561, 1981.)

standardized. Some recommend ulnar lengthening early in the disease. Hulten has described a condition known as the ulna-minus variant. He found in 78% of patients with Kienböck's disease that the ulna was shorter than the radius at their distal articulation. This was true in only 23% of normal wrists. In no patient with Kienböck's disease was the ulna longer than the radius at the distal articulation, but 16% of the control group had a so-called ulna-plus variant.

Persson in 1970 reported a series of patients in whom he lengthened the ulna for this disease. These patients were observed for several years by Moberg and Axelsson. They found 16 who had been operated on some 20 years previously, and all but one had been able to continue with manual labor after the operation. Even in one who had pain, the disease process appeared to have been halted. Because of these findings, Armistead et al. have performed the ulnar lengthening operation for Kienböck's disease, reporting 20 cases in 1982 (Fig. 4-56). The technique is detailed below. The ulna should not be lengthened enough to

impair ulnar deviation of the wrist; usually most wrist movement has been retained. Strong plate fixation of the distal ulna is recommended to avoid nonunion.

In addition to ulnar lengthening, the radius has been shortened by some to accomplish an even distal radioulnar articular surface to the lunate. Shortening of the radius consists of making a transverse osteotomy about 3 inches (7.6 cm) proximal to the distal articular surface, shortening the radius by 2 mm, and fixing the bone with a compression plate.

Others have advised the conservative measure of simple casting if the disease is considered to be quite early, that is, before sclerosis, fragmentation, or collapse occurs. However, this treatment has generally been unacceptable, since it requires 4 or more months of immobilization with an uncertain outcome.

In late cases in which the lunate has collapsed but secondary arthritic changes are absent, the ulnar lengthening operation is still advocated by Armistead et al. Stark, Zemel, and Ashworth recommend use of a hand-carved sili-

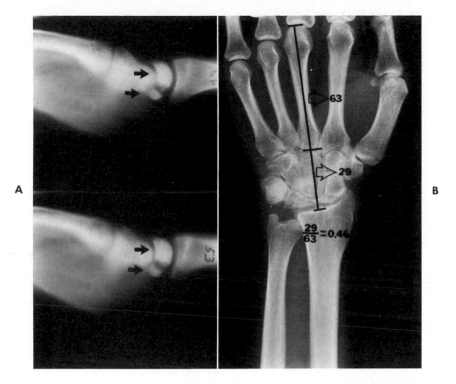

Fig. 4-55. A, Common fracture pattern in Kienböck's disease is so-called anterior-pole type, isolating anterior pole of lunate from remaining portion of bone. Distraction of fracture caused by compressive force exerted by capitate diminishes likelihood of fracture healing. This detail is usually not visible on routine roentgenograms because radial styloid process is superimposed on fracture gap. As dorsal portion of lunate collapses further, anterior pole may be extruded volarly. **B,** Ratio of height of carpus to length of third metacarpal is reduced in this patient with Kienböck's disease. Youm et al. determined that this ratio in normal wrists is 0.54 ± 0.03 and that significantly reduced ratios indicate overall carpal collapse. (From Armistead, R.B., et al.: J. Bone Joint Surg. **64-A:**170, 1982. By permission of Mayo Foundation.)

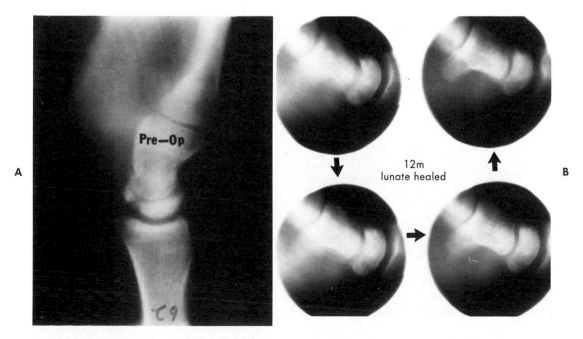

Fig. 4-56. A, Lateral tomogram of wrist, showing typical anterior-pole fracture. **B,** Lunate shows no further collapse 12 months after ulnar lengthening, and early healing is suggested. (From Armistead, R.B., et al.: J. Bone Joint Surg. **64-A:**170, 1982. By permission of Mayo Foundation.)

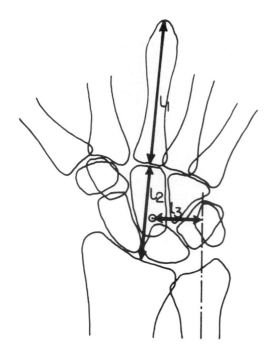

Fig. 4-57. Three kinematic indices: center of rotation, carpal height *(L₂)*, and carpal-ulnar distance *(L₃)*. K_1, length of third metacarpal. Carpal height ratio is L_2/L_1 and the carpal-ulnar distance ratio is L_3/L_1. (From McMurtry, Y., et al.: J. Bone Joint Surg. **60-A**:955, 1978.)

cone rubber spacer in the absence of significant alteration in the shape of the bone, including absence of collapse as measured by the three kinematic indices of McMurtry et al. (Fig. 4-57). The carved prosthetic device is substituted for the lunate, which is excised through a dorsal approach. Both Swanson and Lichtman and their associates advocate replacement with a previously molded lunate-shaped silicone block followed by careful repair of the capsule to avoid the potential dislocation of the block. This ligamentous and capsular reconstruction is extremely important as emphasized by many authors.

When secondary arthritic changes have developed throughout the wrist, the choice of treatment is usually between proximal row resection and arthrodesis of the wrist.

TECHNIQUE OF ULNAR LENGTHENING (ARMISTEAD ET AL.). Make a longitudinal incision over the medial border of the distal ulna. Reflect the extensor carpi ulnaris and flexor carpi ulnaris tendons, and expose the distal third of the ulna subperiosteally. Then make a transverse osteotomy through the medial three fourths of the ulna. Place a slotted plate with four or more holes over the bone centered at the site of the osteotomy. Insert four screws completely so that each is located at the end of the respective slotted hole nearest the center of the plate. Complete the osteotomy through the ulna with a cervical laminectomy spreader and distract the fragments without rotation. The amount of distraction needed is determined by preoperative roentgenograms. It should be 1 or 2 mm more than the negative ulnar length. Insert a cortical iliac graft of the predetermined width into the osteotomy gap. Next, loosen the two screws in the proximal fragment so that the tension of the

surrounding soft tissues will compress the fragments. Now retighten all four screws. Trim any projection from the periphery of the graft. Check the new length of the ulna by roentgenograms. Finally, close the wound over suction drains and apply a padded palm-to-axilla dressing with an external plaster splint.

AFTERTREATMENT. The drains are removed at 24 hours. At 2 weeks the sutures are removed, and a long arm cast is applied. At 4 to 6 weeks the cast is removed, and a splint is worn until there is roentgenographic evidence of healing. The plates and screws are not removed for at least 1 year.

TECHNIQUE FOR INSERTION OF SILICONE IMPLANT (LICHTMAN). Make a dorsal transverse skin incision and identify and retract the superficial veins and nerves. Now enter the wrist joint between the third and fourth extensor compartments. Detach the periosteum on the dorsal tip of the radius by sharp dissection and create a distally based rectangular flap of the wrist capsule. Identify the lunate and its supporting ligaments, which are usually friable. Remove the lunate with an osteotomy and small rongeurs. Preserve the volar wrist ligament, and if necessary, leave small chips of the volar cortex to maintain the integrity of the ligament. Should a gap be made in the volar ligament, close it before the final closure.

Next, make a small hole in the radial side of the triquetrum to accept the stem of the lunate implant. Select an implant of proper size. If the implant is under too much pressure, it will tend to dislodge easily. If the implant is too small, lateral shifting will take place and the stem will tend to pop out of the prepared hole. Move the wrist to determine the fit. When the proper size has been selected, flexion and extension of the wrist do not generally displace the prosthesis even before the capsule is closed. Now close the capsule carefully with appropriate sutures. Again check the stability of the implant.

AFTERTREATMENT. The wrist is immobilized in a forearm splint. Later a cast is applied and worn for 6 weeks after surgery. This stabilizes the wrist while the capsule heals.

ARTHRODESIS OF WRIST

Fusion of the wrist is most often undertaken for tuberculosis, for ununited or malunited fractures of the carpal scaphoid with associated radiocarpal traumatic arthritis, and for severely comminuted fractures of the distal end of the radius; less commonly it is done for Volkmann's ischemic paralysis and for stabilization of the wrist in poliomyelitis and in cerebral palsy of the spastic type.

The wrist should be fused in a position that will not be fatiguing and that will allow maximum grasping strength in the hand. This is usually 10 to 20 degrees extension, the long axis of the third metacarpal shaft being aligned with the long axis of the radial shaft. Clinically it is determined by the position that the wrist normally assumes with the fist strongly clenched.

Of the many techniques that have been described, most include the use of a bone graft. In some the graft bridges from the radius to the proximal carpal bones, but in others it extends distally to the base of the third metacarpal. The carpometacarpal joints may be preserved and thus some motion in the wrist will be retained. However, Haddad and

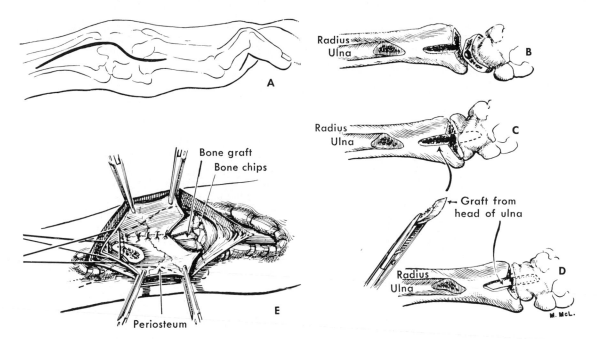

Fig. 4-58. Smith-Petersen arthrodesis of wrist joint. **A,** Medial curvilinear incision. **B,** Distal end of ulna has been resected, joint excised, and bed prepared for ulnar graft. **C** and **D,** Graft inserted. **E,** Bone chips have been packed in defect about grafts; incision closed in layers. (Modified from Smith-Petersen, M.N.: J. Bone Joint Surg. **22:**122, 1940.)

Riordan recommend that the second and third carpometacarpal joints always be included in the fusion to prevent development of any painful motion in them. Furthermore, the disease of the wrist frequently extends into these joints making a complete fusion necessary. The radial or lateral approach permits placement of the bone graft without disturbing the bed of the extensor tendons.

Since the distal radial epiphysis does not close until approximately the seventeenth year of age, care should be taken not to damage it in patients under this age. After partial destruction of the plate by disease or trauma, however, the remaining part may be excised to prevent unequal growth. Fusion of the wrist in children is difficult to secure because of the preponderance of cartilage in the joint. When practicable, operation should be postponed until the patient is 10 to 12 years of age.

In conjunction with many other useful procedures for rheumatoid arthritis of the upper extremity Smith-Petersen reported a method of fusing the wrist suggested by the exposure of the wrist after resection of the distal end of the ulna. This technique, of course, should not be used unless there is disease or derangement of the distal radioulnar joint.

TECHNIQUE (SMITH-PETERSEN). Make a 6.3 cm medial longitudinal skin incision parallel with the distal end of the ulna and extend it distally and anteriorly toward the base of the fifth metacarpal. After subperiosteal exposure, resect the distal 2.5 cm of the ulna. Incise the periosteum and capsule of the wrist joint and continue the dissection to expose the ulnar aspect of the distal end of the radius, the radiocarpal joint, and the carpus. Then create a slot in the radius and carpus to receive a graft fashioned from the resected part of the ulna. Pack in the remnants of the ulnar segment as supplementary grafts (Fig. 4-58).

AFTERTREATMENT. A cast is applied from the upper arm to the tips of the fingers and thumb, with the elbow at a right angle, the forearm in neutral position, and the wrist at 10 to 15 degrees extension. The fingers and thumb are slightly flexed. To allow for swelling, the dorsum of the cast is windowed. Three weeks after surgery a short cast (below the elbow to just proximal to the metacarpophalangeal joints) is applied, checking the correct position of the wrist. Support is continued until firm fusion is present, ordinarily 10 weeks.

• • •

Seddon has modified Smith-Petersen's method of arthrodesis of the wrist to eliminate thickening that results from shortening after resection of the articular surfaces of the radiocarpal joint. Triangular wedges are resected from the distal end of the radius and proximal aspect of the carpus, the resulting defect being diamond-shaped in the sagittal plane when the wrist is in neutral position. The distal end of the ulna is resected, and from it a triangular graft is obtained; with its base facing dorsally it is impacted into the prepared bed while the wrist is held at about 25 degrees extension. The wrist is not shortened, since the only resection is the preparation of the bed. MacKenzie reviewed 34 patients treated by this method, and the result was satisfactory in all.

According to Stein, in 1923 Gill devised a method of wrist arthrodesis in which a graft from the dorsum of the radius is turned about and inserted into the carpus. In 1958 Stein reviewed 15 patients treated by Gill with this procedure between 1923 and 1946, and all were satisfactory.

TECHNIQUE (GILL-STEIN). Expose the lower 5 to 7.5 cm of the radius and the carpus to the bases of the metacarpals through a longitudinal dorsal incision centered over Lister's tubercle. Denude the dorsal surfaces of the radius and carpal bones, and excise any fibrous or granulation tissue. Remove the cortical bone from the dorsal aspect of the proximal carpal row and make a cleft in the capitate transversely in the coronal plane. Then remove a broad plate of bone from the dorsal aspect of the distal end of the radius and rotate it 180 degrees in the coronal plane. Drive the sharp proximal end of the graft into the cleft in the capitate, maintaining the cancellous surface of the graft in contact with the cancellous surfaces of the denuded carpal and distal radial fusion bed (Fig. 4-59). On slight dorsiflexion of the wrist the bone plate locks in the capitate and firmly contacts the dorsal surface of the radius as well. Suture the ligaments over the graft and close the wound in layers.

• • •

Haddad and Riordan have described a technique of arthrodesis of the wrist through a radial or lateral approach. It has these advantages: the distal radioulnar joint is not entered, the extensor tendons to the digits are not so disturbed, and since dorsal thickening is avoided, the appearance of the wrist is not altered. They report only one failure in 24 wrists using this technique.

TECHNIQUE (HADDAD AND RIORDAN). Begin a J-shaped skin incision 2.5 to 3.8 cm proximal to the radial styloid on the

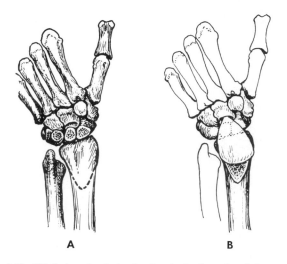

A   **B**

Fig. 4-59. Gill-Stein arthrodesis of wrist. **A,** Outline of graft from distal radius. Proximal row of carpals has been denuded, and cleft has been created in capitate. **B,** Graft is in place; it is locked in capitate by dorsiflexing wrist. (From Stein, I.: Surg. Gynecol. Obstet. **106:**231, 1958.)

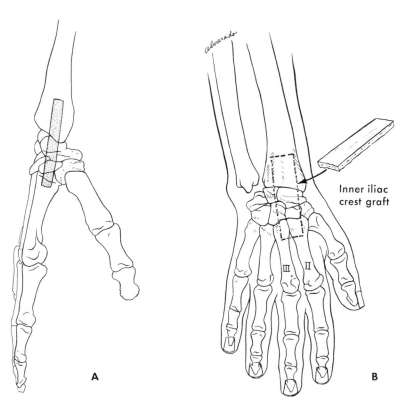

Inner iliac
crest graft

A   **B**

Fig. 4-60. Haddad and Riordan arthrodesis of wrist. **A,** Radial view showing slot cut in distal radius, carpal bones, and bases of second and third metacarpals. **B,** Dorsal view showing shape of graft and its final position, *broken line,* in slot. (From Haddad, R.J., Jr., and Riordan, D.C.: J. Bone Joint Surg. **49-A:**950, 1967.)

midlateral aspect of the forearm, extend it distally across the styloid, and then curve it dorsally to end at the base of the second metacarpal. Now mobilize and retract the superficial branch of the radial nerve. Identify the interval between the first and second dorsal compartments and incise the dorsal carpal ligament in this interval, leaving it attached to the volar aspect of the radius. Mobilize subperiosteally and retract the abductor pollicis longus, extensor pollicis brevis, and the wrist and finger extensors. Next divide the extensor carpi radialis longus tendon just proximal to its insertion on the base of the second metacarpal, leaving a stump distally so that it can be sutured later. Remove the capsule from the radiocarpal, the intercarpal, and the second carpometacarpal joints. Now locate the dorsal branch of the radial artery and ligate and divide its dorsal branch to the dorsal carpal arch. Then denude the radiocarpal joint of articular cartilage and subchondral bone. Using an electric saw, obtain from the medial table of the iliac crest a graft about 3.8 cm long and 2.5 cm wide. With the wrist in 15 degrees dorsiflexion, cut a slot, still using an electric saw, in the distal end of the radius, the carpal bones, and the bases of the second and third metacarpals. Do not cut through the medial cortex of the radius and enter the distal radioulnar joint. Then place the graft in the prepared bed (Fig. 4-60). If the wrist is unstable, insert a nonthreaded Kirschner wire obliquely or longitudinally to engage the base of the second metacarpal and the distal radius; cut off the wire under the skin at the palm, to be removed 6 to 8 weeks later. Close the dorsal carpal ligament deep to the abductor pollicis longus and extensor pollicis brevis. Suture the extensor carpi radialis longus tendon and close the wound.

REFERENCES
General references

Alexander, A.H., and Lichtman, D.M.: Irreducible distal radioulnar joint occuring in a Galeazzi fracture: case report, J. Hand Surg. **6:**258, 1981.

Allen, P.R.: Idiopathic avascular necrosis of the scaphoid: a report of two cases, J. Bone Joint Surg. **65-B:**333, 1983.

Belsole, R.: Physiological fixation of displaced and unstable fractures of the hand, Orthop. Clin. North Am. **11:**393, 1980.

Bilos, Z.J., Pankovich, A.M., and Yelda, S.: Fracture-dislocation of the radiocarpal joint: a clinical study of five cases, J. Bone Joint Surg. **59-A:**198, 1977.

Blalock, H.S., et al.: An instrument designed to help reduce and percutaneously pin fractured phalanges, J. Bone Joint Surg. **57-A:**792, 1975.

Bora, F.W., Jr., and Didizian, N.H.: The treatment of injuries to the carpometacarpal joint of the little finger, J. Bone Joint Surg. **56-A:**1459, 1974.

Borden, J.: Complications of fractures and ligamentous injuries of the hand, Orthop. Rev. **1:**29, November, 1972.

Bowers, W.H.: Mallet deformity of a finger after phalangeal fracture: case report of treatment by the Fowler procedure, J. Bone Joint Surg. **59-A:**525, 1977.

Carter, P.R., Eaton, R.G., and Littler, J.W.: Ununited fracture of the hook of the hamate, J. Bone Joint Surg. **59-A:**583, 1977.

Clinkscales, G.S., Jr.: Complications in the management of fractures in hand injuries, South. Med. J. **63:**704, 1970.

Cooney, W.P., III, Lucca, M.J., Chao, E.Y.S., and Linscheid, R.L.: The kinesiology of the thumb trapeziometacarpal joint, J. Bone Joint Surg. **63-A:**1371, 1981.

Cowen, N.J., and Kranik, A.D.: An irreducible juxta-epiphyseal fracture of the proximal phalanx: report of a case, Clin. Orthop. **110:**42, 1975.

Eaton, R.G., and Malerich, M.M.: Volar plate arthroplasty of the proximal interphalangeal joint: a review of ten years' experience, J. Hand Surg. **5:**260, 1980.

Egawa, M., and Asai, T.: Fracture of the hook of the hamate: report of six cases and the suitability of computerized tomography, J. Hand Surg. **8:**393, 1983.

Gedda, K.O.: Studies on Bennett's fracture: anatomy, roentgenology, and therapy, Acta Chir. Scand. (Suppl. 193), 1954.

Gingrass, R.F., Fehring, B., and Matloub, H.: Intraosseous wiring of complex hand fractures, Plast. Reconstr. Surg. **66:**383, 1980.

Gladstone, H.: Rupture of the extensor digitorum communis tendons following severely deforming fractures about the wrist, J. Bone Joint Surg. **34-A:**698, 1952.

Green, D.P., and Anderson, J.R.: Closed reduction and percutaneous pin fixation of fractured phalanges, J. Bone Joint Surg. **55-A:**1651, 1973.

Green, D.P., and O'Brien, E.T.: Fractures of the thumb metacarpal, South Med. J. **65:**807, 1972.

Griffiths, J.C.: Fractures at the base of the first metacarpal bone, J. Bone Joint Surg. **46-B:**712, 1964.

Hunter, J.M., and Cowen, N.J.: Fifth metacarpal fractures in a compensation clinic population: a report of one hundred and thirty-three cases, J. Bone Joint Surg. **52-A:**1159, 1970.

Ikuta, Y., and Tsuge, K.: Micro-bolts and micro-screws for fixation of small bones in the hand, Hand **6:**261, 1974.

James, J.I.P.: A case of rupture of flexor tendons secondary to Kienböck's disease, J. Bone Joint Surg. **31-B:**521, 1949.

Lane, C.S.: Detecting occult fractures of the metacarpal head: the Brewerton view, J. Hand Surg. **2:**131, 1977.

Lee, M.L.H.: Intra-articular and peri-articular fractures of the phalanges, J. Bone Joint Surg. **45-B:**103, 1963.

Leonard, M.H.: Open reduction of fractures of the neck of the proximal phalanx in children, Clin. Orthop. **116:**176, 1976.

Leonard, M.H., and Dubravcik, P.: Management of fractured fingers in the child, Clin. Orthop. **73:**160, 1970.

Lewis, R.C., Jr. and Hartman, J.T.: Controlled osteotomy for correction of rotation in proximal phalanx fractures, Orthop. Rev. **2:**11, November 1973.

Lilling, M., and Weinberg, H.: The mechanism of dorsal fracture dislocation of the fifth carpometacarpal joint, J. Hand Surg. **4:**340, 1979.

Littler, J.W.: Metacarpal reconstruction, J. Bone Joint Surg. **29:**723, 1947.

McElfresh, E.C., and Dobyns, J.H.: Intraarticular metacarpal head fracture, J. Hand Surg. **8:**383, 1983.

McElfresh, E.C., et al.: Management of fracture-dislocation of the proximal interphalangeal joints by extension-block splinting, J. Bone Joint Surg. **54-A:**1705, 1972.

Moberg, E.: Fractures and ligamentous injuries of the thumb and fingers, Surg. Clin. North Am. **40:**297, 1960.

Murakami, Y., and Todani, K.: Traumatic entrapment of the extensor pollicis longus tendon in Smith's fracture of the radius: case report, J. Hand Surg. **6:**238, 1981.

Nemethi, C.E.: Phalangeal fractures treated by open reduction and Kirschner-wire fixation, Industr. Med. **23:**148, 1954.

Nevaiser, R.J.: Proximal row carpectomy for posttraumatic disorders of carpus, J. Hand Surg. **8:**301, 1983.

Nunley, J.A., and Urbaniak, J.R.: Partial bony entrapment of the median nerve in a greenstick fracture of the ulna, J. Hand Surg. **5:**557, 1980.

Palmer, A.K.: Trapezial ridge fractures, J. Hand Surg. **6:**561, 1981.

Pieron, A.P.: Correction of rotational malunion of a phalanx by metacarpal osteotomy, J. Bone Joint Surg. **54-B:**516, 1972.

Pollen, A.G.: The conservative treatment of Bennett's fracture—subluxation of the thumb metacarpal, J. Bone Joint Surg. **50-B:**91, 1968.

Pratt, D.R.: Exposing fractures of the proximal phalanx of the finger longitudinally through the dorsal extensor apparatus, Clin. Orthop. **15:**22, 1959.

Pritsch, M., Engel, J., and Farin, I.: Manipulation and external fixation of metacarpal fractures, J. Bone Joint Surg. **63-A:**1289, 1981.

Riordan, D.C.: Fractures about the hand, South. Med. J. **50:**637, 1957.

Rychak, J.S., and Kalenak, A.: Injury to the median and ulnar nerves secondary to fracture of the radius: a case report, J. Bone Joint Surg. **59-A:**414, 1977.

Seddon, H.J.: Reconstructive surgery of the upper extremity. In Poliomyelitis, Second International Poliomyelitis Congress, Philadelphia, 1952, J.B. Lippincott Co.

Smith-Peterson, M.N.: A new approach to the wrist joint, J. Bone Joint Surg. **22**:122, 1940.

Smith, R.J.: Boutonnière deformity of the fingers, Bull. Hosp. Joint Dis. **27**:27, 1966.

Spangberg, O., and Thoren, L.: Bennett's fracture: a method of treatment with oblique traction, J. Bone Joint Surg. **45-B**:732, 1963.

Stark, H.H., et al.: Fracture of the hook of the hamate in athletes, J. Bone Joint Surg. **59-A**:575, 1977.

Swanson, A.B.: Fractures involving the digits of the hand, Orthop. Clin. North Am. **2**:261, 1970.

Torisu, T.: Fracture of the hook of the hamate by a golfswing, Clin. Orthop. **83**:91, 1972.

Trevor, D.: Rupture of the extensor pollicis longus tendon after Colles' fracture, J. Bone Joint Surg. **32-B**:370, 1950.

Vance, R.M., Gelberman, R.H., and Evans, E.F.: Scaphocapitate fractures: patterns of dislocation, mechanism of injury, and preliminary results of treatment, J. Bone Joint Surg. **62-A**:271, 1980.

Van Demark, R.E., and Cottam, G.I.W.: Translocation tenorraphy of the extensor pollicis longus after spontaneous rupture in Colles' fracture, Clin. Orthop. **31**:106, 1963.

Vom Saal, F.H.: Intramedullary fixation in fractures of the hand and fingers, J. Bone Joint Surg. **35-A**:5, 1953.

Wagner, C.J.: Method of treatment of Bennett's fracture dislocation, Am. J. Surg. **80**:230, 1950.

Wilson, C.H., and Deitz, J.R.: Skeletal traction for fractures of the hand, South. Surg. **13**:253, 1947.

Carpal bones, fractures, and dislocations

Adams, J.D., and Leonard, R.D.: Fracture of the carpal scaphoid, N. Engl. J. Med. **198**:401, 1928.

Agerholm, J.D., and Lee, M.: The acrylic scaphoid prosthesis in the treatment of the ununited carpal scaphoid fracture, Acta Orthop. Scand. **37**:67, 1966.

Agner, O.: Treatment of ununited fractures of the carpal scaphoid by Bentzon's operation, Acta Orthop. Scand. **33**:56, 1963.

Aitken, A.P., and Nalebuff, E.: Volar transnavicular perilunar dislocation of the carpus, J. Bone Joint Surg. **42-A**:1051, 1960.

Alho, A., and Kankaanpaa, U.: Management of fractured scaphoid bone: a prospective study of 100 fractures, Acta Orthop. Scand. **46**:737, 1975.

Armstrong, J.R.: Closed technique for fixation of fractured carpal scaphoid, Lancet **1**:537, 1941.

Bannerman, M.M.: Fractures of the carpal scaphoid bone, Arch. Surg. **53**:164, 1946.

Barnard, L., and Stubbins, S.G.: Styloidectomy of the radius in the surgical treatment of nonunion of the carpal navicular, J. Bone Joint Surg. **30-A**:98, 1948.

Barr, J.S., et al.: Fracture of the carpal navicular (scaphoid) bone, J. Bone Joint Surg. **35-A**:609, 1953.

Baumann, J.E., and Campbell, R.D.: Significance of architectural types of fractures of the carpal scaphoid and relation to timing of treatment, J. Trauma **2**:431, 1962.

Beckenbaugh, R.D., et al.: Kienböck's disease: the natural history of Kienböck's disease and consideration of lunate fractures, Clin. Orthop. **149**:98, 1980.

Bentzon, P.G.K., and Randlove-Madsen, A.: On fracture of the carpal scaphoid, Acta Orthop. Scand. **16**:30, 1945.

Boegeskov, S., et al.: Fractures of the carpal bones, Acta Orthop. Scand. **37**:276, 1966.

Brown, P.E., and Dameron, T.B.: Surgical treatment for nonunion of the scaphoid, South. Med. J. **68**:415, 1975.

Bryan, R.S., and Dobyns, J.H.: Fractures of the carpal bones other than lunate and navicular, Clin. Orthop. **149**:107, 1980.

Buck-Gramcko, D.: Denervation of the wrist joint, J. Hand Surg. **2**:54, 1977.

Burnett, J.H.: Further observation on treatment of fracture of the carpal scaphoid (navicular), J. Bone Joint Surg. **19**:1099, 1937.

Campbell, R.D., et al.: Indications for open reduction of lunate and perilunate dislocations of the carpal bones, J. Bone Joint Surg. **47-A**:915, 1965.

Cave, E.F.: Retrolunar dislocations of the capitate with fracture or subluxation of the navicular bone, J. Bone Joint Surg. **23**:830, 1941.

Childress, H.M.: Fracture of a bipartite carpal navicular, J. Bone Joint Surg. **25**:446, 1943.

Cleveland, M.: Fracture of the carpal scaphoid, Surg. Gynecol. Obstet. **84**:769, 1947.

Cobey, M.C., and White, R.K.: An operation for nonunion of fractures of the carpal navicular, J. Bone Joint Surg. **28**:757, 1946.

Cole, W.H., and Williamson, G.A.: Fractures of the carpal navicular bone, Minn. Med. **18**:81, 1935.

Cooney, W.P., III, Dobyns, J.H., and Linscheid, R.L.: Fractures of the scaphoid: a rational approach to management, Clin. Orthop. **149**:90, 1980.

Cooney, W.P., III, Dobyns, J.H., and Linscheid, R.L.: Nonunion of the scaphoid: analysis of the results from bone grafting, J. Hand Surg. **5**:343, 1980.

Crittenden, J.J., et al.: Bilateral rotational dislocation of the carpal navicular, Radiology **94**:629, 1970.

Danyo, J.J.: Open navicular bone grafting and styloidectomy, Orthop. Rev. **4**:21, July, 1975.

Davidson, A.J., and Horwitz, M.T.: An evaluation of excision in the treatment of ununited fracture of the carpal scaphoid (navicular) bone, Ann. Surg. **108**:291, 1938.

Dawkins, A.L.: The fractured scaphoid: a modern view, Med. J. Aust. **1**:332, 1967.

Dickson, J.C., and Shannon, J.G.: Fractures of the carpal scaphoid in the Canadian Army, Surg. Gynecol. Obstet. **79**:225, 1944.

Dobyns, J.H., et al.: Traumatic instability of the wrist. In American Academy of Orthopaedic Surgeons: Instructional course lectures, vol. 24, St. Louis, 1975, The C.V. Mosby Co.

Dwyer, F.C.: Excision of the carpal scaphoid for ununited fracture, J. Bone Joint Surg. **31-B**:572, 1949.

Edelstein, J.M.: Treatment of ununited fractures of the carpal navicular, J. Bone Joint Surg. **21**:902, 1939.

Farquharson, E.L.: A splint for fracture of the carpal navicular, J. Bone Joint Surg. **24**:922, 1942.

Fenton, R.L.: The naviculocapitate fracture syndrome, J. Bone Joint Surg. **38-A**:681, 1956.

Ferguson, L.K.: Fractures of the carpal scaphoid, Surg. Clin. North Am. **17**:1603, 1937.

Fisk, G.R.: An overview of injuries of the wrist, Clin. Orthop. **149**:137, 1980.

Foster, E.J., Palmer, A.K., and Levinsohn, E.M.: Hamate erosion: an unusual result of ulnar artery constriction, J. Hand Surg. **4**:536, 1979.

Friedenberg, Z.B.: Anatomic considerations in the treatment of carpal navicular fractures, Am. J. Surg. **78**:379, 1949.

Ganel, A., et al.: Bone scanning in the assessment of fractures of the scaphoid, J. Hand Surg. **4**:540, 1979.

Goeringer, C.F.: Follow-up results of surgical treatment for nonunion of the carpal scaphoid bone, Arch. Surg. **58**:291, 1949.

Green, D.P., and O'Brien, E.T.: Classification and management of carpal dislocations, Clin. Orthop. **149**:55, 1980.

Graner, O., et al.: Arthrodesis of the carpal bones in the treatment of Kienböck's disease, painful ununited fractures of the navicular and lunate bones with avascular necrosis, and old fracture-dislocation of carpal bones, J. Bone Joint Surg. **48-A**:767, 1966.

Hartwig, R.H., and Louis, D.S.: Multiple carpometacarpal dislocations: a report of four cases, J. Bone Joint Surg. **61-A**:906, 1979.

Hill, N.A.: Fractures and dislocations of the carpus, Orthop. Clin. North Am. **1**:275, 1970.

Hopkins, F.S.: Fractures of the scaphoid in athletes, N. Engl. J. Med. **209**:687, 1933.

Howard, F.M., et al.: Rotatory subluxation of the navicular, Clin. Orthop. **104**:134, 1974.

Hudson, T.M., et al.: Isolated rotatory subluxation of the carpal navicular, Am. J. Roentgenol. Radium Ther. Nucl. **126**:601, 1976.

Hull, W.J., et al.: The surgical approach and source of bone graft for symptomatic nonunion of the scaphoid, Clin. Orthop. **115**:241, 1976.

Jaekle, R.F., and Clark, A.G.: Acute fractures of the carpal scaphoid, Surg. Gynecol. Obstet. **68**:820, 1939.

Johnson, R.P.: The acutely injured wrist and its residuals, Clin. Orthop. **149**:33, 1980.

Johnson, R.W.: A study of the healing processes in injuries to the carpal scaphoid, J. Bone Joint Surg. **9**:482, 1927.

Kauer, J.M.G.: Functional anatomy of the wrist, Clin. Orthop. **149**:9, 1980.

Lette, R.R.: Vitallium prosthesis in the treatment of fracture of the carpal navicular, West. J. Surg. **59**:468, 1951.

Linscheid, R.L., et al.: Traumatic instability of the wrist: diagnosis, classification, and pathomechanics, J. Bone Joint Surg. **54-A:**1612, 1972.

Luck, J.V., et al.: Orthopedic surgery in the Army Air Forces during World War II. Arch Surg. **57:**801, 1948.

Marsh, A.P., and Lampros, P.: The naviculocapitate fracture syndrome, Am. J. Roetgenol. Radium Ther. Nucl. Med. **82:**255, 1959.

Mathison, G.W., and MacDonald, R.I.: Irreducible transcapitate fracture and dislocation of the hamate: report of a case, J. Bone Joint Surg. **57-A:**1166, 1975.

Mayfield, J.K.: Mechanism of carpal injuries, Clin. Orthop. **149:**45, 1980.

Mayfield, J.K., Johnson, R.P., and Kilcoyne, R.K.: Carpal dislocations: pathomechanics and progressive perilunar instability, J. Hand Surg. **5:**266, 1980.

Maudsley, R.H., and Chen, S.C.: Screw fixation in the management of the fractured carpal scaphoid, J. Bone Joint Surg. **54-B:**432, 1972.

Mazet, R., and Hohl, M.: Radial styloidectomy and styloidectomy plus bone graft in the treatment of old ununited carpal scaphoid fractures, Ann. Surg. **152:**296, 1960.

Mazet, R., and Hohl, M.: Conservative treatment of old fractures of the carpal scaphoid, J. Trauma **1:**115, 1961.

Mazet, R., and Hohl, M.: Fractures of the carpal navicular, J. Bone Joint Surg. **45-A:**82, 1963.

McDonald, G., and Petrie, D.: Un-united fracture of the scaphoid, Clin. Orthop. **108:**110, 1975.

McGauley, F.F.: Injuries to carpal bones: fracture of the scaphoid and dislocation of the semilunar, Arch. Surg. **10:**764, 1925.

McLaughlin, H.L.: Fracture of the carpal navicular (scaphoid) bone, J. Bone Joint Surg. **36-A:**765, 1954.

McLaughlin, H.L., and Parkes, J.C.: Fracture of the carpal navicular (scaphoid) bone: Gradations in therapy based upon pathology, J. Trauma **9:**311, 1969.

McMillan, F.B.: Injuries to the carpal bones, Am. J. Surg. **42:**633, 1938.

Meekison, D.M.: Some remarks on three common fractures, J. Bone Joint Surg. **27:**80, 1945.

Meuli, H.C.: Arthroplasty of the wrist, Clin. Orthop. **149:**118, 1980.

Meyn, M.A., and Roth, A.M.: Isolated dislocation of the trapezoid bone, J. Hand Surg. **5:**602, 1980.

Meyers, M.H., et al.: Naviculo-capitate fracture syndrome: review of the literature and a case report, J. Bone Joint Surg. **53-A:**1383, 1971.

Monahan, P.R.W., and Galasko, C.S.B.: The scapho-capitate fracture syndrome: a mechanism of injury, J. Bone Joint Surg. **54-B:**122, 1972.

Mulder, J.D.: The results of 100 cases of pseudarthroses in the scaphoid bone treated by the Matti-Russe operation, J. Bone Joint Surg. **50-B:**110, 1968.

Murray, G.: End results of bone-grafting for nonunion of the carpal navicular, J. Bone Joint surg. **28:**749, 1946.

Neviaser, R.J.: Proximal row carpectomy for posttraumatic disorders of carpus, J. Hand Surg. **8:**301, 1983.

Obletz, B.E.: Fresh fractures of the carpal scaphoid, Surg. Gynecol. Obstet. **78:**83, 1944.

Obletz, B., and Halbstein, B.: Nonunion of fractures of the carpal navicular, J. Bone Joint Surg. **20:**424, 1938.

Palmer, A.K., Dobyns, J.H., and Linscheid, R.L.: Management of post-traumatic instability of the wrist secondary to ligament rupture, J. Hand Surg. **3:**507, 1978.

Peimer, C.A., Smith, R.J., and Leffert, R.D.: Distraction-fixation in the primary treatment of metacarpal bone loss, J. Hand Surg. **6:**111, 1981.

Perey, O.: A re-examination of cases of pseudarthrosis of the navicular bone operated on according to Bentzon's technique, Acta Orthop. Scand. **23:**26, 1954.

Perkins, G.: Fractures of the carpal scaphoid, Br. Med. J. **1:**536, 1950.

Primiano, G.A., and Reef, T.C.: Disruption of the proximal carpal arch of the hand, J. Bone Joint Surg. **56-A:**328, 1974.

Robertson, J.M., and Wilkins, R.D.: Fractures of the carpal scaphoid, Br. Med. J. **1:**685, 1944.

Rothberg, A.S.: Ununited fractures of the carpal scaphoid, Am. J. Surg. **56:**611, 1942.

Ruedi, T., et al.: Stable internal fixation of fractures of the hand, J. Trauma **2:**381, 1971.

Russe, O.: Fracture of the carpal navicular, J. Bone Joint Surg. **42-A:**759, 1960.

Russell, T.B.: Inter-carpal dislocations and fracture without dislocations, J. Bone Joint Surg. **31-B:**524, 1949.

Sashin, D.: Treatment of fractures of the carpal scaphoid, Arch. Surg. **52:**445, 1946.

Sherwin, J.M., et al.: Bipartite carpal navicular and the diagnostic problem of bone partition, J. Trauma **11:**440, 1971.

Smith, E.H.: Autogenous bone dowel for relief of fracture of scaphoid bone of wrist, Med. Rec. **139:**655, 1934.

Smith, L., and Friedman, B.: Treatment of ununited fracture of the carpal navicular by styloidectomy of the radius, J. Bone Joint Surg. **38-A:**368, 1956.

Snodgrass, L.E.: End results of carpal scaphoid fractures, Ann. Surg. **97:**209, 1933.

Soto-Hall, R., and Haldeman, K.O.: Treatment of fractures of the carpal scaphoid, J. Bone Joint Surg. **16:**822, 1934.

Soto-Hall, R., and Haldeman, K.O.: The conservative and operative treatment of fractures of the carpal scaphoid, J. Bone Joint surg. **23:**841, 1941.

Soto-Hall, R., et al.: Fractures of the carpal scaphoid, JAMA **129:**335, 1945.

Speed, K.: The unresolved fracture, Surg. Gynecol. Obstet. **60:**341, 1935.

Speed, K.: Fractures of the carpal navicular bone, J. Bone Joint Surg. **7:**682, 1925.

Speed, K.: Fractures of the carpus, J. Bone Joint Surg. **17:**965, 1935.

Speed, K.: Fractures and dislocations of the carpus, Calif. Med. **72:**93, 1950.

Spotoft, J.: Fracture of the navicular bone, Acta Chir. Scand. **125:**524, 1963.

Squire, M.: Carpal mechanics and trauma, J. Bone Joint Surg. **41-B:**210, 1959.

Srivastava, K.K., and Kochhar, V.L.: Congenital absence of the carpal scaphoid: a case report, J. Bone Joint Surg. **54-A:**1782, 1972.

Stark, W.A.: Recurrent perilunar subluxation, Clin. Orthop. **73:**152, 1970.

Stecher, W.R.: Roentgenography of carpal navicular bone, Am. J. Roentgenol. Radium Ther. Nucl. Med. **37:**704, 1937.

Stein, F., et al.: Naviculocapitate fracture syndrome, J. Bone Joint Surg. **51-A:**391, 1969.

Stewart, M.J.: Fractures of the carpal navicular (scaphoid): a report of 436 cases, J. Bone Joint Surg. **36-A:**998, 1954.

Taleisnik, J.: Post-traumatic carpal instability, Clin. Orthop. **149:**73, 1980.

Taleisnik, J., and Kelly, P.J.: The extraosseous and intraosseous blood supply of the scaphoid bone, J. Bone Joint Surg. **48-A:**1125, 1966.

Thomaidis, V.T.: Elbow-wrist-thumb immobilization in the treatment of fractures of the carpal scaphoid, Acta Orthop. Scand. **44:**679, 1973.

Thompson, J.E.: Fractures of the carpal navicular and triquetrum bones, Am. J. Surg. **21:**214, 1907.

Thompson, T.C., et al.: Primary and secondary dislocation of the scaphoid bone, J. Bone Joint Surg. **46-B:**73, 1964.

Thorndike, A., and Garrey, W.E.: Fractures of the carpal scaphoid, N. Engl. J. Med. **222:**827, 1940.

Törngren, S., and Sandqvist, S.: Pseudarthrosis in the scaphoid bone treated by grafting with autogenous bone-peg: a follow-up study, Acta Orthop. Scand. **45:**82, 1974.

Tullos, H.S., et al.: Isolated subluxation of the carpal scaphoid associated with secondary displacement of the capitate, South. Med. J. **66:**568, 1973.

Uematsu, A.: Intercarpal fusion for treatment of carpal instability: a preliminary report, Clin. Orthop. **144:**159, 1979.

Vance, R.M., Gelberman, R.D., and Braun, R.M.: Chronic bilateral scapholunate dissociation without symptoms, J. Hand Surg. **4:**178, 1979.

Verdan, C.: Fractures of the scaphoid, Surg. Clin. North Am. **40:**461, 1960.

Volz, R.G., and Benjamin, J.: Biomechanics of the wrist, Clin. Orthop. **149:**112, 1980.

Wagner, C.J.: Fractures of the carpal navicular, J. Bone Joint Surg. **34-A:**774, 1952.

Wagner, C.J.: Fracture-dislocations of the wrist, Clin. Orthop. **15:**181, 1959.

Watson, H.K.: Limited wrist arthrodesis, Clin. Orthop. **149:**126, 1980.

Watson-Jones, R.: Inadequate immobilization and nonunion of fractures, Br. Med. J. **1:**937, 1934.

Waugh, R.L., and Reuling, L.: Ununited fractures of the carpal scaphoid, Am. J. Surg. **67:**184, 1945.

Waugh, R.L., and Sullivan, R.F.: Anomalies of the carpus, J. Bone Joint Surg. **36-A:**682, 1950.

Weber, E.R.: Biomechanical implications of scaphoid waist fractures, Clin. Orthop. **149:**83, 1980.

Weber, E.R., and Chao, E.Y.: An experimental approach to the mechanism of scaphoid waist fractures, J. Hand Surg. **3:**142, 1978.

Weeks, P.M., Young, V.L., and Gilula, L.A.: A cause of painful clicking wrist: a case report, J. Hand Surg. **4:**522, 1979.

Wesely, M.S., and Barenfeld, P.A.: Trans-scaphoid, transcapitate, transtriquetral, perilunate fracture-dislocation of the wrist: a case report, J. Bone Joint Surg. **54-A:**1073, 1972.

Woods, R.S.: Union of carpal scaphoid (by immobilization) after six weeks delay in treatment, Br. Med. J. **2:**1119, 1937.

Youm, Y., and Flatt, A.E.: Kinematics of the wrist, Clin. Orthop. **149:**21, 1980.

Ziter, F.M.H., Jr.: A modified view of the carpal navicular, Radiology **108:**706, 1973.

Kienböck's disease

Agerholm, J.C., and Goodfellow, J.W.: Avascular necrosis of the lunate bone treated by excision and prosthetic replacement, J. Bone Joint Surg. **45-B:**110, 1963.

Armistead, R.B., Linscheid, R.L., Dobyns, J.H., and Beckenbaugh, R.D.: Ulnar lengthening in the treatment of Kienböck's disease, J. Bone Joint Surg. **64-A:**170, 1982.

Barber, H.M., and Goodfellow, J.W.: Acrylic lunate prostheses: a long-term follow-up, J. Bone Joint Surg. **57-B:**706, 1974.

Beckenbaugh, R.D., Shives, T.C., Dobyns, J.H., and Linscheid, R.L.: Kienböck's disease: the natural history of Kienböck's disease and consideration of lunate fractures, Clin. Orthop. **149:**98, 1980.

Chan, K.P., and Huang, P.: Anatomic variations in radial and ulnar lengths in the wrists of Chinese, Clin. Orthop. **80:**17, 1971.

Crabbe, W.A.: Excision of the proximal row of the carpus, J. Bone Joint Surg. **46-B:**708, 1964.

Doran, A.: The results of treatment in Kienböck's disease, J. Bone Joint Surg. **31-B:**518, 1949.

Dornan, A.: Results of treatment in Kienböck's disease, J. Bone Joint Surg. **31-B:**518, 1949.

Gelberman, R.H., et al.: The vascularity of the lunate bone and Kienböck's disease, J. Hand Surg. **5:**272, 1980.

Gelberman, R.H., et al.: Ulnar variance in Kienböck's disease, J. Bone Joint Surg. **57-A:**674, 1975.

Gillespie, H.S.: Excision of the lunate bone in Kienböck's disease, J. Bone Joint Surg. **43-B:**245, 1961.

Graner, O., et al.: Arthrodesis of the carpal bones in treatment of Kienböck's disease, painful ununited fractures of the navicular and luante bones with avascular necrosis, and old fracture-dislocations of carpal bones, J. Bone Surg. **48-A:**767, 1966.

Hultén, O.: Über anatomische Variationen der Handgelenkknochen, Acta Radiol. **9:**155, 1928.

Inglis, A.E., and Jones, E.C.: Proximal row carpectomy for disease of the proximal row, J. Bone Joint Surg. **51-A:**460, 1977.

Jorgensen, E.C.: Proximal-row carpectomy: an end-result study of twenty-two cases, J. Bone Joint Surg. **51-A:**1104, 1969.

Kienböck, R.: Uber traumatische Malazie des Mondbeins und ihre Folgezustände, Entartungsformen und Kompressionsfrakturen, Fortschr. Geb. Röntgenstr. **16:**77, 1910-1911.

Lichtman, D.M., et al.: Kienböck's disease: the role of silicone replacement arthroplasty, J. Bone Joint Surg. **59-A:**899, 1977.

Lippman, E.M., and McDermott, L.J.: Vitallium replacement of lunate in Kienböck's disease, Milit. Surg. **105:**482, 1949.

Lowry, W.E., and Cord, S.A.: Traumatic avascular necrosis of the capitate bone: case report, J. Hand Surg. **6:**245, 1981.

Marek, F.M.: Avascular necrosis of the carpal lunate, Clin. Orthop. **10:**96, 1957.

Match, R.M.: Nonspecific avascular necrosis of the pisiform bone: a case report, J. Hand Surg. **5:**341, 1980.

McMurtry, R.Y., Youm, Y., Flatt, A.E., and Gillespie, T.E.: Kinematics of the wrist. Part II. Clinical applications, J. Bone Joint Surg. **60A:**955, 1978.

Mouat, T.B., Wilkie, J., and Harding, H.E.: Isolated fracture of the carpal semilunar and Kienböck's disease, Br. J. Surg. **19:**577, 1932.

Nahigian, S.H., et al.: The dorsal flap arthroplasty in the treatment of Kienböck's disease, J. Bone Joint Surg. **52-A:**245, 1970.

Persson, M.: Pathogenese und Behandlung der Kienböckschen Lunatummalazie, die Frakturtheorie im Lichte der Erfolge operativer Radiusverkürzung (Hultén) und einer neuen Operationsmethode—Ulnaverlängerung, Acta Chir. Scand. **92** (Suppl. 98): 1, 1945.

Persson, M.: Causal treatment of lunatomalacia: further experiences of operative ulna lengthening, Acta Chir. Scand. **100:**531, 1950.

Phemister, D.B., Brunschwig, A., and Day, L.: Streptococcal infections of epiphyses and short bones, their relation to Kohler's disease of the tarsal navicular, Legg-Perthes' disease and Kienböck's disease of the os lunatum, JAMA **95:**995, 1930.

Ramakrishna, B., D'Netto, D.C., and Sethu, A.U.: Long-term results of silicone rubber implants for Kienböck's disease, J. Bone Joint Surg. **64-B:**361, 1982.

Ringsted, A.: Doppelseitiger Mb. Kienboeck bei 2 Brüdern, Acta Chir. Scand. **69:**185, 1932.

Roca, J., et al.: Treatment of Kienböck's disease using a silicone rubber implant. J. Bone Joint Surg. **58-A:**373, 1976.

Ståhl, F.: On lunatomalacia (Kienböck's disease): a clinical and roentgenological study, especially on its pathogenesis and the late results of immobilization treatment, Acta Chir. Scand. (Suppl. 126), 1947.

Stark, H.H., Zemel, N.P., and Ashworth, C.R.: Use of a hand-carved silicone-rubber spacer for advanced Kienböck's disease, J. Bone Joint Surg. **63-A:**1359, 1981.

Swanson, A.B.: Silicone rubber implants for the replacement of the carpal scaphoid and lunate bones, Orthop. Clin. North Am. **1:**299, 1970.

Swanson, A.B., and Swanson, G. deG.: Flexible implant resection arthroplasty: a method for reconstruction of small joints in the extremities. In AAOS: Instructional course lectures, vol. 27, 1978, The C.V. Mosby Co.

Tauma, T.: An investigation of the treatment of Kienböck's disease, J. Bone Joint Surg. **48-A:**1649, 1966.

Therkelsen, F., and Andersen, K.: Lunatomalacia, Acta Chir. Scand. **97:**503, 1949.

Arthodesis of wrist

Allende, B.T.: Wrist arthrodesis, Clin. Orthop. **142:**164, 1979.

Beckenbaugh, R.D., and Linscheid, R.L.: Total wrist arthroplasty: a preliminary report, J. Hand Surg. **2:**337, 1977.

Clendenin, M.B., and Green, D.P.: Arthrodesis of the wrist: complications and their management, J. Hand Surg. **6:**253, 1981.

Haddad, R.J., Jr., and Riordan, D.C.: Arthrodesis of the wrist: a surgical technique, J. Bone Joint Surg. **49-A:**950, 1967.

MacKenzie, I.G.: Arthrodesis of the wrist in reconstructive surgery, J. Bone Joint Surg. **42-B:**60, 1960.

Makin, M.: Wrist arthrodesis in paralyzed arms of children. J. Bone Joint Surg. **59-A:**312, 1977.

Seddon, H.J.: Reconstructive surgery of the upper extremity. In Poliomyelitis, Second International Poliomyelitis Congress, Philadelphia, 1952, J.B. Lippincott Co.

Smith-Peterson, M.N.: A new approach to the wrist joint, J. Bone Joint Surg. **22:**122, 1940.

Stein, I.: Gill turnabout radial graft for wrist arthrodesis, Surg. Gynecol. Obstet. **106:**231, 1958.

Watson, H.K., Goodman, M.L., and Johnson, T.R.: Limited wrist arthrodesis. II. Intercarpal and radiocarpal combinations, J. Hand Surg. **6:**223, 1981.

Watson, H.K., and Hempton, R.F.: Limited wrist arthrodesis. I. The triscaphoid joint, J. Hand Surg. **5:**320, 1980.

CHAPTER 5

Nerve injuries

EVALUATION

Injuries of nerves are even more difficult to evaluate than those of tendons and are often overlooked during the preliminary examination. Digital nerve injuries are frequently overlooked because these nerves are so small or because they are not thought of even when the wound is inspected during the operation. When a flexor tendon in the finger is cut, at least one digital nerve is usually cut also.

Sensibility may be checked by stimulation for pain with a sharp pin. This is frequently neither accurate nor conclusive, but stimulation of a normal area for comparison may help the patient's interpretation. Two-point discrimination is better to assess damaged sensory nerves; however, the surgeon must remember that normal two-point discrimination at the finger pulp is 4 mm or less (Fig. 5-1). An experienced examiner may be able to confirm suspected nerve damage by observing for sweating. Since a denervated area will be dry or nonsweating within 30 minutes after nerve injury, careful comparison of the normal and suspected abnormal areas by palpation with a dry fingertip may be diagnostic. The areas of the hand supplied by the major nerves are well known and will not be given here.

Only a few tests for motor function of the nerves will be mentioned here; these are given to emphasize some of the pitfalls that are likely in attempting the more commonly used tests. When testing for motor function of the *radial nerve,* the examiner must not simply support the palm of the hand and interpret extension of the two distal finger joints as caused by function of this nerve. One must test for active extension of the metacarpophalangeal joints and of the wrist and for extension and abduction of the thumb to rule out motor paralysis of the radial nerve.

When the motor function of the *median nerve* is checked, the tip of the thumb should be actively opposed to the pulp of the middle finger, and contraction of the abductor pollicis brevis muscle should be examined both visually and by palpation; one must remember that occasionally the innervation of this muscle varies. Careful comparison of the opposite normal side should be done, observing the active contraction of the abductor pollicis brevis, since the superficial flexor pollicis brevis may simulate it. The latter is frequently innervated by the ulnar nerve.

The motor function of the *ulnar nerve* may be easily tested by checking active abduction of the middle finger from the ulnar to the radial side with the palm resting on a flat surface to prevent active flexion and extension of the fingers (Fig. 5-2). If the long flexor tendons are allowed to act, the fingers will tend to converge and thus prevent accurate interpretation of the function of the volar interosseus muscles; if the long extensor tendons are allowed to act, the fingers tend to diverge, and an accurate test for function of the dorsal interosseus muscles is impossible. To test the adductor of the thumb (which is supplied by the ulnar nerve), ask the patient to place the tip of the thumb on the base of the little finger without flexing the interphalangeal joint of the thumb; this is possible only when the nerve supply to this muscle is intact.

• • •

In 1984 the Clinical Assessment Committee of the American Society for Surgery of the Hand issued a report on nerve function evaluation. The committee recommended four areas to evaluate in the progress of peripheral nerve injury and repair: sensiblity testing, motor testing, subjective evaluation, and pseudomotor function.

Sensibility evaluation

The basic minimum tests recommended are the stationary two-point discrimination and moving two-point discrimination tests.

TECHNIQUE. The hand should be warm and the instrument at room temperature. Apply a blunt caliper distally over the digital pad until blanching occurs. Test each area three times, going progressively from longer distances to shorter

119

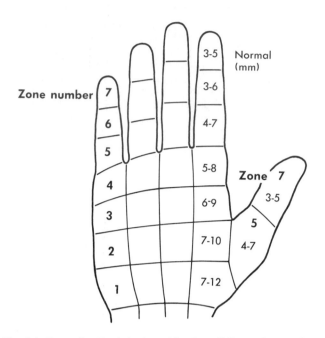

Fig. 5-1. Two-point discrimination of hand sensibility, palmar surface. Dorsal surface averages from 7 mm distally to 12 mm proximally.

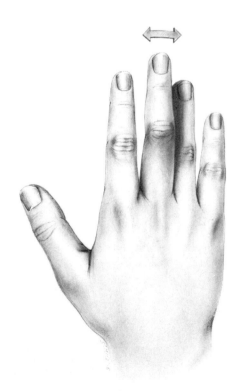

Fig. 5-2. Egawa sign. With patient's palm flat on table, he is asked to raise extended middle finger and abduct it from ulnar side to radial side.

distances. The moving two-point discrimination test is done in a similar fashion. The caliper is applied in an axial direction and moved from proximal to distal aspects along the digital pad. Two out of three correct answers are considered proof of perception with either test.

Motor testing

The committee recommends three basic minimum tests for motor function: Jamar grip tester (all five positions), pinch meter (key pinch), and pinch meter (three-jaw pinch). The Jamar grip tester should be used and the results recorded at all five positions because this reflects the overall integrated function of the hand, in addition to areas of extrinsic and intrinsic muscle deficits.

Subjective evaluation

This is the patient's evaluation of his current status and includes symptoms such as the presence of pain, cold intolerance, dysesthesias, and functional disabilities.

Pseudomotor function

A statement relative to pseudomotor function should be included in the evaluation.

SUTURING OF NERVES
Indications for primary nerve suture

In the past opinions have differed as to the best time to suture a divided nerve; however, in civilian practice in which the necessary technical skill and facilities are available, primary nerve repair in clean incised wounds is now usually advised. That results are better after delayed repair has not been proved. In fact, repairs carried out after 3 months become less successful with each additional month of delay. Therefore early suture provides a better chance for a satisfactory functional result; it also avoids another operation and another period of immobilization. However, well-demarcated and clean bundles of nerve fibers should be seen at both ends of the severed nerve before primary suture is considered. Suture at this time is more difficult than secondary suture because the perineurium is thin; 2 or 3 weeks later the perineurium is thick and will hold small No. 8-0 nylon sutures better. Tension at the suture line must be avoided. Many failures after nerve suture are the result of too much tension that causes later separation of the nerve ends; scar tissue then fills the gap. An incompletely divided nerve should be sutured, and the intact part should be left undisturbed.

Secondary nerve repair must be considered when a severe crushing injury involves a segment of undetermined length. Nerve ends that are damaged can be later identified by the presence of fibrosis. Damage is not easily detectable at the time of primary treatment, but some evidence of the extent of the injury may be obtained from the history and from the condition of the surrounding tissues. A nerve should not be sutured primarily when severe flexion of the wrist and fingers of the acutely injured hand is necessary to avoid tension on the repair; in these instances the proximal end of the nerve should be anchored under normal tension with a suture to prevent its retraction. When a defect in a nerve is large enough to require a nerve graft or extensive dissection for a nerve transfer to gain needed length, primary repair should not be done.

TECHNIQUE OF SUTURE OF NERVES OF HAND. The principles that apply to the suture of the other peripheral nerves also apply to those of the hand. However, the prognosis is better after suture of a digital nerve because it is small and purely sensory and because the distance is short from the point of division to the end organ. The return of sensation is usually proportional to the accuracy of the suture, but the age of the patient has the greatest effect on the result. A digital nerve can be sutured as far distally as the distal volar finger crease.

TECHNIQUE FOR SUTURE OF DIGITAL NERVES. Both digital nerves run anterolaterally in the finger and thus can be exposed through the same midlateral incision when necessary. Dissect the digital nerves and arteries from their common encircling sheath (a part of Cleland's ligament) (Fig. 5-3). Resect the ends of the nerves as necessary to remove crushed edges or neuromas. Approximate the nerve ends without tension; flex the finger joints if necessary. Use a No. 8-0 or 9-0 monofilament nylon suture on

an atraumatic noncutting curved needle (Fig. 5-4). When necessary, the nerve ends may be temporarily held in place by passing the smallest straight Bunnell needle transversely through them into the adjoining soft tissue; thus tension is avoided while the sutures are more accurately tied. Now pass a suture through the neurilemma of the nerve a fraction of a millimeter from its edge and again in a similar manner through the neurilemma of the other end of the nerve; tie the knot with at least five loops to prevent its slipping or untying. Place a second suture on the exact opposite side of the nerve. These first two sutures are left long so that they can be used to rotate the nerve 180 degrees, making accessible all of its surfaces. Place a total of four sutures. After the repair, slowly extend the joints and observe the suture line for tension; note the optimum position of the joints for this purpose and maintain it by splinting after closure. Always suture any divided tendons before suturing any nerve to avoid disrupting the delicate repair.

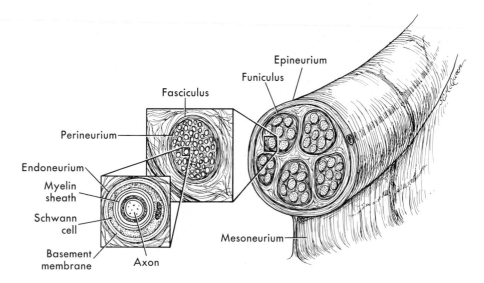

Fig. 5-3. Basic anatomy of peripheral nerves.

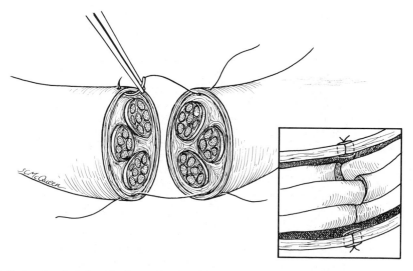

Fig. 5-4. Basic suture technique for laceration of peripheral nerves should result in no tension at suture line, and each small fascicle should be aligned to match opposing, mirror image.

AFTERTREATMENT. After 3 weeks the finger joints are allowed gradual active extension beyond the optimum position noted at surgery; when the defect in the nerve is large, active extension cannot be permitted before 4 weeks. Even though the suture line must be protected, active finger motion must be started as soon as possible to avoid stiffness.

SECONDARY NERVE REPAIR

When multiple tissues must be reconstructed, the repair of nerves is second in the order of priority and is indicated only after good skin coverage has been obtained. All wounds must be completely healed, and the nutritional status of the hand must be satisfactory. When a digital or palmar nerve has not been sutured at the time of laceration, it should usually be sutured as a secondary procedure 2 or 3 weeks later. If at all possible, secondary repair should be delayed no longer than 3 months because the success of repair is jeopardized by further delay. On the other hand, a secondary repair should be attempted even after 12 to 18 months if repair before this time either has not been attempted or has been impossible because of a complication (for example, an infection); such an attempt is highly desirable in children, who have extraordinary regenerative powers. In adults, however, such a delayed repair is very unlikely to result in any useful function.

A secondary nerve repair is indicated not only when the nerve has not been sutured primarily, but also sometimes even after primary suture. Occasionally when a surgeon first sees an injured hand with a nerve deficit a few days or a few weeks after injury, he may learn that a nerve was primarily sutured but cannot learn how expertly it was done; he must then consider an exploration and possible secondary repair. This immediate exploration may sometimes prove secondary suture unnecessary, but if the surgeon accepts prolonged delay while awaiting a result that may not be forthcoming, he jeopardizes the result of any later secondary repair. Exploration is a small price to pay for assurance, since skillful repair of nerves, more than any other one factor, determines the ultimate return of function.

While major nerves are regenerating after repair, the hand may assume an unnatural posture because of change in muscle balance. Even when the nerve lesion is above the wrist, the hand suffers most and may incur fixed contractures before nerve function returns. Proper splinting (p. 21) is therefore necessary to prevent contractures during this period. The patient should be warned that until sensation returns, the anesthetic skin may become infected after even minor trauma or may be burned or frostbitten unless properly protected. He should be instructed to inspect the insensitive areas routinely and to avoid extremes of heat and cold.

Distal to the wrist, it is the digital nerves that are most frequently severed, and in recent years the importance of their repair has been emphasized, and the techniques have been described often. Less often it has been noted that the sensory nerves on the dorsum of the hand may also be repaired by a similar technique if the surgeon knows their anatomy well enough to locate the proximal and distal ends.

Regeneration

After the repair of a digital or other sensory nerve or sensory elements of a mixed nerve, not only does the area of anesthesia decrease in size as regeneration proceeds, but the quality of sensation also changes. At 2 or 3 months the entire area may become paresthetic. Then it may become hyperesthetic when stimulated by light touch or cold; firm pressure is usually less painful. Gradually, the hyperesthesia subsides and relatively normal sensation is reestablished. The quality of sensation may improve for as long as 18 months or more, but fully normal sensation, with two-point discrimination, cannot be expected.

Hypesthesia of a fingertip will sometimes disappear almost immediately after a digital nerve is released by careful dissection from a scarred area, even though the nerve may have been narrowed at the point of compression.

Even though digital nerves regenerate better than more proximal or mixed nerves such as the ulnar nerve, regeneration is poor especially after the age of 50 years. At times a separate operation may not be worthwhile to repair one digital nerve in a noncritical area. The more critical areas of sensory innervation of the fingers are the ulnar side of the thumb, the radial side of the index and middle fingers, and the ulnar side of the little finger. These critical areas are very important in pinch and ulnar border contact. Repair of nerves to these areas should be given priority when a choice must be made because of the limitations of surgical time in relation to multiple tissue involvement or segmental nerve damage.

Digital nerves

A digital nerve may be repaired as far distally as the flexor crease of the distal interphalangeal joint. When repaired secondarily, the line of suture must lie in a vascular bed free of scar. Before surgery the proximal end of the nerve can often be located by passing a firm object, such as a paper clip, distally along the course of the nerve; when it reaches the neuroma at the end of the nerve, the patient will be quick to respond.

TECHNIQUE. Approach the nerve through a midlateral incision (p. 10). Begin proximally and dissect a normal segment of the nerve from its investing fascia; then proceed distally to the scar at the site of injury. Next begin distal to the site of injury and dissect proximally to the scar. With a sharp knife or scissors remove the neuroma from the proximal end of the nerve and the glioma from the distal end (for the details of suture see p. 121).

AFTERTREATMENT. Aftertreatment is discussed on p. 122 following the discussion of the technique for suture of digital nerves.

Superficial radial nerve

Disability after interruption of the superficial radial nerve at the wrist is less than that after interruption of sensory nerves on the volar surface of the hand; there is anesthesia over a variable area on the dorsum of the thumb and index finger. Sometimes the ulnar side of the area of pinch of the thumb receives its major innervation from this nerve. Neuromas caught in dorsal scars are particularly painful because they are stimulated not only by direct touch but also by stretching of the surrounding skin, nerve, and scar when the wrist and fingers are flexed.

Unless there is some unusual reason for repairing the nerve or one of its branches, it should be resected proximal to its site of severance to permit it to lie in an area of minimal scar. It is so common to have a painful and at times disabling neuroma after repair that the small area of lost sensibility is a small disability in comparison.

TECHNIQUE. Technique is as described for digital nerves. Locate the nerve proximally and dissect it distally to the scar; a consistent anatomic landmark proximally is the exit of the nerve from beneath the tendon of the brachioradialis muscle, usually about 5 cm proximal to its insertion into the radial styloid. Then locate the nerve distally and dissect it proximally toward the scar. At the base of the thumb the nerve has usually already divided into two major branches; each is larger than a digital nerve and when severed can be repaired (for the technique of suture see p.121). If the wrist must be extended to appose the nerve ends, it should be maintained in this position for 4 or 5 weeks to prevent tension on the repair. When the distal branch or branches cannot be found, take care to release the nerve proximally from the scar to relieve pain; resect some of it if necessary.

Dorsal branch of ulnar nerve

This nerve is large enough at the wrist and just distal to it to be repaired like a digital nerve. It crosses the ulnar styloid superficially, even though it may have branched from the trunk 5 cm or more proximal to the wrist. If extra length is needed to appose the ends, it may be made to branch from the main trunk more proximally by stripping and is then routed more directly to the dorsum of the hand. The wrist should then be held in extension for 3 to 4 weeks after surgery.

Ulnar nerve at wrist

When the ulnar artery and the tendon of the flexor carpi ulnaris are severed at the wrist, the ulnar nerve is usually severed too. At this level it is both motor and sensory, and therefore proper rotational alignment of the ends is important at the time of suture.

TECHNIQUE. Expose the proximal and distal segments of the nerve but do not yet remove them from their normal beds. With a suture through the epineurium, mark exactly the most anterior aspect of each segment some distance from the scarred area. Now free each segment from the surrounding soft tissues. With clean transverse cuts, excise the neuroma from the proximal segment and the glioma from the distal segment; now look at each cut end for a pattern consisting of large and small bundles. By matching these patterns and using the two epineurial sutures just described, proper rotational alignment should be possible. When further length is needed for suture without tension, dissect and mobilize the nerve more proximally or even, if necessary, transplant it anteriorly from behind the medial epicondyle of the humerus. It should be realized, however, that extensive freeing of a nerve may damage its blood supply. When advancing the nerve distally take care not to strip its branches to the muscles in the proximal forearm.

Flex the elbow as necessary to avoid tension. Suture the nerve as described on p. 121 but use No. 6-0 silk or No. 8-0 or 9-0 nylon and place more than four sutures as needed for a secure repair.

When the ulnar nerve is severed near but just distal to its division into its volar (palmar) superficial sensory branch and its deep motor branch, identify the two small proximal segments and strip them apart in a proximal direction for ease of mobilization; suture each branch separately.

Deep branch of ulnar nerve

Boyes has noted the feasibility of repairing this important branch of the ulnar nerve, which supplies those intrinsic muscles of the hand not supplied by the median nerve: two lumbricals, all interossei, the hypothenar muscles, and the adductor pollicis. These are among those most responsible for the quick and skillful movements of the fingers. Many tendon transfers have been devised to restore motor function lost by interruption of this nerve, but when possible, direct repair of the nerve is desirable.

The first dorsal interosseus muscle is usually supplied by the ulnar nerve, but it should be remembered that in at least 10% of hands it is supplied in part or completely by the median nerve. Occasionally the posterior interosseous or superficial radial branches of the radial nerve may supply the first dorsal and possibly also the second and third dorsal interosseus muscles.

TECHNIQUE (BOYES). Expose the nerve from its origin as a branch of the main trunk at the wrist to its midpalmar part through a curved incision distal to but parallel to the thenar crease; extend it over the hook of the hamate to the flexion crease of the wrist, cross the crease obliquely, and proceed to the ulnar aspect of the distal forearm. Reflect the skin, divide the palmaris brevis muscle at its insertion and reflect it ulnarward so as not to disturb its nerve supply. Retract the ulnar vessels toward the thumb and divide the origins of the abductor digiti quinti, flexor digiti quinti, and opponens digiti quinti muscles. Elevate the tendons of the flexor digitorum. The course of the nerve is now exposed from its origin at the pisiform to the midpalm (Fig. 5-5, *A*). When necessary, the nerve may be further exposed distally by extending the incision to the index metacarpal and by elevating the flexor tendons with the lumbrical muscles. When these are displaced ulnarward, the nerve can be picked up where it passes through the transverse fibers of the adductor pollicis.

When the nerve has been divided by a sharp instrument, gently free it from proximally and distally to the point of damage. This usually allows enough length for suture without tension. In gunshot wounds or others in which nerve substance has been lost, reroute the nerve as follows (Fig. 5-5, *C* and *D*). Split its motor component from the trunk well into the distal forearm. Then divide the volar carpal ligament and free from the ulnar side of the carpus the ulnar bursa that lines the carpal tunnel; displace the proximal end of the nerve into the tunnel. Bring the proximal end to the midpalm by flexing the wrist. In some instances when branches to the hypothenar muscles are still intact, gentle dissection and splitting off of the bun-

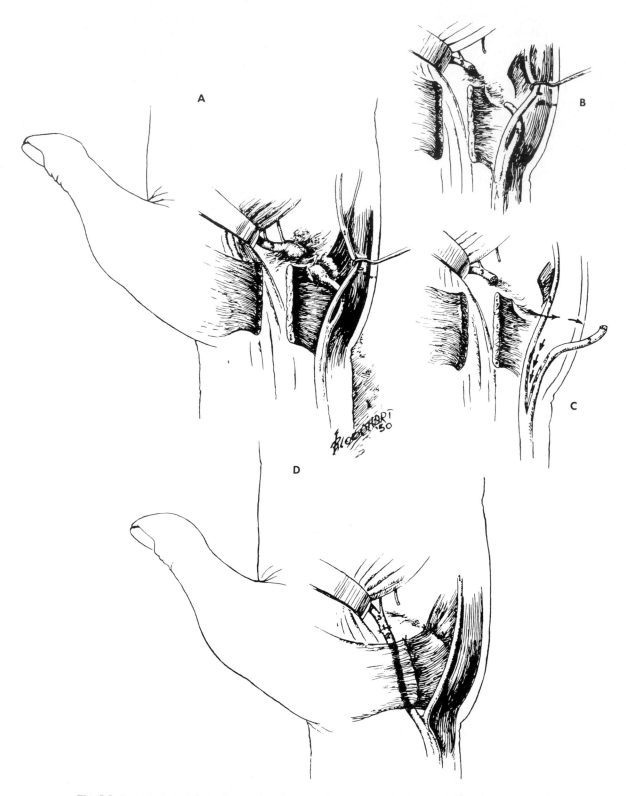

Fig. 5-5. Boyes technique of repairing deep branch of ulnar nerve. **A,** Main trunk and deep branch of ulnar nerve have both been exposed, and volar carpal ligament has been divided. **B,** Ends of deep branch have been freshened. **C,** Deep branch has been split intraneurally into distal forearm. **D,** Deep branch has been rerouted through carpal tunnel, and its ends have been sutured. (From Boyes, J.H.: J. Bone Joint Surg. **37-A:**920, 1955.)

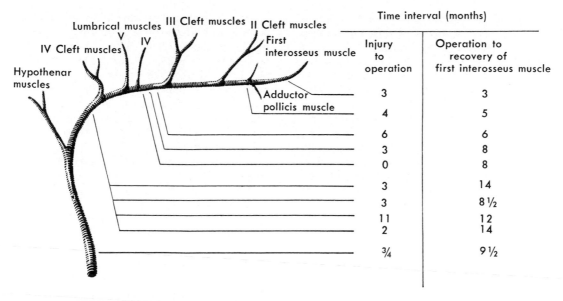

Fig. 5-6. Rate of recovery of voluntary function of first dorsal interosseus muscle after repair of deep branch of ulnar nerve in 10 patients. (Modified from Boyes, J.H.: J. Bone Joint Surg. **37-A**:920, 1955.)

dles will allow branches to be saved and yet permit the nerve to be rerouted. Freshen the ends of the nerve (Fig. 5-5, *B*) and approximate them and their sheath with interrupted No. 8-0 nylon sutures on small curved cutting needles. Suture the volar carpal ligament, replace the insertion of the palmaris brevis, and close the wound.

According to Boyes, the results are proportional to the accuracy of the approximation and inversely proportional to the scarring and fibrosis. Regeneration takes place in an orderly way; the recovery of nerve function may be tested by noting voluntary activity of the first dorsal interosseus muscle (Fig. 5-6).

Median nerve at wrist

Division of the median nerve at the wrist is not unusual, and the vital sensory function of the hand depends on its successful repair. Again we emphasize that (1) the neuroma must be carefully excised from the proximal end to provide healthy fibers and the glioma must be excised from the distal end, (2) surrounding scar must be excised to provide a vascular bed, (3) the repair must be accurate, with the ends in proper rotation, for the nerve contains motor as well as sensory fibers, and (4) tension on the repair must be avoided.

The following points of technique are suggested. A vessel usually lies on the anterior surface of the median nerve parallel with its fibers; this vessel may be of help in securing proper rotational alignment, or it may be obliterated by scar when the repair is late. An epineurial suture in each segment as described for the ulnar nerve at the wrist (p. 123) may aid in obtaining proper rotation. Tension may be reduced by dissecting and mobilizing the nerve proximally in the forearm and by flexing the wrist and elbow. The nerve should be repaired with No. 8-0 or 9-0 nylon on a noncutting atraumatic curved needle as described for digi-

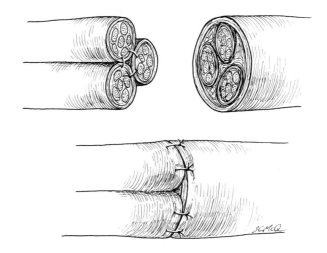

Fig. 5-7. Bundle suture for segmental gap (see text).

tal nerves (p. 121), but more than four sutures should be used for a secure repair.

A nerve graft (p. 127) is indicated when the gap between the nerve ends is so wide that the ends cannot be brought together without excessive tension. When a tendon and the median nerve are both sutured secondarily, the volar carpal ligament should be released to help prevent constriction from scarring.

Median nerve in palm

When the median nerve is divided where it branches in the palm, it may sometimes be repaired with a bundle suture (Fig. 5-7). This suture gathers the several branches of the nerve into a single trunk so that it in turn can be sutured to the proximal segment of the nerve.

Every effort should be made to repair the recurrent branch of the median nerve. It may be hard to find because of surrounding fascia and scar tissue, but once it is seen it can be readily identified by its yellow fibers running transversely toward the base of the thumb. This branch usually projects from the main trunk radially and superficially, passing just over the distal margin of the transverse carpal ligament. It courses slightly posteriorly and laterally to innervate the thenar muscles. There are several important anatomic variations so that this recurrent branch may be represented by two branches instead of one; it may come off the ulnar side of the trunk, and it may perforate the distal portion of the transverse carpal ligament. It is repaired with the technique described for digital nerves (p. 121); the prognosis is good if careful attention is given to anatomic detail.

When the median nerve cannot be repaired, a neurovascular island graft as described next may be indicated.

Neurovascular island grafts

That any digit deprived of sensibility is selectively and unconsciously avoided during use of the hand has become increasingly clear. Therefore, restoration of sensibility to a selected area of a given digit by transfer of a neurovascular island graft is useful at times. In permanent nerve damage, sensibility can be restored to critical areas, especially on the thumb or index finger. In osteoplastic reconstruction of the thumb (p. 199), transfer of a neurovascular island graft is essential; in fact, this reconstruction, once discarded, is now useful when combined with such a transfer. However, sensibility in the graft is never normal after transfer. In those grafts critically examined some time after surgery, sensibility was found to be abnormal in all; in more than half the skin was persistently hyperesthetic, and in all, precise sensory reorientation was incomplete. Reorientation, although it need not be normal for a good functional result, seems to improve with time and with use of the part.

Transfer of a neurovascular island graft may be indicated in permanent sensory deficit on the radial side of an otherwise normal index finger or on the area of pinch of the thumb. Before decision for surgery is made, the following factors must be considered: (1) the dominance of the involved hand, (2) the presence of any scarring in the palm through which an incision must be made for channeling of the neurovascular bundle, (3) the status of the ipsilateral ulnar nerve, (4) the condition of the opposite hand, (5) the age of the patient, and (6) the experience of the surgeon.

Early descriptions of the operation suggested transfer of skin only from the ulnar side of the distal phalanx of the ring finger. However, experience has shown that skin from an entire side of the donor finger should be included in the transfer. This larger transfer increases the area of sensitive skin on the recipient digit and causes no wider sensory loss on the donor digit; of little consequence is the larger free graft required to cover the donor area.

The operation is not so difficult that death of the transferred graft is likely; yet even temporary impairment of the circulation usually causes a permanent and disabling sensory deficit in the graft and thus a partial failure of the operation. Therefore in handling the neurovascular bundle, several points in technique must be emphasized: (1) the bundle, including all veins, should be dissected from proximally to distally so that any anomalies of the vessels may be properly treated; (2) the bundle should not be completely freed from the surrounding fatty tissue, especially at the base of the finger, but should be transferred along with some attached tissue; and (3) the bundle should be channeled through an incision large enough to show the entire bundle and any kinking, twisting, or stretching of the nerve or vessels.

TECHNIQUE. Using a skin pencil, accurately outline the area of sensory deficit on the thumb and prepare to remove skin from a similar area on the ulnar side of the ring finger (or if desired, use as the donor area the radial side of the litte finger or, in the absence of median nerve damage, the ulnar side of the middle finger). If the entire palmar surface of the thumb is insensitive, outline on the ring finger the maximum donor area for transfer. Shape the donor area to include most of the ulnar side of the finger, with darts to near the midline on the palmar and dorsal surfaces between the finger joints. The area thus outlined will include skin supplied by the dorsal branch of the proper digital nerve and will be shaped to prevent tension on the resulting scars during finger movements.

Now inflate a tourniquet on the arm. Then beginning proximally near the base of the palm, make a zigzagged incision distally to the fourth web (Fig. 5-8). Identify and dissect free, along with some surrounding tissue, the common volar digital artery and nerve to the ring and little fingers and the proper digital artery and nerve to the ulnar side of the ring finger. Ligate and divide the proper digital artery to the radial side of the little finger. Next carefully split proximally from the common volar digital nerve the proper digital nerve to the ulnar side of the ring finger. Continue the dissection distally and excise, with this attached neurovascular bundle, the area of skin from the ring finger previously outlined; take special care not to damage the arterial tree and to preserve as many veins as possible. Divide and ligate any small branches of the artery as necessary. Now free the composite graft and carry the island graft across the palm to the recipient area on the thumb; be sure the neurovascular bundle is long enough to permit the transfer without tension on the bundle. The island graft should cover most of the pulp area on the palmar aspect of the thumb and should extend to the ulnar aspect of the digit but not to the distal edge of the nail. Now beginning at the proximal end of the original excision and proceeding to the thumb, make a zigzagged incision across the palm conforming to the skin creases. Excise from the thumb the previously outlined area of skin, and if large enough, save it to be used later as a free graft on the donor finger. Now suture the island graft in place on the thumb. Carefully check the entire neurovascular bundle for stretching, kinking, or twisting and close the palmar incisions. Next cover the donor area of the finger with a full-thickness graft from the recipient thumb, free of fat, or with a thick split graft obtained elsewhere; cover this graft with a stent dressing. Now release the tourniquet and hold the wrist in slight flexion and the thumb in the best position to eliminate tension on the transferred bundle. Carefully observe the island graft for evidence of return of circulation. Remember that

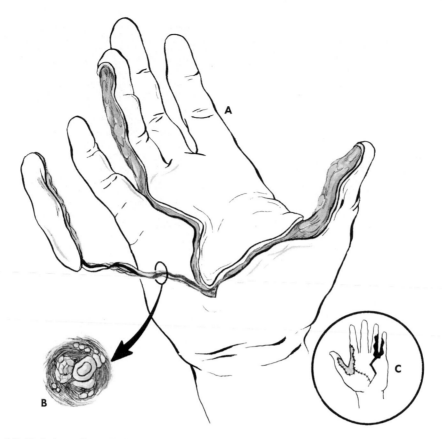

Fig. 5-8. Technique of transferring neurovascular island graft. **A,** Palmar incision has been made, neurovascular island graft has been excised from ulnar surface of ring finger and its bundle has been freed proximally, and insensitive skin has been excised from palmar surface of thumb (see text). **B,** Cross section of neurovascular bundle. **C,** Alternative technique in which neurovascular island graft includes adjacent surfaces of ring and little fingers and area covered by it is larger as shown.

vascular spasm may cause ischemia of the graft for a few minutes. The graft eventually should become cherry pink; if it does not, check again the positions of the wrist and thumb and, if necessary, reopen part of the palmar incision and explore the transferred bundle for kinking.

This procedure may be altered as necessary to meet other given requirements. For instance, in complete median nerve paralysis, if sensibility on the ulnar edge of the thumb pulp is reasonably good as a result of overlap of innervation from the radial nerve, transfer of the island graft to the radial side of the proximal and middle phalanges of the index finger may be desirable. This area of the finger is used especially in strong pinch.

AFTERTREATMENT. A bulky dressing and a dorsal plaster splint are applied that hold the wrist, thumb, and fingers in flexion. The hand is elevated constantly for 4 or 5 days after surgery.

Nerve grafts

Sometimes so much of a nerve has been destroyed that the defect cannot be overcome by mobilizing the nerve, by flexing adjacent joints, or by rerouting the nerve. A nerve graft is then the only recourse; an exception is the use of a neurovascular island graft (just described) when the median nerve has been destroyed. The nerve graft must be autogenous and must be taken from a nerve whose loss of

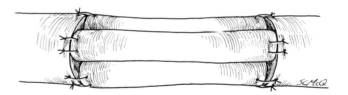

Fig. 5-9. Cable graft is used with bundle suture technique (see text).

function will result in only minimal disability. To survive, the graft must be small in diameter; otherwise central necrosis will result from insufficient nutrition. A graft from the sural nerve is satisfactory and may be obtained through one or more transverse incisions over the proximal and middle calf; a digital nerve from an amputated finger is also satisfactory. A nerve segmental defect in a single digit may be overcome by using a segment of the severed nerve on the opposite, but less critical, side. For instance, should both nerves be cut in the long finger and retracted so as to make it necessary to graft both nerves, the graft may be removed from the ulnar side of the digit to replace the more critical radial side. Obviously this gives no potential for return of sensiblity to the ulnar side; however, it does not involve another area of surgery for the donation of the graft. When the nerve to be grafted is thicker than the nerve intended for the graft, a cable graft may be constructed by making a bundle of two or three strands of the graft; the ends of these grafts are sutured to the ends of the larger nerve (Fig. 5-9). Each small strand is then more likely to obtain sufficient nutrition than would one large graft. A nerve graft must be placed in a satisfactory bed free of scar. It must not be encased in a vein, a plastic tube, or any other material, for this occludes its nutrition. Of course, a nerve graft requires a suture line at each end, and this fact, combined with the chance of necrosis of the graft, reduces the likelihood of an excellent result. But when the proper technique is used, useful function can sometimes be restored. It should be kept in mind that a nerve stripped of more than 15 cm of its mesoneurium has lost its vascular supply and functions as a free graft.

In recent years it has been emphasized that nerve repairs, including nerve grafts, should be done with minimal or no tension. A nerve graft should be of sufficient length to permit the extremity to be placed in extension without tension. Millesi believes that tension is a great detriment to nerve regeneration.

REFERENCES

Adamson, J.E., Horton, C.E., and Crawford, H.H.: Sensory rehabilitation of the injured thumb, Plast. Reconstr. Surg. **40**:53, 1967.

Aschan, W., and Moberg, E.: The Ninhydrin finger printing test used to map out partial lesions to hand nerves. Acta Chir. Scand. **123**:365, 1962.

Berger, A., and Millesi, H.: Nerve grafting, Clin. Orthop. **133**:49, 1978.

Bora, W.F., Pleasure, D.E., and Didizian, N.A.: A study of nerve regeneration and neuroma formation after nerve suture by various techniques, J. Hand Surg. **1**:138, 1976.

Boyes, J.H.: Repair of the motor branch of the ulnar nerve in the palm, J. Bone Joint Surg. **37-A**:920, 1955.

Bralliar, F.: Electromyography: its use and misuse in peripheral nerve injuries, Orthop. Clin. North Am. **12**:229, 1981.

Bromage, P.R.: Nerve physiology and control of pain, Orthop. Clin North Am. **4**:897, 1973.

Broudy, A.S., Leffert, R.D., and Smith, R.J.: Technical problems with ulnar nerve transposition at the elbow: findings and results of reoperation, J. Hand Surg. **3**:85, 1978.

Burkhalter, W.E., and Carneiro, R.S.: Correction of the attritional boutonniere deformity in high ulnar-nerve paralysis, J. Bone Joint Surg. **61-A**:131, 1979.

Cabaud, H.E., Rodkey, W.G., and McCarroll, H.R., Jr.: Peripheral nerve injuries: studies in higher nonhuman primates, J. Hand Surg. **5**:201, 1980.

Cabaud, H.E., et al.: Epineurial and perineurial fascicular nerve repairs: a critical comparison, J. Hand Surg. **1**:131, 1976.

Chacha, P.B., Krishramurti, A., and Soin, K.: Experimental sensory reinnervation of the median nerve by nerve transfer in monkeys, J. Bone Joint Surg. **59-A**:386, 1977.

Dellon, A.L.: Reinnervation of denervated Meissner corpuscles: a sequential histologic study in the monkey following fascicular nerve repair, J. Hand Surg. **1**:98, 1976.

Dellon, A.L.: The moving two-point discrimination test: clinical evaluation of the quickly adapting fiber/receptor system, J. Hand Surg. **3**:474, 1978.

Dellon, A.L.: Clinical use of vibratory stimuli to evaluate peripheral nerve injury and compression neuropathy, Plast. Reconstr. Surg. **65**:466, 1980.

Dobyns, J.H., et al.: Bowler's thumb: diagnosis and treatment: a review of seventeen cases, J. Bone Joint Surg. **54-A**:751, 1972.

Dolich, B.M., Olshansky, K.J., and Babar, A.H.: Use of a cross-forearm neurocutaneous flap to provide sensation and coverage in hand reconstruction, Plast. Reconstr. Surg. **62**:550, 1978.

Doyle, J.R., Semenza, J., and Gilling, B.: The effect of succinylocholine on denervated skeletal muscle, J. Hand Surg. **6**:40, 1981.

Dunham, W., Haines, G., and Spring, J.M.: Bowler's thumb (ulnovolar neuroma of the thumb), Clin. Orthop. **83**:99, 1972.

Edgerton, M.T.: Cross-arm nerve pedicle flap for reconstruction of major defects of the median nerve, Surgery **64**:248, 1968.

Eisen, A.A.: Electromyography and nerve conduction as a diagnostic aid, Orthop. Clin. North Am. **4**:885, 1973.

Finseth, F., Constable, J.D., and Cannon, B.: Interfascicular nerve grafting: early experiences at the Massachusetts General Hospital, Plast. Reconstr. Surg. **56**:492, 1975.

Flynn, J.E., and Flynn, W.F.: Median and ulnar nerve injuries: a long-range study with evaluation of the Ninhydrin test, sensory and motor returns, Ann. Surg. **156**:1002, 1962.

Frykman, G.K., Adams, J., and Bowen, W.W.: Neurolysis, Orthop. Clin. North Am. **12**:325, 1981.

Frykman, G.K., and Waylett, J.: Rehabilitation of peripheral nerve injuries, Orthop. Clin. North Am. **12**:361, 1981.

Frykman, G.K., Wolf, A., and Coyle, T.: An algorithm for management of peripheral nerve injuries, Orthop. Clin. North Am. **12**:239, 1981.

Gellis, M., and Pool, R.: Two-point discrimination distances in the normal hand and forearm: application to various methods of fingertip reconstruction, Plast. Reconstr. Surg. **59**:57, 1977.

Gordon, L., and Buncke, H.J.: Heterotopic free skeletal muscle autotransplantation with utilization of a long nerve graft and microsurgical techniques: a study in the primate, J. Hand Surg. **4**:103, 1979.

Gordon, L., et al.: Predegenerated nerve autografts as compared with fresh nerve autografts in freshly cut and precut motor nerve defects in the rat, J. Hand Surg. **4**:42, 1979.

Grabb, W.C.: Management of nerve injuries in the forearm and hand, Orthop. Clin. North Am. **2**:419, 1970.

Hakstian, R.W.: Perineural neurorrhaphy, Orthop. Clin. North Am. **4**:945, 1973.

Hill, H.L., Vasconez, L.O., and Jurkiewicz, M.J.: Method for obtaining a sural nerve graft, Plast. Reconstr. Surg. **61**:177, 1978.

Holevich, J.: A new method of restoring sensibility to the thumb, J. Bone Joint Surg. **45-B**:496, 1963.

Howell, A.E., and Leach, R.E.: Bowler's thumb: perineural fibrosis of the digital nerve, J. Bone Joint Surg. **52-A**:379, 1970.

Inglis, A.E., Straub, L.R., and Williams, C.S.: Median nerve neuropathy at the wrist, Clin. Orthop. **83**:48, 1972.

Jabaley, M.E., et al.: Comparison of histologic and functional recovery after peripheral nerve repair, J. Hand Surg. **1**:119, 1976.

Jones, E.T., and Louis, D.S.: Median nerve injuries associated with supracondylar fractures of the humerus in children, Clin. Orthop. **150**:181, 1980.

Kleinert, H.E., and Griffin, J.M.: Technique of nerve anastomosis, Orthop. Clin. North Am. **4**:907, 1973.

Kleinert, H.E., et al.: Post-traumatic sympathetic dystrophy, Orthop. Clin. North Am. **4**:917, 1973.

Krag, C., and Rasmussen, K.B.: The neurovascular island flap for defective sensibility of the thumb, J. Bone Joint Surg. **57-B**:495, 1975.

Kutz, J.E., Shealy, G., and Lubbers, L.: Interfascicular nerve repair, Orthop. Clin. North Am. **12**:277, 1981.

Laing, P.G.: The timing of definitive nerve repair, Surg. Clin. North Am. **40**:363, 1960.

Leonard, M.H.: Return of skin sensation in children without repair of nerves, Clin. Orthop. **95**:273, 1973.

Levin, S., Pearsall, G., and Ruderman, R.J.: Von Frey's method of measuring pressure sensibility in the hand: an engineering analysis of Weinstein-Semmes pressure aesthesiometer, J. Hand Surg. **3**:211, 1978.

Lewin, M.L.: Repair of digital nerves in lacerations of the hand and the fingers, Clin. Orthop. **16**:227, 1960.

Louis, D.S.: Nerve function evaluation, Am. Soc. Surg. Hand News **3**(Suppl. B):1, 1984.

Lundborg, G., and Hansson, H.A.: Nerve regeneration through preformed synovial tubes, J. Hand Surg. **5**:35, 1980.

Mackenzie, I.G.: Causes of failure after repair of the median nerve, J. Bone Joint Surg. **43-B**:465, 1961.

Markley, J.M., Jr.: The preservation of close two-point discrimination in the interdigital transfer of neurovascular island flaps, Plast. Reconstr. Surg. **59**:812, 1977.

McFarlane, R.M., and Mayer, J.R.: Digital nerve grafts with the lateral antebrachial cutaneous nerve, J. Hand Surg. **1**:169, 1976.

Millesi, H.: Surgical management of brachial plexus injuries, J. Hand Surg. **2**:367, 1977.

Millesi, H.: Interfascicular nerve grafting, Orthop. Clin. North Am. **12**:287, 1981.

Minkow, F.V., and Bassett, F.H., III: Bowler's thumb, Clin. Orthop. **83**:115, 1972.

Moberg, E.: Evaluation of sensibility in the hand, Surg. Clin. North Am. **40**:357, 1960.

Moberg, E.: Aspects of sensation in reconstructive surgery of the upper extremity, J. Bone Joint Surg. **46-A**:817, 1964.

Moberg, E.: Evaluation and management of nerve injuries in the hand, Surg. Clin. North Am. **44**:1019, 1964.

Moore, J.R., and Weiland, A.J.: Bilateral attritional rupture of the ulnar nerve at the elbow, J. Hand Surg. **5**:358, 1980.

Morrison, W.A., et al.: Neurovascular free flaps from the foot for innervation of the hand, J. Hand Surg. **3**:235, 1978.

Nahai, F., and Wolf, S.L.: Percutaneous recordings from peripheral nerves (preliminary communication), J. Hand Surg. **3**:168, 1978.

Narakas, A.: Surgical treatment of traction injuries of the brachial plexus, Clin. Orthop. **133**:71, 1978.

Narakas, A.: Brachial plexus surgery, Orthop. Clin. North Am. **12**:303, 1981.

Nicolle, F.V., Chir, B., and Woolhouse, F.M.: Restoration of sensory function in severe degloving injuries of the hand, J. Bone Joint Surg. **48-A**:1511, 1966.

O'Connor, R.L.: Digital nerve compression secondary to palmar aneurysm, Clin. Orthop. **83**:149, 1972.

Omer, G.E., Jr.: Injuries to nerves of the upper extremity, J. Bone Joint Surg. **56-A**:1615, 1974.

Omer, G.E., Jr.: Sensation and sensibility in the upper extremity, Clin. Orthop. **104**:30, 1974.

Omer, G.E., Jr.: Physical diagnosis of peripheral nerve injuries, Orthop. Clin. North Am. **12**:207, 1981.

Omer, G.E., Jr., and Spinner, M.: Peripheral nerve testing and suture techniques. In American Academy of Orthopaedic Surgeons: Instructional course lectures, vol. 24, St. Louis, 1975, The C.V. Mosby Co.

Omer, G.E., Jr., and Thomas, S.R.: The management of chronic pain syndromes in the upper extremity, Clin. Orthop. **104**:37, 1974.

Omer, G.E., Jr., et al.: Neurovascular cutaneous island pedicles for deficient median-nerve sensibility: new technique and results of serial functional tests, J. Bone Joint Surg. **52-A**:1181, 1970.

Orgel, M.G., and Terzis, J.K.: Epineural versus perineural repair: an ultrastructural and electrophysiological study of nerve regeneration, Plast. Reconstr. Surg. **60**:80, 1977.

Peacock, E.E., Jr.: Restoration of sensation in hands with extensive median nerve defects, Surgery, **54**:576, 1963.

Poppen, N.K., et al.: Recovery of sensibility after suture of digital nerves, J. Hand Surg. **4**:212, 1979.

Rodkey, W.G., Cabaud, H.E., and McCarroll, H.R., Jr.: Neurorrhaphy after loss of a nerve segment: comparison of epineurial suture under tension versus multiple nerve grafts, J. Hand Surg. **5**:366, 1980.

Rydevik, B., Lundborg, G., and Bagge, U.: Effects of graded compression on intraneural blood flow, J. Hand surg. **6**:3, 1981.

Schuler, F.A., III, and Adamson, J.R.: Pacinian neuroma, an unusual cause of finger pain, Plast. Reconstr. Surg. **62**:576, 1978.

Starkweather, R.J., et al.: The effect of devascularization on the regeneration of lacerated peripheral nerves: an experimental study, J. Hand Surg. **3**:163, 1978.

Stromberg, W.B., Jr., et al.: Injury of the median and ulnar nerves: one hundred and fifty cases with an evaluation of Moberg's Ninhydrin test, J. Bone Joint Surg. **43-A**:717, 1961.

Stromberg, B.V., Vlastou, C. and Earle, A.S.: Effect of nerve graft polarity on nerve regeneration and function, J. Hand Surg. **4**:444, 1979.

Sunderland, S.: The restoration of median nerve function after destructive lesions which preclude end-to-end repair, Brain **97**:1, 1974.

Sunderland, S.: The pros and cons of funicular nerve repair, J. Hand Surg. **4**:20, 1979.

Sunderland, S.: The anatomic foundation of peripheral nerve repair techniques, Orthop. Clin. North Am. **12**:245, 1981.

Swanson, A.B.: Ulnar nerve compression due to an anomalous muscle in the canal of Guyon, Clin. Orthop. **83**:64, 1972.

Swezey, F.L., Bjarnason, D., and Austin, E.S.: Nerve conduction studies in resorptive arthropathies: opera-glass hand, J. Bone Joint Surg. **55-A**:1680, 1973.

Snyder, C.C.: Epineurial repair, Orthop. Clin. North Am. **12**:267, 1981.

Taylor, G.I.: Nerve grafting with simultaneous microvascular reconstruction, Clin. Orthop. **133**:56, 1978.

Terzis, J.K., Dykes, R.W., and Hakstian, R.W.: Electrophysiological recordings in peripheral nerve injury: a review, J. Hand Surg. **1**:52, 1976.

Terzis, J.K., and Strauch, B.: Microsurgery of the peripheral nerve: a physiological approach, Clin. Orthop. **133**:39, 1978.

Uriburu, I.J.F., Morchio, F.J., and Marin, J.C.: Compression syndrome of the deep motor branch of the ulnar nerve (piso-hamate hiatus syndrome). J. Bone Joint Surg. **58-A**:145, 1976.

Wilgis, E.F.S., and Maxwell, G.P.: Distal digital nerve grafts: clinical and anatomical studies, J. Hand Surg. **4**:439, 1979.

Wilson, R.L.: Management of pain following peripheral nerve injuries, Orthop. Clin. North Am. **12**:343, 1981.

Winsten, J.: Island flap to restore stereognosis in hand injuries, N. Engl. J. Med. **268**:124, 1963.

CHAPTER 6

Dislocations and ligamentous injuries

Most dislocations of the joints of the hand can be treated by closed means. Exceptions are (1) unstable dislocations of the carpometacarpal joints of the thumb or fingers, (2) dislocation of the metacarpophalangeal joint of the thumb with a *complete* rupture of the ulnar collateral ligament, which renders pinch weak and unstable, (3) dislocation in which a tendon is trapped, (4) old undiagnosed dislocations, and (5) buttonhole dislocations.

DISLOCATIONS OF CARPOMETACARPAL JOINTS
Carpometacarpal dislocation of thumb

Carpometacarpal dislocation of the thumb is usually accompanied by fracture of the hook of the first metacarpal and is included in the discussion of Bennett's fracture (p. 76). When it occurs without fracture and is recognized early, the dislocation should be reduced and the joint should be immobilized for 4 to 6 weeks to prevent recurrent dislocation. Open reduction and repair of the dorsal and radial ligaments ensure better joint stability. Immobilization for 6 weeks is still indicated after the repair.

Recurrent dislocation of carpometacarpal joint of thumb

In recurrent dislocation or subluxation of the carpometacarpal joint of the thumb, either idiopathic or traumatic,

construction of a ligament to reinforce the deep capsule of the joint may be indicated. The operation is most helpful, of course, when the joint is unstable and painful and when degeneration of its articular surfaces is minimal. This procedure should not be done to relieve symptoms or subluxations of this joint from osteoarthritis.

TECHNIQUE (EATON AND LITTLER). Make a dorsoradial incision along the proximal half of the first metacarpal and curve its proximal end ulnarward around the base of the thenar eminence parallel with the distal flexor crease of the wrist. Next expose the carpometacarpal joint of the thumb subperiosteally and the volar aspect of the greater multangular extraperiosteally. Isolate the distal part of the flexor carpi radialis tendon from its position on the ulnar aspect of the crest of the greater multangular. Now in the distal forearm expose the same tendon through a longitudinal incision and split from its radial side a strip of tendon 6 cm long; free the strip proximally, continue the split distally, and leave the strip attached to the base of the second metacarpal (Fig. 6-1).

Before proceeding further, reduce the first metacarpal on the greater multangular and pass a Kirschner wire through this joint while holding it in appropriate orientation. Care should be taken in placing the wire so as not to interfere with the site where the transverse hole will be drilled through the first metacarpal and through which the tendon transfer will eventually pass.

Now reroute the strip of tendon previously raised from behind the crest of the greater multangular and pass it directly from the base of the second metacarpal to that of the first. Then beginning at the normal site of attachment of the deep capsule of the carpometacarpal joint of the thumb, drill a hole dorsally through the base of the first metacarpal to emerge on the ulnar side of the extensor pollicis brevis tendon. Now pass the strip of tendon through this hole, loop it back deep to the abductor pollicis longus tendon, draw it tight, and suture it to the periosteum near its exit. Finally, loop the tendon strip around the flexor carpi radialis near its insertion and suture it to the base of the first metacarpal.

AFTERTREATMENT. The thumb is immobilized for 4 weeks in extension and abduction.

Carpometacarpal dislocations of finger rays

Since carpometacarpal dislocations are usually associated with fractures, they are discussed on p. 81.

131

DORSAL DISLOCATION OF
METACARPOPHALANGEAL JOINT OF THUMB

Dislocation of any of the metacarpophalangeal joints is possible from hyperextension injuries, but dorsal dislocation of the metacarpophalangeal joint of the thumb is the most common (Fig. 6-2). Early closed reduction may be easy, provided the thumb is maintained in adduction to relax its intrinsic muscles. One method is to use minimal if any tension, hyperextend the metacarpophalangeal joint, and then with the examiner's thumb push forward the proximal end of the proximal phalanx over the end of the metacarpal head. This tends to diminish the buttonhole effect on the metacarpal neck that traction accentuates. If this method is not successful, repeated attempts are contraindicated; open reduction should be done to disengage the head of the metacarpal from a buttonhole slit in the anterior capsule and from the flexor pollicis brevis muscle.

TECHNIQUE. Make a transverse curved incision over the radial and volar aspects of the joint, exposing the articular surfaces of the phalanx and metacarpal. The base of the phalanx lies on the dorsal aspect of the head and neck of the metacarpal, and the head protrudes through the anterior capsule. Disengage the flexor pollicis brevis muscle, releasing the head of the metacarpal. Flex the thumb and push the head through the rent in the capsule to complete the reduction.

AFTERTREATMENT. The thumb is held in moderate flexion by a plaster splint. After 3 weeks the splint is removed and active motion is begun.

• • •

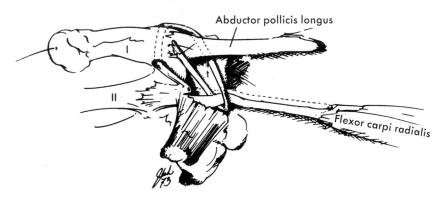

Fig. 6-1. Volar and radial ligament reconstruction with a strip from tendon of flexor carpi radialis, which is left attached at its insertion at base of second metacarpal. Course of tendon strip creates a reinforcement of volar, dorsal, and radial aspects of joint. (From Eaton, R.G., and Littler, J.W.: J. Bone Joint Surg. **55-A:**1655, 1973.)

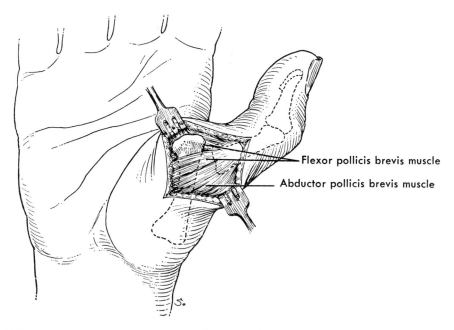

Fig. 6-2. Dislocation of metacarpophalangeal joint of thumb. Metacarpal head has penetrated joint capsule in such a way that were traction applied to thumb, metacarpal neck would be caught by capsule, and reduction would be impossible. Traction should not be applied; rather, metacarpal should be adducted and dislocated joint should be hyperextended while proximal end of proximal phalanx is pushed against and then over metacarpal head.

Farabeuf in 1876 recommended a dorsal surgical approach in irreducible dislocation of the metacarpophalangeal joint of the thumb. The main obstacle to reduction is the dorsally displaced volar plate, and the dorsal approach provides access to the plate, which is tethered tightly over the metacarpal head and neck. The plate is a fibrocartilaginous structure similar in appearance to articular cartilage; it is split in the midline of the thumb so that it will slip around the metacarpal head and permit reduction of the phalanx. Motion is started within a few days after surgery.

IRREDUCIBLE PALMAR DISLOCATION OF METACARPOPHALANGEAL JOINT OF THUMB

Palmar dislocation of the proximal phalanx of the thumb is rare, but sometimes it may be irreducible. Gunther and Zielinski reported a case requiring an open reduction (Figs. 6-3 to 6-5) in which the metacarpal head was trapped between the extensor pollicis longus and extensor pollicis brevis tendons. This required opening the dorsal aponeurosis to relocate the extensor tendon. A rupture of the ulnar collateral ligament was also present and apparently the extensor pollicis longus tendon was trapped at the proximal ligamentous stump on the metacarpal head. This ligament was also repaired after reduction was accomplished.

COMPLETE RUPTURE OF ULNAR COLLATERAL LIGAMENT OF METACARPOPHALANGEAL JOINT OF THUMB

Incomplete rupture of the ulnar collateral ligament of the thumb is common and needs only proper rest for restoration of function, although pain and swelling may persist for several weeks. A thumb spica cast may be indicated.

Complete rupture of the ulnar collateral ligament is more common than that of the radial and of course is more disabling because it renders pinch unstable. It may be suspected clinically by comparing the stability of the joint with that of the opposite thumb. Pathologic rotation of the thumb is also evident. The rupture can be demonstrated by an anteroposterior roentgenogram of the joint while under lateral stress as compared with a similar view of the opposite normal joint.

The ligament may avulse a bony fragment from the phalanx. When this fragment is not widely displaced, it may be treated by applying a thumb spica for 6 weeks.

If a complete rupture is acute, the ligament should be repaired (Fig. 6-6), but when the diagnosis is delayed for a month or longer, fibrosis makes identification and repair of the ligament more difficult but still possible at times. Then a repair might be done by dissecting out the ligament from within the fibrotic mass and reattaching it appropriately. At that time the detached tendinous insertion of the adductor muscle might be advanced and reattached to furnish a dynamic reinforcement. When the repair is done several months after the injury, a graft may be used to replace the ligament. This may be either a boxlike graft with a strip of fascia or palmaris longus tendon passed through both the proximal and distal attachments of the ligament or a graft from the extensor pollicis brevis tendon, either split or in total, threaded through bone and attached by pull-out sutures to reconstruct the ligament. Arthrodesis of the metacarpophalangeal joint may be indicated when there are arthritic changes within the joint or gross disruption of the joint.

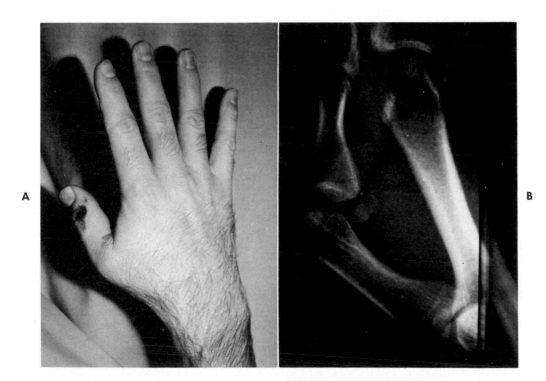

Fig. 6-3. A, Skin abrasion and metacarpophalangeal joint dislocation. **B,** Roentgenogram reveals palmar dislocation of proximal phalanx and widened joint space. (From Gunther, S.F., and Zielinski, C.J.: J. Hand Surg. **7:**515, 1982.)

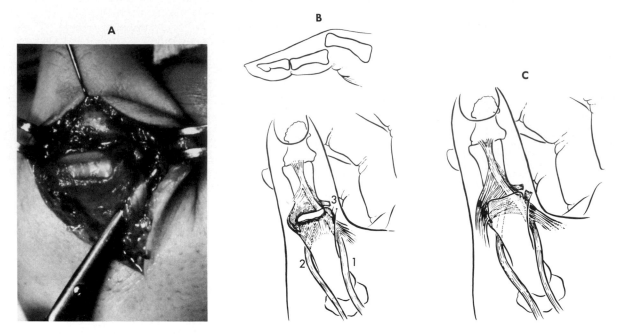

Fig. 6-4. Same patient as in Fig. 6-3. **A,** Surgical findings after incision of skin and subcutaneous fat. Dorsal aspect of metacarpal head is visible through rent in capsule. Whole extensor mechanism is trapped beneath metacarpal head. Hemostat is delivering, but cannot reduce, portion of extensor pollicis longus tendon. **B,** Visible findings: *1,* extensor pollicis longus tendon; *2,* extensor pollicis brevis tendon; and *3,* ruptured ulnar collateral ligament. **C,** Invisible anatomy that is hidden beneath metacarpal head. Bone has buttonholed proximal and dorsal to aponeurotic expansion that connects abductor pollicis brevis to extensor pollicis longus tendon. (From Gunther, S.F., and Zielinski, C.J.: J. Hand Surg. **7:**515, 1982.)

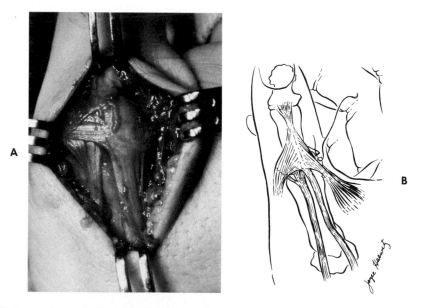

Fig. 6-5. Same patient as in Fig. 6-3. **A,** Appearance after reduction showing normal anatomy of extensor mechanism. **B,** Same showing single suture in ulnar collateral ligament. (From Gunther, S.F., and Zielinski, C.J.: J. Hand Surg. **7:**515, 1982.)

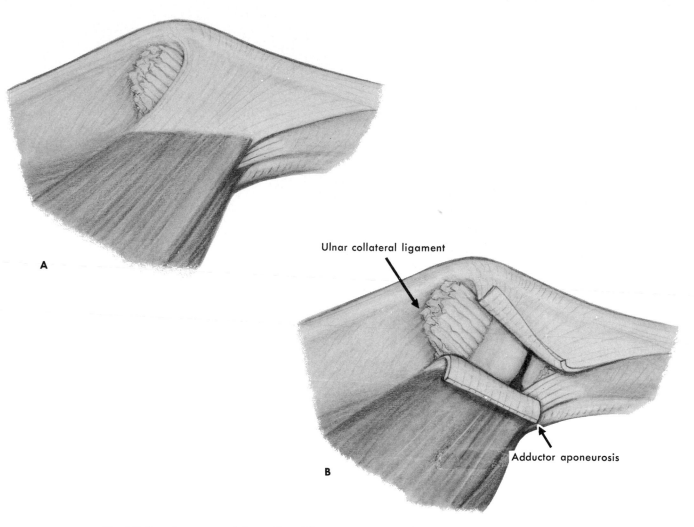

Fig. 6-6. Complete rupture of ulnar collateral ligament of metacarpophalangeal joint of thumb. **A,** Ligament is ruptured distally and is folded back so that its distal end points proximally. **B,** Adductor aponeurosis has been divided, exposing ligament and joint. (From Stener, B.: J. Bone Joint Surg. **44-B:**869, 1962.)

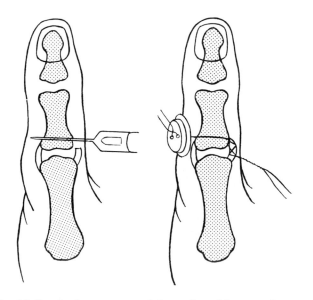

Fig. 6-7. Repair of acute rupture of ulnar collateral ligament of metacarpophalangeal joint of thumb (see text).

TECHNIQUE OF REPAIR BY SUTURE. To repair an acute rupture of the ulnar collateral ligament, make a slightly curved longitudinal incision convex dorsally over the dorsoulnar aspect of the metacarpophalangeal joint or a bayonet-shaped incision with the transverse segment at the joint level. Protect the terminal branches of the superficial radial nerve, which innervate the lateral margins of the thumb pulp. Identify them as they pass distally on each side at the dorsolateral aspect of the metacarpophalangeal joint deep to the subcutaneous fat. Identify the ligament, and at the site of the avulsion drill a small hole through the proximal end of the proximal phalanx, using a Kirschner wire of the smallest diameter as a drill point (Fig. 6-7). Place a Bunnell pull-out suture through the avulsed end of the ligament, pass the ends of the suture through the phalanx, and while holding the joint in slight flexion, tie them over a padded button on the radial side. Pass the twisted pull-out wire loop through the skin near the incision before closure. The same technique is used when a small bone fragment is avulsed by the ligament if the tear is complete and the bone fragment is displaced.

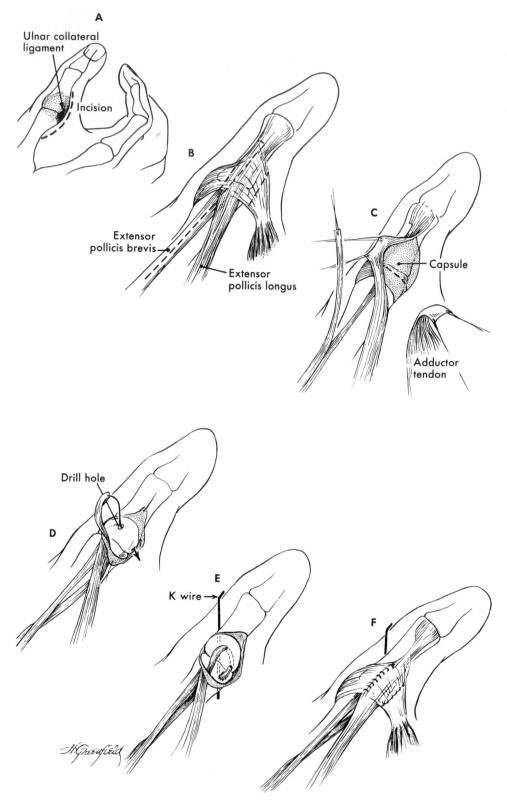

Fig. 6-8. Steps in procedure to reconstruct ulnar collateral ligament using one half of tendon of extensor pollicis brevis. (From Sakellarides, H.T., and DeWeese, J.W.: J. Bone Joint Surg. **58-A**:106, 1976.)

Repair of old ulnar collateral ligament rupture using extensor pollicis brevis tendon (Sakellarides et al.)

TECHNIQUE (FIG. 6-8). Make a Z-shaped incision over the metacarpophalangeal joint, *A*, so as to expose the dorsal as well as the ulnar side of this joint, *B*. Isolate the extensor pollicis brevis tendon and detach it approximately 5 cm proximal to its insertion into the base of the proximal phalanx, *C*. Open the expansion of the adductor tendon and inspect the joint, excising the fibrosed area of the torn ligament so as to expose adequately the ulnar aspect of the joint, *D*. Now drill a hole beginning radial to the insertion of the extensor pollicis brevis and exiting on the volar ulnar aspect of the proximal phalanx at the site of the normal ligamentous attachment, *D*. Pass the tendon through this hole and then drill another hole transversely across the neck of the metacarpal at the site of attachment of the collateral ligament. Now pass the tendon proximally across the joint into this hole and attach it to the bone with a wire pull-out suture over a button on the radial side, *E*. Carefully measure the tendon length and attach the pull-out suture before inserting it in the hole. Close the capsule and the expansion of the adductor muscle, close the wound, *F*, and maintain the thumb in a cast for 4 weeks. Following cast removal, protect the thumb for another 5 weeks in a soft protective dressing or a removable splint.

Repair of old rupture of ulnar collateral ligament of thumb (Neviaser et al.)

This procedure is applicable when the rupture is at least a month old and when difficulty in identifying the torn ligament is expected because of scar tissue. When there is crepitus or pain on a grinding type of manipulation of the joint, an arthrodesis should be done.

TECHNIQUE. Make a V-shaped or chevron incision over the ulnar aspect of the thumb (Fig. 6-9, *A*). Take care to protect the dorsal and volar sensory nerves. Reflect the adductor aponeurosis. The original tear in the capsule and the ulnar collateral ligament may be obliterated by scar (Fig. 6-9, *B*). Detach the tendon of the adductor pollicis from the sesamoid; shape a flap of the collateral ligament in the form of a U based proximally to be used to reef the scarred ligament when the joint is reduced. Drill a hole through the ulnar cortex of the proximal phalanx, approximately 1 cm distal to the metacarpophalangeal joint. After attaching the tightened capsule, advance and reinsert the adductor tendon distally through the hole using a pull-out wire technique (Fig. 6-9, *C*). Close the wound and immobilize the hand in plaster for a minimum of 4 weeks. Start progressive exercise when the plaster is removed.

Repair of old rupture of ulnar collateral ligament of metacarpophalangeal joint of thumb (R.J. Smith)

R.J. Smith points out in his series that the ulnar collateral ligament was ruptured in 77% of the cases and the radial collateral ligament in 23% of the cases. In 3 of the 21 cases, the ulnar collateral ligament was found to be superficial and proximal to the adductor apparatus as described previously by Stener. Because of a volar attachment of the ulnar collateral ligament to the proximal phalanx, Smith advises that there is also a factor of volar

instability permitting subluxation of the proximal phalanx. Early surgical treatment is, of course, preferable, but 3 weeks after injury or longer, a tendon graft by his technique described below is recommended.

TECHNIQUE (FIG. 6-10). Make a midlateral incision, 3 cm long, centered over the ulnar side of the metacarpophalangeal joint of the thumb, *A;* identify the sensory branches of the radial nerve and retract them dorsally. Expose the dorsal expansion of the extensor mechanism and transect it in the midlateral line to expose the joint capsule, *B* and *C*. A remnant of the ruptured ulnar collateral ligament is frequently found attached to the metacarpal neck. Make a 2.8 mm hole transversely across the base of the proximal phalanx, taking care to exit on the volar ulnar side of the phalanx, *D*. Obtain a tendon graft from the palmaris longus, attach a figure-of-eight suture to one end, pass the suture through the hole in the phalanx, *E*, and fix it so that the end of the tendon is fastened in the hole and the other end is free on the ulnar side. Pull the free end of the graft to the area of the metacarpal head, *F*, make parallel incisions longitudinally in the remnant of the ulnar collateral ligament, weave the graft through the portion of the ligament between the incisions, and suture it securely, *G*. When the collateral ligament remnant is not firm, attach the graft proximally through two adjacent drill holes connected by a tunnel in the metacarpal head. Pass the remaining free portion of the graft back distally on itself volar to the first strand and suture the parallel segments of the graft to each other, *G*. The volar attachment of the graft helps to counteract the tendency of the phalanx to subluxate volarward. Repair the capsule by overlapping and suturing, using the double-breasted technique, *H*. Suture the adductor expansion. Hold the thumb in plaster for 4 weeks and follow with a removable splint for an additional 5 weeks.

DISLOCATION OF METACARPOPHALANGEAL JOINT OF FINGERS

These injuries are less common than interphalangeal dislocations. They occur most often in the index finger, and Kaplan's description of it in this finger is vivid (see Fig. 6-12).

"The fibrocartilaginous plate breaks away in the region of its weakest attachment, at the neck of the volar aspect of the second metacarpal; and the flexor tendons in the vaginal ligament and the pretendinous band or the midpalmar fascia, which adheres to the vaginal ligament, are violently displaced to the ulnar side of the metacarpal head. Following this, the fibrocartilaginous plate of the joint is displaced over the head of the metacarpal, landing on the dorsum of this bone, where it becomes wedged between the base of the proximal phalanx and the head. The lateral collateral ligaments, which are now abnormally displaced, lock the phalanx in the abnormal position typical of this dislocation. At the same time the two groups of transverse fibers of the palmar fascia hold the head of the metacarpal, the distal group (the natatory ligament, which moves with the phalanx) applying pressure to the dorsum of the metacarpal head, while the proximal group (the superficial transverse ligament, which extends across the volar aspect of the metacarpal neck) applies pressure to the volar as-

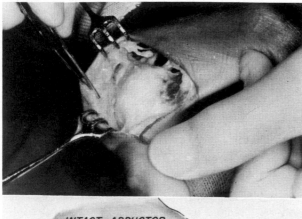

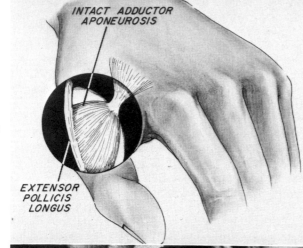

INTACT ADDUCTOR APONEUROSIS

EXTENSOR POLLICIS LONGUS

A

pect.''* Thus the dislocated metacarpal head lies between the natatory ligament and the superficial transverse ligament of the palmar fascia. The flexor tendons are on one side and the lumbrical muscle on the other.

When the dislocation is incomplete, reduction by manipulation is easy. When complete, with the head of the metacarpal displaced volarward and the base of the phalanx dorsalward, open reduction is often required. The major obstruction preventing reduction of the metacarpophalangeal joint is the displaced volar fibrocartilaginous plate lying dorsal to the metacarpal head. Sometimes, however, manipulation alone is successful. We find that 50% can be reduced closed. The joint is hyperextended, the articular surface of the proximal phalanx is forced firmly against the metacarpal neck, and while this force is maintained, the joint is flexed. Sometimes this maneuver will trap the displaced fibrocartilaginous plate and carry it to its normal position anterior to the metacarpal head. The following is Kaplan's technique for open reduction.

TECHNIQUE (KAPLAN). ''In open reduction the incision is started in the thenar crease of the hand at the radial base

---

*From Kaplan, E.B.: Dorsal dislocation of the metacarpophalangeal joint of the index finger, J. Bone Joint Surg. **39-A:**1081, 1957.

Fig. 6-9. Repair of old rupture of ulnar collateral ligament of thumb. (See text.) (From Neviaser, R.J., et al.: J. Bone Joint Surg. **53-A:**1357, 1971).

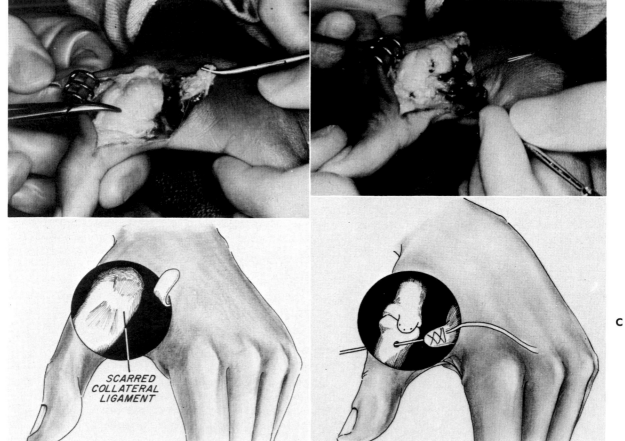

B

SCARRED COLLATERAL LIGAMENT

XXI

C

of the index finger and is continued into the proximal crease of the hand [Fig. 6-11]. . . . To reduce the dislocation following exposure, it is necessary to divide all the constricting bands. The first incision is made to free the constriction of the cartilaginous plate [Fig. 6-12]. This incision is placed parallel and radial to the vaginal ligament and extends from the free edge of the torn ligament to the junction of the periosteum with the proximal phalanx. The incision must penetrate the entire thickness of the plate. Division of the plate alone is not sufficient, however. The transverse fibers of the taut natatory ligament must also be completely divided, and following this, another longitudinal incision should be made through the transverse fibers of the superficial transverse metacarpal ligament. This third incision, which should extend to the ulnar side of the first lumbrical muscle, releases the constriction below the metacarpal head.''*

This triple incision ''frees the base of the proximal phalanx, which then returns to its normal place over the metacarpal head. This, in turn, permits the immediate replacement of the second metacarpal head in line with the other metacarpal heads, following which the flexor tendons, the vaginal ligament, and the nerves and vessels are restored to their normal positions. The wound is then closed in the accepted manner, and the finger is immobilized in functional position for about one week.''*

---

*From Kaplan, E.B.: Dorsal dislocation of the metacarpophalangeal joint of the index finger, J. Bone Joint Surg. **39-A:**1081, 1957.

TECHNIQUE (BECTON ET AL.). Becton et al. believe the dorsal approach has several advantages over the volar approach. The dorsal approach provides full exposure of the fibrocartilaginous volar ligament, which is the structure blocking reduction. The digital nerves are not as likely to be cut, and should there be an occult fracture of the metacarpal head, this can be reduced and fixed more easily (Fig. 6-13).

Over the metacarpophalangeal joint, make a 4 cm midline incision cutting the underlying extensor tendon and joint capsule as well. The fibrocartilaginous liagment may be difficult to identify because it has the same color as the articular cartilage, and its torn margin may not be visible. Make a small incision to ensure the tissue is in fact the fibrocartilaginous ligament; then complete the longitudinal incision (Fig. 6-14). Flex the wrist volarward to release the tension on the flexor tendons; then place traction on the finger and flex the metacarpophalangeal joint, reducing the dislocation. Observe to see if there is any free cartilage missing from the metacarpal head. This may be lodged in the joint. Suture the extensor tendon and skin and splint the finger for 3 weeks.

INTERPHALANGEAL DISLOCATIONS

Most interphalangeal dislocations are reduced immediately by the patient himself or by a bystander (Fig. 6-15). If a collateral ligament is not completely ruptured, joint motion can be reestablished when pain and swelling subside. In a young adult if one or both of the collateral ligaments are completely ruptured, they should be repaired,

Text continued on p. 143.

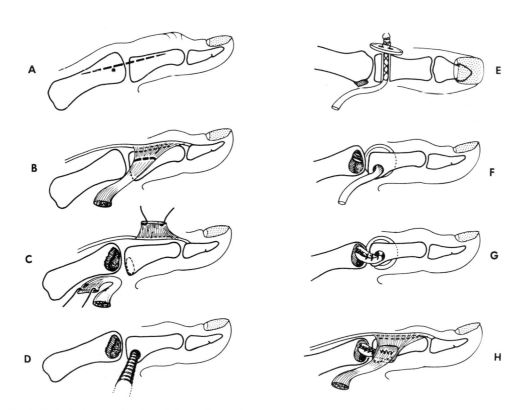

Fig. 6-10. Steps in reconstruction of ulnar collateral ligament for late posttraumatic instability. (From Smith, R.J.: J. Bone Joint Surg. **59-A:**13, 1977).

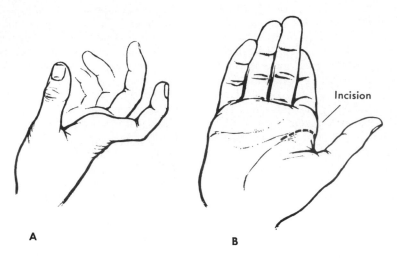

Fig. 6-11. Dislocation of second metacarpophalangeal joint. **A,** Deformity as seen from lateral side. **B,** Skin incision, *broken line,* used in open reduction. (From Kaplan, E.B.: J. Bone Joint Surg. **39-A:**1081, 1957.)

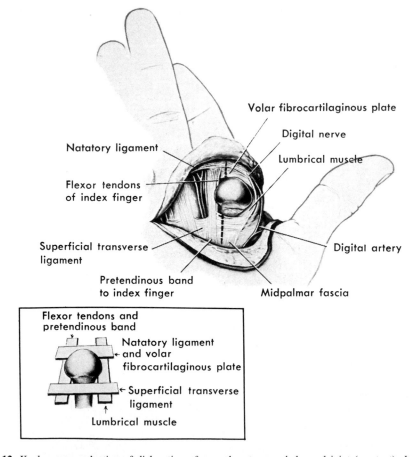

Fig. 6-12. Kaplan open reduction of dislocation of second metacarpophalangeal joint (see text). *Inset,* Diagram of four structures that surround and constrict metacarpal head. (Modified from Kaplan, E.B.: J. Bone Joint Surg. **39-A:**1081, 1957.)

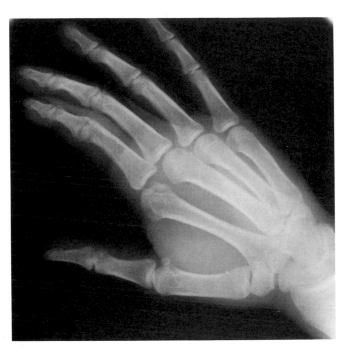

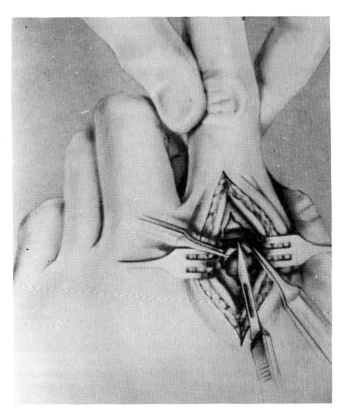

Fig. 6-13. Fracture of metacarpal head with irreducible dislocation. This fracture-dislocation of metacarpophalangeal joint of index finger is ideal for dorsal approach described by Becton et al. since direct access to fracture area is provided.

Fig. 6-14. Dorsal surgical approach to dislocated metacarpophalangeal joint. Volar plate that is caught over dorsal area of metacarpal head is incised longitudinally, and reduction is easily achieved. (From Becton, J.L., Christian, J.D., Jr., Goodwin, H.N., and Jackson, J.G.: J. Bone Joint Surg. **57-A:**698, 1975.)

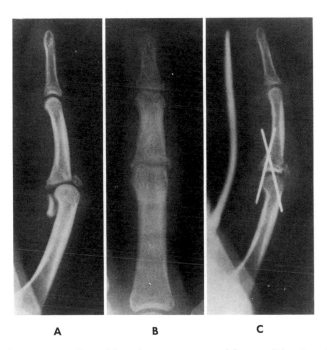

A   B   C

Fig. 6-15. A and **B,** Anteroposterior and lateral roentgenograms of fracture-dislocation after patient's self-reduction. **C,** Lateral roentgenogram after operation. (From Baugher, W.H., and McCue, F.C., III: J. Bone Joint Surg. **61-A:**779, 1979.)

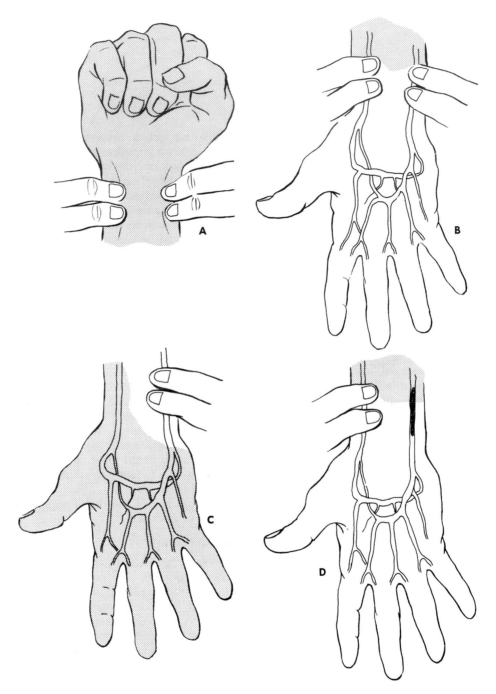

Fig. 7-1. Allen test for patency of radial and ulnar arteries. **A,** Patient elevates hand and makes fist while examiner occludes both radial and ulnar arteries. **B,** Patient extends fingers, and blanching of hand is seen. **C,** Radial artery alone is released, and color of hand returns to normal. **D,** In thrombosis of ulnar artery, test is positive (hand remains blanched) when this artery alone is released.

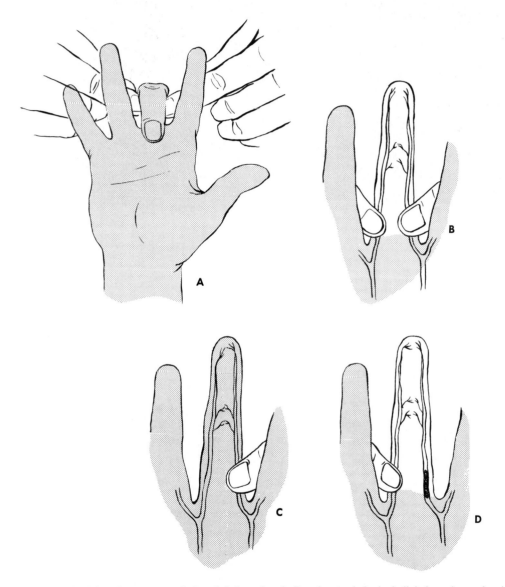

Fig. 7-2. Allen test as applied to digital arteries. **A,** Examiner occludes both digital arteries, and patient flexes finger. **B,** Patient extends finger, and blanching of finger is seen. **C,** When either artery is patent and it alone is released, color of finger returns to normal. **D,** When either artery is thrombosed and it alone is released, finger remains blanched. (From Ashbell, T.S., Koonce, O.E., and Clinard, H.E.: Plast. Reconstr. Surg. **39:**411, 1967.)

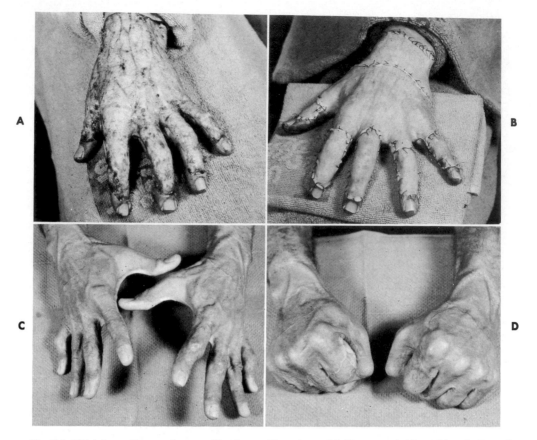

Fig. 7-3. Third-degree burn on dorsum of hand treated by primary debridement and skin grafting. **A,** Burned skin has been excised with care to preserve paratenon and venous network. Lines of excision extend into palm in all web spaces and along midlateral line of each digit with deep darts over joints. **B,** Two days after surgery. Thick split grafts have been applied and carefully anchored by fine sutures. **C** and **D,** Appearance and function of hands after recovery. (From Moncrief, J.A., Switzer, W.E., and Rose, L.R.: Plast. Reconstr. Surg. **33:**305, 1964.)

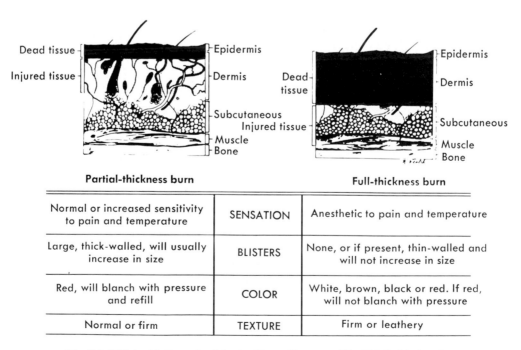

	SENSATION	
Normal or increased sensitivity to pain and temperature	SENSATION	Anesthetic to pain and temperature
Large, thick-walled, will usually increase in size	BLISTERS	None, or if present, thin-walled and will not increase in size
Red, will blanch with pressure and refill	COLOR	White, brown, black or red. If red, will not blanch with pressure
Normal or firm	TEXTURE	Firm or leathery

Fig. 7-4. Differential diagnosis of depth of burn. (Courtesy of Dr. Hal G. Bingham.)

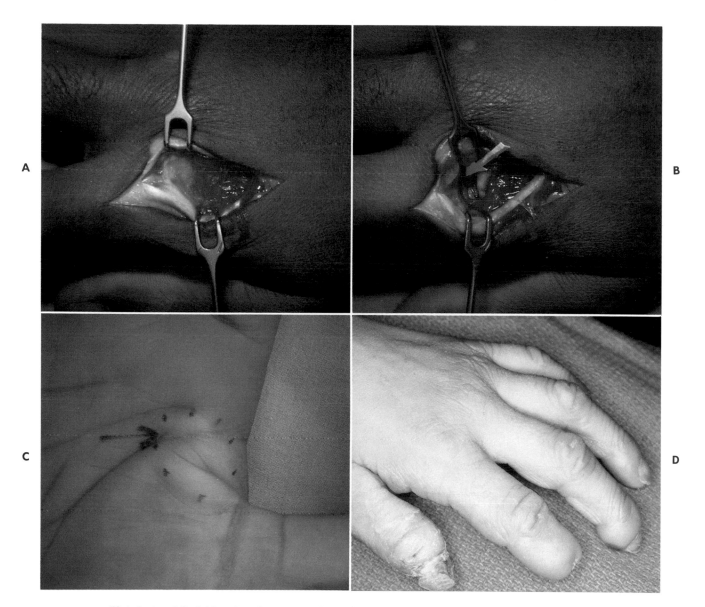

Plate 1. A and **B,** Subluxation of extensor tendon with rupture of metacarpophalangeal joint capsule of right long finger. **C,** Prominence of second metacarpal head dislocated palmarward at metacarpophalangeal joint. Arrow indicates site of digital nerve now stretched and lying just under skin. **D,** Hand of child with congenital insensitivity to pain.

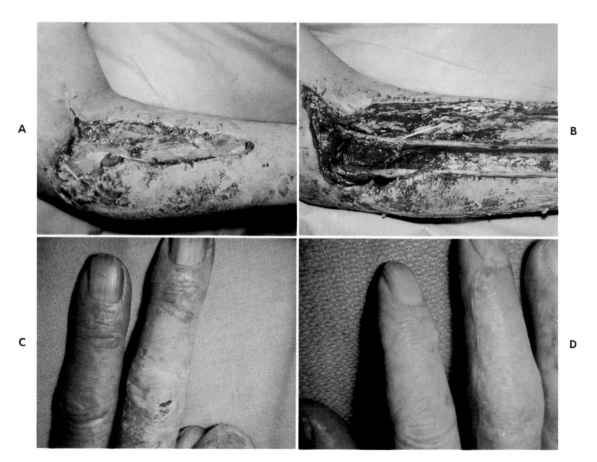

Plate 3. **A** and **B,** Electrical burn of elbow and forearm. **A,** After initial debridement. **B,** After second debridement. Usually two or more debridements are necessary before viable tissue is reached. Amputation above elbow was eventually necessary. **C,** Radiation burns of fingers of physician as a result of use of inefficient lead gloves. **D,** After excision of affected skin and split-thickness skin grafting. Grafts must be relaxed enough to allow full flexion of fingers.

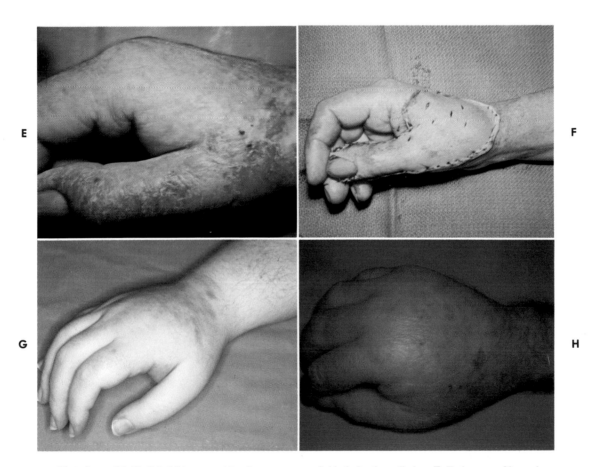

Plate 3, cont'd. E, Painful burn resulting from treatment of skin lesion by radiation. **F,** Entire area of burned skin has been excised and replaced by split-thickness skin graft, relieving pain. However, breakdown of grafted skin may be expected within few years. **G,** This patient had seen several physicians and had multiple diagnostic procedures, including lymphograms, performed before her examination by us. Secretin's disease was suspected. Forearm cast was applied to cover involved area and, as is typical, patient did not return. **H,** Self-inflicted repeated trauma to back of hand.

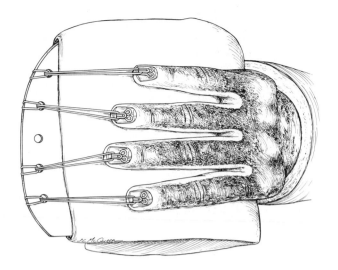

Fig. 7-5. Splinting for burned hand in antideformity position. Rubber band traction on metallic dress hooks "glued" to fingernails achieves positioning without occlusive dressing. (From Salisbury, R.E., and Pruitt, B.A.: Burns of the upper extremity, Philadelphia, 1976, W.B. Saunders Co.)

from scarring or from a severe fibrous reaction produced by prolonged edema or are destroyed from exposure, then restoration of full function later is almost impossible.

Immediate application of mafenide acetate (Sulfamylon) or sulfamethazine helps prevent local infection.

Moncrief et al. have had much experience in treating burns of the hand and the following remarks outline the treatment reported by them. Debridement of the burn should be delayed until 3 days after injury; during this time the patient's general condition is evaluated, and the edema that obliterates the tissue planes and makes accurate excision of the damaged area difficult is allowed to decrease. A general anesthesia is given to the patient and a tourniquet is used. The wound and nails are cleaned, and all loose tissue is removed. The burn is inspected, its boundaries are located, and if desired, they are outlined with a skin pencil. The area to be excised should be patterned to conform to the functional skin creases as described for placing hand incisions (p. 9), thus avoiding tension lines and excessive scarring; this applies not only to the junction between normal skin and grafts but also to any junction between graft and graft. This proper placement of suture lines is especially important in the webs. If the webs require grafting, they should be fully covered and darts of skin should extend distally to the palm; otherwise the darts should extend distally far enough to break any suture line that crosses their proximal borders dorsally. To prevent adduction contracture of the thumb, special care must be used in locating any suture lines on the thumb web. The venous and lymphatic networks are carefully preserved to help prevent edema and to improve nutrition of the extensor tendons. Furthermore, the peritenon, so important in tendon excursion, is also preserved. After the damaged skin has been excised as necessary, small vessels are tied with No. 5-0 plain catgut sutures. Then a finely woven gauze is applied to the raw surface, the hand and arm are wrapped in a bulky compression dressing, the tourniquet

is released, and the extremity is elevated fully for 45 minutes to help control bleeding. This elevation is continued later in bed. After 48 hours the patient is returned to surgery and again with the patient under general anesthesia the wound is cleaned and all bleeding is controlled. A skin graft 0.4 to 0.45 mm thick is obtained from the abdomen or medial aspect of the thigh and is sutured in place; during suture, the graft is stretched almost not at all so that it will be elastic when finger and wrist movements are begun. A dressing is applied with the metacarpophalangeal and interphalangeal joints extended (Fig. 7-5); the wrist is supported on a volar plaster splint. The limb is elevated and the tourniquet is removed. After 72 additional hours the patient is carried to surgery a third time and is anesthetized; the grafts are inspected, any hematomas are evacuated, and any areas on which the grafting has failed are covered by new grafts. Finally a dressing is applied and the patient is kept in bed for a week. Then the dressings are removed and active motion is begun gradually and increased as tolerated.

ELECTRICAL BURNS

Electrical burns, unfortunately, almost always involve the upper extremity (Plate 1). The extent of injury is determined by the characteristics of the injuring current, the duration of contact with it, and the patient's susceptibility to it. Evaluating the extent of injury is always difficult; the first assessment of the severity and depth of tissue damage is usually much too conservative because later more and more apparently normal tissue becomes necrotic.

In a large series, Hunter experienced an overall amputation rate of 43%, including the upper and lower extremity or parts thereof, Initial debridement should begin within 48 hours to help avoid the lethal complication of sepsis associated with the presence of necrotic tissue.

Clinically the burn at first appears as a gray area of skin underlying blistered or charred superficial epithelium; its center is painless because of destruction of sensory nerves. Encircling the area is a line of hyperemia that later becomes edematous. At 7 to 10 days after injury the center of the damaged area begins to slough and necrosis becomes unexpectedly extensive. Sometimes necrosis is quite deep and spreads in all directions beneath the skin. It apparently is caused not only by the burn itself but also by thrombosis of vessels as well. At about 10 days the wound is invaded by bacteria, and softening of tissue and drainage result. To prevent increasing destruction by bacteria, some surgeons recommend debriding the wound early and closing it by either a split-thickness graft or a pedicle graft. Certainly this treatment is ideal if the extent of necrosis can be reasonably determined and if no viable tissue is sacrificed, but accuracy in these matters is difficult. Other surgeons are more conservative and recommend repeated debridement of obviously nonviable tissue, attempting to save all viable structures that may be useful in reconstruction later (Plates 1 to 3, A, F). However, the excision of nonviable tissue may uncover bones, tendons, and nerves, and unless they are soon covered they too become necrotic, probably from bacterial invasion. Obviously the wound must be closed in some manner before any reconstructive surgery. In this type of burn the danger of necrosis of major vessels is great; a larger artery may rupture spontaneously during the night and consequently a tourniquet should be placed by the bed for immediate use should this complication occur.

RADIATION BURNS

In radiation burns or dermatitis caused by overexposure to roentgen rays, the skin becomes pale, dry, atrophic, and wrinkled and scattered keratoses develop; the fingernails split longitudinally. The skin may become increasingly painful and eventually narcotics may be required. Multiple squamous cell carcinomas may develop and cause ulceration. In the past such burns have caused the loss of many digits of physicians and even now are occurring in some medical professionals, typically on the dorsum of the fingers of the left hand, presumably caused by holding cassettes or using the fluoroscope without protection. When breakdown of tissue, pain, or malignant change makes resurfacing the hand necessary, the damaged skin is excised and split-thickness grafts are applied at the same time (Plate 1, *D* and *F*). The area of excision should be generous, including even questionably involved skin; usually all dorsal skin from the wrist distally should be replaced.

CIRCUMFERENTIAL CHEMICAL BURNS

Chemical burns of the hand are usually splash burns. Circumferential burns of the hand are unusual. However, with the increasing availability and use of dimethyl sulf-

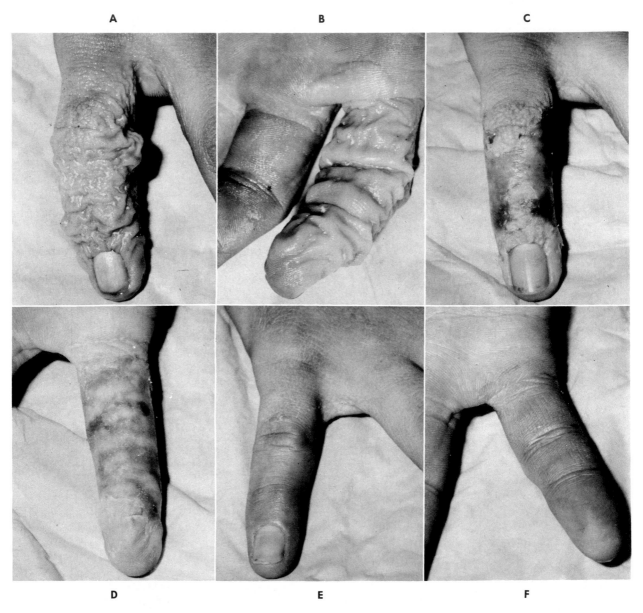

Fig. 7-6. Circumferential chemical burn. **A** and **B,** Dorsal and volar aspects of little finger showing circumferential burn from DMSO. **C,** Finger 5 days after injury. **D,** Volar aspect showing skin blanching on full extension. **E** and **F,** Dorsal and volar views of finger 3 weeks after injury. (From Walker, F.W., and Weinstein, M.A.: J. Hand Surg. **8:**330, 1983.)

oxide (DMSO) as a "pain reliever," particularly by arthritic patients, circumferential burns may become more common. Walker and Weinstein report on a horse-handler and exerciser who experienced circumferential blistering of the entire small finger after wrapping it in a cloth soaked in DMSO for treatment of a minor sprain. Recovery was prompt following early debridement and a therapy program designed to prevent joint stiffness and contractures (Fig. 7-6).

FROSTBITE

In frostbite, tissue is damaged by anoxia caused early by vascular constriction and later by vascular thrombosis. In order of increasing degrees of damage the following develop: erythema, edema, vesiculation, necrosis of skin, necrosis of deeper soft tissue, and necrosis of bone. The immediate basic care of frostbite, whether the tissue is blistered or discolored, is warming and cleaning followed by minimal debridement and watchful waiting for necrosis. After warming, the hand should be washed daily. (Some physicians use Hubbard tanks.) Blisters should not be debrided unless infected. Active motion should be encouraged and frequent washings continued. Amputation should be delayed until there is definite demarcation; this may require several weeks or a few months. In contrast to thermal burns, there is no place for early excision and grafting in treatment of frostbite.

Bigelow and Ritchie reported epiphyseal arrest in several children with severe frostbite (Fig. 7-7); the index and little fingers were involved more frequently than the middle and ring fingers, and the thumb least of all. Disturbances in growth develop gradually, of course, and are probably caused by derangement of the blood supply to the involved epiphyses.

PAINT-GUN INJURIES

Paint-gun injuries are usually caused by wiping the jet opening of a high-pressure gun with the index fingertip. The stream of paint strikes the part with such pressure that it penetrates the skin and spreads widely throughout the underlying fascial planes and tendon sheaths. The resulting distention of tissues and inflammatory reaction cause marked ischemia of tissue; fever and leukocytosis follow. Stark et al. strongly recommend immediate incision and drainage of the injured part with the patient under general anesthesia to relieve pressure and to remove as much of the foreign material as possible; delay in such treatment may result in loss of the part.

GREASE-GUN INJURIES

Grease-gun injuries (Figs. 7-8 and 7-9), like paint-gun injuries just described, are caused by penetration of the tissues by grease or diesel fuel under high pressure. The grease or fuel balloons the soft tissues and follows the planes of least resistance; it causes ischemia and chemical irritation, but the inflammation is not as severe as that caused by paint. Treatment consists of relieving ischemia by decompression and, if possible, preventing infection;

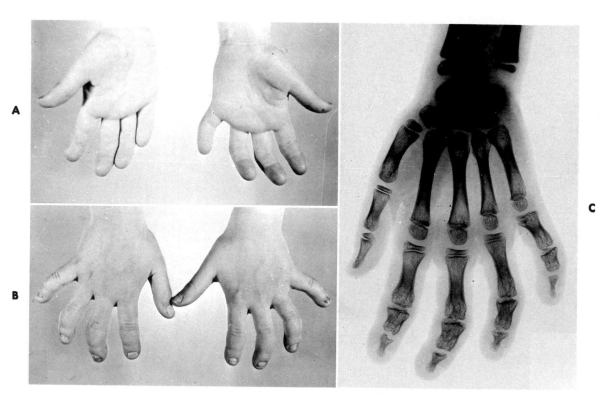

Fig. 7-7. Deformities of fingers of 12-year-old girl caused by frostbite incurred at age of 2 years. **A** and **B,** Note shortening and angulation of various fingers. **C,** Left hand. Note destruction of epiphyses of middle and distal phalanges of all fingers and deformity of epiphysis of proximal phalanx of little finger. Osseous changes in right hand were similar. (From Bigelow, D.R., and Ritchie, G.W.: J. Bone Joint Surg. **45-B:**122, 1963.)

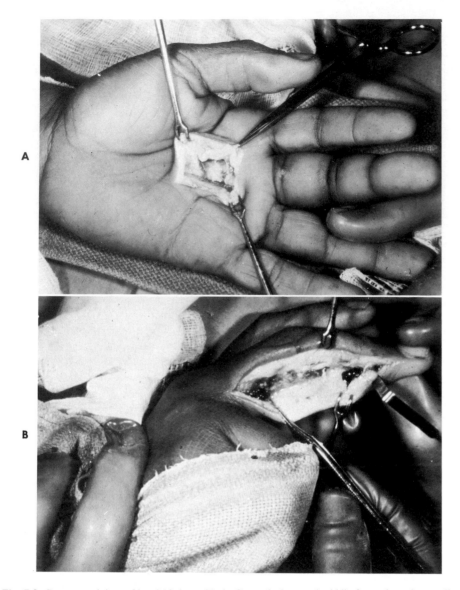

Fig. 7-8. Grease-gun injury of hand 12 days old. **A,** Grease had entered middle finger through a small wound on volar surface at proximal finger crease. Tendon sheath in palm was distended with grease. **B,** Tendon sheath in finger was also distended. All grease was contained within sheath. (From Stark, H.H., Wilson, J.N., and Boyes, J.H.: J. Bone Joint Surg. **43-A:**485, 1961.)

the distended tissues are opened immediately through bold incisions that follow the principles given for placing hand incisions (p. 9), and the foreign material is evacuated. The incisions are closed loosely, if at all, antibiotics are administered, and the hand is immobilized and elevated.

SHOTGUN INJURIES

Shotgun injuries are low velocity missile wounds, are multiple, and often are contaminated by such foreign material as clothing and wadding from the shotgun shell. The wadding is no longer made of horsehair that in the past sometimes contaminated the wound by tetanus spores; it is now made of paper or plastic but is still a dangerous contaminant. In the upper extremity such injuries are usually from close range and the clustered shots cause destruction of multiple tissues. Often the skin surrounding the wound is burned by powder.

The wound should be thoroughly debrided of foreign material, devitalized muscle, fat, and skin, but nerves, even though damaged, should not be excised, Removing every shot is unnecessary but attempts should be made to remove any lodged within joints; any lying just beneath the skin often erode it, are painful, and require removal later. All free osseous fragments should be removed and any segmental defects in bones should be bridged by Kirschner wires to prevent collapse of the bony architecture. When the patient's condition permits and when joints, nerves, and tendons are exposed, the wound may be closed primarily on rare occasions, but an abdominal flap is often necessary. A filleted finger is useful (see Fig.

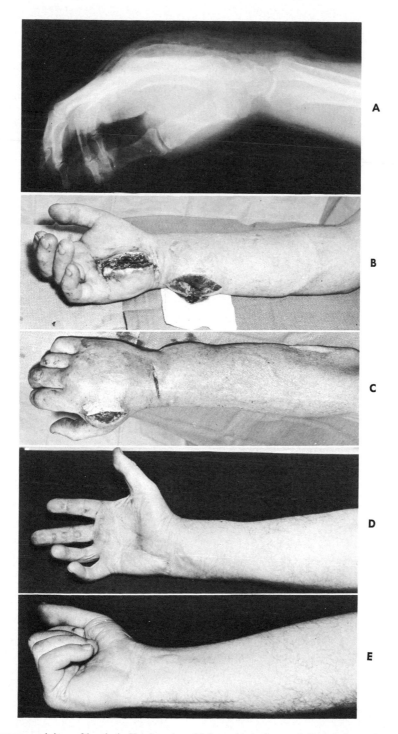

Fig. 7-9. Grease-gun injury of hand. **A,** Hand on day of injury. Air is dispersed through tissues impregnated with grease. **B** and **C,** Hand and forearm 9 days after incision and drainage. Palmar wound was incurred at time of injury, and in it ulnar nerve was found irreparably damaged. **D** and **E,** Hand and forearm 17 months after injury. Function is good. (From Tanzer, R.C.: Surg. Clin. North Am. **43:**1277, 1963.)

2-10). The wound may be left open for a few days but should not be allowed to fill in slowly by granulation tissue and heal spontaneously. Certainly good coverage is necessary before any reconstructive surgery is possible.

WRINGER INJURIES

The term "wringer injury" was first used by Mac-Collum in 1938 to designate a crushing injury of the upper extremity caused by its passage between the rollers of the wringer on an electric washing machine. About 50% of such injuries occur in children under the age of 5 years. Early examination may reveal only abrasions or tears of the skin or occasionally a fracture. However, this first examination is often misleading because hours later there may be severe swelling caused by hemorrhage and edema. When the injury is severe, the skin and deep tissues are burned by the rollers, often at one level where the extremity is blocked from entering farther between the rollers: usually at the base of the thumb, the antecubital fossa, or the axilla. Some of the skin avulsion may be caused by vigorous attempts of the patient to free his limb while the rollers are still in motion. We have seen bursting of the skin at the thumb web and the thenar muscles protruding through the opening like toothpaste squeezed from a tube.

Some surgeons advise hospitalization for 48 hours after injury even when the skin is not broken. The limb is cleaned with soap and water and any open wounds are debrided and closed loosely, if at all. A pressure dressing that includes the entire hand is applied immediately, care being taken that pressure is evenly distributed. First the area is covered by finely woven nonadherent gauze and flat gauze pads are applied; then large masses of cotton and an elastic bandage are rolled on evenly. The extremity is elevated and is kept so throughout treatment. At 24 hours the dressing is removed, the wound is inspected for blisters, hematomas, and necrosis, and the dressing is reapplied; this is repeated every 24 hours until the injury becomes stabilized. Then if necessary, any devitalized tissue is excised and the wound is closed appropriately.

PSYCHOFLEXED AND PSYCHOEXTENDED HANDS

There are at least two typical postures of the hand associated with severe psychiatric depression and other mental disorders. One is the psychoflexed hand as reported by Frykman et al. (Fig. 7-10). They described the posture as one in which the ulnar three digits are severely flexed and contracted, often causing maceration in the palm; it is almost uncorrectable permanently. It interferes with the hygiene of the hand and may cause an offensive odor. In addition, secondary infection can occur from implanting the fingernails into the palm. There is no predilection for the minor or dominant hand. The psychoflexed hand should be carefully differentiated from such disorders as Dupuytren's contracture, arthrogryposis multiplex congenita, and certain spastic hand deformities secondary to stroke or cerebral palsy. These conditions are usually easily distinguished by the experienced orthopaedic surgeon.

Simmons and Vasile (Fig. 7-11) have reported the clenched-fist syndrome in which the entire fist is clenched. However, the ulnar three digits are more predominantly involved.

Spiegel and Chase reported a patient who was eventually cured of severe contractures of all digits by exercises and self-hypnosis. Their method emphasizes and enhances the patient's control over the disability rather than questioning its cause.

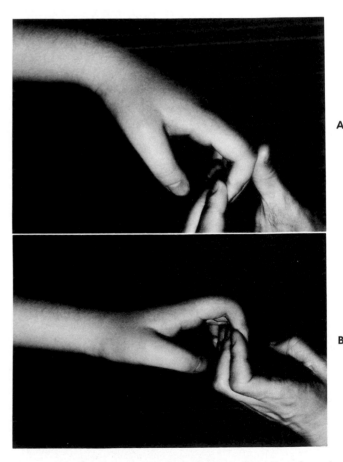

A

B

Fig. 7-11. Clenched-fist syndrome. **A** and **B,** Finger flexion is unchanged by wrist motion (paradoxical stiffness). (From Simmons, B.P. and Vasile, R.G.: J. Hand Surg. **5:**420, 1980.)

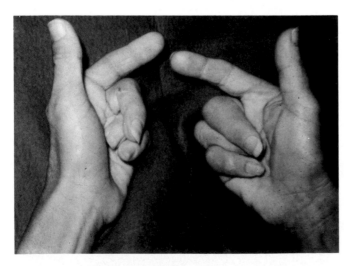

Fig. 7-10. Psychoflexed hands. Patient had flexion contractures of ulnar three fingers of both hands with palmar maceration. (From Frykman, G. et al.: Clin. Orthop. **174:**156, 1983.)

The second posture is the psychoextended hand. It is similar to the psychoflexed hand except that the ulnar three digits are held in rigid hyperextension at the proximal interphalangeal joints and in flexion at the metacarpophalangeal joints. This seems to permit a partially functioning hand consisting of a pinch mechanism preserved between the thumb and index finger. The index finger metacarpophalangeal joint is held in flexion but active flexion and extension are preserved at the proximal interphalangeal joint, thus permitting opposition to the thumb pulp. We have seen several patients who would permit passive extension at the metacarpophalangeal joint and passive flexion at the proximal interphalangeal joint, but after release the posture quickly recurred. Increased hyperextension is eventually possible at the proximal interphalangeal joints by persistent stretching. These patients express only a casual concern over the posture; they neither demand treatment nor are overly distressed by their problem. However, they may permit surgery to be performed, but the surgeon must be aware that almost nothing seems to be of lasting help, including casting, amputation of fingertips, or fixation of joints with Kirschner wires. Psychiatric management should be the initial treatment.

SELF-INDUCED INJURIES

Self-induced injury should be suspected when there is a history of prolonged edema, lack of wound healing, or a deformity without a plausible explanation.

Further suspicion should be aroused when the patient gives a history of having seen several competent physicians who were unable to establish an organic diagnosis after multiple diagnostic procedures such as venograms, lymphograms, and EMGs. Casting the edematous part or wound is helpful in diagnosis, since often the patient does not return or returns without the cast with some excuse for having removed it.

Always inspect the entire extremity for evidence of some type of constricting band proximally. The edema will vary in severity depending on the length of time and the frequency with which and how recently the limb has been constricted. The constriction is usually applied when the person is alone. Plate 3, *G* and *H* shows a self-induced injury.

REFERENCES

Arterial injuries (vascular affections)

Allen, E.V.: Thromboangiitis obliterans: methods of diagnosis of chronic occlusive arterial lesions distal to the wrist with illustrative cases, Am. J. Med. Sci. **178:**237, 1929.

Allen, E.V., Barker, N.W., and Hines, E.A., Jr.: Peripheral vascular diseases, ed. 3, Philadelphia, 1962, W.B. Saunders Co.

Ashbell, T., Koonce, O.E., and Clinard, N.E.: The digital Allen test, Plast. Reconstr. Surg. **39:**411, 1967.

Barker, N.W., and Hines, E.A., Jr.: Arterial occlusion in the hands and fingers associated with repeated occupational trauma, Mayo Clin. Proc. **19:**345, 1944.

Butsch, J.L., and Jones, J.M.: Injuries of the superficial palmar arch, J. Trauma, **3:**505, 1963.

Cameron, B.N.: Occlusion of the ulnar artery with impending gangrene of the fingers: relieved by section of the volar carpal ligament, J. Bone Joint Surg. **36-A:**406, 1964.

Caneirio, R.S., and Mann, R.J.: Occlusion of the ulnar artery associated with an anomalous muscle: a case report, J. Hand Surg. **4:**412, 1979.

Conn, J., Jr., Bergan, J.J., and Bell, J.L.: Hypothenar hammer syndrome: posttraumatic digital ischemia, Surgery **68:**1122, 1970.

Costigan, D.C., Riley, J.N., and Coy, F.E.: Thrombo-fibrosis of the ulnar artery in the palm, J. Bone Joint Surg. **41-A:**702, 1959.

Diaz, J.E., Jones, R.S., and Ciceric, W.F.: Perforation of the deep palmar arch produced by surgical wire after tenorrhaphy: a case report and review of the literature, J. Bone Joint Surg. **57-A:**1150, 1975.

Eaton, R.G., and Green, W.T.: Epimysiotomy and fasciotomy in the treatment of Volkmann's ischemic contracture, Orthop. Clin. North Am. **3:**175, 1972.

Flatt, A.E.: Digital artery sympathectomy, J. Hand Surg. **5:**550, 1980.

Flatt, A.E.: Tourniquet time in hand surgery, Arch. Surg. **104:**190, 1972.

Gardner, C.: Traumatic vasospasm and its complications, Am. J. Surg. **83:**468, 1952.

Gelberman, R.H., et al.: Forearm arterial injuries, J. Hand Surg. **4:**401, 1979.

Gibbon, J.H., Jr., and Landis, E.M.: Vasodilatation in the lower extremities in response to immersing forearms in warm water, J. Clin. Invest. **11:**1019, 1932.

Goren, M.L.: Palmar intramural thrombosis in the ulnar artery, Calif. Med. **89:**424, 1958.

Herndon, W.A., et al.: Thrombosis of the ulnar artery in the hand: report of five cases, J. Bone Joint Surg. **57-A:**994, 1975.

Imparato, A.M.: Management of vascular injuries of the upper extremities, Orthop. Clin. North Am. **2:**383, 1970.

Jackson, J.P.: Traumatic thrombosis of the ulnar artery in the palm, J. Bone Joint Surg. **36-B:**438, 1954.

Kartchner, M.M., and Wilcox, W.C.: Thrombolysis of palmar and digital arterial thrombosis by intra-arterial Thrombolysin, J. Hand Surg. **1:**67, 1976.

Kleinert, H.E., and Volianitis, G.J.: Thrombosis of the palmar arterial arch and its tributaries: etiology and newer concepts in treatment, J. Trauma **5:**447, 1965.

Koman, L.A., and Urbaniak, J.R.: Ulnar artery insufficiency: a guide to treatment, J. Hand Surg. **6:**16, 1981.

Lanz, U.: Anatomical variations of the median nerve in the carpal tunnel, J. Hand Surg. **2:**44, 1977.

Lawrence, R.R., and Wilson, J.N.: Ulnar artery thrombosis in the palm: case reports, Plast. Reconstr. Surg. **36:**604, 1965.

Leddy, J.P., and Packer, J.W.: Avulsion of the profundus tendon insertion in athletes, J. Hand Surg. **2:**66, 1977.

Leriche, R., Fontaine, R., and Dupertius, S.M.: Arterectomy with follow-up studies on 78 operations, Surg. Gynecol. Obstet. **64:**149, 1937.

Lowrey, C.W., Chadwick, R.O., and Waltman, E.N.: Digital vessel trauma from repetitive impact in baseball catchers, J. Hand Surg. **1:**236, 1976.

Martin, A.F.: Ulnar artery thrombosis in the palm: a case report, Clin. Orthop. **17:**373, 1960.

McCormack, L.J., Cauldwell, E.W., and Anson, B.J.: Brachial and antebrachial arterial patterns: a study of 750 extremities, Surg. Gynecol. Obstet. **96:**43, 1953.

Middleton, D.S.: Occupational aneurysms of palmar arteries, Br. J. Surg. **21:**215, 1933.

Neviaser, R.J., and Adams, J.P.: Vascular lesions in the hand: current management, Clin. Orthop. **100:**111, 1974.

Pickering, G.W., and Hess, W.: Vasodilatation of the hands and feet in response to warming the body, Clin. Sci. **1:**213, 1934.

Poirier, R.A., and Stansel, H.C., Jr.: Arterial aneurysms of the hand, Am. J. Surg. **124:**72, 1972.

Smith, J.W.: True aneurysms of traumatic origin in the palm, Am. J. Surg. **104:**7, 1962.

Spittel, J.A., Jr.: Aneurysms of the hand and wrist, Med. Clin. North Am. **42:**1007, 1958.

Suzuki, K., Takahashi, S., and Hakagawa, T.: False aneurysm in a digital artery, J. Hand Surg. **5:**402, 1980.

Teece, L.C.: Thrombosis of the ulnar artery, Aust. N.Z.J. Surg. **19:**156, 1949.

Thio, R.T.: False aneurysm of the ulnar artery after surgery employing a tourniquet, Am. J. Surg. **123:**604, 1972.

Tompkins, D.G.: Exercise myopathy of the extensor carpi ulnaris muscle: report of a case, J. Bone Joint Surg. **59-A:**407, 1977.

Trevaskis, A.E., et al.: Thrombosis of the ulnar artery in the hand: a case report, Plast. Reconstr. Surg. **33:**73, 1964.

Tsuge, K.: Treatment of established Volkmann's contracture, J. Bone Joint Surg. **57-A:**925, 1975.

Watson, H.K. et al.: Post-traumatic interosseus-lumbrical adhesions: a cause of pain and disability in the hand, J. Bone Joint Surg. **56-A:**79, 1974.

Wilgis, E.F.S.: Observations on the effects of tourniquet ischemia, J. Bone Joint Surg. **53-A:**1343, 1971.

Zuckerman, I.C., and Procter, S.E.: Traumatic palmar aneurysm, Am. J. Surg. **72:**52, 1946.

Special hand injuries

Adams, J.P., and Fowler, F.D.: Wringer injuries of the upper extremity: a clinical, pathological, and experimental study, South. Med. J. **52:**798, 1959.

Adamson, J.E.: Treatment of the stiff hand. Orthop. Clin. North Am. **2:**467, 1970.

Allen, J.E., Beck, A.R., and Jewett, T.C., Jr.: Wringer injuries in children, Arch. Surg. **97:**194, 1968.

Bigelow, D.R., and Ritchie, G.W.: The effects of frostbite in childhood, J. Bone Joint Surg. **45-B:**122, 1963.

Brown, H.: Closed crush injuries of the hand and forearm, Orthop. Clin. North Am. **2:**253, 1970.

Browne, E.Z., Jr., Teague, M.A., and Snyder, C.C.: Burn syndactyly, Plast. Reconstr. Surg. **62:**92, 1978.

Caldwell, E.H., and McCormack, R.M.: Acute radiation injury of the hands: report of a case with a twenty-one year follow-up, J. Hand Surg. **5:**568, 1980.

Cannon, B., and Zuidema, G.D.: The care and the treatment of the burned hand, Clin. Orthop. **15:**111, 1959.

Condon, K.C., and Kaplan, I.J.: A method of diagnosis and management of the burned hand, Br. J. Plast. Surg. **12:**129, 1959.

Crow, M.L., and McCoy, F.J.: Volume increase Z-plasty to the finger skin: its application in electrical ring burns, J. Hand Surg. **2:**402, 1977.

Drake, D.A., et al.: An unusual ring injury, J. Hand Surg. **2:**111, 1977.

Dupertuis, S.M., and Musgrave, R.H.: Burns of the hand. Surg. Clin. North Am. **40:**321, 1960.

El-Adwar, L., and Arafa, A.G.: A rare injury of the thumb similar to degloving, J. Bone Joint Surg. **57-A:**998, 1975.

Elton, R.C., and Bouzard, W.C.: Gunshot and fragment wounds of the metacarpus, South. Med. J. **68:**833, 1975.

Flagg, S.V., Finseth, F.J., and Krizek, T.J.: Ring avulsion injury, Plast. Reconstr. Surg. **59:**241, 1977.

Frykman, G., et al.: The psychoflexed hand, Clin. Orthop. **174:**156, 1983.

Gant, T.D.: The early enzymatic debridement and grafting of deep dermal burns to the hand, Plast. Reconstr. Surg. **66:**185, 1980.

Gelberman, R.H., et al.: High-pressure injection injuries of the hand, J. Bone Joint Surg. **57-A:**935, 1975.

Given, K.S., Puckett, C.L., and Kleinert, H.E.: Ulnar artery thrombosis, Plast. Reconstr. Surg. **61:**405, 1978.

Goldner, J.L.: Reconstructive surgery of the hand following thermal injuries, Clin. Orthop. **13:**98, 1959.

Grace, T.G., and Omer, G.E.: The management of upper extremity pit viper wounds, J. Hand Surg. **5:**168, 1980.

Hardin, C.A., and Robinson, D.W.: Coverage problems in the treatment of wringer injuries, J. Bone Joint Surg. **36-A:**292, 1954.

Hawkins, L.G., Lischer, C.G., and Sweeney, M.: The main line accidental intra-arterial drug injection: a review of seven cases, Clin. Orthop. **94:**268, 1973.

Horner, R.L., Wiedel, J.D., and Brailliar, F.: The orthopaedist and epidermolysis bullosa, Orthop. Rev. **1:**21 August 1972.

Horner, R.L., et al.: Involvement of the hand in epidermolysis bullosa, J. Bone Joint Surg. **53-A:**1347, 1971.

Hunter, J.M.: Salvage of the burned hand, Surg. Clin. North Am. **47:**1059, 1967.

Iritani, R.I., and Siler, V.E.: Wringer injuries of the upper extremity, Surg. Gynecol. Obstet. **113:**677, 1961.

Kleinert, H.E., and Williams, D.J.: Blast injuries of the hand, J. Trauma **2:**10, 1962.

Larmon, W.A.: Surgical management of tophaceous gout, Clin. Orthop. **71:**56, 1970.

Leonard, L.G., Munster, A.M., and Su, C.T.: Adjunctive use of intravenous fluorescein in tangential excision of burns of the hand, Plast. Reconstr. Surg. **66:**30, 1980.

Lewis, G.K.: Electrical burns on the upper extremities, J. Bone Joint Surg. **40-A:**27, 1958.

Louis, D.S., and Renshaw, T.: Injuries to the upper extremity inflicted by the mechanical cornpicker, Clin. Orthop. **92:**231, 1973.

Lynn, H.B., and Reed, R.C.: Wringer injuries, JAMA **174:**500, 1960.

MacCollum, D.W.: Wringer arm: report of 26 cases, N. Engl. J. Med. **218:**549, 1938.

McKay, D., et al.: Infections and sloughs in the hands in drug addicts, J. Bone Joint Surg. **55-A:**741, 1973.

Moncrief, J.A., Switzer, W.E., and Rose, L.R.: Primary excision and grafting in the treatment of third-degree burns of the dorsum of the hand, Plast. Reconstr. Surg. **33:**305, 1964.

Moseley, T., and Hardman, W.W., Jr.: Treatment of wringer injuries in children, South. Med. J. **58:**1372, 1965.

Neviaser, R.J., et al.: The puffy hand of drug addiction: a study of the pathogenesis, J. Bone Joint Surg. **54-A:**629, 1972.

Peterson, R.A.: Electrical burns on the hand. Treatment by early excision, J. Bone Joint Surg. **48-A:**407, 1966.

Posch, J.L., and Weller, C.N.: Mangle and severe wringer injuries of the hand in children, J. Bone Joint Surg. **36-A:**57, 1954.

Poticha, S.M., Bell, J.L., and Mehn, W.H.: Electrical injuries with special reference to the hand, Arch. Surg. **85:**852, 1962.

Poulos, E.: The open treatment of wringer injuries in children, Am. Surg. **24:**458, 1958.

Pulvertaft, R.G.: Twenty-five years of hand surgery: personal reflections, J. Bone Joint Surg. **55-B:**32, 1973.

Ramos, H., Posch, J.L., and Lie, K.K.: High-pressure injection injuries of the hand, Plast. Reconstr. Surg. **45:**221, 1970.

Robertson, D.C.: The management of the burned hand. J. Bone Joint Surg. **40-A:**625, 1958.

Robinson, D.W., Masters, F.W., and Forrest, W.J.: Electrical burns: a review and analysis of 33 cases, Surgery **57:**385, 1965.

Robson, M.C., and Heggers, J.P.: Evaluation of hand frostbite blister fluid as a clue to pathogenesis, J. Hand Surg. **6:**43, 1981.

Salisbury, R.E., and Pruitt, B.A.: Burns of the upper extremity, Philadelphia, 1976, W.B. Saunders Co.

Sanguinetti, M.V.: Reconstructive surgery of roller injuries of the hand, J. Hand Surg. **2:**134, 1977.

Simmons, B.P., and Vasile, R.G.: The clenched fist syndrome, J. Hand Surg. **5:**420, 1980.

Smith, J.R., and Gomez, N.H.: Local injection therapy of neuromata of the hand with triamcinolone acetonide: a preliminary study of twenty-two patients, J. Bone Joint Surg. **52-A:**71, 1970.

Spiegel, D. and Chase, R.A.: The treatment of contractures of the hand using self-hypnosis, J. Hand Surg. **5:**428, 1980.

Stark, H.H., Ashworth, C.R., and Boyes, J.H.: Paint-gun injuries of the hand, J. Bone Joint Surg. **49-A:**637, 1967.

Stark, H.H., Wilson, J.N., and Boyes, J.H.: Grease-gun injuries of the hand, J. Bone Joint Surg. **43-A:**485, 1961.

Tanzer, R.C.: Grease-gun type injuries of the hand, Surg. Clin. North Am. **43:**1277, 1963.

Tubiana, R.: Hand reconstruction, Acta Orthop. Scand **46:**446, 1975.

Wakefield, A.R.: Hand injuries in children, J. Bone Joint Surg. **46-A:**1226, 1964.

Walker, F.W., and Weinstein, M.A.: Circumferential finger burn from dimethyl sulfoxide (DMSO), J. Hand Surg. **8:**330, 1983.

Walton, S.: Injection gun injury of the hand with anticorrosive paint and paint solvent: a case report, Clin. Orthop. **74:**141, 1971.

CHAPTER 8

Reconstruction after injury

Before attempting an elective procedure, the surgeon must ask himself if the expected improvement will justify the time, effort, and discomfort involved. When the answer is not clear, surgery should not be done. The patient's occupation, age, emotional maturity, and motivation must all be considered before deciding whether to undertake even a single procedure; should multiple operations be necessary to improve function, emotional maturity is even more important.

Each patient presents an individual problem. For example, a laborer has a severely deformed little finger: amputation leaving a well-padded painless stump might be the wisest choice; multiple reconstructive procedures, although perhaps resulting in a more pleasing appearance, would cost him much more in time lost from work and might produce no better function. Conversely a young woman with a similar deformity might insist on almost any procedure other than amputation, no matter how long the hospitalization and convalescence involved.

Before any elective operation, the patient should be thoroughly informed of the surgeon's plans so that he knows not only what to expect from the operation, but also what is expected of him. He should be told the probable duration of hospitalization and disability, the discomfort involved, and the intended results. He should be advised specifically of his own responsibility for active exercises during convalescence. Unless the psychologic preparation is painstaking and thorough, misunderstanding may occur and may discourage not only the patient but also at times the surgeon. The plan should be not only thought out and talked out, it should be recorded. Photographs and movies are valuable parts of the record; drawings are useful for ready reference. When multiple procedures are considered, such as skin replacement followed by tendon transfers or secondary nerve sutures, two separate examinations with complete notes are advised, the second being at least 2 weeks after the first and without reference to it; the conclusions from each examination should then be compared in detail and carefully considered. After a decision is reached, admitting the patient immediately to the hospital and proceeding with surgery the next day are unwise, even if the patient is willing. Instead operative arrangements should be made for several days later so that he has time to make personal arrangements, to think over the plan, and to adjust his thinking to it.

The surgeon, too, should have time to study the plan. He must see vividly what functions are impaired and why and how they are to be improved; he must know not only the surgical techniques required, but also the sequence in which they should be performed. For example, when multiple procedures are necessary to improve the function of the whole upper extremity, it is usually wise to begin with the hand, for if its function cannot be improved, the more proximal procedures may be futile. When multiple tissues require reconstruction, they must be treated in the proper order of precedence. Good *skin coverage* to reestablish the nutrition of the part is prerequisite to any further reconstruction. Infection must not be present or threatening. Next *nerves* must be intact or must be so repaired that return of sensation is expected. Then and only then is the *bony architecture* restored; otherwise restoration may not be worthwhile. Finally at least passive *joint motion* must be obtained before *tendon grafts or transfers are* indicated. Almost always, placing the wrist in extension and the metacarpophalangeal joints in flexion, either actively or passively, must be possible before a tendon is transferred or grafted. This sequence must be emphasized. First the hand is covered so that its nutrition is reestablished, then its sensation is restored or expected, next its bony archi-

161

tecture is restored, then joint motion is obtained, and finally tendons are repaired. This section is devoted to the indications and techniques involved.

For reconstruction following tendon, nerve, and bone injuries, see individual headings.

SKIN COVERAGE
Granulating areas

A granulating area on the hand should never be left to heal with a scar. When a hand has not been completely covered with skin during the treatment of an acute injury, a split-thickness graft should be applied as soon as the surface is clean enough to support it; even when the entire granulating surface is not clean, any portion that is clean enough should be covered. Exposed tendons, joints, or cortical bone should be covered with flap grafts (p. 15).

Scars

A scar is a poor substitute for skin; it is inelastic, and its sensation is abnormal. The absence of elasticity restricts the motion of otherwise unobstructed underlying joints, interferes with the nutrition of adjacent parts, and limits the motion of joints, tendons, and ligaments when the scar is adherent to them. A scar contracts during healing and will not stretch later. Attempts to stretch a scar may be beneficial only in that normal surrounding skin is stretched. When a linear scar is left spanning a joint, the intermittent stretching caused by active motion will cause it to hypertrophy. Forced passive stretching causes any scar to rupture and fissure, only to heal and become thicker. A scar not only lacks normal sensation, but it also may become painful when it traps and strangulates nerve endings.

Scars cannot be entirely eliminated because the process of healing depends on the production of scar tissue. However, a scar can be partially replaced by tissue of better quality, and the direction or location of its lines can be changed so that they interfere less with function. A scar may be treated surgically (1) to eliminate deformity, (2) to restore joint motion, (3) to provide better skin coverage to nourish vulnerable parts or to permit operation on deeper structures such as tendons or nerves, (4) to relieve pain, and (5) occasionally to improve the appearance of the hand. Sometimes the excision of normal skin is necessary in moving the lines of the scar to a more desirable location.

When possible, a scar should not be replaced until it has matured, usually after a minimum of 3 months. However, it should be treated earlier when it severely limits joint motion; for example, when a metacarpophalangeal joint is held in extension or a proximal interphalangeal joint in flexion, the joint will develop a severe secondary contracture unless the offending scar is treated as soon as possible without awaiting its maturation.

For the purposes of treatment, scars may be classified as *linear* scars and *area* scars; either type, of course, may be volar or dorsal and may or may not involve the deep structures.

METHODS OF CORRECTING LINEAR SCARS

Disabling linear scars usually result from surgical incisions or traumatic lacerations that cross flexor creases. When such a scar on a finger is narrow and is surrounded by normal tissue (Fig. 8-1), it may be released by a Z-plasty (p. 17), but a scar more than 2 mm wide on the volar surface is hard to correct in this way because the skin here is less mobile than that on the dorsum. In some instances the scar must be replaced by a full-thickness free graft (p. 15), a cross finger flap (p. 32), or a local flap (p. 16). On the palm a linear scar may represent loss of skin substance, and in this instance a free thick split graft or a full-thickness graft is indicated (p. 15); correcting a scar contracture here by Z-plasty is difficult. On the dorsum of

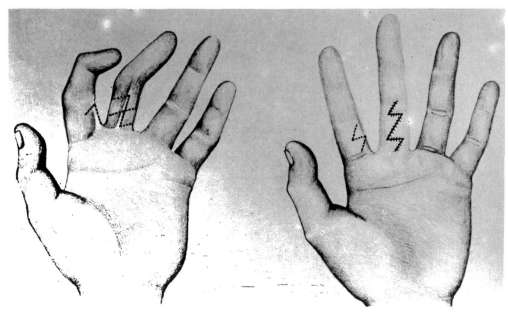

Fig. 8-1. Flexion contractures caused by linear scars may be released by Z-plasties.

the hand most disabling linear scars may be corrected by Z-plasty.

METHODS OF CORRECTING AREA SCARS

An area scar represents a skin loss greater than the area of the final scar, since it has contracted during healing; it must always therefore be replaced by a graft that is larger than the scar (Fig. 8-2). Since the skin for any graft should be as near like the lost skin as possible, a local flap (p. 16) or cross finger flap (p. 32) is preferable when only a small area is lost. When the area is large, when bare bone or tendon is left after excision of the scar, or when a reconstructive procedure is planned, a distant flap containing both skin and subcutaneous fat is necessary. Deeper parts of the scar may be excised when the flap is applied, but tendons or nerves must not be repaired until later. They are then exposed through an incision along the edge of the flap and not through it.

An area scar on the dorsum of the hand involving only the skin may be replaced by a medium or a thick split graft of carefully planned size. The normal adult hand has about 5 cm extra skin longitudinally on the dorsum to allow flexion of the wrist and fingers and about 2.5 cm extra transversely to allow development of the metacarpal arch when making a fist (Fig. 8-3). A graft here, then, must allow for some of this extra skin as well as for previous shrinkage of the scar and later shrinkage of the graft and must be placed while the hand is in the position of function. Otherwise it will be much too tight.

For an area scar on the palm, like skin cannot be used, since it is found only on the sole of the foot (palmar skin is made to withstand friction and shock and is more sen-

sitive than dorsal skin). When the scar is superficial, a thick split graft can be used (p. 14); when deep vulnerable structures are involved, a full-thickness graft is preferable (p. 15), although it is harder to handle, is less likely to survive, and must be limited in size by the fact that it leaves a defect in the donor area that must be closed by suture after its edges are undermined.

For an insensitive large area scar on the radial side of an otherwise normal index finger or on the area of pinch of the thumb, a neurovascular island graft may be indicated (p. 126).

When a graft is applied to the hand, the rules that guide the location and direction of hand incisions (p. 9) must be carefully followed because the graft will heal to the normal skin with a linear scar (Fig. 8-4).

LIMITATION OF JOINT MOTION

When joint motion is limited or absent, an intelligent plan of treatment is possible only on the basis of accurate diagnosis. When there is only a single cause of limitation of motion, the diagnosis may be simple. But since a joint

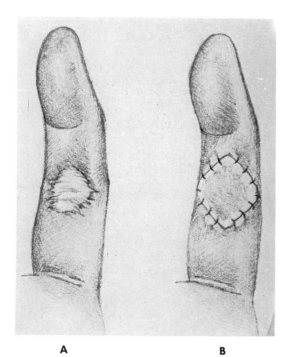

Fig. 8-2. Area scar, **A,** has been replaced by a full-thickness skin graft, **B,** which is larger than scar.

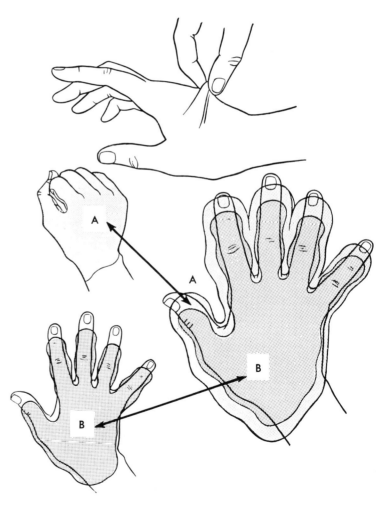

Fig. 8-3. Extra skin on dorsum of normal hand allows flexion of wrist and fingers, and when fist is formed, development of metacarpal arch. When fist is made, all of skin represented by area *A,* is required, but when fingers are extended, only skin represented by area *B* is required.

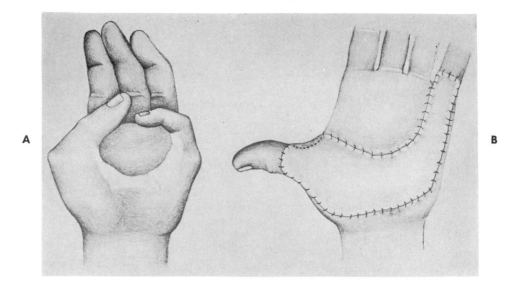

Fig. 8-4. Area scar on palm, **A,** has been replaced by graft, **B,** with margins that follow rules that guide location and direction of hand incisions.

tends to lose motion whenever it is not actively used and since the reasons for its inactivity may be various, diagnosis of the cause of some secondary joint contractures may be complicated; the joint contracture itself may obscure the primary cause of limitation. The most frequent causes are discussed in the paragraphs that follow.

Interruption of tendons

When the continuity of a tendon is interrupted by avulsion or laceration, a specific active movement of one or more joints of the hand is lost. Passive movement in the opposite direction by the examiner will reveal complete loss of the resistance by muscle tone that is ordinarily transmitted through the tendon. The posture of the hand or fingers may also be abnormal. Some days after the injury, edema or secondary joint contracture may limit passive motion. The treatment is tendon repair (see p. 41 for basic techniques).

Adherence of tendons

When a tendon completely adheres to bone, specific active movements of one or more joints distal to the area of adherence are lost. Specific passive movements are also limited because the adherent tendon acts as a checkrein. For example, when a profundus tendon is stuck to the shaft of the proximal phalanx, the two distal finger joints cannot be actively flexed by this tendon. However, the proximal interphalangeal joint may be actively flexed by the sublimis tendon; the metacarpophalangeal joint may also be actively flexed by the profundus and sublimis tendons along with the intrinsics. Adherence of the profundus tendon to the shaft of the proximal phalanx also checks full passive or active extension of the two distal finger joints. Active extension of the distal interphalangeal joint may be increased by passive flexion of the proximal interphalangeal joint; likewise, active extension of the proximal interphalangeal joint may be increased by passive flexion of the

distal interphalangeal joint. Extension of the wrist may initiate flexion of the metacarpophalangeal joint through the adherent tendon.

Adherence of tendon to fracture site

The adherence of a tendon to a fracture site is usually associated with (1) volar angulation of a phalangeal fracture after poor reduction, (2) external pressure against the tendon, forcing it against the fracture during healing, (3) crush injuries, or (4) laceration of the tendon sheath. A sublimis tendon usually adheres to the proximal phalanx, causing a flexion contracture of the proximal interphalangeal joint; a profundus tendon usually adheres to the middle phalanx, causing a flexion contracture of the distal interphalangeal joint. Although rarely involved, an extensor tendon usually adheres to the metacarpal shaft or proximal phalanx. When measurements of motion of adjacent joints demonstrate no progress in loosening the tendon by active exercise, surgery may be indicated.

HOWARD TECHNIQUE TO FREE ADHERENT EXTENSOR TENDON. Make a longitudinal incision parallel to the lateral margin of the involved metacarpal and away from any previous scar. Free the tendon from the bone and smooth the bone with a rasp or osteotome. Remove all scar tissue from the tendon. Place a polyethylene sheet over the bone and anchor it with sutures at its corners (Fig. 8-5). Immobilize the part for 5 days and then begin volunatry motion. Improvement can be expected up to a year.

Paralysis of muscles

Paralyzed muscles cause loss of specific active movements of joints, but passive movement is not limited in any direction unless the joint is secondarily contracted. The posture of the hand at rest may reveal loss of muscle tone. A passive movement that normally would have been resisted by the muscle now paralyzed reveals a decrease in muscle tone; however, there will be some resistance,

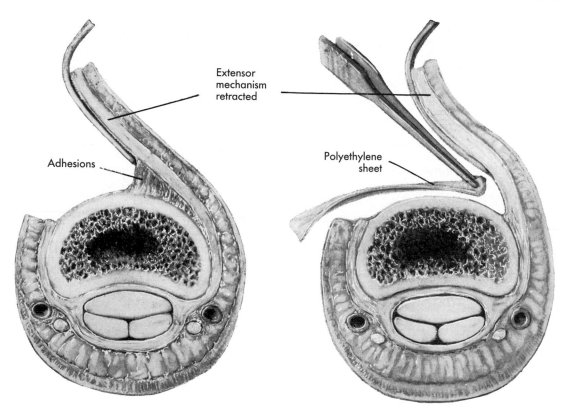

Extensor
mechanism
retracted

Adhesions

Polyethylene
sheet

Fig. 8-5. Howard technique to free adherent extensor tendon (see text).

whereas when a tendon is severed, resistance to passive movement is absent. The treatment consists either of nerve repair or of appropriate tendon transfers. (See discussion of the paralytic hand, p. 215.)

Contracture of muscles

When muscles are contracted, as in Volkmann's ischemic contracture, active motion in the joints controlled by these muscles is limited or absent; passive motion is also limited. When the finger flexors are contracted, the fingers at rest may assume a posture of flexion; passive extension is limited until a more proximal joint or joints are passively flexed. Passive flexion of the wrist may allow complete finger extension (for treatment see p. 299).

Contracture of skin

Skin contracture as a result of a scar may obviously be the cause of limitation of joint motion; when seen late, however, it may be difficult to determine whether this limitation is caused by the skin contracture or a joint contracture or both. Blanching of the skin on the extremes of passive motion is an indication of at least some skin contracture. When the contracture is on the dorsum of the hand, extension of the wrist may release the mobile skin in this area and allow an increase in passive flexion of the metacarpophalangeal joints; however, if the joints are also contracted, no increase in flexion will be gained by this maneuver. When the joint alone is contracted, the skin can be moved over the joint, at least to some extent (see p. 162 for methods of correcting scars).

Contracture of joint

A joint contracture may be caused by any of the causes of limitation of motion just given, or by direct joint damage. When the limitation is produced by bone block from a malunited intraarticular fracture, motion may be equal both actively and passively but is suddenly blocked at a certain position; when the limitation is associated with ligamentous contracture, flexion is rapidly brought to a halt, but in this instance there is a springy sensation rather than a sudden block. The proximal interphalangeal and metacarpophalangeal joints are more often affected by ligamentous contractures. (Correction of joint contracture by capsulotomy is discussed on p. 174.)

Locking of joint

The most frequent cause of locking or triggering of a finger joint (usually of the proximal interphalangeal) is stenosing tenosynovitis at the level of the proximal flexor pulley just distal to the bifurcation of the flexor digitorum sublimis tendon (p. 353). But we have seen triggering of this joint caused by lesions other than tenosynovitis. Locking of the metacarpophalangeal joint caused too by lesions other than tenosynovitis has been reported. Most such lesions have been traumatic and all have involved the fibrocartilaginous volar plate, the volar capsule, or some other structure about the metacarpophalangeal joint. The typical triggering is often absent, and flexion or extension of the joint may be firmly blocked.

As already mentioned, we have seen triggering of the proximal interphalangeal joint unrelated to stenosing tenosynovitis. In a physician with a history of partial division of the flexor digitorum sublimis tendon in the distal palm, typical triggering of the middle finger developed. At surgery a fibrous enlargement was found on the tendon at the site of old injury. In a young woman triggering was caused by a mass attached to the flexor digitorum sublimis tendon; the mass was discolored, had apparently been ruptured, and microscopically was found to be a ganglion. In a man triggering was caused by a traumatic avulsion of one slip of the flexor digitorum sublimis tendon that became folded back into the palm; then the tendon could no longer enter the flexor pulley with ease.

Locking or triggering of a finger joint may occur in children, apparently from a congenital malformation in which there is discrepancy in size between the flexor tendons and the sheath. Frequently it is first seen at the age of about 2 years; sometimes it continues to be bothersome for many years, but many cases will subside within a year or two.

Several lesions have been found responsible for locking of a metacarpophalangeal joint: (1) a tear in the volar capsule, the proximal part of which retracted and rolled up into a band caught around the metacarpal head; (2) a partial tear of the volar plate or the accessory collateral ligaments (Bruner); (3) a transverse tear of the volar plate that formed a pocket as shown in Fig. 8-6 (Yancey and Howard); (4) an osteophyte on the palmar surface of the metacarpal head that caught the fibrous cuff of the volar plate as shown in Fig. 8-7 (Goodfellow and Weaver); (5) an osteophyte on the lateral aspect of the metacarpal head that impinged on the collateral ligament as shown in Fig. 8-8 (Aston); (6) an abnormal configuration of the metacarpal head as shown in Fig. 8-9 (Flatt); (7) a large sesamoid bone that rubbed against a ridge on the radial side of the palmar surface of the metacarpal head as shown in Fig. 8-10 (Flatt); (8) a trapped sesamoid and an abnormal band of tissue as shown in Fig. 8-11 (Bloom and Bryan); (9) an irregular articular surface of the metacarpal head after an undiagnosed fracture as shown in Fig. 8-12 (Dibbell and Field).

Locking or triggering of a finger joint from any cause can be corrected surgically if an accurate diagnosis is made. Release of the flexor pulley, excision of an offending osteophyte, or other appropriate surgery is then indicated.

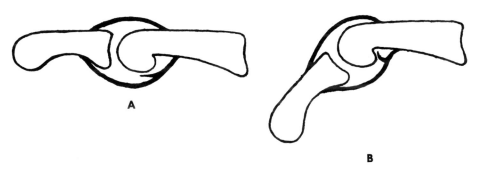

Fig. 8-6. Locking of metacarpophalangeal joint of index finger. **A,** Volar plate had been partially torn. **B,** Extension of joint was blocked when torn edge of plate became caught behind metacarpal head. Torn edge of plate and prominence of metacarpal head were excised. (From Yancey, H.A., Jr., and Howard, L.D., Jr.: J. Bone Joint Surg. **44-A:**380, 1962.)

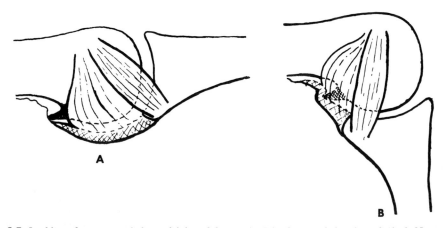

Fig. 8-7. Locking of metacarpophalangeal joint of finger. **A,** Joint is extended and not locked. Note osteophyte on volar margin of metacarpal head. **B,** Osteophyte is caught in fibrous cuff of volar plate, preventing extension of joint. Treatment consisted of release over osteophyte of palmar metacarpophalangeal ligament. (From Goodfellow, J.W., and Weaver, J.P.A.: J. Bone Joint Surg. **43-B:**772, 1961.)

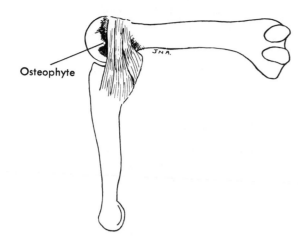

Fig. 8-8. Locking of metacarpophalangeal joint of middle finger. Ulnar collateral ligament was caught on small osteopohyte, preventing extension of joint. Treatment consisted of division of ligament. (From Aston, J.N.: J. Bone Joint Surg. **42-B:**75, 1960.)

Osteophyte

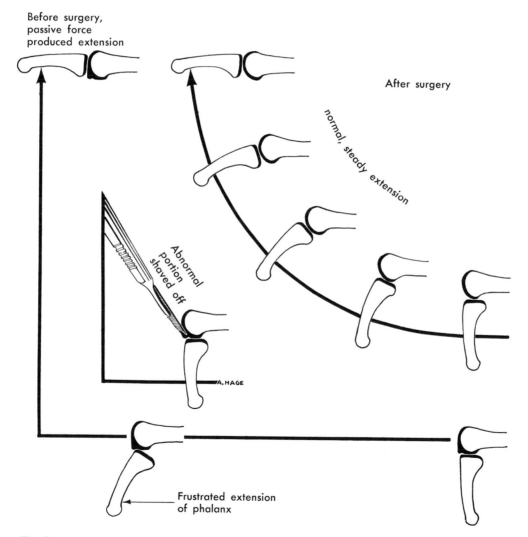

Fig. 8-9. Locking of metacarpophalangeal joint of little finger. Active extension of joint was blocked by abnormal configuration of metacarpal head. Treatment consisted of shaving off abnormal part of head. (From Flatt, A.E.: J. Bone Joint Surg. **43-A:**240, 1961.)

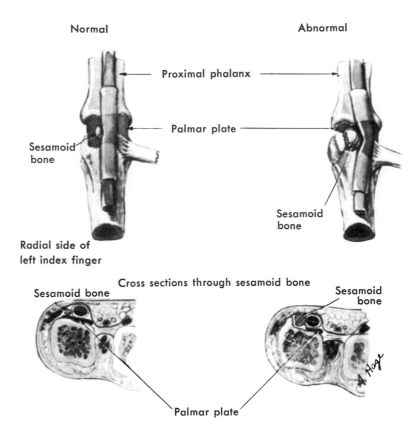

Fig. 8-10. Locking of metacarpophalangeal joint of index finger. Extension of joint was prevented by large sesamoid bone that rubbed against ridge of bone on radial side of palmar surface of metacarpal head. Tendon sheath was displaced ulnarward. Treatment consisted of excision of sesamoid and abnormal ridge of bone and repair of joint capsule. (Modified from Flatt, A.E.: J. Bone Joint Surg. **40-A:**1128, 1958.)

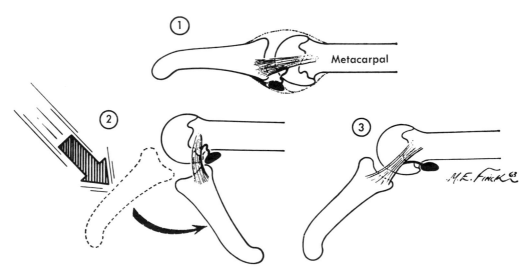

Fig. 8-11. Locking of metacarpophalangeal joint of index finger after injury. *1,* Before injury, loose abnormal band extending from sesamoid to collateral ligament caused no trouble. *2,* Direction, *arrows,* of force of injury caused abnormal flexion of joint and forced sesamoid proximal to volar margin of metacarpal head. *3,* When joint recoiled back into extension, abnormal band became tight, locking sesamoid proximal to metacarpal head and preventing full extension. Treatment consisted of excision of abnormal band and sesamoid. (From Bloom, M.H., and Bryan, R.S.: J. Bone Joint Surg. **47-A:**1383, 1965.)

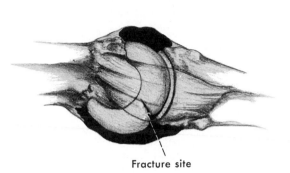

Fracture site

Fig. 8-12. Locking of metacarpophalangeal joint of finger caused by irregularity of metacarpal head after undiagnosed fracture. (From Dibbell, D.G., and Field, J.H.: Plast. Reconstr. Surg. **40:**562, 1967.)

Recurrent hyperextension and locking of proximal interphalangeal joint

In this deformity the finger temporarily assumes a swan-neck posture when an attempt is made to flex it from the fully extended position. The proximal interphalangeal joint is hyperextended, the distal interphalangeal joint is flexed, and flexion of the former joint is impossible unless initiated passively. The lateral bands are subluxated dorsally so that they bowstring across the dorsum of the proximal interphalangeal joint, holding it extended; then because its angle of approach is altered, the flexor digitorum sublimis cannot flex the joint. The volar fibrocartilaginous plate is stretched and may even be detached, but the intrinsic muscles and joint capsule are not contracted. The deformity may be traumatic or may be caused by congenital hyperextensibility of joints and become worse with aging.

We have corrected the deformity and relieved the recurrent locking by transferring a lateral band of the extensor hood volarward to act as a checkrein.

TECHNIQUE. Through a curved dorsal incision centered over the proximal interphalangeal joint expose and define one lateral band of the extensor hood. Then at the junction of the proximal and middle thirds of the proximal phalanx detach the band from the extensor mechanism, leaving it attached to the distal phalanx (Fig. 8-13). Just opposite

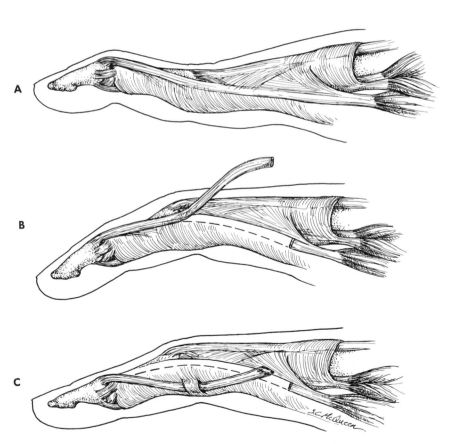

Fig. 8-13. Technique of correcting recurrent hyperextension and locking of proximal interphalangeal joint. **A,** Lateral view of extensor hood and flexor tendon sheath. **B,** One lateral band of hood has been detached proximally. **C,** Detached band has been threaded through small pulley made in flexor tendon sheath opposite proximal interphalangeal joint and has been sutured to hood under enough tension to create slight flexion contracture of joint.

the proximal interphalangeal joint make a small pulley in the flexor tendon sheath. Then thread the detached lateral band from distally to proximally through this pulley and suture it to the extensor hood under just enough tension to create a slight flexion contracture of the joint; later the transferred lateral band will stretch and the joint will extend a few more degrees. Close the incision and apply a bulky dry dressing.

AFTERTREATMENT. At about 2½ weeks the bandage is removed and motion is gradually begun.

• • •

For the same deformity Curtis recommends transferring half of the flexor digitorum sublimis tendon across the volar fibrocartilaginous plate of the proximal interphalangeal joint to reinforce the plate. This gives a stronger tenodesis of the joint such as is needed in athletes but requires more extensive exposure about the joint.

TECHNIQUE (CURTIS). Make a midlateral incision (p. 10) at the level of proximal interphalangeal joint, incise the flexor tendon sheath, and identify the flexor tendons. Then at the proximal bifurcation of the flexor digitorum sublimis divide the ipsilateral half of the tendon (Fig. 8-14). Next drill a hole transversely through the distal end of the proximal phalanx. Carry the freed half of the sublimis tendon deep to the flexor digitorum profundus tendon to the opposite side of the proximal phalanx, thread it through the hole in the phalanx, and anchor it with a pull-out wire suture under enough tension to cause a slight flexion contracture of the proximal interphalangeal joint. Close the incision and splint the finger in a heavy bandage.

AFTERTREATMENT. At 3 weeks the bandage is removed and motion of the joint is begun.

ARTHRODESIS

Arthrodesis of a joint of a finger or thumb may be indicated when the joint has been so damaged by injury or disease that pain, deformity, or instability makes motion a liability rather than an asset. Occasionally a joint must be arthrodesed because the muscles that control the digit are not strong enough to both stabilize and move all joints. Of the finger joints, arthrodesis is indicated for the proximal interphalangeal most often; it is indicated less often for the distal interphalangeal joint because sometimes stabilizing that joint by tenodesis is easier; it is indicated least often for the metacarpophalangeal joint because motion in this joint is more valuable than in the others, and consequently when muscle power is sufficient, arthroplasty is indicated more frequently than arthrodesis. When the muscles that control the finger are weak, procedures to limit their function to the metacarpophalangeal joint are indicated. Of the joints of the thumb, arthrodesis is indicated most often for the metacarpophalangeal joint because it results in little loss of function and for the interphalangeal joint because

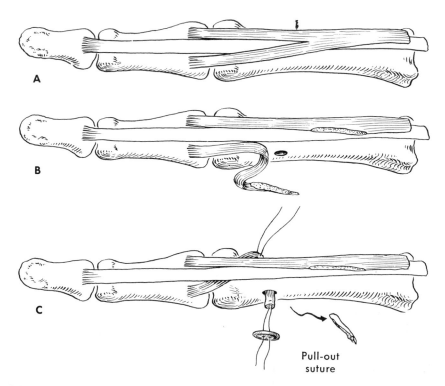

Fig. 8-14. Curtis technique of correcting recurrent hyperextension and locking of proximal interphalangeal joint. **A,** Palmar view of flexor tendons. **B,** One half of flexor digitorum sublimis tendon has been divided at bifurcation of tendon. Hole has been drilled in proximal phalanx. **C,** Freed half of tendon has been carried deep to flexor digitorum profundus tendon to opposite side of proximal phalanx, threaded through hole in bone, and anchored with pull-out suture under enough tension to cause slight flexion contracture of proximal interphalangeal joint. (From Curtis, R.M.: In Flynn, J.E., editor: Hand surgery, Balitmore, 1966, The Williams & Wilkins Co.)

it results in partial loss of discrete pinch only; it is indicated least often for the carpometacarpal joint because this is the most useful joint of the thumb. If the carpometacarpal joint of the thumb is to be arthrodesed, the function of the two distal joints should be satisfactory.

The following are the recommended degrees of flexion in arthrodesis of the various joints: in the fingers, the metacarpophalangeal joints should be in flexion of 20 to 30 degrees, the proximal interphalangeal joints 40 to 50 degrees (there should be less flexion in this joint on the radial than on the ulnar side of the hand), and the distal interphalangeal joints 15 to 20 degrees; in the thumb, the interphalangeal joint in flexion of 20 degrees, the metacarpophalangeal joint 25 degrees, and the carpometacarpal joint with the metacarpal in opposition.

Proximal interphalangeal joint

TECHNIQUE. Open the joint through either a midlateral incision (p. 10) or a dorsal incision in the form of an inverted V (Fig. 8-16, *A*) if the extensor mechanism of the distal interphalangeal joint is destroyed. With a thin osteotome, square off the proximal end of the middle phalanx; shape the distal end of the proximal phalanx by resecting its condyles until the proper angle for arthrodesis is obtained. Do not disturb the volar cartilaginous plate; in a flexion contracture the plate stabilizes and compresses the ends of the bones anteriorly as the joint is forced into extension. Stabilize the joint with parellel or crossed Kirschner wires (Fig. 8-15); be sure the bones are compressed as

the wires are inserted. Use any resected bone as small grafts about the joint. Immobilize the joint with a splint until roentgenograms show solid fusion.

For gross lateral instability of proximal interphalangeal joint see p. 84 in the section on fractures.

Distal interphalangeal joint

TECHNIQUE. The technique of distal interphalangeal joint arthrodesis is the same as described for a proximal interphalangeal joint (Fig. 8-16).

Metacarpophalangeal joint of thumb

Arthrodesis of the metacarpophalangeal joint of the thumb is frequently useful in rheumatoid arthritis and is described in the discussion of the rheumatoid hand (p. 283).

Carpometacarpal joint of thumb

Arthrodesis of the carpometacarpal joint of the thumb may be indicated for degenerative arthritis after malunited Bennett's fracture, subluxation of the joint, osteoarthritis, or rheumatoid arthritis. Careful consideration should be given to other causes of pain in and about the carpometacarpal area of the thumb. Signs of de Quervain's disease can be demonstrated by adduction of the thumb and ulnar deviation of the wrist (Finkelstein's test). When this maneuver produces pain, the diagnosis can be substantiated by local tenderness over the tunnel of the abductor pollicis longus and extensor pollicis brevis.

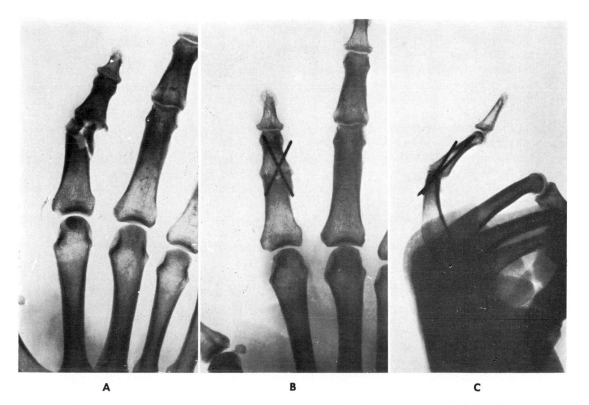

A   **B**   **C**

Fig. 8-15. Arthrodesis of proximal interphalangeal joint. **A,** Comminuted fracture involving joint. **B** and **C,** Joint has been arthrodesed. Small fragments were used as grafts, and joint was fixed with two Kirschner wires. Placing wires parallel is preferred to crossing them as shown here.

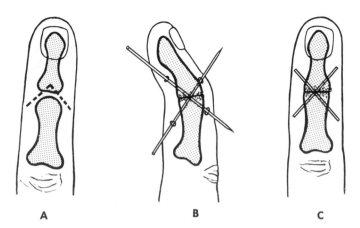

Fig. 8-16. Arthrodesis of distal interphalangeal joint. **A,** Inverted V incision. **B,** Joint surfaces have been resected, small chips of bone have been placed about joint, and joint has been fixed with two Kirschner wires. **C,** Wires have been cut off beneath skin.

The axial compression test, when painful, helps to confirm that symptoms are mechanical and located in the carpometacarpal joint of the thumb. This test is performed by rotating the metacarpal as the joint is compressed by the metacarpal. Local tenderness is usually maximum on the radial aspect. Carpal tunnel syndrome may accompany degenerative changes at the base of the thumb joint and be accentuated postoperatively; therefore careful evaluation should be made preoperatively for this cause of pain. Old fractures of the scaphoid should be ruled out by roentgenograms, along with arthritic changes in adjacent joints.

The advantages of arthrodesis versus arthroplasty of the joint by resecting the greater multangular must be considered: arthrodesis relieves pain, provides stability, and possibly increases strength; arthroplasty relieves pain, increases mobility, and probably decreases strength. Activity following resection arthroplasty usually causes sufficient pain to require protection and immobilization for several weeks after operation. When movement is started, pinch will be extremely weak and possibly painful until a buildup of heavy scar tissue takes place over several months. Activities such as opening car doors, cutting with scissors, and winding watches will eventually be possible with practice and patience but cannot be expected to be easy within the first few weeks. See p. 286 for discussion of implant arthroplasty of this joint.

Arthrodesis requires cast immobilization for several weeks following operation with any technique; however, once fusion is accomplsihed, the joint is pain free. Degenerative changes in surrounding joints do not necessarily contraindicate arthrodesis to relieve pain at the carpometacarpal joint.

Regardless of the technique used for arthrodesis of the carpometacarpal joint of the thumb, the position of the first metacarpal should be that of normal maximum abduction and opposition. This permits pulp-to-pulp pinch with the index finger and, with flexion of the interphalangeal thumb joint, with the middle finger. Extension of the thumb's metacarpophalangeal joint clears the thumb to permit a full fist with unobstructed flexion of all fingers.

TECHNIQUE (STARK, MOORE, ASHWORTH, AND BOYES). Expose the metacarpophalangeal joint through a curved volar in-

cision at the base of the thumb at the level of the insertion of the abductor pollicis longus tendon, being careful to avoid the sensory branches of the superficial radial and lateral antebrachial cutaneous nerves. Divide this tendon and the origin of the opponens muscle at the base of the first metacarpal; open the joint capsule through a transverse incision. With an osteotome, remove all the articular cartilage and the subchondral cortical bone on both joint surfaces. Place the thumb as described above so that the index and middle finger pulps can easily reach the thumb pulp but allowing sufficient room in extension to permit full finger flexion into the palm. Compress the bony surfaces in the position desired (see above), and maintain this by 2 or 3 small Kirschner wires. If more bony contact is needed, add an iliac bone graft. Repair the capsule and the abductor pollicis longus tendon, and suture the skin.

AFTERTREATMENT. The thumb is maintained in a thumb spica cast for 2 weeks, at which time the skin sutures are removed. The thumb spica is reapplied and healing is checked periodically by roentgenograms until fusion is obtained, usually at 12 weeks.

• • •

Another excellent method for arthrodesis of the carpometacarpal joint of the thumb consists of denuding the articular surfaces of the joint and using a bone peg for internal fixation (Fig. 8-17, *A* and *B*); thus fixation with Kirschner wires that must be removed later and the interposition of a bone graft between the osseous surfaces as required in some methods are unnecessary.

TECHNIQUE. Begin a curved incision on the dorsoradial aspect of the first metacarpal and curve it volarward at the distal wrist crease (Fig. 4-10). Then expose the carpometacarpal joint and remove its articular cartilage. Hold the thumb in the desired position in relation to the greater multangular, and starting on the base of the first metacarpal, introduce a small drill obliquely across the joint. Then check the position of both the joint and the drill by roentgenograms, remove the drill, and enlarge the hole to make it about 3 mm in diameter. Now obtain from the proximal ulna a square bone peg large enough to fit tightly in the hole; a square peg provides better fixation than a round

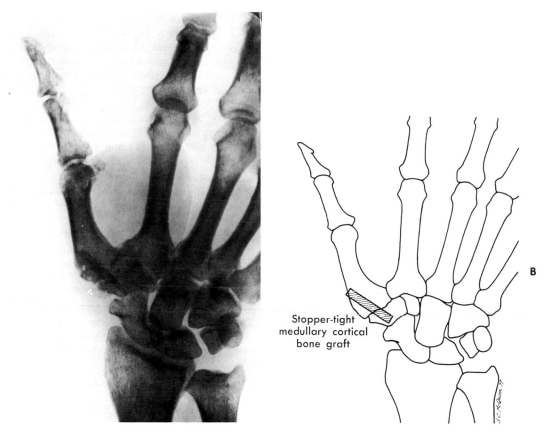

A

B

Stopper-tight
medullary cortical
bone graft

Fig. 8-17. A, Arthrodesis of carpometacarpal joint of thumb for osteoarthritis using square bone peg for internal fixation (see text). **B,** Drawing of roentgenogram shown in **A.** At times it is more convenient to slot graft across joint and add Kirschner wire internal fixation.

one. Drive the peg through the hole and across the joint, taking care that its proximal end remains within the greater multangular. Then place small chips of bone in any crevices about the joint and close the wound. Apply a cast as is used for immobilizing a fractured scaphoid.

AFTERTREATMENT. The cast is changed as necessary but is worn until roentgenograms demonstrate osseous union.

ARTHROPLASTY

Of the finger joints, the metacarpophalangeal is treated by arthroplasty most often; the operation is indicated when degenerative changes in the joint of an otherwise useful finger result in less than 30 degrees of motion. Furthermore, the muscles that control the joint must be of functional strength; otherwise the arthroplasty is of no value. Arthroplasty of the proximal interphalangeal joint is indicated only when a joint with lateral instability of 30 degrees and motion in flexion and extension of 60 degrees or more is preferable to one arthrodesed in the position of function. Again the muscles that control the joint must be of functional strength. The two central digits are more suitable for this arthroplasty because the lateral instability is controlled somewhat by the fingers on each side. Arthroplasty of the distal interphalangeal joint of a finger and the interphalangeal joint of the thumb is never indicated.

The techniques of arthroplasty using an interpositional Silastic prosthesis for the carpometacarpal joint of the

thumb and the metacarpophalangeal joints of the fingers are described in Chapter 12.

Carpometacarpal joint of thumb

Resecting the greater multangular may be indicated for painful degenerative arthritis of the first carpometacarpal joint or for an adduction contracture and loss of mobility of the joint caused by direct injury or contracture of the surrounding soft tissue (see discussion of the adducted thumb).

Resection of the greater multangular usually is considered more strongly for a woman past middle age. If strong pinch is desired, an arthrodesis of the carpometacarpal joint should be carried out. After resection, pinch and grasp maneuvers can be accomplished with considerable force but only after several months of practice and rehabilitation. Insertion of a Silastic prosthesis seems to have advantages over simple resection arthroplasty. (See p. 286 for Swanson technique.)

TECHNIQUE OF EXCISING GREATER MULTANGULAR (GOLDNER AND CLIPPINGER). Make an incision parallel with the abductor pollicis longus tendon and extend it into the web space as far as necessary to release the soft tissues. Divide the superficial fascia and release the fascia over the abductor, the adductor insertion, and that part of the adductor originating on the third metacarpal. Reflect dorsalward the dorsal branch of the radial artery and the sensory branch of

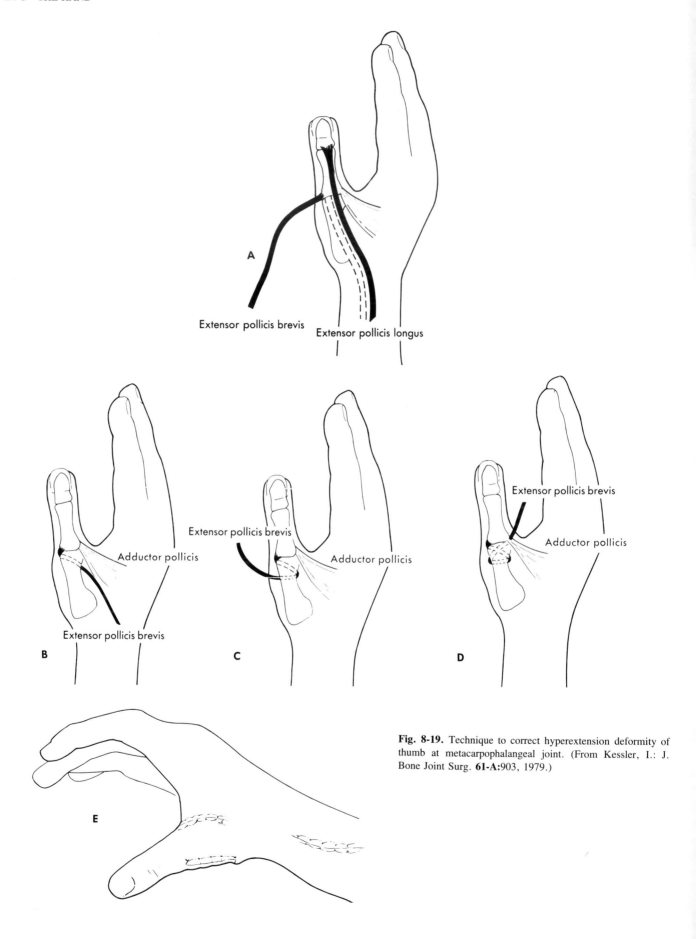

Extensor pollicis brevis   Extensor pollicis longus

A

Adductor pollicis

Extensor pollicis brevis

B

Extensor pollicis brevis

Adductor pollicis

C

Extensor pollicis brevis

Adductor pollicis

D

E

Fig. 8-19. Technique to correct hyperextension deformity of thumb at metacarpophalangeal joint. (From Kessler, I.: J. Bone Joint Surg. **61-A:**903, 1979.)

retinacular ligament. Expose the collateral ligament by approaching the joint from the base of the middle phalanx and elevating the transverse retinacular ligament; preserve this ligament to stabilize the joint after excision of the collateral ligament. Starting at its distal attachment excise en bloc as much of the collateral ligament as possible (Fig. 8-18, *D*). Repeat the procedure on the opposite side of the joint.

When the contracture is of long duration, the volar synovial pouch may have been obliterated; if so, restore it with a small curved elevator or by forcing the base of the phalanx into flexion. When the interosseus muscle is contracted, lengthen its tendon by tenotomy and suture (Fig. 8-18, *E*). If necessary, free the extensor tendon over the dorsum of the finger through the same approach.

Satisfactory passive motion must be demonstrated during surgery because no further motion can be anticipated after surgery. It is also important that the flexor tendons not be adherent in the palm; should they be, make a palmar incision to release them.

Inject a steroid preparation into the joint and close the wound. Apply a dressing of compressed wet cotton to prevent swelling.

AFTERTREATMENT. At 2 to 3 days motion is begun under supervision. The joint is splinted alternately in flexion and extension. Splinting is continued until the range of motion obtained at surgery is possible both actively and passively; it may be necessary at least part of the time for as long as 3 or 4 months. (Also see lumbrical plus finger, p. 69.)

HYPEREXTENSION DEFORMITY OF METACARPOPHALANGEAL JOINT OF THUMB

Hyperextension deformity of the thumb may result from trauma, rheumatoid arthritis, or paralysis. If there is a usable joint surface with satisfactory active flexion, arthrodesis may be avoided by using the technique of Kessler. He uses the extensor pollicis brevis tendon to tether the metacarpophalangeal joint in slight flexion. In his experience this tendon transfer does not have the disadvantage of stretching out, as is seemingly true with the capsulorrhaphy technique.

TECHNIQUE (KESSLER). Make a 4 cm incision on each side of the metacarpophalangeal joint and slightly dorsal. Make another incision just proximal to the radial styloid to approach the extensor pollicis brevis tendon at its musculotendinous junction. Sever the tendon at this point. Bring the severed segment out at the incision near the insertion of the tendon (Fig. 8-19, *A*). Thread the end of the tendon radially and volarly across the metacarpophalangeal joint obliquely and superficial to the flexor pollicis longus tendon sheath. Bring it out on the ulnar side of the metacarpal head (Fig. 8-19, *B*). Drill a hole transversely through the neck of the metacarpal and pass the tendon through the hole to the radial side (Fig. 8-19, *C*). Place sufficient tension on the tendon to hold the metacarpophalangeal joint securely at 20 degrees of flexion. Now once again cross the tendon volarly toward the insertion of the adductor pollicis muscle. Suture it to the adductor pollicis so that the metacarpophalangeal joint is held securely in 20 degrees of flexion without permitting any further exten-

sion (Fig. 8-19, *D*). Metacarpophalangeal joint passive flexion of 30 to 40 degrees should be possible. Immobilize the thumb in a short plaster cast for 3 weeks. A Kirschner wire may also be inserted across the metacarpophalangeal joint.

REFERENCES
Reconstruction after injury (skin coverage)

Apfelberg, D.B., et al.: Treatment of colloid milium of the hand by dermabrasion, J. Hand Surg. **3**:98, 1978.

Barton, N.J.: A modified thenar flap, Hand, **7**:150, 1975.

Baudet, J., LeMaire, J.M., and Guimberteau, J.C.: Ten free groin flaps, Plast. Reconstr. Surg. **57**:577, 1976.

Beasley, R.W.: Local flaps for surgery of the hands, Orthop. Clin. North Am. **2**:219, 1970.

Boswick, J.A., Jr.: Management of the burned hand, Orthop. Clin. North Am. **2**:311, 1970.

Boswick, J.A., Jr.: Rehabilitation of the burned hand, Clin. Orthop. **104**:162, 1974.

Brown, H., and Flynn, J.E.: Abdominal pedicle flap for hand neuromas and entrapped nerves, J. Bone Joint Surg. **55-A**:575, 1973.

Brown, H.G.: Electrical and cold injuries of the hand, Orthop. Clin. North Am. **2**:321, 1970.

Carroll, R.E.: Ring injuries in the hand, Clin. Orthop. **104**:175, 1974.

Chase, R.A.: The damaged index digit: a source of components to restore the crippled hand, J. Bone Joint Surg. **50-A**:1152, 1968.

Chase, R.A.: The severely injured upper limb: to amputate or reconstruct: that is the question, Arch. Surg. **100**:382, 1970.

Chase, R.A.: Early salvage in acute hand injuries with a primary island flap, Plast. Reconstr. Surg. **48**:521, 1971.

Chase, R.A., Hentz, V.R., and Apfelberg, D.: A dynamic myocutaneous flap for hand reconstruction, J. Hand Surg. **5**:594, 1980.

Chase, R.A., and Nagel, D.A.: Cosmetic incisions and skin, bone, and composite grafts to restore function of the hand. In American Academy of Orthopaedic Surgeons: Instructional course lectures, vol. 23, St. Louis, 1974, The C.V. Mosby Co.

Dabezies, E.J.: An advancement pedicle flap for the late coverage of pulp injuries of the digits, South. Med. J. **67**:340, 1974.

Dowden, R.V., and McCraw, J.B.: Muscle flap reconstruction of shoulder defects J. Hand Surg. **5**:382, 1980.

Eversmann, W.W., Burkhalter, W.E., and Dunn, C.: Transfer of the long flexor tendon of the index finger to the proximal phalanx of the long finger during index-ray amputation, J. Bone Joint Surg. **53-A**:769, 1971.

Foucher, G., and Braun, J.B.: A new island flap transfer from the dorsum of the index to the thumb, Plast. Reconstr. Surg. **63**:344, 1979.

Frackelton, W.H., and Teasley, J.L.: Neurovascular island pedicle-extension in usage, J. Bone Joint Surg. **44-A**:1069, 1962.

Glanz, S.: Repair of contractures of the hand with pedal full-thickness skin grafts, Am. J. Surg. **100**:412, 1960.

Graham, W.P., III: Incisions, amputations, and skin grafting in the hand, Orthop. Clin. North Am. **2**:213, 1970.

Green, D.P., and Dominguez, O.J.: A transpositional skin flap for release of volar contractures of a finger at the metacarpophalangeal joint, Plast. Reconstr. Surg. **64**:516, 1979.

Hanna, D.C.: Resurfacing the hand in acute injuries, Surg. Clin. North Am. **40**:331, 1960.

Houghland, R.G.: Secondary index finger amputations to improve function and dexterity, Orthop. Rev. **4**:52, July 1975.

Hurwitz, D.J., and White, W.L.: Application of glove designs in resurfacing the dorsum of the hand, Plast. Reconstr. Surg. **62**:385, 1978.

Johnson, R.K., and Iverson, R.E.: Cross-finger pedicle flaps in the hand, J. Bone Joint Surg. **53-A**:913, 1971.

Joshi, B.B.: Sensory flaps for the degloved mutilated hand, Hand **6**:247, 1974.

Kleinman, W.B., and Dustman, J.A.: Preservation of function following complete degloving injuries to the hand: use of simultaneous groin flap and partial-thickness skin graft, J. Hand Surg. **6**:82, 1981.

Lewin, M.L.: Digital flaps in reconstructive and traumatic surgery, Clin. Orthop. **15**:74, 1959.

Lie, K.K., and Posch, J.L.: Island flap innervated by radial nerve for restoration of sensation in an index stump: case report, Plast. Reconstr. Surg. 47:386, 1971.

Lie, K.K., and Posch, J.L.: Island flap innervated by radial nerve for restoration of sensation in an index stump: case report, Plast. Reconstr. Surg. 47:386, 1971.

Littler, J.W.: Neurovascular skin island transfer in reconstructive hand surgery, Trans. Int. Soc. Plast. Surg. 2:175, 1960.

MacDougal, B., Wray, R.C., Jr., and Weeks, P.M.: Lateral-volar finger flap for the treatment of burn syndactyly, Plast. Reconstr. Surg. 57:167, 1976.

Macht, S.D., and Watson, H.K.: The Moberg volar advancement flap for digital reconstruction, J. Hand Surg. 5:372, 1980.

Maquieira, N.O.: An innervated full-thickness skin graft to restore sensibility to fingertips and heels, Plast. Reconstr. Surg. 53:568, 1974.

May, H.: Plastic repair of skin defects of the hand, Clin. Orthop. 15:86, 1959.

May, J.W., Jr., and Barlett, S.P.: Staged groin flap in reconstruction of the pediatric hand, J. Hand Surg. 6:163, 1981.

May, J.W., Jr., and Gordon, L.: Palm of hand free flap for forearm length preservation in nonreplantable forearm amputation: a case report, J. Hand Surg. 5:377, 1980.

McDonald, J., and Webster, J.P.: Early covering of extensive traumatic deformities of the hand and foot, Plast. Reconstr. Surg. 1:49, 1946.

McFarlane, R., and Stromberg, W.B.: Resurfacing of the thumb following major skin loss, J. Bone Joint Surg. 44-A:1365, 1962.

McGarth, M.H., Adelbert, D., and Finseth, F.: The intravenous fluorescein test: use in timing of groin flap division, J. Hand Surg. 4:19, 1979.

McGregor, I.A.: Flap reconstruction in hand surgery: the evolution of presently used methods, J. Hand Surg. 4:1, 1979.

Miura, T., and Nakamura, R.: Use of paired flaps to simultaneously cover the dorsal and volar surfaces of a raw hand, Plast. Reconstr. Surg. 54:286, 1974.

Murray, J.F., Ord, J.V.R., and Gavelin, G.E.: The neurovascular island pedicle flap: an assessment of late results in sixteen cases, J. Bone Joint Surg. 49-A:1285, 1967.

Pho, R.W.H.: Local composite neurovascular island flap for skin cover in pulp loss of the thumb, J. Hand Surg. 4:11, 1979.

Pohl, A.L., Larson, D.L., and Lewis, S.R.: Thumb reconstruction in the severely burned hand, Plast. Reconstr. Surg. 57:320, 1976.

Posner, M.A., and Smith, R.J.: The advancement pedicle flap for thumb injuries, J. Bone Joint Surg. 53-A:1618, 1971.

Rose, E.H., and Buncke, H.J.: Free transfer of a large sensory flap from the first web space and dorsum of the foot including the second toe for reconstruction of a mutilated hand, J. Hand Surg. 6:196, 1981.

Salisbury, R.E., McKeel, D.W., and Mason, A.D., Jr.: Ischemic necrosis of the intrinsic muscles of the hand after thermal injuries, J. Bone Joint Surg. 56-A:1701, 1974.

Schlenker, J.D.: Transfer of a neurovascular island pedical flap based upon the metacarpal artery: a case report, J. Hand Surg. 4:16, 1979.

Scott, J.E.: Amputation of the finger, Br. J. Surg. 61:574, 1974.

Shaw, D.T., et al.: Interdigital butterfly flap in the hand (double-opposing Z-plasty), J. Bone Joint Surg. 55-A:1677, 1973.

Smith, R.C., and Furnas, D.W.: The hand sandwich: adjacent flaps from opposing body surfaces, Plast. Reconstr. Surg. 57:351, 1976.

Thompson, R.V.S.: Closure of skin defects near the proximal interphalangeal joint—with special reference to the patterns of finger circulation, Plast. Reconstr. Surg. 59:77, 1977.

Tubiana, R., and DuParc, J.: Restoration of sensibility in the hand by neurovascular skin island transfer, J. Bone Joint Surg. 43-B:474, 1961.

White, W.L.: Flap grafts to the upper extremity, Surg. Clin. North Am. 40:389, 1960.

Reconstruction after injury (bone and joint)

Adams, J.P.: Correction of chronic dorsal subluxation of the proximal interphalangeal joint by means of a criss-cross volar graft, J. Bone Joint Surg. 41-A:111, 1959.

Alldred, A.: A locked index finger, J. Bone Joint Surg. 36-B:102, 1954.

Alldred, A.J.: Rupture of the collateral ligament of the metacarpo-phalangeal joint of the thumb, J. Bone Joint Surg. 37-B:443, 1955.

Allende, B.T., and Engelem, J.C.: Tension-band arthrodesis in the finger joints, J. Hand Surg. 5:269, 1980.

Aston, J.N.: Locked middle finger, J. Bone Joint Surg. 42-B:75, 1960.

Bloom, M.H., and Bryan, R.S.: Locked index finger caused by hyperflexion and entrapment of sesamoid bone. J. Bone Joint Surg. 47-A:1383, 1965.

Braun, R.M.: Trephine techniques for small bone grafts, Hand 6:103, 1974.

Bruner, J.M.: Use of single iliac-bone graft to replace multiple metacarpal loss in dorsal injuries of the hand, J. Bone Joint Surg. 39-A:43, 1957.

Bruner, J.M.: Recurrent locking of the index finger due to internal derangement of the metacarpophalangeal joint, J. Bone Joint Surg. 43-A:450, 1961.

Buch, V.I.: Clinical and functional assessment of the hand after metacarpophalangeal capsulotomy, Plast. Reconstr. Surg. 53:452, 1974.

Burton, R.I.: Basal joint arthrosis of the thumb, Orthop. Clin. North Am. 4:331, 1973.

Campbell, C.S.: Gamekeeper's thumb, J. Bone Joint Surg. 37-B:148, 1955.

Carroll, R.E., and Hill, N.A.: Arthrodesis of the carpo-metacarpal joint of the thumb, J. Bone Joint Surg. 55-B:292, 1973.

Carroll, R.E., and Taber, T.H.: Digital arthroplasty of the proximal interphalangeal joint, J. Bone Joint Surg. 36-A:912, 1954.

Coonrad, R.W., and Goldner, J.L.: A study of the pathological findings and treatment in soft-tissue injury of the thumb metacarpophalangeal joint: with a clinical study of the normal range of motion in one thousand thumbs and a study of post mortem findings of ligamentous structures in relation to function, J. Bone Joint Surg. 50-A:439, 1968.

Curtis, R.M.: Capsulectomy of the interphalangeal joints of the fingers, J. Bone Joint Surg. 36-A:1219, 1954.

Curtis, R.M.: Joints of the hand. In Flynn, J.E., editor: Hand surgery, Baltimore, 1966, The Williams & Wilkins Co.

Curtis, R.M.: Management of the stiff hand. In The practice of hand surgery, Oxford, 1981, Blackwell Scientific Publications, Ltd.

Dibbell, D.G., and Field, J.H.: Locking metacarpal phalangeal joint, Plast. Reconstr. Surg. 40:562, 1967.

Eaton, R., and Littler, J.W.: A study of the basal joint of the thumb: treatment of its disability by fusion, J. Bone Joint Surg. 51-A:661, 1969.

Eaton, R.G., and Littler, J.W.: Ligament reconstruction for the painful thumb carpometacarpal joint, J. Bone Joint Surg. 55-A:1655, 1973.

Flatt, A.E.: Recurrent locking of an index finger, J. Bone Joint Surg. 40-A:1128, 1958.

Flatt, A.E.: A locking little finger, J. Bone Joint Surg. 43-A:240, 1961.

Fowler, S.B.: Mobilization of metacarpophalangeal joint, J. Bone Joint Surg. 29:193, 1947.

Frank, W.E., and Dobyns, J.: Surgical pathology of collateral ligamentous injuries of the thumb, Clin. Orthop. 83:102, 1972.

Froimson, A.I.: Tendon arthroplasty of the trapeziometacarpal joint, Clin. Orthop. 70:191, 1970.

Gervis, W.H.: Excision of the trapezium for osteoarthritis of the trapeziometacarpal joint, J. Bone Joint Surg. 31-B:537, 1949.

Gervis, W.H.: A review of excision of the trapezium for osteoarthritis of the trapezio-metacarpal joint after twenty-five years, J. Bone Joint Surg. 55-B:56, 1973.

Goldner, J.L., and Clippinger, F.W.: Excision of the greater multangular bone as an adjunct to mobilization of the thumb, J. Bone Joint Surg. 41-A:609, 1959.

Goodfellow, J.W., and Weaver, J.P.: Locking of the metacarpophalangeal joints, J. Bone Joint Surg. 43-B:772, 1961.

Gould, J.S., and Nicholson, B.G.: Capsulectomy of the metacarpophalangeal and proximal interphalangeal joints, J. Hand Surg. 4:482, 1979.

Graham, W.C., and Riordan, D.C.: Reconstruction of a metacarpophalangeal joint with a metatarsal transplant, J. Bone Joint Surg. 30-A:848, 1948.

Haraldsson, S.: Extirpation of the trapezium for osteoarthritis of the first carpometacarpal joint, Acta Orthop. Scand. 43:347, 1972.

Harrison, S.H.: The Harrison-Nicolle intramedullary peg: follow-up study of 100 cases, Hand 6:304, 1974.

Howard, L.D., Jr.: Locking proximal finger joint (2 cases), Spectator Correspondence Club Letter, 1961, (mimeographed).

Huffaker, W.H., Wray, R.C., Jr., and Weeks, P.M.: Factors influencing final range of motion in the fingers after fractures of the hand, Plast. Reconstr. Surg. 63:82, 1979.

Kessler, I.: Complete avulsion of the ulnar collateral ligament of the metacarpophalangeal joint of the thumb, Clin. Orthop. **29:**196, 1963.

Kessler, I.: Silicone arthroplasty of the trapezio-metacarpal joint, J. Bone Joint Surg. **55-B:**285, 1973.

Kessler, I.: A simplified technique to correct hyperextension deformity of the metacarpophalangeal joint of the thumb, J. Bone Joint Surg. **61-A:**903, 1979.

Kleinert, H.E., and Kasdan, M.L.: Reconstruction of chronically subluxated proximal interphalangeal finger joint, J. Bone Joint Surg. **47-A:**958, 1965.

Lasserre, C., Pauzat, D., and Derennes, R.: Osteoarthritis of the trapezio-metacarpal joint, J. Bone Joint Surg. **31-B:**534, 1949.

Leach, R.E., and Bolton, P.E.: Arthritis of the carpometacarpal joint of the thumb: results of arthrodesis, J. Bone Joint Surg. **50-A:**1171, 1968.

Lee, B.S.: Degenerative arthritis of the carpometacarpal joint of the thumb, Orthop. Rev. **2:**45, April 1973.

Mack, G.R., Lichtman, D.M., and MacDonald, R.I.: Fibular autografts for distal defects of the radius, J. Hand Surg. **4:**576, 1979.

Massengill, J.B., et al.: Mechanical analysis of Kirschner wire fixation in a phalangeal model, J. Hand Surg. **4:**351, 1979.

McGarth, M.H., and Watson, H.K.: Late results with local bone graft donor sites in hand surgery, J. Hand Surg. **6:**234, 1981.

Moberg, E.: Arthrodesis of finger joints, Surg. Clin. North Am. **40:**465, 1960.

Moberg, E.: Biological problems in the evolution of hand prostheses, Orthop. Clin. North Am. **4:**1161, 1973.

Moberg, E., and Henrikson, B.: Technique for digital arthrodesis: a study of 150 cases, Acta Chir. Scand. **118:**331, 1960.

Moberg, E., and Stener, B.: Injuries to the ligaments of the thumb and fingers: diagnosis, treatment, and prognosis, Acta Chir. Scand. **106:**166, 1953.

Morris, H.D.: Tendon and mortise grafts for bridging metacarpal defects due to gunshot wounds, Surgery **20:**364, 1946.

Müller, G.M.: Arthrodesis of the trapezio-metacarpal joint for osteoarthritis, J. Bone Joint Surg. **31-B:**540, 1949.

Murley, A.H.: Excision of the trapezium in osteoarthritis of the first carpo-metacarpal joint, J. Bone Joint Surg. **42-B:**502, 1960.

Murray, R.A.: The injured or abnormal thumb: recommendations for treatment, South. Med. J. **52:**845, 1959.

Palmer, A.K., and Louis, D.S.: Assessing ulnar instability of the metacarpophalangeal joint of the thumb, J. Hand Surg. **3:**542, 1978.

Peacock, E.E., Jr.: Reconstructive surgery of hands with injured central metacarpophalangeal joints, J. Bone Joint Surg. **38-A:**291, 1956.

Potenza, A.D.: A technique for arthrodesis of finger joints, J. Bone Joint Surg. **55-A:**1534, 1973.

Pratt, D.R.: Exposing fractures of the proximal phalanx of the finger longitudinally through the dorsal extensor apparatus, Clin. Orthop. **15:**22, 1959.

Redler, I., and Williams, J.T.: Rupture of a collateral ligament of the proximal interphalangeal joint of the fingers: analysis of eighteen cases, J. Bone Joint Surg. **49-A:**322, 1967.

Sakellarides, H.T., and DeWeese, J.W.: Instability of the metacarpophalangeal joint of the thumb: reconstruction of the collateral ligaments using the extensor pollicis brevis tendon, J. Bone Joint Surg. **58-A:**106, 1976.

Sims, C.D., and Bentley, G.: Carpometacarpal osteo-arthritis of the thumb, Br. J. Surg. **57:**442, 1970.

Slocum, D.B.: Stabilization of the articulation of the greater multangular and the first metacarpal, J. Bone Joint Surg. **25:**626, 1943.

Smillie, I.S.: Intermetacarpal fusion, J. Bone Joint Surg. **35-B:**256, 1953.

Souter, W.A.: The boutonnière deformity: a review of 101 patients with division of the central slip of the extensor expansion of the fingers, J. Bone Joint Surg. **49-B:**710, 1967.

Stark, H.H., et al.: Fusion of the first metacarpotrapezial joint for degenerative arthritis, J. Bone Joint Surg. **59-A:**22, 1977.

Swanson, A.B., and Herndon, J.H.: Flexible (silicone) implant arthroplasty of the metacarpophalangeal joint of the thumb, J. Bone Joint Surg. **59-A:**362, 1977.

Turkell, J.H.: Tenodesis as adjunct to fixation in fusion of interphalangeal joints, Bull. Hosp. Joint Dis. **20:**103, 1959.

Watson, H.K., Light, T.R., and Johnson, T.R.: Checkrein resection for flexion contracture of the middle joint, J. Hand Surg. **4:**67, 1979.

Watson, H.K., and Shaffer, S.R.: Concave-convex arthrodeses in joints of the hand, Plast. Reconstr. Surg. **46:**368, 1970.

Weckesser, E.C.: Rotational osteotomy of the metacarpal for overlapping fingers, J. Bone Joint Surg. **47-A:**751, 1965.

Weeks, P.M.: The chronic boutonnière deformity: a method of repair, Plast. Reconstr. Surg. **40:**248, 1967.

Weeks, P.M., Wray, R.C., Jr., and Kuxhaus, M.: The results of nonoperative management of stiff joints in the hand, Plast. Reconstr. Surg. **61:**58, 1978.

Weiland, A.J., et al.: Free vascularized bone grafts in surgery of the upper extremity, J. Hand Surg. **4:**129, 1979.

Weilby, A., and Søndorf, J.: Results following removal of silicone trapezium metacarpal implants, J. Hand Surg. **3:**154, 1978.

Weinman, D.T., and Lipscomb, P.R.: Degenerative arthritis of the trapeziometacarpal joint: arthrodesis or excision? Mayo Clin. Proc. **42:**276, 1967.

Wilson, J.N.: Basal osteotomy of the first metacarpal in the treatment of arthritis of the carpometacarpal joint of the thumb, Br. J. Surg. **60:**854, 1973.

Yancey, H.A., Jr., and Howard, L.D., Jr.: Locking of the metacarpophalangeal joint, J. Bone Joint Surg. **44-A:**380, 1962.

Young, V.L., Wray, R.C., Jr., and Weeks, P.M.: The surgical management of stiff joints in the hand, Plast. Reconstr. Surg. **62:**835, 1978.

reconstructive procedure. Skin from an otherwise useless digit may be employed as a free graft. Skin and deeper soft structures may be useful as a filleted graft (p. 35); if desired, the bone may be removed primarily and the remaining flap suitably fashioned during a secondary procedure. Skin well supported by one or more neurovascular bundles but not by bone may be saved and used as a neurovascular island graft (p. 126). Segments of nerves may be useful as autogenous grafts. A musculotendinous unit, especially a flexor digitorum sublimis or an extensor indicis proprius, may be saved for transfer to improve function in a surviving digit, as for example, to improve adductor power of the thumb when the third metacarpal shaft has been destroyed or to improve abduction when the recurrent branch of the median nerve has been destroyed. Tendons of the flexor digitorum sublimis of the fifth finger, the extensor digiti quinti, and the extensor indicis proprius may be useful as free grafts. Bones may be used as peg grafts or for filling osseous defects. Under certain circumstances even joints may be useful.

5. Every effort, of course, should be made to salvage the thumb.

PRINCIPLES OF AMPUTATION OF FINGERS

Whether an amputation is carried out primarily or secondarily, certain principles must be observed to obtain a painless and useful stump. The volar skin flap should be semicircular at its end and, to assure adequate coverage, almost as long as a phalanx; it must be long enough to cover the volar surface and tip of the stump and to join the dorsal flap slightly dorsally. The ends of the digital nerves should be carefully dissected from the volar flap and resected at least 6 mm proximal to its end; during this resection tension on the nerves should not be sufficient to rupture axons more proximally, since this might cause discomfort later. Neuromas at the nerve ends are inevitable but they should be allowed to develop only in padded areas where they are less likely to be painful. The digital arteries should be cauterized. When scarring or a skin defect makes the fashioning of a classic flap impossible, then one of a different shape may be improvised, but the end of the bone must be padded well. Flexor and extensor tendons should be drawn distally, divided, and allowed to retract proximally. When an amputation is through a joint, the flares of the osseous condyles should be resected to avoid

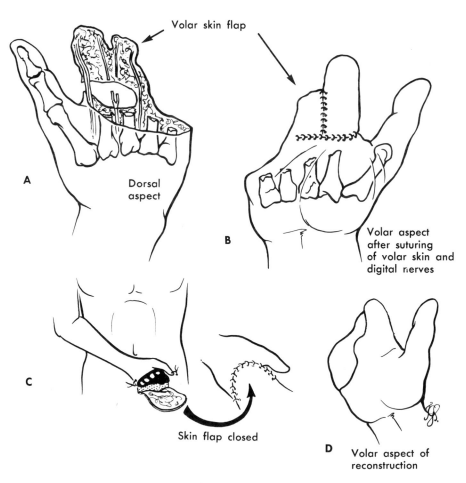

Fig. 9-2. Use of skin supported by neurovascular bundles. **A,** Dorsum of hand amputated through four ulnar metacarpals. Part of volar skin of index and middle fingers remains, is partially detached, but is viable. **B,** Volar surface after repair of severed nerves and suture of the skin. **C,** Application of an abdominal flap to the dorsum of the hand. **D,** Volar surface after application of abdominal flap.

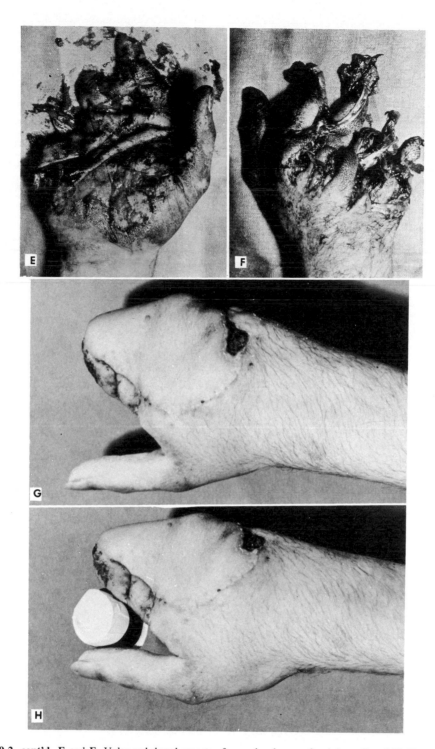

Fig. 9-2, cont'd. E and **F,** Volar and dorsal aspects of same hand soon after injury. **G** and **H,** Hand after reconstruction. Sensibility is normal in skin preserved from index finger. Bone graft to remnant of third metacarpal has provided stability for pinch and grasp. (From Entin, M.A.: Surg. Clin. North Am. **48:**1063, 1968.)

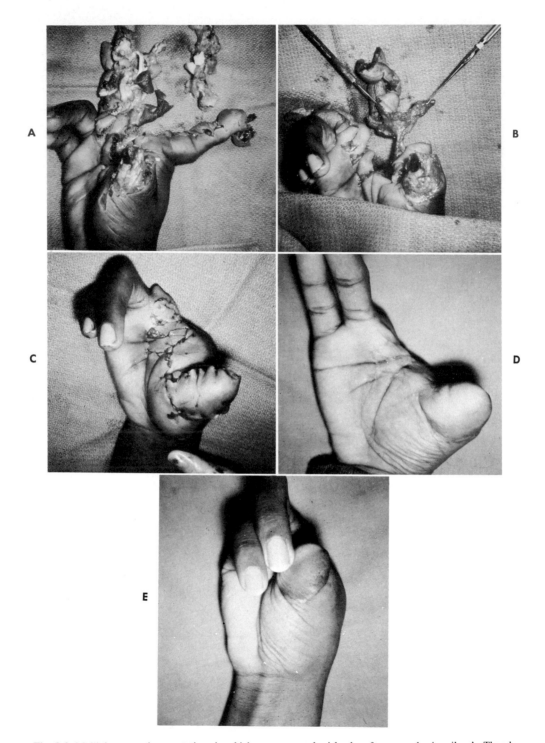

Fig. 9-3. Multiple traumatic amputations in which a neurovascular island graft was used primarily. **A,** Thumb was partially amputated, including its entire area of pinch and grasp. Index and middle fingers were damaged beyond repair. **B,** Index and middle fingers were amputated, but skin of radial side of index finger and its intact neurovascular bundle were salvaged. **C,** Neurovascular island graft was sutured on thumb, providing not only immediate coverage but also sensibility. **D** and **E,** Function after healing without additional operations.

clubbing of the stump. Before the wound is closed, the tourniquet should be released and any bleeding controlled because fingers are quite vascular and any hematoma is painful and delays healing. The flaps should be stitched together with small interrupted sutures; little consideration should be given to "dog-ears" at each end of the suture line because they tend to disappear and because excising them requires narrowing the base of the volar flap, which might endanger its blood supply.

The principles of handling tissues are described in the section on open hand injuries (p. 29).

AMPUTATIONS OF FINGERTIP

Amputations of the fingertips vary markedly depending on the amount of skin lost, the depth of the soft tissue defect, and whether the phalanx has been exposed or even partially amputated (Fig. 9-4). Proper treatment is determined by the exact type of injury and whether other digits have been injured too.

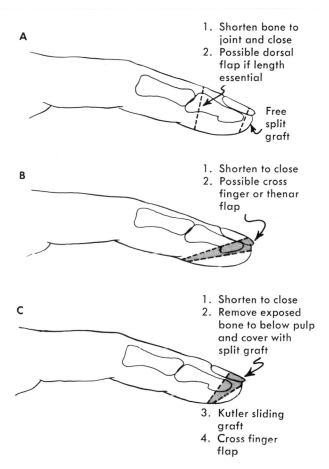

A

1. Shorten bone to joint and close
2. Possible dorsal flap if length essential

Free split graft

B

1. Shorten to close
2. Possible cross finger or thenar flap

C

1. Shorten to close
2. Remove exposed bone to below pulp and cover with split graft
3. Kutler sliding graft
4. Cross finger flap

Fig. 9-4. Techniques useful in closing amputations of fingertip. **A,** For amputations at more distal level, free split graft is applied; at more proximal level, bone is shortened to permit closure or, if length is essential, dorsal flap may be used. **B,** For amputations through *stippled area,* bone may be shortened to permit closure or cross finger or thenar flap may be used. **C,** For amputations through *stippled area,* bone may be shortened to permit closure, exposed bone may be resected and split-thickness graft is applied, Kutler advancement flaps may be used, or cross finger flap may be applied. In small children, fingertip will heal without a graft nearly as fast as with a graft.

When skin alone has been lost, only it requires replacement and this is accomplished by a free graft. However, when the soft tissue defect is deep and the phalanx is exposed, then deeper tissues as well as skin must be replaced. Several methods of coverage are available as follows. *Reamputation* of the finger at a more proximal level provides ample skin and other soft tissues for closure but requires shortening the finger. Yet it may be indicated when other parts of the hand are severely injured or when the entire hand would be endangered by keeping a finger in one position for a long time as is required for a flap; this is especially true for patients with arthritis or for those over 50 years of age. For small children reamputation is not indicated because in them nature will cover the exposed bone in a remarkably short time even if the surgeon does not. A *free skin graft* may be used for coverage but normal sensibility is never restored. A split-thickness graft is often sufficient when the bone is only slightly exposed and its end is nibbled off beneath the fat. Such a graft contracts during healing and eventually becomes about half its original size. Sometimes a full-thickness graft is available from other injured parts of the hand but the fat should be removed from its deep surface. Occasionally the amputated part of the fingertip is recovered and replacing it as a free graft is tempting; although it usually survives in children, it rarely does so in adults. The medial aspect of the arm just inferior to the axilla is a convenient area from which to obtain a small full-thickness graft. Any free graft should always be secured by a stent dressing tied over the end of the finger.

When deeper tissues as well as skin must be replaced to cover exposed bone, one of several different flaps or grafts may be used. The *Kutler V-Y* or *Atasoy triangular advancement flaps* involve the injured finger alone but provide no additional skin and sometimes impair sensibility. The *bipedicle dorsal flap* is useless unless the finger has been amputated proximal to the nail bed. However, when further shortening is unacceptable, this type of flap can be raised from the dorsum of the injured finger and carried distally without involving another digit. The *cross finger flap* provides excellent coverage but may be followed by stiffness not only of the involved finger but of the donor finger as well. This type of coverage requires operation in two stages and a split-thickness graft to cover the donor site. The *thenar flap* also requires operation in two stages. Furthermore, it usually will not cover as large a defect as will a cross finger flap and it sometimes is followed by tenderness of the donor site. But it has the advantage of involving only one finger directly. A *local neurovascular island graft* shifted distally seems ideal in that a good pad with normal sensibility is provided. However, to obtain the desired result, an experienced surgeon with almost perfect technique is required. A *flap from a distant area* such as the abdomen or subpectoral region should be avoided because of all the procedures mentioned it provides the poorest coverage and is followed by the most complications. Such flaps usually are too thick and are unstable, hyperpigmented, and hypersensitive.

Reamputation of fingertip

The technique of reamputation follows the same principles as described for amputation of a finger, see above.

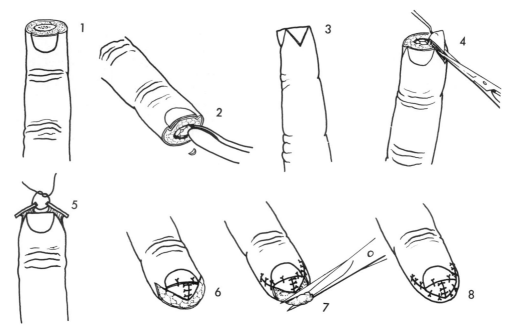

Fig. 9-5. Kutler V-Y advancement flaps (see text). (From Fisher, R.H.: J. Bone Joint Surg. **49-A:**317, 1967.)

Free skin graft

The techniques of applying free skin grafts are described on p. 14.

Kutler V-Y or Atasoy triangular advancement flaps

This type of fingertip coverage is appealing because it involves the injured finger alone. However, it provides only limited coverage and does not consistently result in normal sensibility.

TECHNIQUE (KUTLER; FISHER). Anesthetize the finger by digital block at the proximal phalanx and apply a rubber catheter as a tourniquet. Debride the tip of the finger of uneven edges of soft tissue and any protruding bone (Fig. 9-5). Then develop two triangular flaps, one on each side of the finger with the apex of each directed proximally and centered in the midlateral line of the digit. Avoid making the flaps too large; their sides should each measure about 6 mm and their bases about the same or slightly less. Develop the flaps farther by incising deeper toward the nail bed and volar pulp. Take care not to pinch them with thumb forceps or hemostats. Rather, near the base of each insert a skin hook and apply slight traction in a distal direction. Now with a pair of small scissors and at each apex divide the pulp just enough (usually not more than half its thickness) to allow the flaps to be mobilized toward the tip of the finger. Avoid dividing any pulp distally. Round off the sharp corners of the remaining part of the distal phalanx and reshape its end to conform with the normal tufts. Next approximate the bases of the flaps and stitch them together with small interrupted nonabsorbable sutures; then stitch the dorsal sides of the flaps to the remaining nail or nail bed and close the defect volar to the flaps. Apply Xeroform gauze and a routine dressing.

TECHNIQUE (ATASOY ET AL.). Under tourniquet control and appropriate anesthesia, cut a triangle in the remaining pulp skin area with the base equal in width to the cut edge of the nail (Fig. 9-6). Develop a full-thickness flap with nerves and blood supply preserved. Carefully separate the fibrofatty subcutaneous tissue from the periosteum and flexor tendon sheath using sharp dissecting instruments and cutting the vertical septa that hold the flap in place. Mobilize and distally advance the flap, and suture it in place with interrupted sutures. A few millimeters of the phalanx may be removed when the bone protrudes.

Bipedicle dorsal flaps

A bipedicle dorsal flap is useful when a finger has been amputated proximal to its nail bed and when preserving all its remaining length is essential but attaching it to another finger is undesirable. When this flap can be made wide enough in relation to its length, one of its pedicles may be divided, leaving it attached only at one side (Fig. 9-7).

TECHNIQUE. Beginning distally at the raw margin of the skin and proceeding proximally, elevate the skin and subcutaneous tissue from the dorsum of the finger. Then at a more proximal level make a transverse dorsal incision to create a bipedicle flap long enough, when drawn distally, to cover the bone and other tissues on the end of the stump. Suture the flap in place and cover the defect created on the dorsum of the finger by a split-thickness skin graft. The flap may be made more mobile by freeing one of its pedicles but this decreases its vascularity.

Cross finger flaps

The technique of applying cross finger flaps is described on p. 32.

Thenar flap

TECHNIQUE. With the thumb held in abduction, flex the injured finger so that its tip touches the middle of the

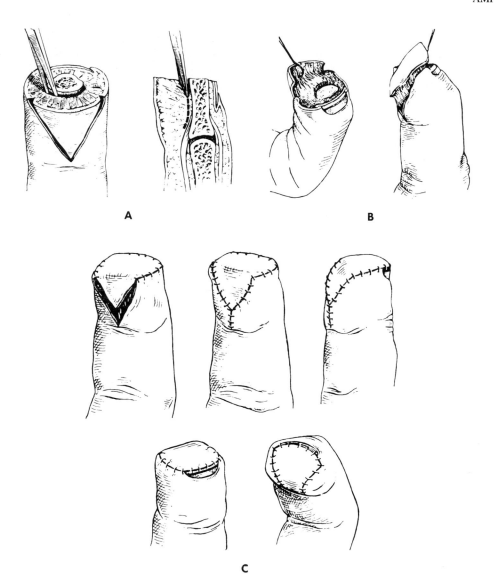

Fig. 9-6. Atasoy V-Y technique. **A,** Skin incision and mobilization of triangular flap. **B,** Advancement of triangular flap. **C,** Suturing of base of triangular flap to nail bed and closure of defect, V-Y technique. (From Atasoy, E., et al.: J. Bone Joint Surg. **52-A:**921, 1970.)

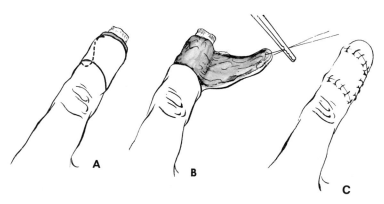

Fig. 9-7. Dorsal pedicle flap useful for amputations proximal to nail when preserving length is essential. It may have two pedicles or, as illustrated here, only one. **A,** Flap has been outlined. **B,** Flap has been elevated, leaving only a single pedicle. **C,** Flap has been sutured in place over end of stump and remaining defect on dorsum of finger has been covered by split-thickness skin graft.

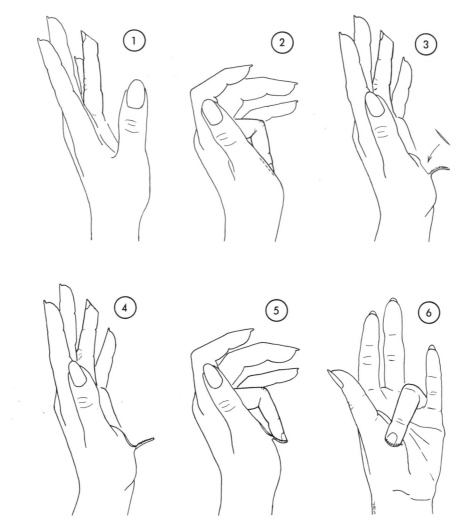

Fig. 9-8. Thenar flap for amputation of fingertip. **1,** Tip of ring finger has been amputated. **2,** Finger has been flexed so that its tip touches middle of thenar eminence, and thenar flap has been outlined. **3,** Split-thickness graft is to be sutured to donor area before flap is attached to finger. **4,** Split-thickness graft is in place. **5** and **6,** End of flap has been attached to finger by sutures passed through nail and through tissue on each side of it.

thenar eminence (Fig. 9-8). Outline on the thenar eminence a flap that when raised will be large enough to cover the defect and will be properly positioned; pressing the bloody stump of the injured finger against the thenar skin will outline by bloodstain the size of the defect to be covered. With its base proximal, raise the thenar flap to include most of the underlying fat; handle the flap with skin hooks to avoid crushing it even with small forceps. Make the flap sufficiently wide that when sutured to the convex fingertip it will not be under tension. Furthermore, make its length no more than twice its width. By gentle undermining of the skin border at the donor site the defect can be closed directly without resorting to a graft. Attach the distal end of the flap to the trimmed edge of the nail by sutures passed through the nail. The lateral edges of the flap should fit the margins of the defect but, to avoid impairing circulation in the flap, suture only their most distal parts, if any, to the finger. Prevent the flap from folding back on itself and strangulating its vessels. Finally, control all bleeding, check the positions of the flap and finger, and

apply wet cotton gently compressed to follow the contours of the graft and the fingertip. Now hold the finger in the proper position by gauze and adhesive tape and splint the wrist.

AFTERTREATMENT. At 4 days the graft is dressed and thereafter kept as dry as possible by dressing it every 1 or 2 days and by leaving it partially exposed. At 2 weeks the base of the flap is detached, and the free skin edges are sutured in place. The contours of the fingertip and the thenar eminence will improve with time.

Local neurovascular island graft

A limited area of the touch pad may be resurfaced by a local neurovascular island graft. This graft provides satisfactory padding and normal sensibility to the most important working surface of the digit.

TECHNIQUE. Make a midlateral incision on each side of the finger (or thumb) beginning distally at the defect and extending proximally to the level of the proximal interphalangeal joint. On each side and beginning proximally,

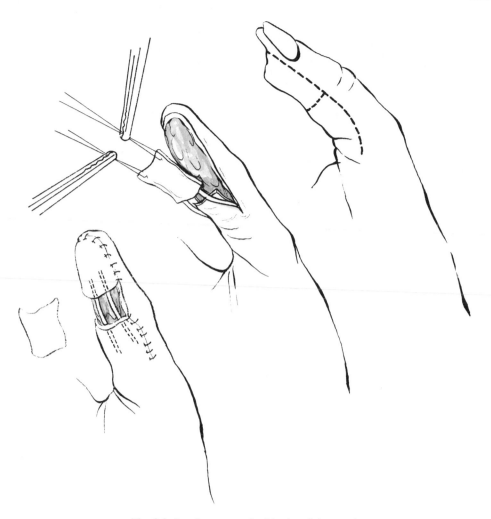

Fig. 9-9. Local neurovascular island graft (see text).

carefully dissect the neurovascular bundle distally to the level selected for the proximal margin of the graft (Fig. 9-9). Here make a transverse volar incision through the skin and subcutaneous tissues but carefully protect the neurovascular bundles. Then if necessary, make another transverse incision at the margin of the defect, thus freeing a rectangular island of the skin and underlying fat to which is attached the two neurovascular bundles. Carefully draw this island or graft distally and place it over the defect, avoiding too much tension on the bundles; should tension be sufficient to embarrass the circulation in the graft, then dissecting the bundles more proximally or flexing the distal interphalangeal joint or both may be necessary. Suture the graft in place with interrupted small nonabsorbable sutures. Now cover the defect created on the volar surface of the finger by a free full-thickness graft. Place over the grafts wet cotton carefully shaped to fit the contour of the area and to prevent pressure on the neurovascular bundles. Then apply a compression dressing.

AMPUTATIONS OF SINGLE FINGER
Index finger

When the index finger is amputated at its proximal interphalangeal joint or at a more proximal level, the remaining stump is useless and may hinder pinch between the thumb and middle finger. Therefore, in most instances, when a primary amputation must be at such a proximal level, any secondary amputation should be through the base of the second metacarpal. This index ray amputation is especially desirable in women for cosmetic reasons. However, because it is a more extensive operation than amputation through the finger, it may cause stiffness of the other fingers and is contraindicated in arthritic hands and in men past middle age. Unless knowledge of anatomy is precise, the branch of the median nerve to the second web may be damaged. Furthermore, improper technique may result in a sunken scar on the dorsum of the hand or in anchoring the first dorsal interosseus to the extensor mechanism rather than to the base of the proximal phalanx, causing intrinsic overpull.

TECHNIQUE OF INDEX RAY AMPUTATION. With a skin pencil outline the incision as follows. First make a mark on the palmar edge of the second web near the base of the index finger; this is the distal point on which the skin flap will be rotated. Then outline the dorsal part of the incision that extends from this point proximally over the second metacarpal shaft and, curving radially, ends over its base. Next, from the same point on the web, outline the rest of the incision that extends first distally to the proximal interphalangeal joint of the index finger near the ulnar side of its volar surface, then radially across the volar surface of the finger, and finally proximally and ulnarward to join the dorsal part of the incision just proximal to the second metacarpophalangeal joint. Now make the incision as just outlined. Ligate and divide the dorsal veins and at a more proximal level divide the branches of the superficial radial nerve. Divide too the tendons of the extensor digitorum communis of the index finger and the extensor indicis proprius and allow them to retract. Detach the tendinous insertion of the first dorsal interosseus and dissect the muscle proximally from the second metacarpal shaft. Then detach the volar interosseus from the same shaft and divide the transverse metacarpal ligament that connects the second and third metacarpal heads; take care not to damage the radial digital nerve of the middle finger. With bone-cutting forceps carefully divide the second metacarpal about 1.9 cm distal to its base; do not disarticulate the bone at its proximal end. Smooth any rough edges on the remaining part of the metacarpal. Divide both flexor tendons of the index finger and allow them to retract. Then divide both digital nerves of the finger proximal to the skin incision; carefully divide the ulnar digital nerve distal to the common digital nerve to prevent damaging the radial digital nerve of the middle finger. Next anchor the tendinous insertion of the first dorsal interosseus to the base of the proximal phalanx of the middle finger; do not anchor it to the extensor tendon or its hood because this might cause intrinsic overpull. With a running suture approximate the muscle bellies in the area previously occupied by the second metacarpal shaft. Now shape the skin edges as desired, ligate or cauterize all obvious bleeders, and remove the tourniquet. Close the skin with small interrupted nonabsorbable sutures and drain the wound dorsally. Apply a well-molded wet dressing that conforms to the wide new web between the middle finger and the thumb, and support the wrist by a large bulky dressing or a plaster splint.

AFTERTREATMENT. The hand is elevated immediately after surgery for 48 hours. At 24 hours the drain is removed.

Middle or ring finger

In contrast to the proximal phalanx of the index finger, this phalanx of either the middle or ring finger is important functionally. The absence of either makes a hole through which small objects can drop when the hand is used as a cup or in a scooping maneuver; furthermore, it makes the remaining fingers tend to deviate toward the midline of the hand. In multiple amputations the length of either the middle or ring finger becomes even more important. The third and fourth metacarpal heads are important too because they help stabilize the metacarpal arch by providing attachments for the transverse metacarpal ligament.

In a child or woman, when the middle finger has been amputated proximal to the proximal interphalangeal joint and especially when amputated proximal to the metacarpal head, transposing the index ray ulnarward to replace the third ray may be indicated. This operation results in more natural symmetry, removes any conspicuous stump, and makes the presence of only three fingers less obvious. It must be remembered, however, that excising the third metacarpal shaft removes the origin of the adductor pollicis and thus weakens pinch. Therefore the index ray

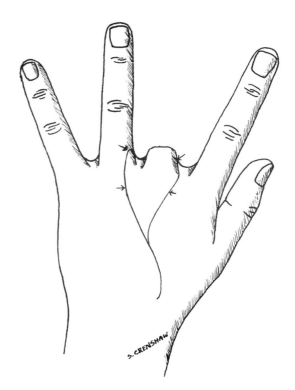

Fig. 9-10. Peacock technique of transposing index ray. Dorsal incision is shown; arrows indicate points along skin edges that will be brought together. Similar palmar incision is made (see text).

should not be transposed unless this adductor can be reattached elsewhere; furthermore, the operation is contraindicated when the hand is needed for heavy manual labor.

In the same circumstances when the ring finger has been similarly amputated, transposing the fifth ray radialward to replace the fourth is rarely indicated. Resection of the fourth metacarpal at its base or at the carpometacarpal joint and closure of the skin so as to create a common web will permit a "folding-in" of the fifth digit to close the gap without actually transposing the fifth metacarpal.

TECHNIQUE OF TRANSPOSING INDEX RAY (PEACOCK). Plan the incision so that a wedge of skin will be removed from both the dorsal and volar surfaces of the hand (Fig. 9-10). Plot in the region of the transverse metacarpal arch the exact points that must be brought together to form a smooth arch across the dorsum of the hand when the second and fourth metacarpal heads are approximated. Curve the proximal end of the dorsal incision slightly toward the second metacarpal base so that the base can be easily exposed. Fashion the distal end of the incision so that a small triangle of skin will be excised from the ring finger to receive a similar triangle of skin from the stump or the area between the fingers; transferring this triangle is important to prevent the suture line from passing through the depths of the reconstructed web. After the dorsal and volar wedges of skin have been removed and the flaps elevated, expose the third metacarpal through a longitudinal incision in its periosteum. The index ray will be the right length when its metacarpal is moved directly to the third metacarpal base. Therefore with an oscillating saw divide transversely the third metacarpal as close to its base as possible. Then excise the third metacarpal shaft and the interosseus muscles to the middle finger. Take care not to damage interosseus muscles to the remaining fingers. Next identify the neurovascular bundles to the middle finger; individually ligate the arteries and veins, and divide the digital nerves between the metacarpals deep within the substance of the interosseus muscle mass. While the wrist is held flexed, draw the flexor tendons distally as far as possible and divide them. Next retract the extensor tendons of the index finger, expose the second metacarpal at its base, and divide the bone at the same level as the third metacarpal. From the radial side of the second metacarpal gently dissect the intrinsic muscles just enough to allow this metacarpal to be placed on the base of the third without undue tension on the muscles. Then bevel obliquely the second metacarpal base to produce a smooth contour on the side of the hand. From the excised third metacarpal fashion a key graft to extend from one fragment of the reconstructed metacarpal to the other. Then insert a Kirschner wire longitudinally through the metacarpophalangeal joint of the transposed ray and bring it out on the dorsum of the flexed wrist; draw it proximally through the metacarpal until its distal end is just proximal to the metacarpophalangeal joint. With the wrist flexed, cut off the proximal part of the wire and allow the remaining end to disappear beneath the skin. Next flex all the fingers simultaneously to assure correct rotation of the transposed ray and insert a Kirschner wire transversely through the necks of the fourth and the transposed metacarpals. Now close the skin and insert a rubber drain. Apply a soft pressure dressing; no additional external support is needed.

AFTERTREATMENT. At 2 days the rubber drain is removed, and at 8 to 10 days the entire dressing and the sutures are removed. Then a light volar plaster splint is applied to keep the wrist in the neutral position and support the transposed ray; however, the splint is removed daily for cleaning the hand and exercising the small joints. At about 5 weeks when the metacarpal fragments have united, the Kirschner wires are removed with the patient under local anesthesia.

RING FINGER AVULSION INJURIES

The soft tissue of the left ring finger is usually forcefully avulsed at its base when a metal ring worn on that finger catches on a nail or hook. The force is usually sufficient to cause separation of the skin and nearly always damages the vascular supply to the distal tissue. Fractures and ligamentous damage may occur also, but the tendons seem to be the last to separate. Early evaluation of an incomplete avulsion may be disarming as to the severity of damage, but after 48 hours, all soft tissues usually are clearly nonviable. Attempts at salvage routinely fail unless the vascular supply can be reestablished. Amputation of the fourth ray with closure of the web is the procedure of choice in a child or woman. By resecting the fourth ray at its base or at the carpometacarpal joint, the fifth ray will close without having to be surgically transposed. Simple amputation of the finger itself should be done in the face of necrosis and infection and, if indicated, the ray amputation is done later as an elective procedure.

Little finger

As much of the little finger as possible should be saved provided all the requirements for a painless stump are satisfied. Often this finger survives when all others have been destroyed, and then it becomes important in forming pinch with the thumb. But when the little finger alone is amputated and when the appearance of the hand is important or the amputation is at the metacarpophalangeal joint, the fifth metacarpal shaft is divided obliquely at its middle third; then the insertion of the abductor digiti quinti is transferred to the proximal phalanx of the ring finger just as the first dorsal interosseus is transferred to the middle finger in the index ray amputation already described. This smooths the ulnar border of the hand and is used most often as an elective procedure for a contracted or painful little finger.

AMPUTATIONS OF THUMB

In partial amputation of the thumb, in contrast to one of a single finger, reamputation at a more proximal level to obtain closure should not be considered because the thumb should never be shortened. Therefore the wound should be closed primarily by a free graft, an advancement pedicle flap (described later), or a local or distant flap.

When a flap is necessary, taking it from the dorsum of either the hand or the index or middle finger is preferable (Fig. 2-6). A flap from one of these areas provides a touch pad that is stable but that will not regain normal sensibility.

Covering the volar surface of the thumb with an abdominal flap is contraindicated; even when thin, abdominal skin and fat provide a poor surface for pinch because they lack fibrous septa and will roll or shift under pressure. Furthermore, skin of the abdomen is dissimilar in appearance to that of the hand and its digits. When the skin and pulp, including all neural elements, have been lost from a significant area of the thumb, a neurovascular island graft (p. 126) may be indicated. However, the defect should be closed primarily by a split-thickness graft; then the neurovascular island graft or, if feasible, a local neurovascular island graft or advancement flap as described for fingertip amputations (p. 188) is applied secondarily.

When the thumb has been amputated so that a useful segment of the proximal phalanx remains, the only surgery necessary, if any, except for primary closure of the wound, is deepening the thumb web by Z-plasty (p. 18). When amputation has been at the metacarpophalangeal joint or at a more proximal level, then reconstruction of the thumb may be indicated (p. 195).

Advancement pedicle flap for thumb injuries

Advancement flaps for fingertip injuries usually will survive if the volar flap incisions are not brought proximal to the proximal interphalangeal joint. In the thumb, however, the venous drainage is not as dependent on the volar flap, and thus this technique is safer and the flap can be longer (Fig. 9-11).

Under tourniquet control and appropriate anesthesia, a midlateral incision is made on each side of the thumb from the tip to the metacarpophalangeal joint. The flap created contains both neurovascular bundles and should be elevated without disturbing the flexor tendon sheath (Fig. 9-12). Flexion of both the joints will allow the flap to be advanced and carefully sutured over the defect with interrupted sutures. The joints should be maintained in flexion for 3 weeks postoperatively. This rather large flap is used only when a large area of thumb pulp is lost.

AMPUTATIONS OF MULTIPLE DIGITS

In *partial amputations of all fingers,* preserving the remaining length of the digits is much more important than in a single finger amputation (Figs. 9-13 and 9-14). Be-

Fig. 9-11. Thumb tip amputation levels. Acceptable procedures by level are: *1,* split-thickness graft; *2,* cross finger flap or advancement flap; *3,* advancement flap, cross finger flap, or shorten thumb and close; *4,* split-thickness skin graft; *5,* shorten bone and split-thickness skin graft, advancement flap, or cross finger flap; *6,* advancement flap or cross finger flap; *7,* advancement flap and removal of nail bed remnant.

cause of the natural hinge action between the first and fifth metacarpals, any remaining stump of the little finger must play an important role in prehension with the intact thumb; this hinge action may be increased about 50% by dividing the transverse metacarpal ligament between the fourth and fifth rays.

In *complete amputation of all fingers,* if the intact thumb cannot easily reach the fifth metacarpal head, then phalangization of the fifth metacarpal is helpful. In this operation the fourth metacarpal is resected and the fifth is osteotomized, rotated, and separated from the rest of the palm. Lengthening of the fifth metacarpal is also helpful.

TECHNIQUE OF PHALANGIZATION OF FIFTH METACARPAL. Over the fourth metacarpal make dorsal and volar longitudinal incisions that join distally. Expose and resect the transverse metacarpal ligament on each side of the fourth metacarpal head. Then divide proximally the digital nerves to the ring finger and ligate and divide the corresponding vessels. Next resect the fourth metacarpal shaft just distal to its carpometacarpal joint. Through the same incision osteotomize the fifth metacarpal near its base. Slightly abduct and flex the distal fragment and rotate it toward the thumb, and fix the fragments with a Kirschner wire. Next cover the raw surfaces between the third and fifth metacarpals with split-thickness grafts, creating a web at the junction of the proximal and middle thirds of the bones. Be sure the padding over the fifth metacarpal head is good and, if possible, sensation is normal at its point of maximum contact with the thumb.

• • •

In *partial amputation of all fingers and the thumb,* function may be improved by lengthening the digits relatively and by increasing their mobility. Function of the thumb may be imporved by deepening its web by Z-plasty (p. 18)

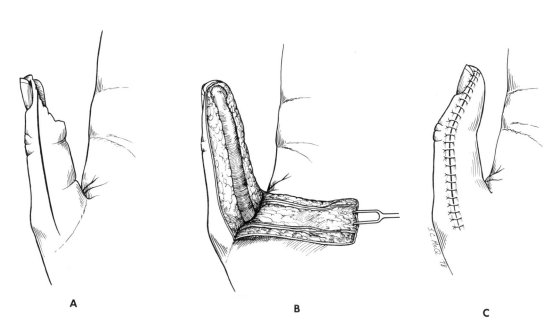

Fig. 9-12. Advancement pedicle flap for thumb injuries. **A,** Deep thumb pad defects exposing bone may be covered with an advancement pedicle flap. **B,** Advancement of neurovascular pedicle. **C,** Flexion of the distal joint of the thumb is necessary to permit placement of the flap (see text).

and by osteotomizing both the first and fifth metacarpals and rotating their distal fragments toward each other (Fig. 9-15) and at the same time, if helpful, by tilting the fifth metacarpal toward the thumb. When the first carpometacarpal joint is functional but the first metacarpal is quite short, the second metacarpal may be transposed to the first to lengthen it and to widen and deepen the first web.

In *complete amputation of all fingers and the thumb* in which the amputation has been transversely through the metacarpal necks, phalangization of selected metacarpals may improve function. The fourth metacarpal is resected to increase the range of motion of the fifth, and function

of the fifth is further improved by osteotomy of the metacarpal in which the distal fragment is rotated radialward and flexed. The second metacarpal is resected at its base but, to preserve the origin of the adductor pollicis, the third metacarpal is not. The thumb should not be lengthened by osteoplastic reconstruction (p. 195) unless sensibility can be added to its volar surface. When the amputation has been through the middle of the metacarpal shafts, prehension probably cannot be restored, but hook can be accomplished by flexing the stump at the wrist. Furthermore, this motion at the wrist can be made even more useful by fitting an artificial platform to which the palmar surface of the stump can be actively opposed.

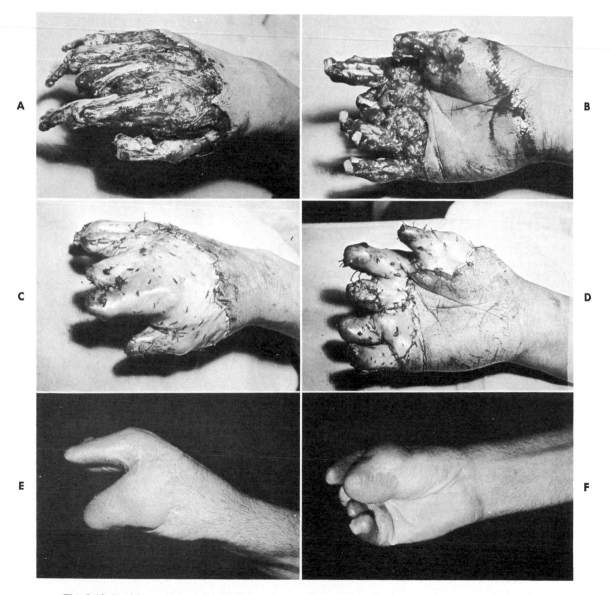

Fig. 9-13. Partial amputation of multiple digits and massive avulsion of skin and neurovascular bundles from digits and hand. **A** and **B,** Dorsal and palmar views soon after injury. **C** and **D,** Dorsal and palmar views after primary operation. Index, ring, and little fingers have been shortened as little as possible and whole area has been covered by split-thickness skin grafts. **E** and **F,** After secondary repair with pedicle graft to thumb and first web. Function is good. (From Matev, I.: J. Bone Joint Surg. **49-B:**722, 1967.)

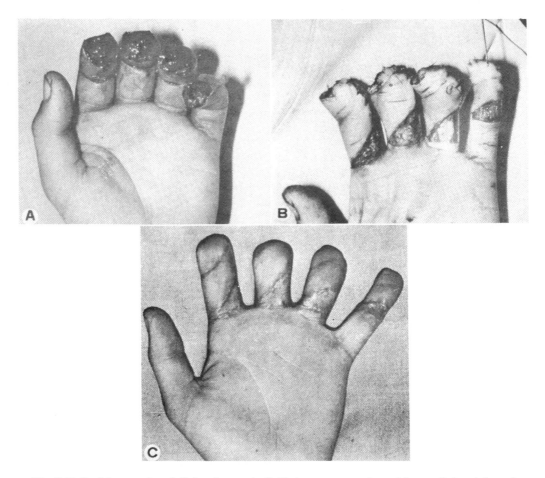

Fig. 9-14. Partial amputation of all four fingers. **A,** Guillotine-type amputations of fingers. **B,** Local flaps of volar skin were transposed distally, preserving remaining length of fingers. Each flap was elevated anterior to ulnar neurovascular bundle but posterior to radial neurovascular bundle. **C,** After healing. In each finger motion was normal, and sensibility was retained in end. (From Hueston, J.: Plast. Reconstr. Surg. **37:**349, 1966.)

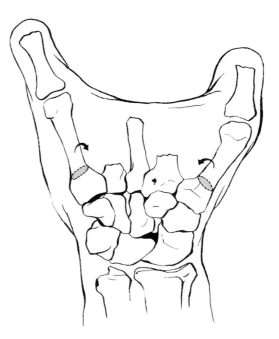

Fig. 9-15. In multiple amputations including the thumb, function may be improved by osteotomizing first and fifth metacarpals and rotating their distal fragments toward each other (see text).

PAINFUL AMPUTATION STUMP

Amputation stumps are often painful enough to require revision; in fact such revision is probably the most frequent elective operation in hand surgery. A neuroma located in an unpadded area near the end of the stump is the usual cause of pain. It is diagnosed by carefully pressing the stump with a small firm object such as the blunt end of a pencil; a well-localized area of extreme tenderness no more than 1 or 2 mm in diameter is always found, usually in line with a digital nerve. When painful, a neuroma should be excised; it and the attached nerve are freed from the scar, and the nerve is divided at a more proximal level where its end will be covered by sufficient padding. Another neuroma will develop but should be painless when located in a padded area.

Pain in an amputation stump may also be caused by bony prominences covered only by thin skin such as a split-thickness graft or by skin made tight by scarring. In these instances excising the thin skin or scar, shortening the bone, and applying a sufficiently padded graft may be indicated. Amputation stumps that are painful because of thin skin coverage at the pulp and nail junction can be improved by using a limited advancement flap as described under thumb amputations. In the finger, proximal dissection to develop these flaps should not extend proximal to the proximal interphalangeal joint.

Finally, painful cramping sensations in the hand and forearm may be caused by flexion contracture of a stump resulting from overstretching of extensor tendons or adherence of flexor tendons; release of any adherent tendons is helpful.

RECONSTRUCTIONS AFTER AMPUTATION
Reconstruction after amputation of hand

Amputation of both hands is, of course, extremely disabling. In selected patients the Krukenberg operation is helpful. It converts the forearm to forceps in which the radial ray acts against the ulnar ray. Swanson compares function of the reconstructed limb with the use of chopsticks. Normal sensibility between the tips of the rays is assured by proper shifting of skin during closure of the wound. The operation, because it provides not only prehension but also sensibility at the terminal parts of the limb, is especially helpful in blind patients with bilateral amputation. But it is helpful too in other patients with similar amputations, especially in surroundings where modern prosthetic services are unavailable. According to Swanson, children with bilateral congenital amputation find the reconstructed limb much more useful than a mechanical prosthesis; they transfer dominance to this limb when a prosthesis is used on the opposite one (Fig. 9-16). In children the appearance of the limb after surgery has not been distressing and, furthermore, the operation does not prevent the wearing of an ordinary prosthesis when desired.

TECHNIQUE (KRUKENBERG; SWANSON). Make a longitudinal incision on the flexor surface of the forearm slightly toward the radial side (Fig. 9-17, A); make a similar one on the dorsal surface slightly toward the ulnar side, but on this surface elevate a V-shaped flap to form a web at the junction of the rays (Fig. 9-17, B). Now separate the forearm muscles into two groups (Fig. 9-17, C and D): on the radial side carry the radial wrist flexors and extensors, the radial half of the flexor digitorum sublimis, the radial half of the extensor digitorum communis, the brachioradialis, the palmaris longus, and the pronator teres; on the ulnar side carry the ulnar wrist flexors and extensors, the ulnar half of the flexor digitorum sublimis, and the ulnar half of the extensor digitorum communis. If they make the stump too bulky or the wound hard to close, resect as necessary the pronator quadratus, the flexor digitorum profundus, the flexor pollicis longus, the abductor pollicis longus, and the extensor pollicis brevis; take care here not to disturb the pronator teres. Next incise the interosseous membrane throughout its length along its ulnar attachment, taking care not to damage the interosseous vessel and nerve. The radial and ulnar rays can now be separated 6 to 12 cm at their tips depending on the size of the forearm; motion at their proximal ends occurs at the radiohumeral and proximal radioulnar joints. The opposing ends of the rays should touch; if not, osteotomize the radius or ulna as necessary. Now the adductors of the radial ray are the pronator teres, the supinator, the flexor carpi radialis, the radial half of the flexor digitorum sublimis, and the palmaris longus; the abductors of the radial ray are the brachioradialis, the extensor carpi radialis longus, the extensor carpi radialis brevis, the radial half of the extensor digitorum communis, and the biceps. The adductors of the ulnar ray are the flexor carpi ulnaris, the ulnar half of the flexor digitorum sublimis, the brachialis, and the anconeus; the abductors of the ulnar ray are the extensor carpi ulnaris, the ulnar half of the extensor digitorum communis, and the triceps.

Remove the tourniquet, obtain hemostasis, and observe the circulation in the flaps. Now excise any excess fat, rotate the skin around each ray, and close the skin over each so that the suture line is not on the opposing surface of either (Fig. 9-17, E and F). Excise any scarred skin at the ends of the rays and, if necessary to permit closure, shorten the bones; in children the skin is usually sufficient for closure and the bones must not be shortened because growth at the distal epiphyses will still be incomplete. Preserve any remaining rudimentary digit. Next suture the flap in place at the junction of the rays and apply any needed split-thickness graft. Insert small rubber drains and, with the tips of the rays separated 6 cm or more, apply a compression dressing.

AFTERTREATMENT. The limb is constantly elevated for 3 to 4 days. The sutures are removed at the usual time. At 2 to 3 weeks rehabilitation to develop abduction and adduction of the rays is begun.

Reconstruction after amputation of multiple digits

Several reconstructive operations are useful after amputation of multiple digits at various levels. These are discussed along with the primary treatment of these injuries (p. 192).

Reconstruction of thumb

Absence of the thumb, either traumatic or congenital, causes a severe deficiency in hand function; in fact, such an absence usually is considered to constitute a 40% disa-

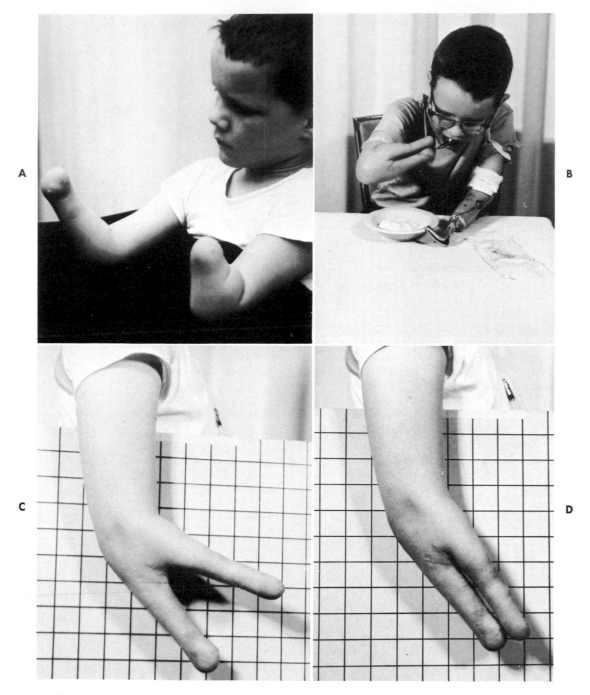

Fig. 9-16. Krukenberg operation for bilateral amputation of hand. **A,** Boy 7 years old who was born with left upper partial hemimelia and right upper archeiria. **B,** After Krukenberg operation child uses reconstructed limb as his dominant one. Secondary osteotomy of radius was necessary to improve apposition at tips of rays. **C,** Tips of rays can be separated about 7.5 cm. **D,** Pinch and grasp are excellent. (From Swanson, A.B.: J. Bone Joint Surg. **46-A:**1540, 1964.)

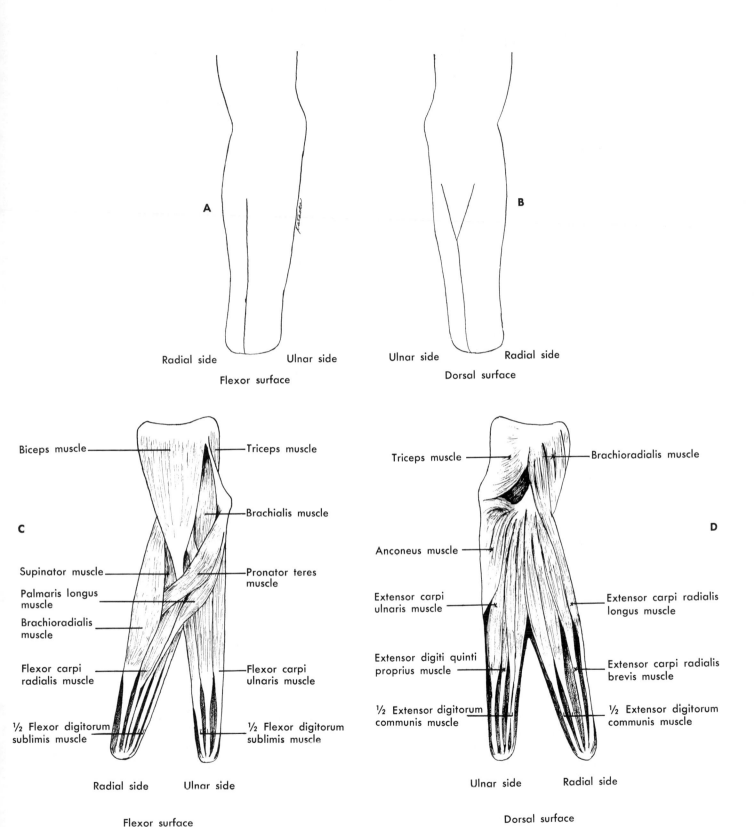

Fig. 9-17. Krukenberg operation. **A,** Incision on flexor surface of forearm. **B,** Incision on dorsal surface (see text). **C** and **D,** Forearm muscles have been separated into two groups (see text). *Continued.*

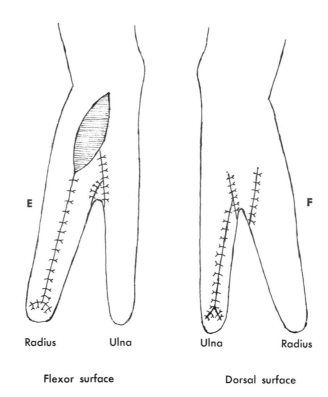

Radius Ulna Ulna Radius

Flexor surface Dorsal surface

Fig. 9-17, cont'd. E, Closure of skin on flexor surface of forearm; parallel lines indicate location of any needed split-thickness skin graft. **F,** Closure of skin on dorsal surface (see text). (Modified from Swanson, A.B.: J. Bone Joint Surg. **46-A:**1540, 1964.)

bility of the hand as a whole. Thus, when the thumb is partially or totally absent, reconstructive surgery is appealing. But before any decision for surgery is made, several factors must be considered: the length of any remaining part of the thumb, the condition of the rest of the hand, the occupational requirements and age of the patient, and the knowledge and experience of the surgeon. When the opposite thumb is normal, some surgeons question the need for reconstructing even a totally absent thumb; at least reconstruction here is not mandatory. However, function of the hand can surely be improved by a suitable operation carefully planned and skillfully executed, especially in a young patient.

Usually the thumb should be reconstructed only when amputation has been at the metacarpophalangeal joint or at a more proximal level. When this joint and a useful segment of the proximal phalanx remain, the only surgery necessary, if any, is deepening of the thumb web by Z-plasty (p. 18). Furthermore, when amputation has been through the interphalangeal joint, the distal phalanx, or the pulp of the thumb, only appropriate coverage by skin is necessary unless sensibility in the area of pinch is grossly impaired; in this latter instance, a more elaborate coverage as by a neurovascular island transfer may be indicated (p. 126).

A reconstructed thumb must meet five requirements. First, sensibility, although not necessarily normal, should be painless and sufficient for recognition of objects held in the position of pinch. This is probably the most important requirement. Second, the thumb should have sufficient stability so that pinch pressure does not cause the thumb joints to deviate or collapse or cause the skin pad to shift. Third, there should be sufficient mobility to enable the hand to flatten and the thumb to oppose for pinch. Fourth, the thumb should be of sufficient length to enable the opposing digital tips to touch it. Sometimes amputation or stiffness of the remaining digits may require greater than normal length of the thumb to accomplish prehension. Fifth, the thumb should be cosmetically acceptable since, if it is not, it may remain hidden and not be used.

Several reconstructive procedures are possible, and the choice depends on the length of the stump remaining and the sensibility of the remaining thumb pad (Fig. 9-18). The thumb may be lengthened by a short bone graft and transferred local skin for sensibility. The nonopposing surface then is skin grafted as in the Gillies-Millard "cocked hat" procedure. Another possibility is pollicizing a digit. Another promising possibility is direct free transfer of a toe to the hand with anastomosis of both the vessels and the nerves by microsurgical technique. In this procedure, nerve restoration is never normal. The osteoplastic technique with a bone graft and tube pedicle skin graft supplemented by a neurovascular pedicle is now rarely recommended.

For congenital absence of the thumb, pollicization of the index finger is the most used technique. Congenital ab-

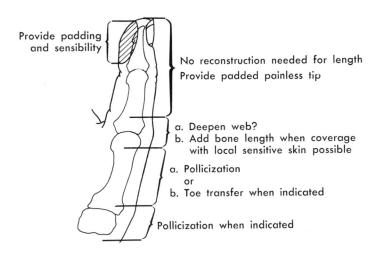

Provide padding and sensibility

No reconstruction needed for length
Provide padded painless tip

a. Deepen web?
b. Add bone length when coverage with local sensitive skin possible

a. Pollicization
 or
b. Toe transfer when indicated

Pollicization when indicated

Fig. 9-18. Thumb reconstruction at various levels. Basic needs are sensibility, stability, mobility, and length.

sence of the thumb is frequently associated with other congenital malformations such as congenital absence of the radius and occasionally metabolic disorders including blood dyscrasias. These latter should be well assessed before elective procedures for thumb reconstruction are performed. These reconstructive procedures are usually done after the first year or two of life.

The so-called floating thumb, a congenital anomaly in which the distal segment of the thumb has no major attachment except a narrow soft tissue pedicle and appears to dangle from a skin thread, is not considered useful enough to attempt reconstruction. The skin of this digit may be used for a skin graft if it is needed, but as a rule it should be detached during the first few months of life.

LENGTHENING OF METACARPAL AND TRANSFER OF LOCAL FLAP

When amputation of the thumb has been at the metacarpophalangeal joint or within the condylar area of the first metacarpal, the thenar muscles are able to stabilize the digit. In these instances, lengthening of the metacarpal by bone grafting and transfer of a local skin flap may be indicated. Furthermore, the technique as described by Gillies and Millard may be completed in one stage and thus the time required for surgery and convalescence is less than in some other reconstructions. But it has disadvantages: the bone graft may resorb and become shortened or its end, after contraction of the flap, may perforate the skin. This procedure requires that there be minimal scarring of the amputated stump.

TECHNIQUE (GILLIES AND MILLARD, MODIFIED). Make a curved incision around the dorsal, radial, and volar aspects of the base of the thumb (Fig. 9-19). Undermine the skin distally but stay superficial to the main veins to prevent congestion of the flap. Now continue the undermining until a hollow flap has been elevated and has been slipped off the end of the stump; the blood supply to the flap is from a source around the base of the index finger in the thumb web. (If desired, complete elevation of the flap may be

delayed until a second operation, as described by Gillies and Millard.) Next attach an iliac bone graft or a phalanx excised from a toe to the distal end of the metacarpal by tapering the graft and fitting it into a hole in the end of the metacarpal. Fix the graft to the bone by a Kirschner wire and place iliac chips about its base. Be sure the graft is small enough that the flap can be easily placed over it. Now cover the raw area at the base of the thumb by a split-thickness skin graft.

AFTERTREATMENT. The newly constructed thumb is immobilized by a supportive dressing and a volar plaster splint is applied to the palm and forearm. The Kirschner wire is removed when the graft has united with the metacarpal. Minor Z-plasties may be necessary later to relieve the volar and dorsal web formed by advancing the flap.

OSTEOPLASTIC RECONSTRUCTION AND TRANSFER OF NEUROVASCULAR ISLAND GRAFT

Verdan recommends osteoplastic reconstruction, especially when the first carpometacarpal joint has been spared and is functional. It is a useful method when the remaining part of the first metacarpal is short. As in the technique of Gillies and Millard, no finger is endangered and all are spared to function against the reconstructed thumb. Transfer of a neurovascular island graft supplies discrete sensibility to the new thumb but precise sensory reorientation is always lacking. For this reconstruction to be successful, the surgeon must be experienced in the use of tubed pedicle grafts and other skin grafting techniques, and the reconstructed thumb must be shorter than normal, never long enough for its end to lie opposite the proximal interphalangeal joint of the index finger.

TECHNIQUE (VERDAN). Raise from the abdomen, the subpectoral region, or some other appropriate area a tubed pedicle graft that contains only moderate subcutaneous fat. Next excise the skin and subcutaneous tissue over the distal end of the first metacarpal; make this area for implantation of the tubed graft a long oval and as large as possible so that the graft may include many vessels and nerves

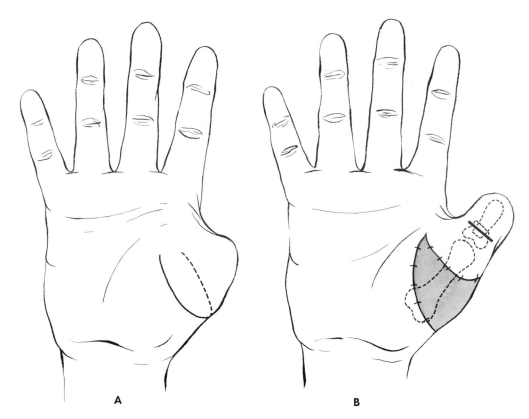

Fig. 9-19. Reconstruction of thumb by technique of Gillies and Millard, modified. **A,** Curved incision around dorsal, radial, and volar aspects of base of thumb has been outlined. **B,** Hollow flap has been undermined and elevated, iliac bone graft has been fixed (this time to base of proximal phalanx), and raw area at base of thumb has been covered by split-thickness skin graft.

and will not constrict later (Fig. 9-20, *A*). Insert into the end of the first metacarpal an iliac bone graft shaped like a palette to imitate the normal thumb. Do not place the graft in line with the first metacarpal but rather place it at an obtuse angle in the direction of opposition. Be sure the graft is not too long. Then place the end of the tubed pedicle over the bone graft and suture it to its prepared bed on the thumb (Fig. 9-20, *C*). Immobilize the hand and tubed pedicle so as to allow normal motion of the fingers and some motion of the shoulder and elbow. After 3 to 4 weeks free the tubed pedicle. Then close the skin over the distal end of the newly constructed thumb, or transfer a neurovascular island graft from an appropriate area to the volar aspect of the thumb to assist in closure and to improve sensation and circulation in the digit (Fig. 9-20, *D* to *G*).

AFTERTREATMENT. A supportive dressing and a volar plaster splint are applied. The newly constructed thumb is protected for about 8 weeks to prevent or decrease resorption of the bone graft. If a neurovascular island graft was not included in the reconstruction (Fig. 9-21), then this transfer must be performed later.

POLLICIZATION

Because pollicization (transposition of a finger to replace an absent thumb) endangers the finger, some sur-

geons recommend transposition only of an already shortened or otherwise damaged one. In this instance, full function of the new thumb can hardly be expected. In fact, full function cannot be expected even after successful transposition of a normal finger. Yet in the hands of an experienced surgeon, pollicization is worthwhile, especially in complete bilateral congenital absence of the thumb or in bilateral traumatic amputation at or near the carpometacarpal joint. When amputation has been traumatic, extensive scarring may require resurfacing by a pedicle skin graft before pollicization.

In the following techniques the index finger is transposed to replace the thumb.

TECHNIQUE (LITTLER). In congenital absence of the thumb with absence of the greater multangular or in traumatic amputation at the first carpometacarpal joint, the repositioned index finger is fixed to its own metacarpal base as described first here. In these instances, the second metacarpophalangeal joint serves as the carpometacarpal joint of the new thumb. Thus placing the new thumb in position for true opposition is impossible.

Make a racquet-shaped incision encircling the base of the index finger. Extend the handle of the racquet proximally and gently curve it first volarly and then dorsally (Fig. 9-22, *A*). Preserve the dorsal vein of the finger. Now free the neurovascular bundles and flexor mechanism of

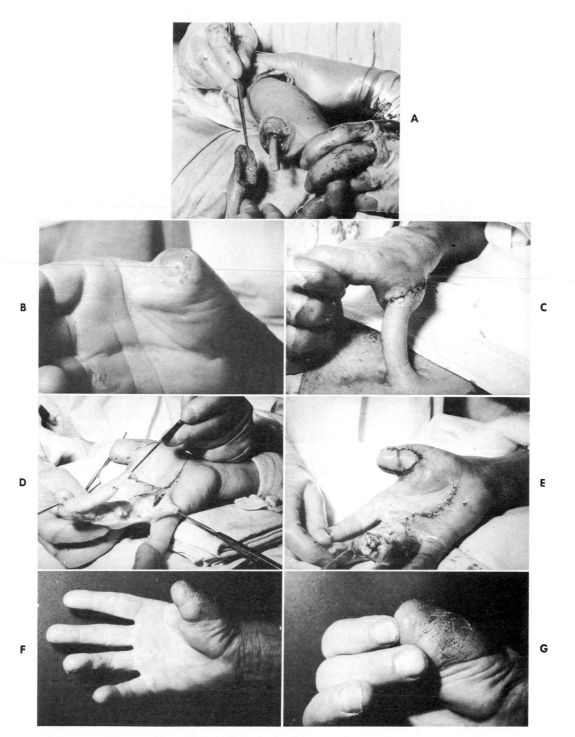

Fig. 9-20. Osteoplastic reconstruction of thumb. **A,** Large area for implantation of tubed pedicle graft has been prepared, iliac bone graft has been inserted into end of first metacarpal, and tubed pedicle graft has been prepared for suture in place over bone graft. **B,** to **G,** Osteoplastic reconstruction of thumb and transfer of neurovascular island graft. **B,** Status of thumb before operation. **C,** First metacarpal has been lengthened by iliac bone graft and tubed pedicle graft has been sutured in place, enclosing bone graft. **D,** Pedicle has been freed and skin has been closed over end of newly constructed thumb; there are trophic changes in the digit and sensibility is markedly deficient. Neurovascular island graft is being raised from ulnar side of ring finger. **E,** Island graft has been transferred to thumb and donor area has been covered by free skin graft. **F** and **G,** Status 3 years after surgery; trophic changes have disappeared and sensibility in the skin of island graft is "normal." (From Verdan, C.: Surg. Clin. North Am. **48:**1033, 1968.)

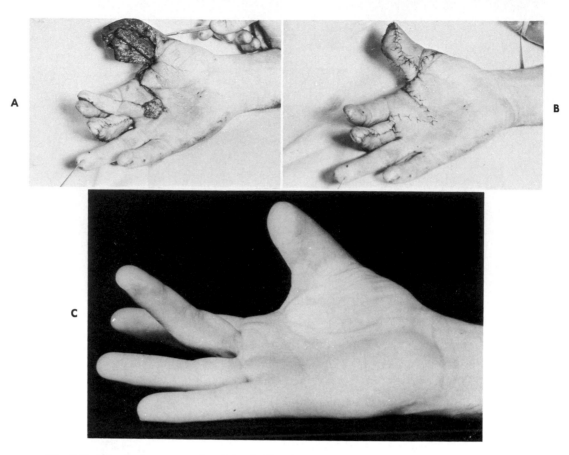

Fig. 9-21. Osteoplastic reconstruction of thumb. Thumb was traumatically amputated through metacarpophalangeal joint with loss of skin dorsally. On day of injury treatment consisted of application of tubed pedicle graft. **A,** One month after injury. Tubed pedicle has been detached from abdomen and opened, iliac bone graft has been inserted into end of first metacarpal, neurovascular island graft has been raised from ulnar side of middle finger, and donor area on finger has been covered by split-thickness skin graft. **B,** Neurovascular island graft has been sutured in place on end and palmar aspect of newly constructed thumb, and palmar incision has been closed. **C,** One year after injury. Thumb has been satisfactorily reconstructed and there is a large area of sensitive skin on its end and palmar surface. (**A** and **B** courtesy Mr. J.T. Hueston; **C** from Hueston, J.: Br. J. Plast. Surg. **18:**304, 1965.)

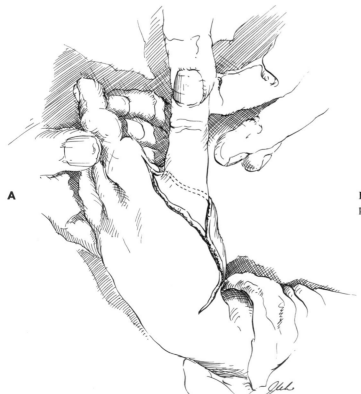

Fig. 9-22. Littler pollicization for congenital absence of thumb or amputation at carpometacarpal joint. **A,** Skin incision.

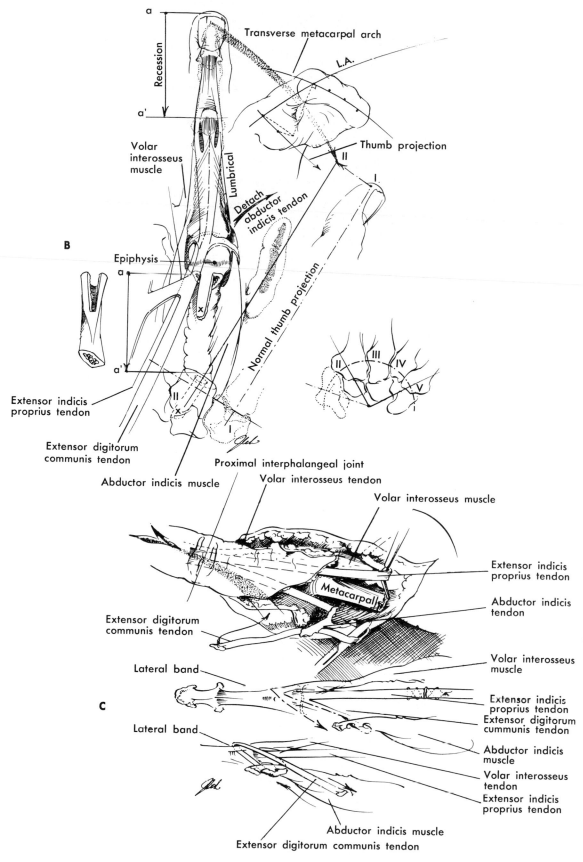

Fig. 9-22, cont'd. B, Detachment of abductor indicis, resection of second metacarpal shaft, and freeing of extensor digitorum communis (see text). **C,** Readjustment of extensor mechanism and fixation of abductor indicis by extensor digitorum communis tendon (see text).

Continued.

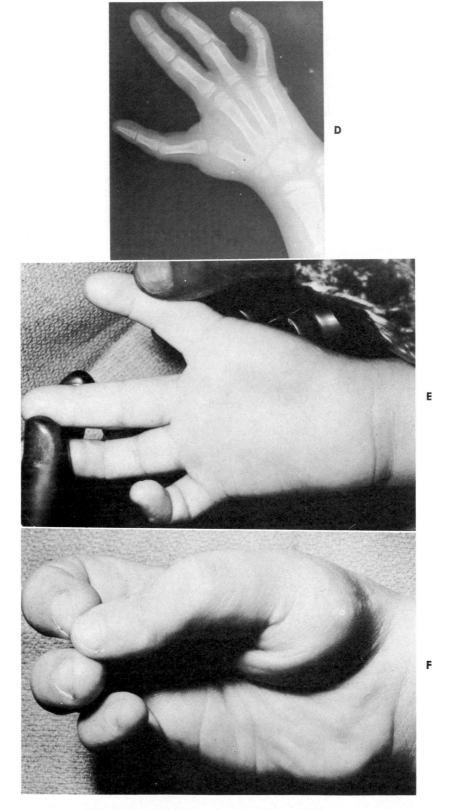

Fig. 9-22, cont'd. D, Left hand after treatment of congenital absence of thumb by limited repositioning of index finger. **E,** Left hand after surgery (function improved). **F,** Right hand after treatment for same anomaly by Littler technique as described in text (function is much better than in the left hand; note pinch). (Modified from Littler, J.W.: In Adams, J.P., editor: Current practice in orthopaedic surgery, vol. 3, St. Louis, 1966, The C.V. Mosby Co.)

the finger by dividing the palmar fascia and septa. Next divide the intermetacarpal ligament and interosseous fascia between the second and third metacarpals, and the proper volar digital artery to the radial side of the middle finger. Detach the insertion of the abductor indicis muscle that in the normal hand is the first dorsal interosseus (Fig. 9-22, *B*). Expose subperiosteally the metacarpal and divide it obliquely at its base at a right angle to the normal projection of the thumb. Section proximally the extensor digitorum communis tendon and separate it distally from the extensor indicis proprius and the radial lateral band to near the proximal interphalangeal joint. Next resect the metacarpal shaft just proximal to the epiphysis, preserving a dorsal strut for better fixation of the remaining head to the metacarpal base. Thus the metacarpophalangeal joint is preserved to act as the carpometacarpal joint of the new thumb. Now fix the metacarpal head to the metacarpal base in the normal thumb projection. Next fix the tendon of insertion of the abductor indicis at the level of the proximal interphalangeal joint by passing the extensor digitorum communis tendon twice through it, then around the ulnar lateral band, and then proximally to suture it to the abductor indicis (Fig. 9-22, *C*). Resect the redundant part of the extensor indicis proprius tendon and suture together the free ends of the tendon under proper tension.

In traumatic amputation through the first metacarpal shaft, the second metacarpophalangeal joint is not needed as a substitute for the carpometacarpal joint of the thumb. Thus the index metacarpal, except for its base, the metacarpophalangeal joint, and the proximal part of the proximal phalanx are discarded, and the retained part of the proximal phalanx is rotated and fixed to the stump of the first metacarpal as described next.

Begin the incision dorsally over the junction of the middle and distal thirds of the second metacarpal, extend it distally to the web between the middle and index fingers, then laterally across the proximal flexion crease, and then proximally to the starting point; from here continue it to the end of the amputated thumb, then proximally along the dorsum of the first metacarpal, and then slightly ulnarward to permit subsequent shifting of the skin (Fig. 9-23, *A*). Be careful to protect the dorsal vein to the index finger. Next reflect the volar flap anteriorly to expose the first dorsal interosseus, lumbrical, and adductor muscles and the radial neurovascular bundle (Fig. 9-23, *D*). Reflect distally the triangular flap from the dorsum of the finger to expose the extensor tendons and dorsal aponeurosis, the intermetacarpal ligament, and the common volar artery with its digital branches to the index and middle fingers. Then divide the juncturae tendinum and the fascia between the tendons of the extensor communis of the index and middle fingers. By further dissection at the base of the index finger carefully isolate the neurovascular pedicles. To allow radial shift of the nerves, vessels, and flexor tendons, section the compartmental septa of the palmar fascia. Locate the bifurcation of the common volar artery at the distal border of the intermetacarpal ligament; divide and ligate here the proper digital artery to the radial side of the middle finger (Fig. 9-23, *E*). The common volar nerve usually divides more proximally, but if necessary, separate it farther. Now section the first dorsal interosseus

and volar interosseus muscles at their musculotendinous junctions. Divide as far proximally as possible the extensor tendons to the index finger and reflect them distally. Remove a bone graft from the dorsal surface of the second metacarpal to be used for medullary fixation of the transposed finger to the first metacarpal (Fig. 9-23, *B* and *C*). Divide the index metacarpal at its base and the proximal phalanx near its base and discard the intervening bony segments; resect only enough bone from the proximal phalanx to make the new thumb of proper length. Next place the bone graft in the medullary canal of the first metacarpal and transfix it with a Kirschner wire. Then transpose the index finger to the thumb by placing its exposed proximal phalanx over the protruding medullary graft in proper pronation; transfix it too with a Kirschner wire to maintain its position after surgery. Now suture the extensor pollicis longus tendon to the extensor communis tendon of the transposed finger by the end-to-end method (Fig. 9-23, *F*). Divide the extensor indicis proprius at its junction with the common extensor, withdraw it proximal to the dorsal carpal ligament, transfer it in line with the extensor pollicis longus, and suture it to the extensor mechanism under proper tension. Now shift the volar flap ulnarward and suture it to the soft tissues at the side of the third metacarpal. Suture the dorsal flap to the margin of this flap and to the triangular flap on the dorsum of the transposed finger. Then close the remaining incisions.

AFTERTREATMENT. The hand and newly constructed thumb are immobilized in the functional position. The sutures are removed at 12 days. Six to 8 weeks later function is gradually resumed. The roentgenograms of a hand before and after this type of pollicization are shown in Fig. 9-24.

• • •

In the Riordan technique, again the index ray is shortened by resection of its metacarpal shaft. To simulate the greater multangular, the second metacarpal head is positioned palmar to the normal plane of the metacarpal bases, and the metacarpophalangeal joint acts as the carpometacarpal joint of the new thumb. The first dorsal interosseus is converted to an abductor pollicis brevis and the first volar interosseus to an adductor pollicis. The technique as described is for an immature hand with congenital absence of the thumb including the greater multangular, but it can be modified appropriately for other hands.

TECHNIQUE (RIORDAN). Beginning on the proximal phalanx of the index finger make a circumferential oval incision (Fig. 9-25, *A* and *B*); on the dorsal surface place the incision level with the middle of the phalanx and on the palmar surface level with the base of the phalanx. From the radiopalmar aspect of this oval extend the incision proximally, radially, and dorsally to the radial side of the second metacarpal head, then palmarward and ulnarward to the radial side of the third metacarpal base in the mid-palm, and finally again radially to end at the radial margin of the base of the palm. Dissect the skin from the proximal phalanx of the index finger, leaving the fat attached to the digit and creating a full-thickness skin flap. Next isolate and free the insertion of the first dorsal interosseus, and strip from the radial side of the second metacarpal shaft the origin of the muscle. Then isolate and free the insertion

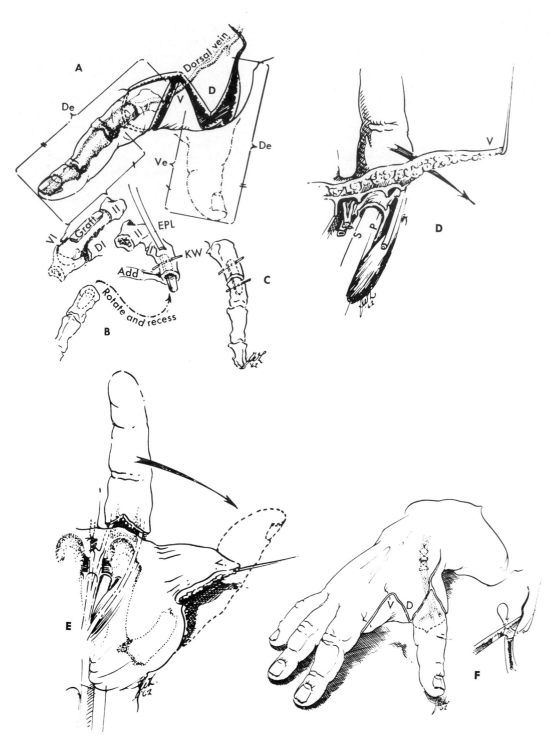

Fig. 9-23. Littler pollicization for amputation of thumb through metacarpal shaft. **A,** Skin incision. Note tenting of incision in anterior midline and preservation of dorsal vein. *Dotted line,* thumb after repositioning of index finger. Note shortening of index ray to simulate natural length of thumb. **B,** Treatment of bone: removal of graft from dorsum of second metacarpal, discard of bone between base of second metacarpal and base of proximal phalanx, and shift of index finger. **C,** Fixation of bone by graft and two Kirschner wires. **D,** Reflection anteriorly of volar flap to expose first dorsal interosseus, lumbrical, and adductor muscles and radial neurovascular bundle. **E,** Ligation of proper volar digital artery to middle finger, and division of intermetacarpal ligament and deep palmar fascia. **F,** Closure of skin flaps. Note special mattress suture used to snug tips of triangular flaps into position. Suture of extensor digitorum communis to extensor pollicis longus is shown. (Courtesy Dr. J. William Littler.)

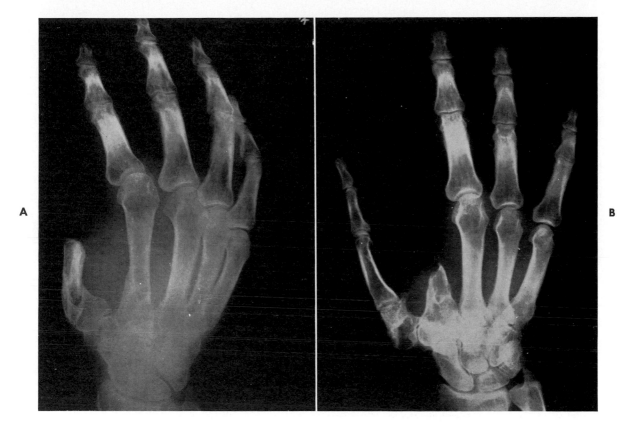

Fig. 9-24. Same as Fig. 9-23. **A,** Before surgery. **B,** After surgery. (Courtesy Dr. J. William Littler.)

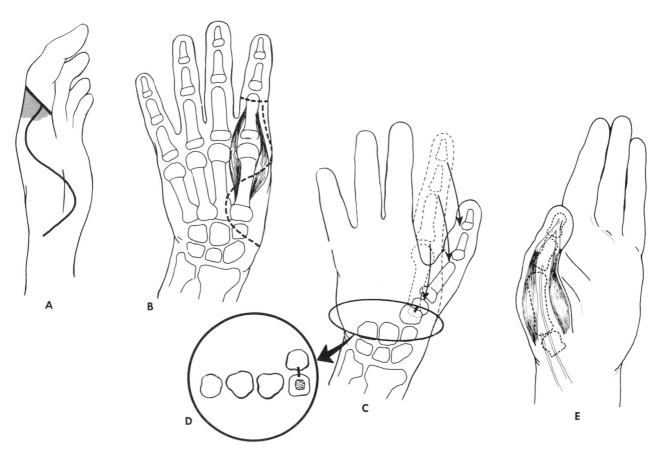

Fig. 9-25. Riordan pollicization for congenital absence of thumb, including greater multangular, in immature hand. **A** and **B,** Incision (see text). Skin of proximal phalanx, stippled area in **A,** is elevated as full-thickness skin flap. **C** and **D,** Second metacarpal has been resected by dividing base proximally and by cutting through epiphysis distally, and finger has been relocated proximally and radially. Second metacarpal head has been anchored palmar to second metacarpal base and simulates greater multangular (see text). **E,** Insertion of first dorsal interosseus has been anchored to radial lateral band of extensor mechanism of new thumb and origin to soft tissues at base of digit; insertion of first volar interosseus has been anchored to opposite lateral band and origin to soft tissues.

of the first volar interosseus and strip from the ulnar side of the metacarpal shaft the origin of this muscle. Take care to preserve the nerve and blood supplies to the muscle in each instance. Now separate the second metacarpal head from the metacarpal shaft by cutting through its epiphysis with a knife; preserve all of its soft tissue attachments. Then divide the second metacarpal at its base, leaving intact the insertions of the extensor carpi radialis longus and flexor carpi radialis; discard the metacarpal shaft. Next carry the index finger proximally and radially and relocate the second metacarpal head palmar to the second metacarpal base so that it simulates a greater multangular (Fig. 9-25, *C*); take care to rotate and angulate it so that the new thumb is properly positioned. Anchor it in this position with a wire suture (Fig. 9-25, *D*). Now anchor the insertion of the first dorsal interosseus to the radial lateral band of the extensor mechanism of the new thumb and its origin to the soft tissues at the base of the digit; this muscle now functions as an abductor pollicis brevis (Fig. 9-25, *E*). Likewise anchor the insertion of the first volar interosseus to the opposite lateral band and its origin to the soft tissues; this muscle now functions as an adductor pollicis. Shorten the extensor indicis proprius by resecting a segment of its tendon; this muscle now functions as an extensor pollicis brevis. Likewise shorten the extensor digitorum communis by resecting a segment of its tendon. Anchor the proximal segment of the tendon to the base of the proximal phalanx; this muscle now functions as an abductor pollicis longus. Trim the skin flaps appropriately; fashion the palmar flap so that when sutured it will place sufficient tension on the new thumb to hold it in opposition. Suture the flaps but avoid a circumferential closure at the base of the new thumb. Apply a pressure dressing of wet cotton and then a plaster cast.

AFTERTREATMENT. At 3 weeks the cast is removed and motion is begun. The thumb is appropriately splinted.

· · ·

Buck-Gramcko has reported experience with 100 operations for pollicization of the index finger in children with congenital absence or marked hypoplasia of the thumb. He emphasizes a reduction in length of the pollicized digit and accomplishes this by removing the entire second metacarpal with the exception of the head, which acts as a new greater multangular. For best results, the index finger has to be initially rotated approximately 160 degrees during the operation so that it is opposite the pulp of the ring finger. This position changes somewhat during the suturing of the muscles and the skin so that at the end of the operation there is rotation of approximately 120 degrees. In addition, the pollicized digit is angulated approximately 40 degrees into palmar abduction.

TECHNIQUE (BUCK-GRAMCKO). Make an S-shaped incision down the radial side of the hand just onto the palmar surface. Begin the incision near the base of the index finger on the palmar aspect and end it just proximal to the wrist. Make a slightly curved transverse incision across the base of the index finger on the palmar surface, connecting at right angles to the distal end of the first incision. Connect both ends of the incision on the dorsum of the hand as shown in Fig. 9-26, *A*. Make a third incision on the dorsum of the proximal phalanx of the index finger from the

proximal interphalangeal joint extending proximally to end at the incision around the base of the index finger (Fig. 9-26, *A*). Through the palmar incision, free the neurovascular bundle between the index and middle fingers by ligating the artery to the radial side of the middle finger. Then separate the common digital nerve carefully into its component parts for the two adjacent fingers so that no tension will be present after the index finger is rotated. Sometimes an anomalous neural ring is found around the artery; split this ring very carefully so that angulation of the artery after transposition of the finger will not occur. When the radial digital artery to the index finger is absent, it is possible to perform the pollicization on a vascular pedicle of only one artery. On the dorsal side, preserve at least one of the great veins.

Now, on the dorsum of the hand, sever the tendon of the extensor digitorum communis at the metacarpophalangeal level. Detach the interosseus muscles of the index finger from the proximal phalanx and the lateral bands of the dorsal aponeurosis. Partially strip subperiosteally the origins of the interosseus muscles from the second metacarpal, being careful to preserve the neurovascular structures.

Now, osteotomize and resect the second metacarpal as follows. If the phalanges of the index finger are of normal length, the whole metacarpal is resected with the exception of its head. When the phalanges are relatively short, the base of the metacarpal must be retained in order to obtain the proper length of the new thumb. When the entire metacarpal is resected except for the head, rotate the head as shown in Fig. 9-26, *C,* and attach it by sutures to the joint capsule of the carpus and to the carpal bones, which in young children can be pierced with a sharp needle. Bony union is not essential, and fibrous fixation of the head is sufficient for good function. When the base of the metacarpal is retained, fix the metacarpal head to its base with one or two Kirschner wires, again in the previously described position. In attaching the metacarpal head, bring the proximal phalanx into complete hyperextension in relation to the metacarpal head for maximum stability of the joint. Unless this is done, hyperextension is likely at the new "carpometacarpal" joint (Fig. 9-26, *C*). Suture the proximal end of the detached extensor digitorum communis tendon to the base of the former proximal phalanx (now acting as the first metacarpal) to become the new "abductor pollicis longus." Section the extensor indicis proprius tendon, shorten it appropriately, and then suture it by end-to-end anastomosis.

Suture the tendinous insertions of the two interosseus muscles to the lateral bands of the dorsal aponeurosis by weaving the lateral bands through the distal part of the interosseus muscle and turning them back distally to form a loop that is sutured to itself. In this way, the first palmar interosseus will become an "adductor pollicis" and the first dorsal interosseus an "abductor brevis" (Fig. 9-26, *B* and *D*).

Close the wound by fashioning a dorsal skin flap to close the defect over the proximal phalanx and fashion the rest of the flaps as necessary for skin closure as in Fig. 9-26, *A*.

AFTERTREATMENT. The hand is immobilized for 3 weeks and then careful active motion is begun.

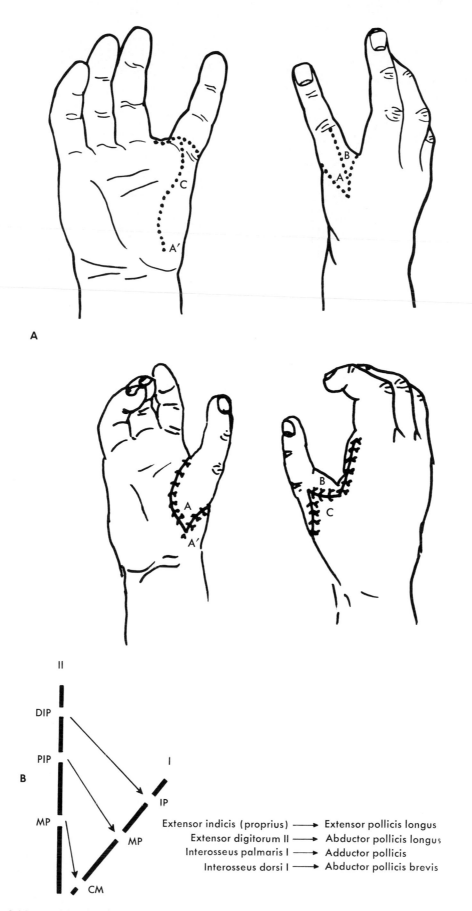

Extensor indicis (proprius) ⟶ Extensor pollicis longus
Extensor digitorum II ⟶ Abductor pollicis longus
Interosseus palmaris I ⟶ Adductor pollicis
Interosseus dorsi I ⟶ Abductor pollicis brevis

Fig. 9-26. A, Skin incisions (above) and appearance of suture line after pollicization. Letters facilitate orientation for shifting of skin flaps. **B,** Diagram of reduction of bones and joints; on right is change in function of muscles.   *Continued.*

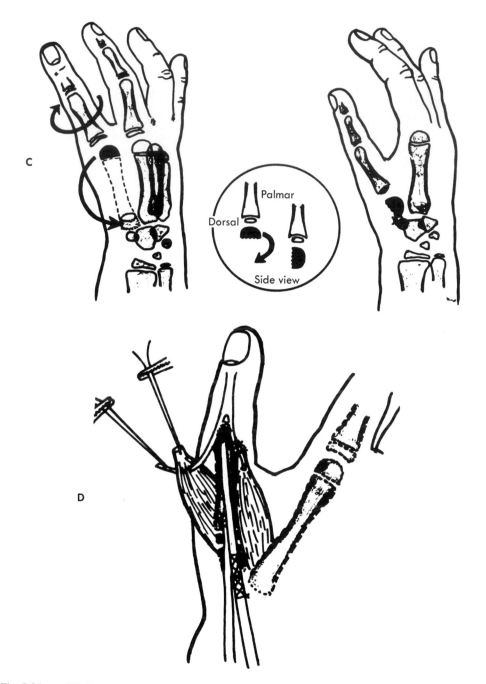

Fig. 9-26, cont'd. C, Diagram of reduction, rotation, and angulation of index finger; diagram in circle shows rotation of metacarpal head, which is done to prevent hyperextension deformity. **D,** Muscle stabilization by two interosseus muscles that are sutured to separated lateral bands of dorsal aponeurosis. Extensor communis tendon is fixed to base of new metacarpal as an abductor pollicis longus; extensor indicis proprius tendon is shortened. (From Buck-Gramcko, D.: J. Bone Joint Surg. **53-A:**1605, 1971.)

REFERENCES

Ahstrom J.P., Jr.; Pollicization in congenital absence of the thumb, Curr. Pract. Orthop. Surg. **5:**1, 1973.

Argamaso, R.V.: Rotation-transposition method for soft tissue replacement on the distal segment of the thumb, Plast. Reconstr. Surg. **54:**366, 1974.

Atasoy, E.: The cross thumb to index finger pedicle, J. Hand Surg. **5:**572, 1980.

Atasoy, E., et al.: Reconstruction of the amputated finger tip with a triangular volar flap: a new surgical procedure, J. Bone Joint Surg. **52-A:**921, 1970.

Bowe, J.J.: Thumb reconstruction by index transposition, Plast. Reconstr. Surg. **32:**414, 1963.

Brent, B.: Replantation of amputated distal phalangeal parts of fingers without vascular anastomoses, using subcutaneous pockets, Plast. Reconstr. Surg. **63:**1, 1979.

Broadbent, T.R., and Woolf, R.M.: Thumb reconstruction with contiguous skin-bone pedicle graft, Plast. Reconstr. Surg. **26:**494, 1960.

Brown, H., and Getty, P.: Leprosy and thumb reconstruction by opponensplasty or phalangizing the first metacarpal, J. Hand Surg. **4:**432, 1979.

Brown, H., et al.: Phalangizing the first metacarpal: case report, Plast. Reconstr. Surg. **45:**294, 1970.

Brown, P.W.: Adduction-flexion contracture of the thumb: correction with dorsal rotation flap and release of contracture, Clin. Orthop. **88:**161, 1972.

Brown, P.W.: Sacrifice of the unsatisfactory hand, J. Hand Surg. **4:**417, 1979.

Buck-Gramko, D.: Pollicization of the index finger: method and results in aplasia and hypoplasia of the thumb, J. Bone Joint Surg. **53-A:**1605, 1971.

Bunnell, S.: Digit transfer by neurovascular pedicle, J. Bone Joint Surg. **34-A:**772, 1952.

Bunnell, S.: Reconstruction of the thumb, Am. J. Surg. **95:**168, 1958.

Butler, B., Jr.: Ring-finger pollicization: with transplantation of nail bed and matrix on a volar flap, J. Bone Joint Surg. **46-A:**1069, 1964.

Button, M., and Stone, E.J.: Segmental bony reconstruction of the thumb by composite groin flap: a case report, J. Hand Surg. **5:**488, 1980.

Carroll, R.E.: Transposition of the index finger to replace the middle finger, Clin. Orthop. **15:**27, 1959.

Carroll, R.E.: Ring injuries in the hand, Clin. Orthop. **104:**175, 1974.

Clarkson, P.: Reconstruction of hand digits by toe transfers, J. Bone Joint Surg. **37-A:**270, 1955.

Clarkson, P.: On making thumbs, Plast. Reconstr. Surg. **29:**325, 1962.

Clarkson, P.: Erratum (on making thumbs), Plast. Reconstr. Surg. **30:**491, 1962.

Clarkson, P., and Chandler, R.: A toe to thumb transplant with nerve graft, Am. J. Surg. **95:**315, 1958.

Clarkson, P., and Furlong, F.: Thumb reconstruction by transfer of big toe, Br. Med. J. **2:**1332, 1949.

Clayton, M.L.: Index ray amputation, Surg. Clin. North Am. **43:**367, 1963.

Cobbett, J.R.: Free digital transfer: report of a case of transfer of a great toe to replace an amputated thumb, J. Bone Joint Surg. **51-B:**677, 1969.

Cuthbert, J.B.: Pollicisation of the index finger, Br. J. Plast. Surg. **1:**56, 1948-1949.

Davis, J.E.: Toe to hand transfers (pedochyrodactyloplasty), Plast. Reconstr. Surg. **33:**422, 1964.

De Oliveira, J.C.: Some aspects of thumb reconstruction, Br. J. Surg. **57:**85, 1970.

Doi, K., et al.: Reconstruction of an amputated thumb in one stage: case report—free neurovascular flap transfer with iliac-bone graft, J. Bone Joint Surg. **61-A:**1254, 1979.

Entin, M.A.: Salvaging the basic hand, Surg. Clin. North Am. **48:**1063, 1968.

Fisher, R.H.: The Kutler method of repair of finger-tip amputation, J. Bone Joint Surg. **49-A:**317, 1967.

Flatt, A.E.: The thenar flap, J. Bone Joint Surg. **39-B:**80, 1957.

Flatt, A.E., and Wood, V.E.: Multiple dorsal rotation flaps from the hand for thumb web contractures, Plast. Reconstr. Surg. **45:**258, 1970.

Freeman, B.S.: Reconstruction of thumb by toe transfer, Plast. Reconstr. Surg. **17:**393, 1956.

Freiberg, A., and Manktelow, R.: The Kutler repair for fingertip amputations, Plast. Reconstr. Surg. **50:**371, 1972.

Gillies, H.: Autograft of amputated digit: suggested operation, Lancet **1:**1002, 1940.

Gillies, H., and Millard, R.D., Jr., editors: The principles and art of plastic surgery, vol. 2, part V, chap. 23, Boston, 1957, Little, Brown & Co.

Gordon, S.: Autograft of amputated thumb, Lancet **2:**823, 1944.

Graham, W.P., III: Incisions, amputations, and skin grafting in the hand, Orthop. Clin. North Am. **2:**213, 1970.

Harrison, S.H.: Restoration of muscle balance in pollicization, Plast. Reconstr. Surg. **34:**236, 1964.

Hentz, V., Jackson, I., and Fogarty, D.: Case report: false aneurysm of the hand secondary to digital amputation, J. Hand Surg. **3:**199, 1978.

Hirshowitz, B., Karev, A., and Rousso, M.: Combined double Z-plasty and Y-V advancement for thumb web contracture, Hand, **7:**291, 1975.

Holm, A., and Zachariae, L.: Fingertip lesions: an evaluation of conservative treatment versus free skin grafting, Acta Orthop. Scand. **45:**382, 1974.

Hueston, J.: The extended neurovascular island flap, Br. J. Plast. Surg. **18:**304, 1965.

Hueston, J.: Local flap repair of fingertip injuries, Plast. Reconstr. Surg. **37:**349, 1966.

Hughes, N.C., and Moore, F.T.: A preliminary report on the use of a local flap and peg bone graft for lengthening a short thumb, Br. J. Plast. Surg. **3:**34, 1950-1951.

Hung-Yin, C., et al.: Reconstruction of the thumb, Chin. Med. J. **79:**541, 1959 (Abstracted by David E. Hallstrand, Int. Abstr. Surg. **111:**177, 1960.)

Irigaray, A.: New fixing screw for completely amputated fingers, J. Hand Surg. **5:**381, 1980.

Jeffery, C.C.: A case of pollicisation of the index finger, J. Bone Joint Surg. **39-B:**120, 1957.

Johnson, H.A.: Formation of a functional thumb post with sensation in phocomelia, J. Bone Joint Surg. **49-A:**327, 1967.

Joshi, B.B.: One-stage repair for distal amputation of the thumb, Plast. Reconstr. Surg. **45:**613, 1970.

Joshi, B.B.: A local dorsolateral island flap for restoration of sensation after avulsion injury of fingertip pulp, Plast. Reconstr. Surg. **54:**175, 1974.

Joyce, J.L.: A new operation of the substitution of a thumb, Br. J. Surg. **5:**499, 1918.

Kaplan, E.B.: Replacement of an amputated middle metacarpal and finger by transposition of the index finger, Bull. Hosp. Joint Dis. **27:**103, 1966.

Kaplan, I.: Primary pollicization of injured index finger following crush injury, Plast. Reconstr. Surg. **37:**531, 1966.

Kaplan, I., and Plaschkes, J.: One stage pollicisation of little finger, Br. J. Plast. Surg. **13:**272, 1960-1961.

Keiter, J.E.: Immediate pollicization of an amputated index finger, J. Hand Surg. **5:**584, 1980.

Kelikian, H., and Bintcliffe, E.W.: Functional restoration of the thumb, Surg. Gynecol. Obstet. **83:**807, 1946.

Kettelkamp, D.B., and Ramsey, P.: Experimental and clinical autogenous distal metacarpal reconstruction, Clin. Orthop. **74:**129, 1971.

Kleinert, H.E.: Finger tip injuries and their management, Am. Surg. **25:**41, 1959.

Lassar, G.N.: Reconstruction of a digit following loss of all fingers with preservation of the thumb, J. Bone Joint Surg. **41-A:**519, 1959.

Leung, P.C., and Kok, L.C.: Use of an intramedullary bone peg in digital replantations, revascularization and toe-transfers, J. Hand Surg. **6:**281, 1981.

Lewin, M.L.: Partial reconstruction of thumb in a one-stage operation, J. Bone Joint Surg. **35-A:**573, 1953.

Lewin, M.L.: Severe compression injuries of the hand in industry: amputation versus rehabilitation, J. Bone Joint Surg. **41-A:**71, 1959.

Lewin, M.L.: Sensory island flap in osteoplastic reconstruction of the thumb, Am. J. Surg. **109:**226, 1965.

Littler, J.W.: Subtotal reconstruction of thumb, Plast. Reconstr. Surg. **10:**215, 1952.

Littler, J.W.: Neurovascular pedicle method of digital transposition for reconstruction of thumb, Plast. Reconstr. Surg. **12:**303, 1953.

Littler, J.W.: Principles of reconstructive surgery of the hand. In Converse, J.M., editor: Reconstructive plastic surgery, vol. 4, Philadelphia, 1964, W.B. Saunders Co.

Littler, J.W.: Digital transposition. In Adams, J.P., editor: Current practice in orthopaedic surgery, vol. 3, St. Louis, 1966, The C.V. Mosby Co.

Littler, J.W.: On making a thumb: one hundred years of surgical effort, J. Hand Surg. 1:35, 1976.

Lobay, G.W., and Moysa, G.L.: Primary neurovascular bundle transfer in the management of avulsed thumbs, J. Hand Surg. 6:31, 1981.

Mansoor, I.A.: Metacarpal lengthening: a case report, J. Bone Joint Surg. 51-A:1639, 1969.

Matev, I.B.: First metacarpal lengthening for thumb reconstruction, Am. Dig. Foreign Orthop. Lit. 1st qtr:10, 1970.

Matev, I.B.: Thumb reconstruction after amputation at the metacarpophalangeal joint by bone-lengthening: a preliminary report of three cases, J. Bone Joint Surg. 52-A:957, 1970.

Matev, I.: Wringer injuries of the hand, J. Bone Joint Surg. 49-B:722, 1967.

Mathes, S.J., Buchannan, R., and Weeks, P.M.: Microvascular joint transplantation with epiphyseal growth, J. Hand Surg. 5:586, 1980.

Matthews, D.: Congenital absence of functioning thumb, Plast. Reconstr. Surg. 26:487, 1960.

May, J.W., Jr., et al.: Free neurovascular flap from the first web of the foot in hand reconstruction, J. Hand Surg. 2:387, 1977.

McCash, C.: Toe pulp-free grafts in finger-tip repair, Br. J. Plast. Surg. 11:322, 1958-1959.

McGregor, I.A., and Simonetta, C.: Reconstruction of the thumb by composite bone-skin flap, Br. J. Plast. Surg. 17:37, 1964.

Metcalf, W., and Whalen, W.P.: Salvage of the injured distal phalanx: plan of care and analysis of 369 cases, Clin. Orthop. 13:114, 1959.

Millender, L.H., et al.: Delayed volar advancement flap for thumb tip injuries, Plast. Reconstr. Surg. 52:635, 1973.

Miller, A.J.: Single finger tip injuries treated by thenar flap, Hand 6:311, 1974.

Miura, T.: An appropriate treatment for postoperative Z-formed deformity of the duplicated thumb, J. Hand Surg. 2:380, 1977.

Miura, T.: Thumb reconstruction using radial-innervated cross-finger pedicle graft, J. Bone Joint Surg. 55-A:563, 1973.

Moore, F.T.: The technique of pollicisation of the index finger, Br. J. Plast. Surg. 1:60, 1948-1949.

Morrison, W.A., O'Brien, B. McC., and MacLeod, A.M.: Thumb reconstruction with a free neurovascular wrap-around flap from the big toe, J. Hand Surg. 5:575, 1980.

Murray, J.F., Carman, W., and MacKenzie, J.K.: Transmetacarpal amputation of the index finger: a clinical assessment of hand strength and complications, J. Hand Surg. 2:471, 1977.

Nemethi, C.E.: Reconstruction of the distal part of the thumb after traumatic amputation: restoration of function and sensation using nerve, tendon, and bone from the amputated portion, J. Bone Joint Surg. 42-A:375, 1960.

O'Brien, B. McC., et al.: Hallux-to-hand transfer, Hand 7:128, 1975.

O'Brien, B. McC., et al.: Microvascular second toe transfer for digital reconstruction, J. Hand Surg. 3:123, 1978.

Ohmori, K., and Harii, K.: Transplantation of a toe to an amputated finger, Hand 7:135, 1975.

Peacock, E.E., Jr.: Metacarpal transfer following amputation of a central digit, Plast. Reconstr. Surg. 29:345, 1962.

Pierce, G.W.: Reconstruction of the thumb after total loss, Surg. Gynecol. Obstet. 45:825, 1927.

Posner, M.A.: Ray transposition for central digital loss, J. Hand Surg. 4:242, 1979.

Prpic, I.: Reconstruction of the thumb immediately after injury, Br. J. Plast. Surg. 17:49, 1964.

Reid, D.A.C.: Reconstruction of the thumb, J. Bone Joint Surg. 42-B:444, 1960.

Reis, N.D.: Gillies "cocked hat" reconstruction for total loss of ulnar four fingers, Hand 5:229, 1973.

Riordan, D.C.: Personal communication, 1969.

Robinson, O.G., Jr.: Primary reconstruction of the thumb using amputated part and tube pedicle flap, South. Med. J. 66:1025, 1973.

Rose, E.H., and Buncke, H.J.: Simultaneous transfer of the right and left second toes for reconstruction of amputated index and middle fingers in the same hand: case report, J. Hand Surg. 5:590, 1980.

Rybka, F.J., and Pratt, F.E.: Thumb reconstruction with a sensory flap from the dorsum of the index finger, Plast. Reconstr. Surg. 64:141, 1979.

Salis, J.G.: Primary pollicisation of an injured middle finger, J. Bone Joint Surg. 45-B:503, 1963.

Schiller, Carl: Nail replacement in finger tip injuries, Plast. Reconstr. Surg. 19:521, 1957.

Schlenker, J.D., Kleinert, H.E., and Tsai, T.-M.: Methods and results of replantation following traumatic amputation of the thumb in sixty-four patients. J. Hand Surg. 5:63, 1980.

Schmauk, B.: On the problem of thumb substitution (Zur Problematik des Daumenersatzes), Med. Welt 9:482, 1960. (Abstracted by Joseph C. Mulier, Int. Abstr. Surg. 111:178, 1960.)

Scott, J.E.: Amputation of the finger, Br. J. Surg. 61:574, 1974.

Shaw, M.H., and Wilson, I.S.P.: An early pollicisation, Br. J. Plast. Surg. 3:214, 1950-1951.

Smith, J.R., and Bom, A.F.: An evaluation of finger-tip reconstruction by cross-finger and palmar pedicle flap, Plast. Reconstr. Surg. 35:409, 1965.

Smith, R.J., and Dworecka, F.: Treatment of the one-digit hand, J. Bone Joint Surg. 55-A:113, 1973.

Snow, J.W.: The use of a volar flap for repair of fingertip amputations: a preliminary report, Plast. Reconstr. Surg. 52:299, 1973.

Snowdy, H.A., Omer, G.E., Jr., and Sherman, F.C.: Longitudinal growth of a free toe phalanx transplant to a finger, J. Hand Surg. 5:71, 1980.

Soiland, H.: Lengthening a finger with the "on the top" method. Acta Chir. Scand. 122:184, 1961.

Stefani, A.E., and Kelly, A.P.: Reconstruction of the thumb: a one-stage procedure, Br. J. Plast. Surg. 15:289, 1962.

Stern, P.J., and Lister, G.D.: Pollicization after traumatic amputation of the thumb, Clin. Orthop. 155:85, 1981.

Sturman, M.J., and Duran, R.J.: Late results of finger-tip injuries, J. Bone Joint Surg. 45-A:289, 1963.

Sullivan, J.G., et al.: The primary application of an island pedicle flap in thumb and index finger injuries, Plast. Reconstr. Surg. 39:488, 1967.

Swanson, A.B.: Restoration of hand function by the use of partial or total prosthetic replacement: I. The use of partial prostheses, J. Bone Joint Surg. 45-A:276, 1963.

Swanson, A.B.: Restoration of hand function by the use of partial or total prosthetic replacement: II. Amputation and prosthetic fitting for treatment of the functionless, asensory hand, J. Bone Joint Surg. 45-A:284, 1963.

Swanson, A.B.: The Krukenberg procedure in the juvenile amputee, J. Bone Joint Surg. 46-A:1540, 1964.

Swanson, A.B.: Levels of amputation of fingers and hand: considerations for treatment, Surg. Clin. North Am. 44:1115, 1964.

Swanson, A.B., Boeve, N.R., and Lumsden, R.M.: The prevention and treatment of amputation neuromata by silicone capping, J. Hand Surg. 2:70, 1977.

Tamai, S., et al.: Hallux-to-thumb transfer with microsurgical technique: a case report in a 45-year-old woman, J. Hand Surg. 2:152, 1977.

Tamai, S., et al.: Traumatic amputation of digits: the fate of remaining blood: an experimental and clinical study, J. Hand Surg. 2:13, 1977.

Tanzer, R.C., and Littler, J.W.: Reconstruction of the thumb, Plast. Reconstr. Surg. 3:533, 1948.

Tegtmeier, R.E.: Thumb-to-thumb transfer following severe electrical burns to both hands, J. Hand Surg. 6:269, 1981.

Tubiana, R., and Roux, J.P.: Phalangization of the first and fifth metacarpals: indications, operative technique, and results, J. Bone Joint Surg. 56-A:447, 1974.

Tubiana, R., Stack, H., and Hakstian, R.W.: Restoration of prehension after severe mutilations of the hand, J. Bone Joint Surg. 48-B:455, 1966.

Usui, M., et al.: An experimental study on "replantation toxemia": the effect of hypothermia on an amputated limb, J. Hand Surg. 3:589, 1978.

Verdan, C.: The reconstruction of the thumb, Surg. Clin. North Am. 48:1033, 1968.

Watman, R.N., and Denkewalter, F.R.: A repair for loss of the tactile pad of the thumb, Am. J. Surg. **97:**238, 1959.

Weckesser, E.C.: Reconstruction of a grasping mechanism following extensive loss of digits, Clin. Orthop. **15:**60, 1959.

Weiland, A.J., et al.: Replantation of digits and hands: analysis of surgical techniques and functional results in 71 patients with 86 replantations, J. Hand Surg. **2:**1, 1977.

Whitaker, L.A., et al.: Retaining the articular cartilage in finger joint amputations, Plast. Reconstr. Surg. **49:**542, 1972.

White, W.F.: Fundamental priorities in pollicisation, J. Bone Joint Surg. **52-B:**438, 1970.

Wilkinson, T.S.: Reconstruction of the thumb by radial nerve innervated cross-finger flap, South. Med. J. **65:**992, 1972.

Winspur, I.: Single-stage reconstruction of the subtotally amputated thumb: a synchronous neurovascular flap and Z-plasty, J. Hand Surg. **6:**70, 1981.

Zancolli, E.: Transplantation of the index finger in congenital absence of the thumb, J. Bone Joint Surg. **42-A:**658, 1960.

CHAPT

Para

Principles
Planning
Evalu
Time
Technica
Restoration
Restorati
Corre
Tendo
Transf
thur
Transf
223
Muscl
Restorati
Transfe
thun
Royle-
Restoration
Restoration
Peripheral
Low radi
High radi
Techni
Low ulna
High ulna
Low med
High med
Combined
Combined
Severe paral
causes, 2

The ha
Motion is
pinch, gra
joints amo
and by the
To be pur
are crosse
selves mov
balanced a
this stabili
extensors,
the strong
nists of the
ute to stab
With the
of function
ing use of

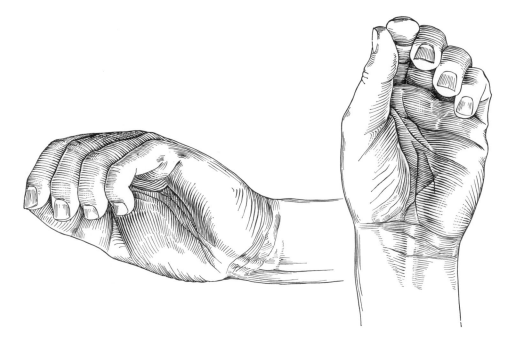

Fig. 10-1. Position of muscle balance. (From White, W.L.: Surg. Clin. North Am. **40:**427, 1960.)

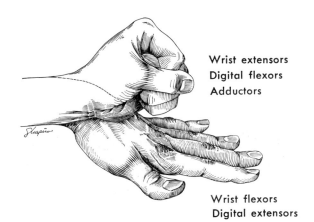

Wrist extensors
Digital flexors
Adductors

Wrist flexors
Digital extensors
Abductors

Fig. 10-2. Synergistic muscles of hand (see text). (From White, W.L.: Surg. Clin. North Am. **40:**427, 1960.)

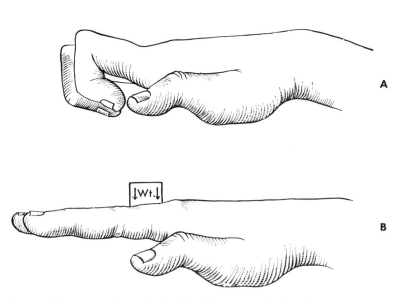

A

↓Wt.↓

B

Fig. 10-3. Clawing of hand caused by paralysis of intrinsic muscles. **A,** Long finger extensors cannot extend interphalangeal joints because metacarpophalangeal joints are hyperextended. **B,** Long finger extensors can extend interphalangeal joints because hyperextension of metacarpophalangeal joints has been prevented.

tensor force is exerted at the metacarpophalangeal joints; without stabilization of the metacarpophalangeal joints in a neutral or slightly flexed position by the intrinsics, the long extensors cannot extend the interphalangeal joints. Finally the wrist is pulled into flexion by the strong finger flexors; this causes a tenodesing effect on the long finger extensors that hyperextends the metacarpophalangeal joints even farther. In addition to deformities of the fingers, the thumb is adducted by its long extensor because this muscle is unopposed by the intrinsic muscles of opposition and abduction. This adducted position is accompanied by extension of the carpometacarpal joint that in turn increases tension on the long thumb flexor tendon that crosses the volar side of the joint; thus the interphalangeal joint is flexed because the long flexor is unopposed by the adductor pollicis and abductor pollicis brevis muscles that normally aid in extending it. The position of the hand just described is known as the *intrinsic minus* position, and in it secondary joint contractures or even subluxations occur. Whether the loss of intrinsic function is caused by disease or by trauma, the results of dynamic muscle imbalance are the same. However, sensation in clawhand varies according to the cause of imbalance: in poliomyelitis sensation is normal, in peripheral nerve lesions the presence or absence of sensation depends on the level of the lesion and the nerve involved, in Hansen's disease sensation is absent, sometimes in a glovelike distribution, and in syringomyelia sensation is partially absent.

Spasticity of muscles can also disrupt the balance of the hand. Muscle tension may be constantly or intermittently increased and may not be controlled and balanced effectively by the opposing normal muscles. Such a situation is sometimes seen in cases of cerebral palsy and it can cause overstretching of muscles and dislocation of joints.

PRINCIPLES OF TENDON TRANSFER

Tendon transfers are useful in restoring functions of the hand lost because of paralysis produced by disease or trauma. However, some basic principles must be followed if transfers are to be successful and if an increase in imbalance and thus in deformity is to be avoided. After these principles are discussed, some frequent patterns of functional loss will be discussed, and specific tendons for transfer will be suggested for each.

Planning tendon transfer

Whether the original cause of imbalance has been traumatic, congenital, infectious, or vascular, the hand must be evaluated in terms of function lost and function retained. Before appropriate muscles for transfer can be selected, those available for transfer must be known; their strength, their amplitude of excursion, the synergistic group to which they belong, and the importance of their present function must all be considered. Sometimes it is helpful to list in one column functions that should be restored and in an opposite column the tendons available for transfer; transfers may be planned with more ease and accuracy by matching these columns, as shown in the accompanying chart of radial nerve palsy.

The two most important points in considering a muscle for transfer are its expendability and its strength. Restoring one major function such as finger extension is contraindicated if done at the expense of another major function such as finger flexion. The strength of a muscle is graded from 0 to 5 as follows:

0 Zero—no contraction
1 Trace—palpable contraction only
2 Poor—moves joint but not against gravity
3 Fair—moves joint against gravity
4 Good—moves joint against gravity and resistance
5 Normal—normal strength

A muscle will usually lose strength by one grade when transferred and therefore should be good or normal if the transfer is to be satisfactory. In addition to expendability and strength, the synergistic group in which the muscle acts and the amplitude of excursion of its tendon should be considered. As previously stated, rehabilitation of a muscle whose tendon has been transferred is less difficult when the transfer is synergistic (for example, a wrist flexor transfer to the finger extensors); therefore, although not essential, transferring a muscle within its own synergistic group is desirable. The amplitude of excursion of the tendon should be sufficient for satisfactory function, although it may not be as great as that of the tendon or tendons it will replace. For example, the brachioradialis, an expendable muscle for transfer, is capable of pulling its tendon through only a short excursion but can sometimes be useful, if not ideal, as a transfer to the long thumb flexor because even limited flexion of the interphalangeal joint of the thumb is useful.

However, as pointed out by Boyes, the excursion of the brachioradialis can be increased by dissecting its tendon proximally and freeing all of its fascial attachments. The muscle is not useful as a transfer to a finger flexor because the limited motion it can produce in finger joints, even after its excursion has been increased, is of little value.

The transfer of tendons is the final step in rehabilitation of the hand. It should not be made until any scar tissue has been satisfactorily replaced because transferred tendons must be surrounded by fat to prevent their adhering to raw bone or subcutaneous scar; consequently a flap graft containing fat is necessary to replace scar. A satisfactory

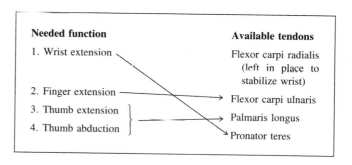

Needed function	Available tendons
1. Wrist extension	Flexor carpi radialis (left in place to stabilize wrist)
2. Finger extension	Flexor carpi ulnaris
3. Thumb extension	Palmaris longus
4. Thumb abduction	Pronator teres

range of passive joint motion is also necessary *before* the transfer; proper splinting or ligamentous release is carried out as needed. Stiffness or contracture of joints cannot be corrected by tendon transfers alone; furthermore, if uncorrected, stiffness or contracture will prevent a transferred tendon from moving at the proper time after surgery so that the tendon becomes permanently adherent to the surrounding tissues. Also malalignment of bone must be corrected by osteotomy, and any necessary bone grafting must be carried out before transfer. Finally, necessary operations to restore any loss of sensibility must precede tendon transfer.

Other factors must sometimes be considered in timing tendon transfers. In poliomyelitis some recovery of muscle power may be expected until 18 months after the acute disease, and consequently this much time must pass before an accurate evaluation is possible; then any further recovery cannot be expected to improve muscle strength more than one grade, if any. During this period of waiting, parts must be properly splinted to improve available muscle function and to prevent fixed deformity. In congenital anomalies the relative muscle strength will not change. In syringomyelia weakness may increase even after transfer. Peripheral nerve injuries must be considered individually; in division of the radial nerve at the midhumerus transfers for finger and thumb extension and for thumb abduction should be delayed for 6 months or longer after neurorrhaphy; however, early transfer to restore wrist extension should be considered. This provides an internal splint for the wrist and immediately improves the function of the hand. Transfer of the pronator teres to the extensor carpi radialis brevis is recommended. In high median nerve lesions, some function should return in the most proximal muscles in 4 months (and in 3 months in low median nerve lesions) or the nerve should be explored or tendon transfers should be considered.

Technical considerations for tendon transfer

The strength of the muscle has been evaluated clinically before surgery, but its color at the time of tendon transfer provides a further check. A muscle suitable for transfer is dark pink or red, indicating satisfactory nutrition and the presence of normal muscle fibers. A weak or paralyzed muscle is pale pink and is smaller than normal, and its amplitude of excursion (Table 10-1) is less than normal when tested at surgery; such a muscle is not suitable for transfer (Fig. 10-4).

A muscle that has been detached from its insertion some time before transfer will have developed a contracture, and consequently its tendon should be anchored under more tension than usual because it will stretch and regain some of its excursion. A muscle and its tendon should not make an acute angle between the origin of the muscle and the new attachment of the tendon—the straighter the muscle the more efficient its action. When an acute angle is necessary, a pulley must be created, but efficiency of the muscle is diminished by friction at the pulley. In freeing a muscle for transfer, care must be taken to avoid stretching or otherwise damaging the neurovascular bundle, which usually enters the proximal third of the muscle belly. A transferred tendon cannot be expected to glide properly

Table 10-1. Amplitude of excursion

Tendons	Amplitude (mm)
Wrist tendons	33
Flexor profundus	70
Flexor sublimis	64
Extensor digitorum communis	50
Flexor pollicis longus	52
Extensor pollicis longus	58
Extensor pollicis brevis	28
Abductor pollicis longus	28

From Curtis, R.M.: Orthop. Clin. North Am. **5**:231, 1974.

when it crosses raw bone, passes through fascia without a sufficient opening, or is buried within scarred tissue; with a few exceptions, transferred tendons should be passed subcutaneously. Should it be necessary to split a transferred tendon and anchor it to two or more separate points, the muscle will act primarily on the slip of tendon under greatest tension and may distort function; thus great care must be taken to equalize tension on the slips at the time of attachment.

The more distal to a given joint a tendon is anchored, the more power the muscle can exert on the joint but also the more is the excursion required of the tendon to provide normal motion. Furthermore, the greater the angle of approach of a tendon to bone, the greater the force the muscle can exert on the bone and across the joint, but this creates a bowstring effect in a pulleyless system. Most muscles lie almost parallel to the bones whose joints they act on, and few approach a bone at near a right angle; the pronator quadratus and the supinator are notable exceptions.

RESTORATION OF PINCH
Restoration of opposition of thumb

Opposition of the thumb is necessary for pinch—one of the three most important functions of the hand. But adduction of the thumb is necessary too (see restoration of adduction of thumb, p. 226). Frequently opposition is either partially or totally lost in poliomyelitis or median nerve palsy. Opposition depends primarily on function of the intrinsic muscles of the thumb, especially the abductor pollicis brevis. Yet, extrinsic muscles are also necessary to stabilize dynamically the metacarpophalangeal and interphalangeal joints of the thumb, or these joints must be stabilized by arthrodesis or tenodesis. At the same time, the carpometacarpal joint of the thumb must be freely movable, unrestricted by contracture of the joint capsule or other structures of the thumb web (see discussion of the adducted thumb).

Opposition of the thumb is a complex motion made by coordination of (1) abduction of the thumb from the palmar surface of the index finger, (2) flexion of the metacarpophalangeal joint of the thumb, (3) internal rotation or pronation of the thumb, (4) radial deviation of the proximal phalanx of the thumb on the metacarpal, and finally (5) motion of the thumb toward the fingers. Although opposition is the result of coordinate function of all the long and short muscles that act on the thumb, the abductor pol-

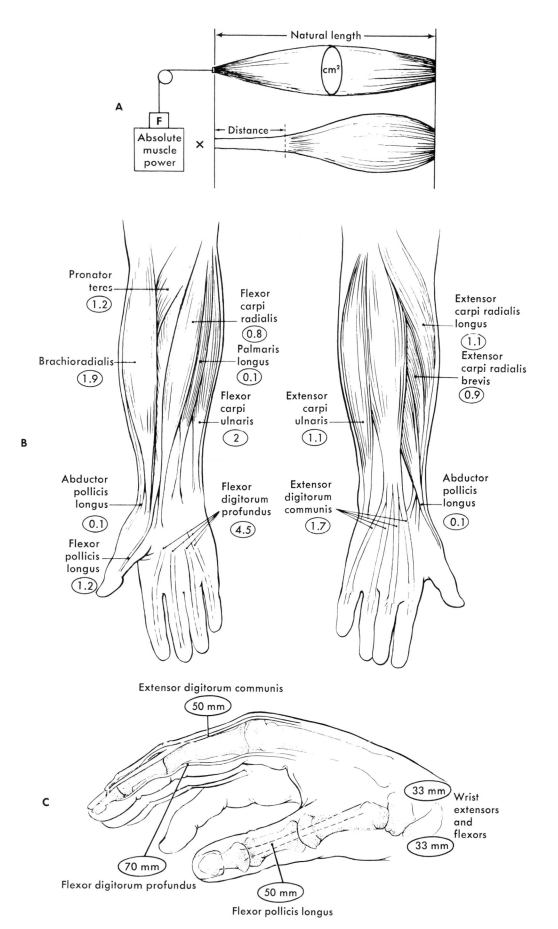

Fig. 10-4. Power of muscle transfer (**A** and **B**). **A,** Working capacity of muscle. $W = Fxd$, , when F (force) = absolute muscle power, 3.65 × cm² of physiologic cross section and d (distance) = amplitude or displacement. **B,** Working capacity of muscle in mkg (meter-kilograms). **C,** Muscle amplitude in millimeters. (From Curtis, R.M.: Orthop. Clin. North Am. **5:**231, 1974.)

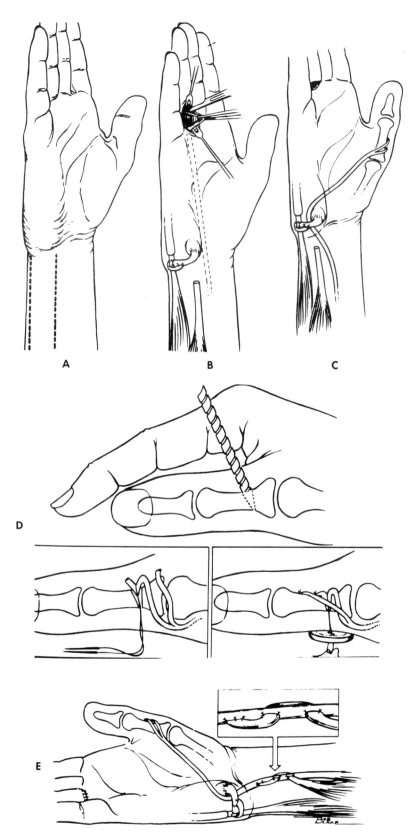

Fig. 10-9. A, Two incisions made at wrist. **B,** Through volar incision, flexor sublimis tendon to ring finger and flexor carpi ulnaris tendon are exposed. Through ulnar incision, extensor carpi ulnaris is exposed. Flexor carpi ulnaris tendon is divided 4 cm from its insertion, and free end of distal segment is sutured to extensor carpi ulnaris. Flexor digitorum sublimis tendon to ring finger is exposed through a transverse incision at proximal flexor crease of finger, and its two slips are divided. **C,** Tendon of flexor digitorum sublimis to ring finger is drawn proximally through volar incision at wrist, threaded through pulley, and passed through sub-cutaneous tissue to metacarpophalangeal joint of thumb. **D,** Securing two slips of transferred tendon to base of proximal phalanx. A hole is made through proximal phalanx in an ulnar-to-radial direction and is made larger on ulnar side to accept a loop of one tendon slip, which is secured with a pull-out suture. **E,** After transfer has been secured to thumb phalanx, tension is adjusted (see text) and proximal segment of flexor carpi ulnaris tendon is sutured to transferred tendon. (From Groves, R.J., and Goldner, J.L.: J. Bone Joint Surg. **57-A:**112, 1975.)

pulley and continue it subcutaneously to the proximal end of the proximal phalanx of the thumb. Here insert one split portion of this tendon into the bone with a pull-out wire and another into the bone by direct attachment. Suture the proximal functioning segment of the flexor carpi ulnaris and its tendon into the sublimis tendon unit under sufficient tension that dorsiflexion of the wrist provides full opposition of the thumb (Fig. 10-9, *E*).

MUSCLE TRANSFER (ABDUCTOR DIGITI QUINTI) TO RESTORE OPPOSITION

When other motors are unavailable or must be transferred elsewhere, the abductor digiti quinti muscle may be transferred as first described in 1921 by Huber and more recently by Littler and Cooley (1963). This muscle, because its mass and excursion are similar to those of the abductor pollicis brevis, is an excellent substitute for it. Cosmetically the transfer is helpful, since it fills the space left by the wasted thenar muscles. It does not require a pulley.

TECHNIQUE (LITTLER AND COOLEY). Make a curved palmar incision along the radial border of the abductor digiti quinti muscle belly extending from the proximal side of the pisiform proximally to the ulnar border of the little finger distally (Fig. 10-10). Free both tendinous insertions of the muscle, one from the extensor expansion and the other

from the base of the proximal phalanx. Lift the muscle from its fascial compartment and carefully expose its neurovascular bundle. Isolate the bundle, taking care not to damage the veins. Next free the origin of the muscle from the pisiform, but retain the origin on the flexor carpi ulnaris tendon; now the muscle can be mobilized enough for its insertion to reach the thumb. Make a curved incision on the radial border of the thenar eminence and create across the palm a subcutaneous pocket to receive the transfer. Now fold the abductor digiti quinti muscle over about 170 degrees (like a page of a book) and pass it subcutaneously to the thumb (Fig. 10-10, *C*). Suture its tendons of insertion to the insertion of the abductor pollicis brevis. Throughout the procedure avoid compression of and undue tension on the muscle and its neurovascular pedicle. Apply a carefully formed light compression dressing and then a volar plaster splint to hold the thumb in abduction and the wrist in slight flexion.

• • •

In 1929 Camitz described a transfer of the palmaris longus tendon to enhance opposition of the thumb. Braun has recently called attention to this useful procedure. He recommends it when the abductor pollicis brevis has weakened and atrophied from a partial median nerve palsy, which happens in carpal tunnel syndrome. An advantage

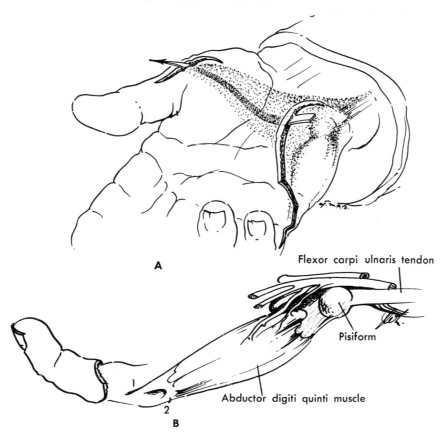

Fig. 10-10. Littler transfer of abductor digiti quinti to restore opposition. **A,** Two skin incisions. Intervening skin (*shaded area*) is undermined, creating pocket to receive transfer. **B,** Anatomy of abductor digiti quinti. Neurovascular bundle is located proximally on deep surface of muscle. Muscle inserts on both proximal phalanx, *1,* and extensor tendon, *2,* of little finger. *Continued.*

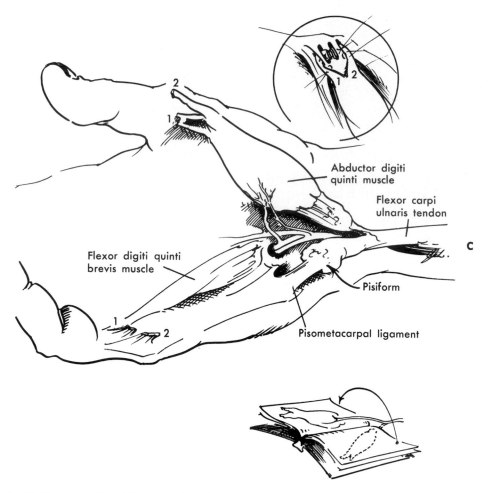

Fig. 10-10, cont'd. C, Origin of muscle is freed from pisiform but not from flexor carpi ulnaris tendon. Muscle is folded over about 170 degrees and is passed subcutaneously to thenar area, and its two tendons of insertion, *1* and *2,* are sutured to abductor pollicis brevis tendon. (Modified from Littler, J.W., and Cooley, S.G.E.: J. Bone Joint Surg. **45-A:**1389, 1963.)

of the operation is its close proximity to the median nerve, which may require repair or release that can be done at the same time without much additional surgery. It does not produce true opposition, but elevates the thumb toward the flexed and abducted position.

TECHNIQUE (CAMITZ). Make a curved incision parallel to the base of the thenar crease and extend it proximally 1½ inches (3.8 cm) up the forearm. Isolate the palmaris longus tendon in the distal forearm and preserve its insertion on the deep palmar fascia. Then incise along parallel lines from the insertion to the palmaris longus into the palmar fascia, obtaining a strip of fascia distally to lengthen the tendon enough to reach the distal part of the abductor pollicis brevis tendon. Pass the lengthened tendon into a small skin incision made over the thumb metacarpal and suture it to the tendon of the abductor pollicis brevis under appropriate tension.

Restoration of adduction of thumb

Adduction of the thumb is as necessary for strong pinch as is opposition. Whereas opposition is the refined unique movement that places the thumb within the flexion arc of the fingers so that the tips of the thumb and fingers can oppose, adduction is the force that stabilizes the thumb in the desired position. When the adductor pollicis is paralyzed, as in ulnar nerve palsy, firm pinch between the pulps of the thumb and the flexed index and middle fingers is impossible; furthermore, the thumb cannot be brought across the palm for pinch with the ring and little fingers. Eventually the interphalangeal joint of the thumb becomes hyperflexed and the metacarpophalangeal joint becomes hyperextended. The flexor pollicis longus can provide some power of adduction when the thumb is held in slight adduction so that the muscle flexes the digit through an arc parallel to the plane of the palm.

Several transfers have been devised to restore adduction. When adduction alone is absent, the brachioradialis or one of the radial wrist extensors may be lengthened by a graft, transferred palmarward through the third interosseous space, and carried across the palm to the tendon of the adductor pollicis. Such a transfer provides adduction only and this in the direction normally provided by the adductor pollicis. It is most often indicated in ulnar nerve palsy because in this instance restoring abduction of the thumb is

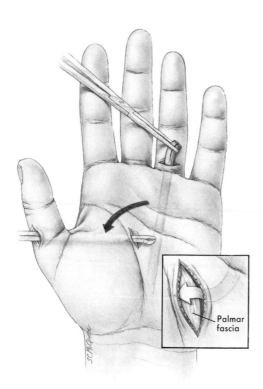

Fig. 10-11. Brand transfer (see text).

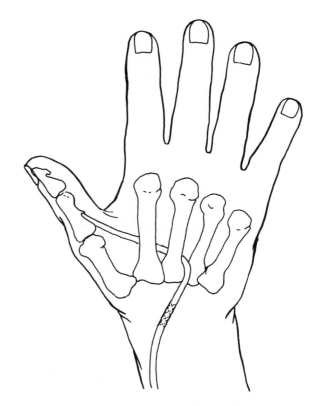

Fig. 10-12. Boyes transfer of brachioradialis or radial wrist extensor to restore thumb adduction (see text).

unnecessary; however, it should be combined with some procedure to restore abduction of the index finger. When both adduction and opposition of the thumb are absent, unless some other provision is made to restore adduction, a single tendon transfer to restore opposition should have its pulley located not near the pisiform but more distally so that some adduction will be restored too. The Royle-Thompson transfer meets this requirement. In it the flexor digitorum sublimis of the ring finger is brought out in the palm distal to the deep transverse carpal ligament that acts as a pulley, is carried across the palm, and is anchored to the tendon of the adductor pollicis. Opposition is only partially restored in that abduction and pronation of the thumb remain limited. To restore abduction of the index finger as well as adduction of the thumb, the sublimis tendon may be split and one slip anchored to the tendon of the adductor pollicis and the other to the insertion of the first dorsal interosseus as reported by Omer.

Several other operations are also available: the Brand transfer (Fig. 10-11) uses the sublimis of the ring finger as its motor. It traverses the palm superficial to the fascia and inserts on the radial aspect of the thumb. The sublimis is sectioned at the proximal phalanx through a short incision and is brought out at the midpalm just ulnar to the thenar crease. This tendon is passed through the natural openings of the fascia between the ring and middle finger at the distal third of the palm. It is then passed subcutaneously to be inserted on the radial side of the thumb at the level of the metacarpophalangeal joint. This tends to ensure pronation of the thumb as well as some restoration of power of adduction.

TRANSFER OF BRACHIORADIALIS OR A RADIAL WRIST EXTENSOR TO RESTORE THUMB ADDUCTION

TECHNIQUE (BOYES). Transfer of the brachioradialis is preferred. Detach the insertion of the muscle and carefully free the tendon proximally of all fascial attachments, thus increasing its excursion. Then anchor a tendon graft (plantaris or palmaris longus) to the adductor tubercle of the thumb by a pull-out wire, or suture the graft to the tendon of insertion of the adductor pollicis. Pass the graft along the adductor muscle belly and through the third interosseous space to the dorsum of the hand (Fig. 10-12). Then pass it subcutaneously in a proximal and radial direction and suture it to the end of the brachioradialis tendon. When a radial wrist extensor is used, pass the tendon graft deep to the extensor digitorum communis tendons and attach it to the wrist extensor. Apply a plaster cast while holding the thumb in adduction and the wrist in extension.

AFTERTREATMENT. At 3 weeks the cast is removed and active exercises are begun.

• • •

Smith reported on 18 patients in whom he transferred the extensor carpi radialis brevis tendon to provide strong thumb adduction. To extend the tendon he used a tendon graft and passed it through the second interosseous space. The operation was performed on patients in whom the power of pinch was 25% of normal or less because of paralysis of the adductor muscle. On the average, pinch was reported to have doubled in strength after the transfer (Figs. 10-13 and 10-14).

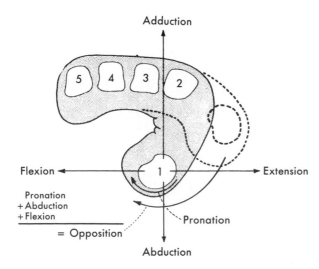

Fig. 10-13. Adduction and abduction of thumb are in plane perpendicular to palm. Flexion and extension of thumb are in palmar plane. Pronation and supination are rotation of the thumb about its longitudinal axis. Opposition is complex of abduction, flexion, and pronation of first metacarpal (as well as flexion/abduction of proximal phalanx and extension of distal phalanx). (From Smith, R.C.: J. Hand Surg. **8**:4, 1983.)

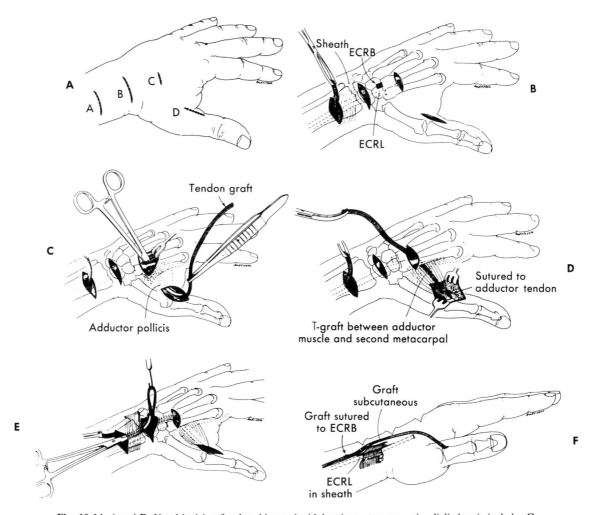

Fig. 10-14. A and **B,** Usual incision for detaching and withdrawing extensor carpi radialis brevis includes **C,** channeling tendon graft through second interspace, and **D,** attaching graft to tendon of adductor pollicis. **B,** Extensor carpi radialis brevis is transected distally and withdrawn proximal to dorsal retinacular ligament ("sheath"). **C,** Tendon graft (palmaris longus or plantaris) is passed deep to the adductor pollicis and between second and third metacarpals. **D,** Tendon graft is sutured to adductor tendon. **E,** Proximal end of tendon graft is passed subcutaneously to proximal incision. **F,** Tendon graft sutured proximally to extensor carpi radialis brevis with thumb adducted and wrist at zero degrees extension. Extensor carpi radialis brevis is at resting length. Graft made slightly longer if thenars are paralyzed. (From Smith, R.C.: J. Hand Surg. **8**:4, 1983.)

TECHNIQUE (SMITH). Make two dorsal transverse incisions over the extensor carpi radialis brevis tendon proximal to its insertion (Fig. 10-14, *A*). Divide the tendon near its insertion on the third metacarpal base and withdraw it through the incision proximal to the dorsal retinaculum (Fig. 10-14, *B*). Now make a third incision between the second and third metacarpals and remove a window of tissue from the paralyzed interosseus muscles. Finally, make a longitudinal incision on the ulnar side of the metacarpophalangeal joint of the thumb. With a curved hemostat, tunnel deep to the adductor pollicis muscle and through the window in the second interosseous space. Secure an appropriate tendon graft (usually the palmaris longus tendon). Draw the graft through the tunnel from the thumb to the dorsum of the hand (Fig. 10-14, *C*), and suture it to the tendon of the adductor pollicis (Fig. 10-14, *D*). Now pass the proximal end of the graft subcutaneously to the most proximal incision (Fig. 10-14, *E*), and suture it to the extensor carpi radialis brevis tendon, taking up all slack, but with no tension, so that the thumb lies just palmar to the index finger with the wrist in neutral position (Fig. 10-14, *F*). Dorsiflex the wrist and note that the thumb is pulled into adduction. Then flex the wrist and note that the thumb lies firmly against the palm.

AFTERTREATMENT. The hand is immobilized in plaster with the thumb in neutral position and with the wrist in 40 degrees dorsiflexion. The plaster is removed in 3 weeks, and active motion is encouraged.

ROYLE-THOMPSON TRANSFER, MODIFIED

TECHNIQUE. Make a midlateral incision (p. 10) over the ulnar aspect of the ring finger and free the insertion of the flexor digitorum sublimis tendon (Fig. 10-15). Then bring the tendon out of the palm through a short transverse incision and split it into two slips. Now make a curved incision on the dorsoradial aspect of the thumb as described for the Riordan transfer (p. 220). Tunnel the slips of the sublimis tendon radially into this incision. Then suture one slip to the extensor pollicis longus tendon distal to the metacarpophalangeal joint; tunnel the other slip dorsally over the metacarpal and suture it on the ulnar side of the thumb to the tendon of insertion of the adductor pollicis (Brand's dual insertion). Close the wounds and apply a cast holding the thumb in adduction and the wrist in moderate flexion.

AFTERTREATMENT. At 3 weeks the cast is removed and active exercises are begun.

RESTORATION OF ABDUCTION OF INDEX FINGER

The index is the finger against which the thumb is brought most frequently in pinch. Therefore if pinch is to be strong, this finger must be stable enough to provide the necessary resistance to the thumb; flexion, extension, abduction, and a stable metacarpophalangeal joint are required. Abduction of the index finger is also especially useful in such activities as playing a piano or using a typewriter. In poliomyelitis, abduction of the index finger is lost so frequently that its restoration is considered here separately from that of the intrinsic functions of the other fingers.

A transfer to restore abduction to this finger provides a substitute chiefly for the first dorsal interosseus muscle;

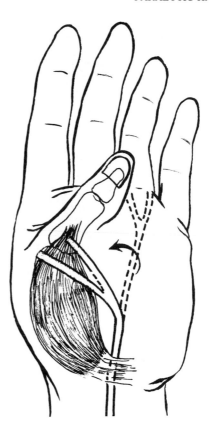

Fig. 10-15. Modified Royle-Thompson transfer to restore thumb adduction (see text).

therefore the transferred tendon is attached to the tendon of insertion of this muscle. The tendons most frequently transferred are those of the extensor indicis proprius, extensor pollicis brevis, and palmaris longus; any of these when transferred will abduct the index finger but will not stabilize it for strong pinch. A sublimis tendon has also been used, but this is generally contraindicated unless the hand is otherwise strong. When opposition must also be restored, the sublimis to the ring finger should be transferred to the thumb (p. 220).

TECHNIQUE. Begin a curved incision at the midlateral point on the radial side of the proximal phalanx of the index finger; carry it proximally over the radial aspect of the metacarpophalangeal joint and then curve it dorsally to end at the middle of the index metacarpal. To add length to the extensor indicis proprius tendon, elevate a small flap of the dorsal expansion over the metacarpophalangeal joint where it is attached to the insertion of the tendon. Withdraw the tendon proximally, free it throughout the wound, and close the defect in the expansion. Then pass the tendon radially in a gentle curve, roughen the tendon of the first dorsal interosseus muscle, and suture the transferred tendon to it with a single mattress suture.

• • •

When fusion of the metacarpophalangeal joint of the thumb is necessary because of an unacceptable hyperextension deformity in ulnar nerve palsy, the extensor pollicis brevis tendon may be transferred to the first dorsal

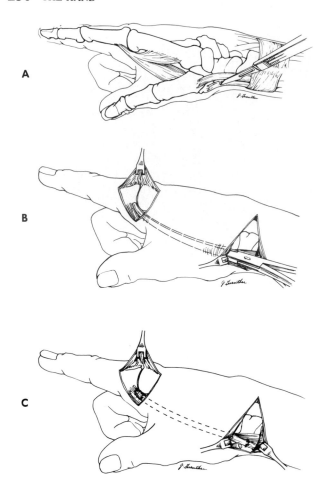

Fig. 10-16. A, Accessory slip inserting into greater multangular is detached distal to retinaculum. Functional slip, inserting into metacarpal, is preserved. **B,** Subcutaneous tunnel is created from radial styloid to insertion of first dorsal interosseus. **C,** Tendon graft is woven into tendon of the first dorsal interosseus and sutured to the accessory slip. (From Neviaser, R.J., Wilson, J.N., and Gardner, M.M.: J. Hand Surg. **5**:53, 1980.)

interosseus since it is no longer useful in its normal position.

Neviaser, Wilson, and Gardner have suggested transfer of a slip of the abductor pollicis longus tendon to replace the first dorsal interosseus muscle. In most patients the abductor pollicis longus tendon consists of two or more slips; in only 20% or less is there a single tendon here. The normal insertion is on the base of the thumb metacarpal and one or more of the extra slips insert on the greater multangular or into the abductor pollicis brevis. One of these extra slips is used in this transfer. The authors reported on 18 hands on which the operation was performed. They did not present comparative measurements of strength but stated that all patients had satisfactory pinch and stability and an increase in strength.

TECHNIQUE (NEVIASER, WILSON, AND GARDNER). Make a transverse incision near the insertion of the abductor pollicis longus. Identify the slips of the abductor tendon at the level of the radial styloid and note their insertions. Take care to avoid the branches of the superficial radial nerve. Apply traction to each of the slips to determine which insert on the metacarpal and which insert elsewhere. Select

a slip that does not insert on the metacarpal and divide it at its insertion. Then make a second incision over the radial side of metacarpophalangeal joint of the index finger and identify the tendon of the first dorsal interosseus muscle. The authors prefer a chevron incision with its base dorsally (Fig. 10-16). Now make a subcutaneous tunnel from the radial styloid to the base of the index finger. Obtain a tendon graft from the palmaris longus or elsewhere and weave it into the first dorsal interosseus tendon distal to the metacarpophalangeal joint.

Pass the graft subcutaneously into the area of the radial styloid without disturbing the first dorsal compartment. At the level of the radial styloid, with both the index finger and the wrist in neutral position, suture the graft to the selected slip of the abductor pollicis longus. Immobilize the wrist for 3 to 4 weeks and then begin active exercise.

RESTORATION OF INTRINSIC FUNCTION OF FINGERS

Loss of intrinsic muscle function of the fingers may result from paralytic disease or low median and ulnar nerve lesions; low lesions of these nerves cause selective paralysis of the intrinsic muscles but spare the long extrinsics to act unopposed (Fig. 10-17) and produce a clawhand. The mechanics of development of this deformity are discussed in the introduction to this chapter.

Loss of intrinsic muscle power may cause hyperextension of the metacarpophalangeal joints in a mobile hand; however, this deformity usually is not the primary or most disabling aspect of this paralysis. It has been shown that with intrinsic paralysis, grasp is diminished 50% or more because of the lack of power of flexion at the metacarpophalangeal joints. Additionally, there is asynchronous movement in flexion of the fingers themselves (Fig. 10-18). The roll-up maneuver of the fingers in the intrinsically paralyzed hand demonstrates this characteristic (Fig. 10-17). The interphalangeal joints must flex first, followed next by the metacarpophalangeal joints and ultimately by full flexion of the fingers. In-phase flexion of the metacarpophalangeal joints is lost with the loss of intrinsic muscle power; thus the hand is unable to grasp a large object. As previously mentioned, it also lacks power of grasp because metacarpophalangeal flexion is dependent entirely on the long flexors in the absence of intrinsics. Power of pinch is also diminished in addition to the effects of paralysis of the thenar muscles since the collateral ligaments of the metacarpophalangeal joints of the fingers are lax in extension and the stabilizing intrinsic musculature that would ordinarily give lateral stability is paralyzed. Divergence of the fingers is automatic with extension produced by the long extensor tendons, and as a result of the alignment of the finger flexors, convergence of the tips on grasping is automatic. However, to stabilize the fingers in extension at the metacarpophalangeal joint, especially for the resistance of the index finger to the pinch pressure of the thumb, the intrinsics are essential.

Many procedures have been devised to block hyperextension of the metacarpophalangeal joints, but stabilizing these joints at a selected position and permitting controlled deviation from side to side requires functioning intrinsic muscles.

The restoration of grasping power should be sought

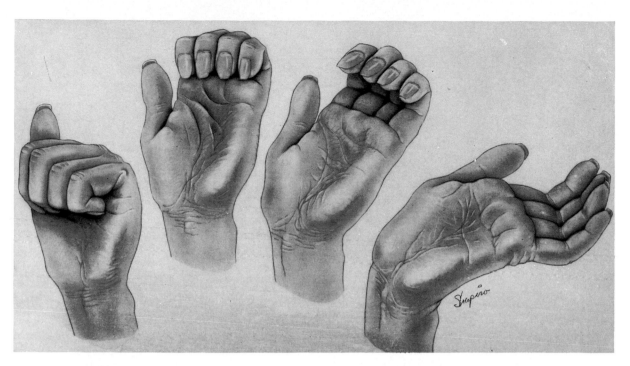

Fig. 10-17. Low median and ulnar nerve palsy. Action of extrinsic flexors of fingers when all of intrinsic muscles are paralyzed. From right to left note that these long flexors first flex distal interphalangeal joints, then flex proximal interphalangeal joints, and finally flex metacarpophalangeal joints. (From White, W.L.: Surg. Clin. North Am. **40:**427, 1960.)

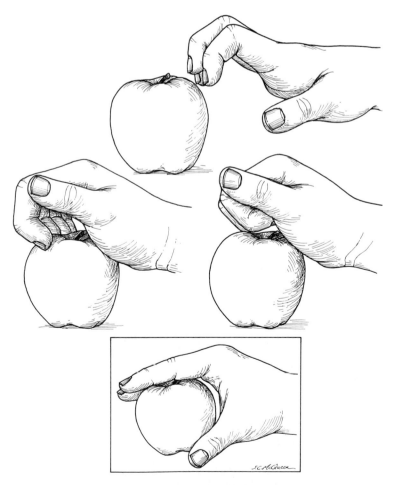

Fig. 10-18. Intrinsic muscle palsy. Flexion of metacarpophalangeal joints occurs only after interphalangeal joints are fully flexed. Fingers thus curl into the hand and push away any large object they wish to grasp. *Inset,* Object is grasped after compensatory manipulation. (Redrawn from Smith, R.J.: In American Academy of Orthopaedic Surgeons: Instructional course lectures, vol. 24, St. Louis, 1975, The C.V. Mosby Co.)

when there are suitable muscles available for the reconstruction, but this will depend on individual circumstances.

In this section detailed knowledge of both the anatomy and function of the intrinsic muscles is assumed and will not be reviewed; however, it must be emphasized here that the interosseus and lumbrical muscles flex the metacarpophalangeal joints and extend the interphalangeal joints of the fingers but that the long finger extensors are capable of extending the interphalangeal joints if the metacarpophalangeal joints are stabilized and cannot hyperextend (Fig. 10-3). This principle (that the long finger extensors can extend interphalangeal joints provided that hyperextension of the metacarpophalangeal joints is prevented) is the basis for many of the operations for intrinsic paralysis. The metacarpophalangeal joints may be stabilized by capsuloplasty (Zancolli), by tenodesis (Fowler; Riordan), by bone block (Howard; Mikhail), by arthrodesis, or by tendon transfers that actively extend the interphalangeal joints as well as flex the metacarpophalangeal joints. The proper operation for a given hand depends on the muscles available for transfer, the amount of passive motion present in the finger and wrist joints, and the opinion and experience of the surgeon. Transfers to replace intrinsic function of the fingers are the most variable, complicated, and surgically difficult ones carried out in the hand. Several different transfers have been devised, but no one has been universally accepted, and rightly so, because each hand is an individual problem, even when the only concern is intrinsic paralysis.

Sir Harold Stiles in 1922 attempted to restore intrinsic function by detaching a sublimis tendon, splitting it, and transferring it to the dorsum of the fingers to the extensor tendons. He did not report on followed patients. Bunnell in 1942 modified this procedure by detaching the sublimis tendon from each finger, splitting it, and passing one slip to each side of the extensor aponeurosis of each finger by way of the lumbrical canals. This transfer removed the powerful flexor of the proximal interphalangeal joints and converted it into an extensor of the same joints. In many instances this transfer has been too strong and has pulled the proximal interphalangeal joints into extension, especially when the hand has been supple before surgery; this complication, which produces an intrinsic plus deformity, has been known to occur several months to many years after the transfer. However, a modification of this procedure in which only one sublimis is transferred to all fingers may be useful in treating clawhands with some restriction of motion in the proximal interphalangeal joints (Fig. 10-19).

Frequently, flexing the wrist in an attempt to extend the interphalangeal joints has become a necessary habit after intrinsic paralysis. Its object is to create a tenodesing effect on the long extensor tendons. If this flexion is too marked, the Bunnell transfer is rendered ineffective, but when the intrinsics are weak but not paralyzed, the transfer may be useful if the wrist extensors are strong enough to prevent flexion of the wrist. When flexing the wrist is a chronic habit and a wrist flexor can be spared, Riordan has transferred the flexor carpi radialis (Fig. 10-23).

Fowler has split the extensor proprius tendons of the index and little fingers to form four slips and has attached

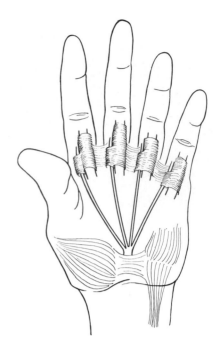

Fig. 10-19. Modification of Bunnell transfer to restore intrinsic function of fingers (see text).

one each to the extensor aponeuroses on the radial side of the index and middle fingers and on the ulnar side of the ring and little fingers. In a later modification, the tendons are split as described, but the slips are passed to the volar side of the deep transverse metacarpal ligament and are attached to the radial side of the extensor aponeurosis of each finger (Figs. 10-20 and 10-21). This is a more efficient transfer and has the advantage of a tenodesing effect when the wrist is flexed. However, the ends of the tendon slips must be advanced about 2.5 cm to reach their destinations on the extensor aponeuroses and are therefore under considerable tension; sometimes an intrinsic overpull or intrisic plus deformity develops. This excessive tension may be avoided as follows. The detached end of the extensor indicis proprius tendon is split into two slips, is passed to the volar side of the deep transverse metacarpal ligament, and is attached to the radial side of the ring and little fingers. One end of a free tendon graft is then attached to the musculotendinous junction of the extensor indicis proprius, and the other end is split into two slips that are passed distally in a similar manner and are attached to the radial side of the middle and index fingers. Riordan has further modified this procedure by attaching the tendon graft to the freed insertion of the palmaris longus tendon instead of to the musculotendinous junction of the extensor indicis proprius (Fig. 10-22).

Brand has had much experience in evaluating transfers for intrinsic paralysis in Hansen's disease. He devised a technique using the extensor carpi radialis brevis tendon lengthened by a free graft from the plantaris tendon (Fig. 10-23); the distal end of the graft is split into four slips or tails, and each tail is passed to the volar side of the deep transverse metacarpal ligament and is attached on the radial side of each proximal phalanx to the extensor aponeu-

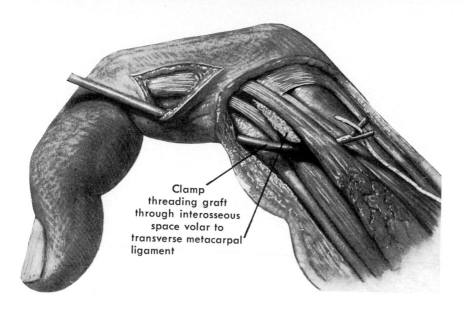

Clamp
threading graft
through interosseous
space volar to
transverse metacarpal
ligament

Fig. 10-20. Any tendons transferred from dorsum of hand to restore intrinsic function of fingers *must pass to volar side* of deep transverse metacarpal ligament.

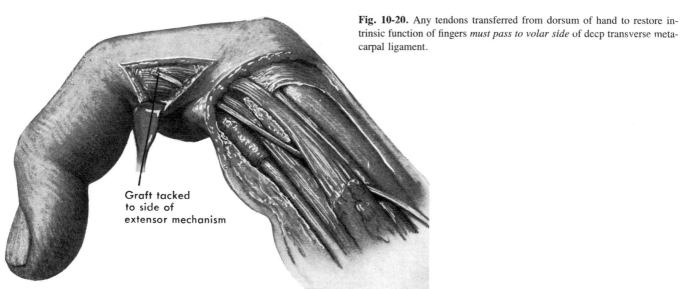

Graft tacked
to side of
extensor mechanism

Tendon being routed volar to deep
transverse metacarpal ligament

Site of extensor indicis proprius tendon

Site of extensor
digiti V proprius tendon

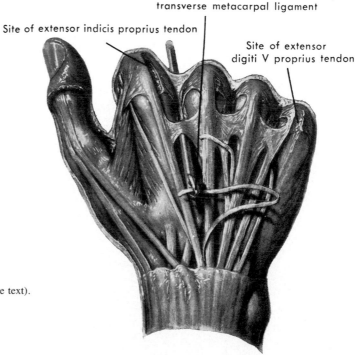

Fig. 10-21. Fowler transfer to restore intrinsic function of fingers (see text).

rosis, except in the index finger where it is attached on the ulnar side. In his opinion, index finger pinch can be secured more firmly when the finger is in adduction rather than in abduction (Fig. 10-24). More recently he has advised transferring the extensor carpi radialis longus or brevis to the volar side of the forearm and extending it by a four-tailed graft through the carpal tunnel and the lumbrical canals and finally to the extensor aponeuroses as before (Fig. 10-25). This transfer crowds the carpal tunnel and may cause symptoms of median nerve compression should the nerve be functioning.

For severe clawing of the hand with flexion of the wrist, Riordan advises freeing the insertion of the flexor carpi radialis and transferring it to the dorsum of the wrist; here it is prolonged with a four-tailed graft, each tail of which is passed volar to the deep transverse metacarpal ligament and is attached to the radial sides of the extensor aponeuroses (Fig. 10-26).

The procedures just described require that muscles strong enough for transfer be available. When they are not, a capsuloplasty or a tenodesing procedure to stabilize the metacarpophalangeal joints may be indicated. Zancolli described a satisfactory capsulodesis (Fig. 10-27). Riordan devised a tenodesing procedure in which the extensor carpi radialis brevis and extensor carpi ulnaris tendons are each cut halfway through at about the level of the junction of the middle and distal thirds of the forearm; a half of each tendon is then stripped distally and is left attached at its insertion on a metacarpal base. Each strand of tendon is split into two strips forming four slips; each slip is then passed through an interosseous space and along the volar side of the deep transverse metacarpal ligament to a finger and is attached to the radial side of its extensor aponeu-

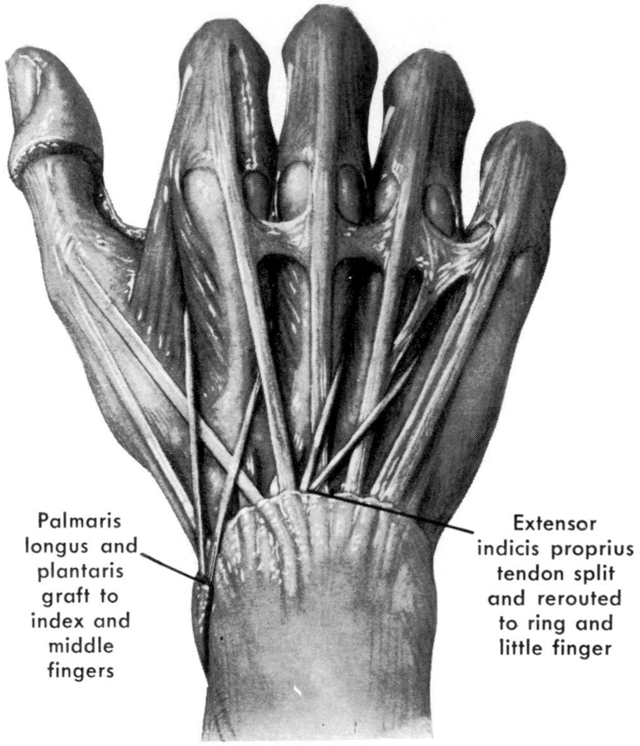

Palmaris longus and plantaris graft to index and middle fingers

Extensor indicis proprius tendon split and rerouted to ring and little finger

Fig. 10-22. Riordan transfer to restore intrinsic function of fingers (see text).

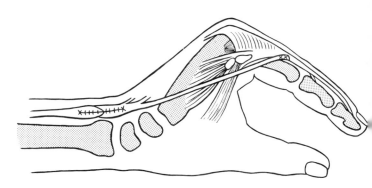

Fig. 10-23. Brand transfer of extensor carpi radialis brevis tendon prolonged with free graft to restore intrinsic function of fingers (see text).

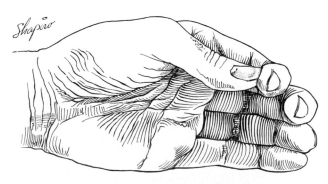

Fig. 10-24. When firm pinch will be more useful than abduction of index finger, transferred tendon is attached to ulnar lateral band of extensor hood rather than to insertion of first dorsal interosseus muscle. Note that during pinch index finger is in adduction rather than in abduction. (From White, W.L.: Surg. Clin. North Am. **40:**427, 1960.)

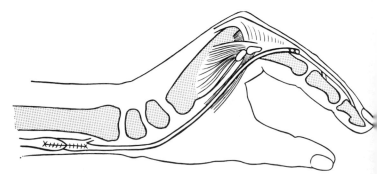

Fig. 10-25. Brand transfer of extensor carpi radialis longus or brevis, first to volar side of forearm and then, after being prolonged with free graft, to extensor aponeuroses to restore intrinsic function of fingers (see text).

rosis. The disadvantage of this tenodesis is that it cannot be activated by motion of the wrist as can the Fowler tenodesis. Fowler uses a free tendon graft attached to the fingers as in the Riordan technique but anchored proximally in the area of the dorsal carpal ligament proximal to the wrist. Thus, when the wrist is flexed, the tenodesis is activated (Fig. 10-28).

When the finger flexors and the wrist flexors and extensors are strong and when there is no habitual flexion of the wrist, the operation of choice to restore function of the finger intrinsics is the modified Bunnell procedure in which the flexor digitorum sublimis of the ring finger is transferred (Fig. 10-19). When flexing the wrist is habitual or there is a flexion contracture of the joint and when a wrist flexor can be spared, the Riordan transfer of the flexor carpi radialis to the dorsum of the wrist prolonged by tendon grafts (Fig. 10-26) is a good choice; however, at least one strong wrist flexor should remain after the transfer. When the wrist extensors are strong and the flexors are weak, the Brand transfer of the extensor carpi radialis longus volarward and prolonged by a free graft through the carpal tunnel (Fig. 10-25) may be indicated. The Brand transfer of the extensor carpi radialis brevis prolonged by a free graft carried between the metacarpals and attached to the extensor aponeuroses (Fig. 10-23) may be complicated by difficulty in reeducation. When a flexor digitorum sublimis or a wrist flexor or extensor is not available for transfer or cannot be spared, the extensor proprius tendons of the index and little fingers may be transferred by the Fowler technique (Figs. 10-20 and 10-21) or the Riordan modification of the Fowler technique in which the palmaris longus tendon is one of the transfers (Fig. 10-22) may be used. When no muscle is available for transfer and when the joints are supple, the Zancolli capsulodesis of the metacarpophalangeal joints (Fig. 10-27), a Fowler tenodesis (Fig. 10-28), or a Riordan tenodesis may be indicated.

The tendency to overload the extensor mechanism by routine attachment of transferred tendons to the lateral bands has been noted; this means that the desirable flexor power to the metacarpophalangeal joints is not obtained; therefore, Brooks and Jones have suggested attaching the transfers to the flexor tendon sheath (Fig. 10-29) and depending on the intact extensor power to extend the proximal interphalangeal joints with the metacarpophalangeal joints stabilized. Likewise, Burkhalter and others have suggested a bony attachment of the transferred tendon to the midportion of the proximal phalanx to provide leverage for flexion at the metacarpophalangeal joint and a better restoration of grip (see Fig. 10-38). However, we have found the Zancolli tendon insertion into the flexor sheath to be much less time consuming than insertion into bone. It also eliminates the tendency for hyperextension of the proximal interphalangeal joint as occurs sometimes when the tendon is inserted into the extensor mechanism.

TECHNIQUE OF TENDON TRANSFER (BUNNELL, MODIFIED). Transfer the sublimis tendon of either the ring or middle finger. Make a midlateral incision (p. 10) about 3.8 cm long on the radial side of the selected finger, beginning at the midshaft of the proximal phalanx and extending distally to beyond the proximal interphalangeal joint. Deepen the incision to the flexor tendon sheath, open the sheath laterally, and identify and divide the sublimis tendon at the level of the proximal interphalangeal joint. Separate the two slips of the tendon so that the tendon can be withdrawn into the

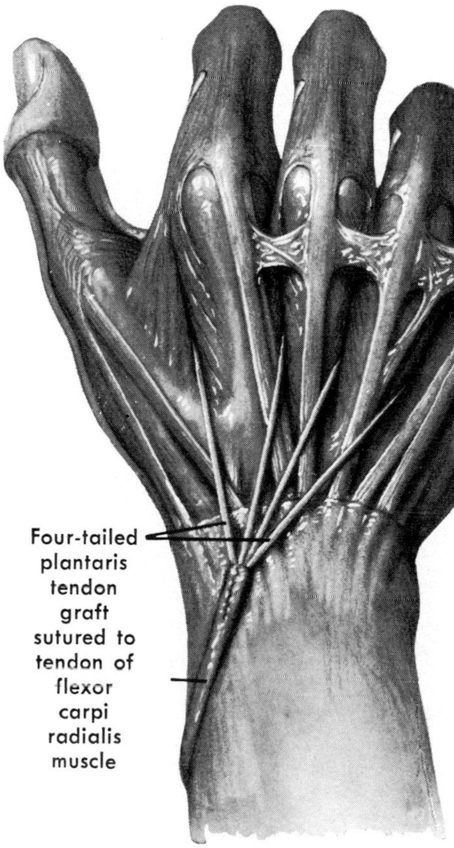

Four-tailed plantaris tendon graft sutured to tendon of flexor carpi radialis muscle

Fig. 10-26. Riordan transfer to restore intrinsic function of fingers (see text).

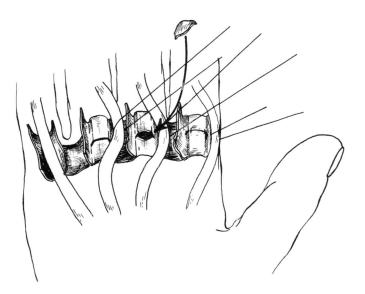

Fig. 10-27. Zancolli capsulodeses for intrinsic paralysis (see text).

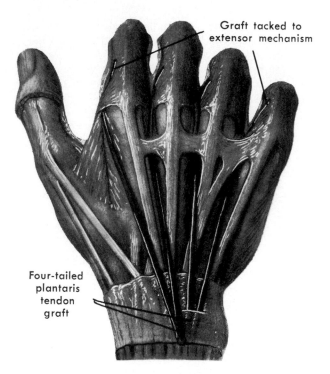

Graft tacked to extensor mechanism

Four-tailed plantaris tendon graft

Fig. 10-28. Fowler tenodesis for intrinsic paralysis (see text).

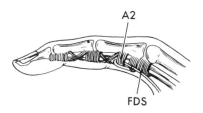

A2

FDS

Fig. 10-29. Flexor digitorum slip *(FDS)* inserted into strip of annular ligament *(A2)* at the middle of the proximal phalanx. (Redrawn from slide supplied by Brooks in Riordan, D.C.: J. Hand Surg. **8:**748, 1983.)

palm. Now make a transverse incision about 3.8 cm long at the level of the proximal palmar crease. Identify the sublimis tendon, withdraw it through the palmar incision, and split it into four equal tails. Now make a longitudinal incision about 2.5 cm long on the radial side (and slightly dorsal) of the proximal phalanx of each finger, except the donor finger, and identify the extensor aponeuroses. Then with either a narrow instrument, a wire loop, or a tendon carrier, pass each tail of tendon through the lumbrical canal of a finger and over the oblique fibers of the extensor aponeurosis to its dorsum (Fig. 10-19). Passage through the lumbrical canals should be easy; if any obstruction is met, redirect the instrument. Now with the metacarpophalangeal joint at 80 or 90 degrees flexion, the interphalangeal joints at neutral, and the wrist at 30 degrees flexion, suture each tail to the aponeurosis under some tension with interrupted sutures and bury its end. Usually 2.5 to 3.8 cm of redundant tendon must be excised. Close the incisions and immobilize the hand with the wrist in neutral position,

the metacarpophalangeal joints in flexion, and the interphalangeal joints in extension.

Arthur Brooks recommends attaching each transfer to the flexor pulley at the level of the proximal phalanx, thus preventing the development of hyperextension deformities of the proximal interphalangeal joints.

AFTERTREATMENT. At 3 weeks the cast is removed, and each finger is splinted with a plaster or plastic gutter splint in a neutral position. Movement of the metacarpophalangeal joints and resisted active extension of the wrist are then encouraged. The finger splints are removed and are reapplied daily until reeducation is complete.

TECHNIQUE OF TENDON TRANSFER (BRAND). Divide the extensor carpi radialis brevis tendon at the distal end of the radius through a short dorsal transverse incision (Fig. 10-30). Then make a second incision 8.9 cm proximal to the first, withdraw the tendon through it, and lay the tendon on a wet towel. Remove a plantaris tendon for a graft (p. 64) and divide it in half or double it on itself to make two grafts. Split open the end of the motor tendon along the natural plane of cleavage, spread it out, and suture the graft to it as shown in Fig. 10-31. Then introduce a tendon-tunneling forceps at the first incision, pass it subcutaneously to the second, grasp the ends of the tendon grafts, and pull them through so that the anastomosis lies under intact skin. Split the end of each graft into two parts to form a total of four slips or tails. Then make a longitudinal dorsoulnar incision over the proximal phalanx of the index finger and dorsoradial incisions over the proximal phalanx of the middle, ring, and little fingers. Identify the lumbrical tendon and lateral band of the extensor aponeurosis in each finger; tunnel from this point on each finger through the palm and appropriate interosseous space; grasp a strand of tendon graft and withdraw it into the finger. Take care to tunnel to the volar side of the deep transverse metacarpal ligament and then between the appropriate metacarpal shafts (Fig. 10-32). When all tendon grafts are in position, suture them one by one under equal tension to the lateral band of the dorsal expansion of each finger, first the index, then the little, and finally the intermediate ones (Fig. 10-33). The transfers should be completely relaxed when the wrist is dorsiflexed 45 degrees, the metacarpophalangeal joints are flexed 70 degrees, and the interphalangeal joints are in neutral. Close the wounds and apply a light plaster cast.

As an *alternative method,* use the extensor carpi radialis longus tendon. Through a dorsal transverse incision free its insertion and withdraw it through a second incision at the middle of the forearm. Then make an incision on the anterior aspect of the forearm 7.5 cm proximal to the wrist, tunnel from the anterior incision deep to the brachioradialis to the proximal incision, and draw the tendon into the anterior incision. Suture the grafts to the motor tendon as decribed previously. Then through a midpalmar incision introduce a tunneler, pass it through the carpal tunnel into the forearm, and draw the grafts into the palm, leaving the anastomosis proximal to the carpal tunnel. Then pass each strand of the graft separately to its finger destination.

AFTERTREATMENT. Aftertreatment is as described for the modified Bunnell technique.

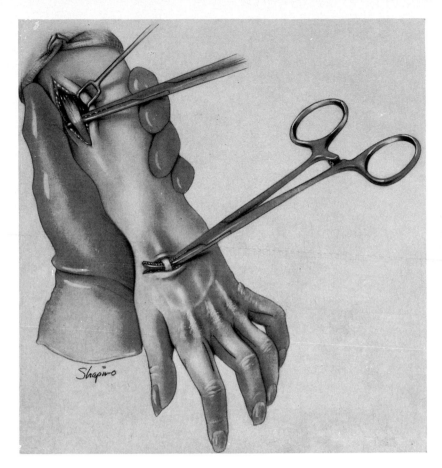

Fig. 10-30. Brand transfer of extensor carpi radialis brevis prolonged with free graft to restore intrinsic function of fingers. Tendon to be transferred has been isolated both proximally and distally. (From White, W.L.: Surg. Clin. North Am. **40:**427, 1960.)

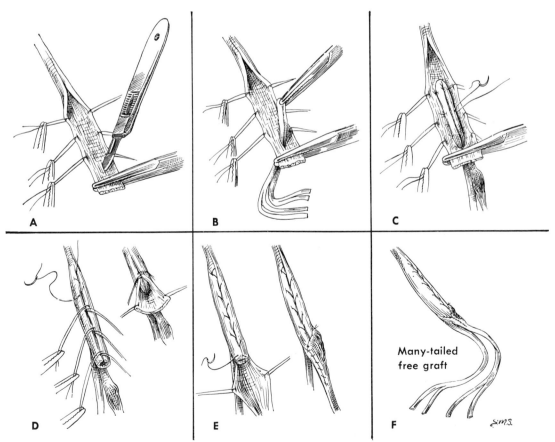

Many-tailed
free graft

A   B   C   D   E   F

Fig. 10-31. Same as Fig. 10-30. **A,** End of extensor carpi radialis brevis tendon has been split open and spread out and is being perforated with knife. **B,** Plantaris tendon (graft) has been doubled on itself and is being pulled through perforation. **C,** Graft is being attached to tendon with interrupted sutures. **D,** Extensor carpi radialis brevis tendon is being closed over graft with running suture. **E,** On *left,* one strand of graft is being spread open to cover end of extensor carpi radialis brevis tendon. On *right,* graft has been sutured over end of tendon. **F,** Two ends of graft have been split into four slips or tails. (From White, W.L.: Surg. Clin. North Am. **40:**427, 1960.)

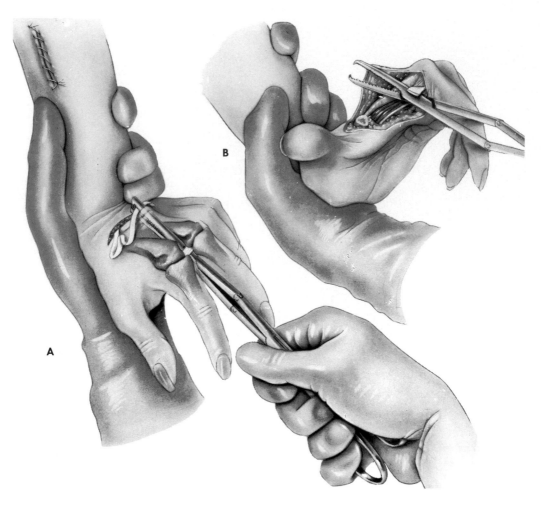

Fig. 10-32. Same as Fig. 10-30. **A,** Extensor carpi radialis brevis along with attached tendon graft has been passed subcutaneously from proximal incision to one at distal radius. One slip of graft is being passed to volar side of deep transverse metacarpal ligament and into index finger on its ulnar side. **B,** For purposes of demonstration, second ray has been removed to show that forceps pass to volar side of ligament. (From White, W.L.: Surg. Clin. North Am. **40:**427, 1960.)

TECHNIQUE OF TENDON TRANSFER (FOWLER). In this transfer, the extensor indicis proprius and the extensor digiti quinti proprius are used as motors (Fig. 10-21). Make a dorsal incision over the radial aspect of the metacarpophalangeal joint of the index finger and identify the extensor indicis proprius tendon. Dissect the tendon from the extensor aponeurosis, obtaining as much length as possible by excising a part of the aponeurosis with it; otherwise the tendon will be too tight after transfer. Then suture the residual defect in the aponeurosis. Split the extensor indicis proprius into two equal parts, pass each volar to the deep transverse metacarpal ligament, and attach one each to the extensor aponeurosis on the radial side of the index and middle fingers, as in the Bunnell technique. Now make a dorsal incision over the little finger, identify the extensor digiti quinti proprius tendon, and free its insertion; split this tendon also into two equal parts, pass each volar to the deep transverse metacarpal ligament, and attach one each to the radial side of the ring and little fingers. Take care that this tendon is not too tight. The Riordan modification of this

operation (Fig. 10-22) does not use the extensor digiti quinti proprius.

AFTERTREATMENT. Aftertreatment is as described for the modified Bunnell technique.

TECHNIQUE OF CAPSULODESIS (ZANCOLLI). Make a transverse incision in the palm at the level of the distal crease. Undermine widely the skin and fat and expose the flexor tendon sheaths; take care not to damage the neurovascular bundles. Now over each metacarpophalangeal joint make a longitudinal incision in the peritendinous fascia and tendon sheath and expose the flexor tendons. Carefully retract the tendons and expose the underlying metacarpophalangeal joint (Fig. 10-27). Resect an elliptic segment of the volar fibrocartilaginous plate including the vertical septum and its deep origin. Resect enough tissue to produce a 10 to 30 degree flexion contracture when the plate is closed. Now close the plate by wire or heavy silk sutures placed laterally in its thickest part, this being at the insertion of the accessory collateral ligaments. If desired to maintain position of the joints, insert transarticular Kirschner wires.

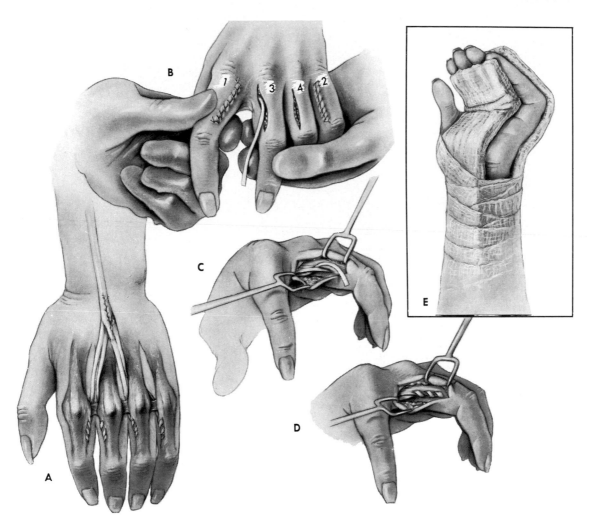

Fig. 10-33. Same as Fig. 10-30. **A,** One slip of tendon graft has been sutured to appropriate lateral band of each finger. **B,** First, one slip of graft is sutured to ulnar lateral band of index finger and then one each to radial lateral band of little, long, and ring fingers in that order. **C and D,** Method of weaving a slip of graft into lateral band. **E,** Wrist has been immobilized in 45 degrees of dorsiflexion and metacarpophalangeal joints in 70 degrees of flexion. (From White, W.L.: Surg. Clin. North Am. **40:**427, 1960.)

Close the wound and apply a dorsal plaster splint holding the metacarpophalangeal joints in flexion and the wrist in extension.

AFTERTREATMENT. Movements of the interphalangeal joints are continued after surgery. At 3 weeks the cast and any Kirschner wires are removed and exercises of the metacarpophalangeal joints are begun.

TECHNIQUE OF TENODESIS (FOWLER). In this operation, a tendon graft is substituted for the finger intrinsics; the graft may be activated by flexing the wrist (Fig. 10-28). Obtain a tendon graft (p. 64) twice as long as the distance from the dorsum of the wrist to the proximal interphalangeal joints. Make a transverse incision on the dorsum of the wrist and expose the dorsal retinaculum of the wrist. Then pass the graft through the retinaculum just distal to its proximal edge. Split each end of the graft into two equal slips. Then transfer each slip to a finger, as in the Fowler transfer described previously. Suture the graft under proper tension so that when the wrist is flexed force will be exerted on the extensor mechanism so as to extend the interphalangeal joints without hyperflexing the metacarpophalangeal joints.

PERIPHERAL NERVE PALSIES
Low radial nerve palsy

In low radial nerve palsy all digital extensors and the long thumb abductor are paralyzed. The radial wrist extensors and the brachioradialis are spared, in contrast to high radial nerve palsy in which these are also lost. Therefore the basic functions to be restored are extension of all digits (including the thumb) and radial abduction of the thumb (Fig. 10-34). The muscles available for transfer are the synergistic wrist flexors, the long finger flexors, and the pronator teres. Murphey in 1914 and Jones in 1916 described transfers that have been successful, but Zachary, in 1946 in a critical survey, pointed out that one strong

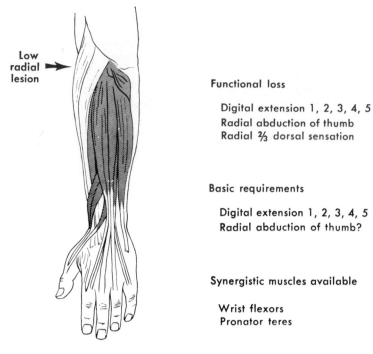

Low radial lesion ➤

Functional loss

Digital extension 1, 2, 3, 4, 5
Radial abduction of thumb
Radial ⅔ dorsal sensation

Basic requirements

Digital extension 1, 2, 3, 4, 5
Radial abduction of thumb?

Synergistic muscles available

Wrist flexors
Pronator teres

Fig. 10-34. Low radial nerve palsy. (Modified from White, W.L.: Surg. Clin. North Am. **40:**427, 1960.)

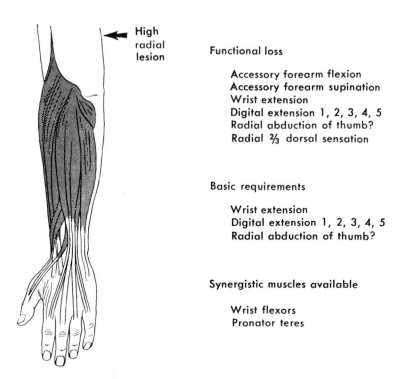

◄ **High radial lesion**

Functional loss

Accessory forearm flexion
Accessory forearm supination
Wrist extension
Digital extension 1, 2, 3, 4, 5
Radial abduction of thumb?
Radial ⅔ dorsal sensation

Basic requirements

Wrist extension
Digital extension 1, 2, 3, 4, 5
Radial abduction of thumb?

Synergistic muscles available

Wrist flexors
Pronator teres

Fig. 10-35. High radial palsy. Boyes has added the sublimi to available synergistic muscles listed here (see text). (From White, W.L.: Surg. Clin. North Am. **40:**427, 1960.)

wrist flexor, usually the flexor carpi radialis (or a strong palmaris longus), should not be transferred but should be left in place to prevent hyperextension of the wrist. This is an important observation because hyperextension certainly occurs in mobile wrists and causes incomplete extension of the metacarpophalangeal joints.

Many combinations of transfers for low radial nerve palsy are possible, but only one will be mentioned here, as follows. The flexor carpi ulnaris is transferred to the long finger extensors. The palmaris longus is transferred to the long thumb extensor after the latter has been transposed from around Lister's tubercle to the radial side of the wrist in line with the first metacarpal (if the palmaris longus is absent, the brachioradialis is transferred instead). The pronator teres is transferred to the long thumb abductor. All requirements are then met because the lost functions are restored, and the wrist extensors and one strong wrist flexor are left undisturbed to stabilize the wrist.

High radial nerve palsy

In high radial nerve palsy the radial wrist extensors and the brachioradialis are paralyzed in addition to those muscles paralyzed in low radial nerve palsy (Fig. 10-35). The synergistic wrist flexors, long finger flexors, and the pronator teres, however, are again available for transfer. A satisfactory plan is to transfer the insertion of the pronator teres to the extensor carpi radialis brevis tendon, the flexor carpi ulnaris to the long finger extensors, and the palmaris longus to the long thumb extensor. The long thumb extensor is transposed from around Lister's tubercle and is rerouted along the radial side of the wrist in line with the first metacarpal; the palmaris longus tendon can then be attached to it in a straight line. Thus both extension and abduction of the thumb are restored (Riordan), and a transfer to the long thumb abductor is not absolutely necessary. The flexor carpi radialis remains undisturbed to prevent hyperextension of the wrist. When the palmaris longus is absent, the flexor carpi ulnaris may be transferred not only to the long finger extensors but to the long thumb extensor as well. Even when the palmaris longus is present Omer uses this transfer. He then shifts the extensor pollicis brevis tendon volarward and ulnarward and sutures the palmaris longus tendon to it; since the insertions of these tendons are not detached, some abduction of the thumb is restored and function of the palmaris longus as a flexor of the wrist is retained. However, to preserve the basic functional movements of the wrist, Boyes advises against transfer of the flexor carpi ulnaris; instead he transfers two flexor digitorum sublimis tendons (see below for the Boyes technique).

TECHNICAL CONSIDERATIONS FOR TRANSFER

The muscle belly of the flexor carpi ulnaris extends almost to the insertion of its tendon, and all along its course the muscle takes origin from the surrounding fascia. Therefore the tendon cannot be easily withdrawn through an incision proximal to its insertion; rather, the incision should be curved and should be made over the entire distal half of the muscle; then its fibers should be dissected from the tendon proximally to the middle of the forearm so that the tendon can be transferred with gradual angulation to the long extensor tendons of the fingers. The level of su-

ture is proximal to the dorsal carpal ligament; this ligament can be partially excised proximally if necessary, but a bowstring effect will be produced if it is completely excised.

The transferred tendons should be tunneled through subcutaneous fat without touching scar, fascia, or muscle. To prevent excessive scarring, the tunnels should be only wide enough to permit the tendons to glide without obstruction. The superficial fascia should be excised from the area in which tendons are attached. In transfers for radial nerve palsy the tendons should be sutured under a little more tension than is usual because the strong flexors tend to stretch the transferred muscles; mattress sutures of wire are preferred.

One way of adjusting tension on the transferred flexor carpi ulnaris is to maintain the wrist in a neutral position while the fingers are completely extended; place maximum pull on the transferred tendon-muscle unit, and suture the transferred tendon to the tendons of the finger extensors. The tendon is simply pulled tight, held over, and then sutured to these tendons without severing them. Through-and-through mattress sutures are used to individually attach each extensor tendon to the larger transferred tendon unit. After suture, the correct tension is verified when, on extending the wrist, full passive finger flexion is possible, and when, on flexing the wrist, there is sufficient tension on the extensors to fully extend the finger.

An additional incision on the middle third of the forearm is needed to attach the pronator teres tendon to the extensor carpi radialis brevis. The pronator teres tendon is detached from its insertion with as much periosteum as possible to make suturing easier. The tab of tendon insertion and the periosteum are inserted in the muscle belly of the extensor carpi radialis brevis and held with one untied suture. This suture is not tied until after the tension testing of the transfer to the finger extensors. An additional short incision may be made about the muscle-tendon junction of the palmaris longus to bring it out proximally for gradual angulation toward the thumb extensor. Its insertion can be reached through the dorsal incision at the wrist that extends volarward for the dissection of the flexor carpi ulnaris.

Many variations of the above are possible, including placing the palmaris longus into the abductor pollicis longus tendon or into the extensor pollicis brevis tendon; both provide some abduction of the thumb. However, if this is done, care should be taken to attach the extensor pollicis longus to the transferred flexor carpi ulnaris so that there will be an active thumb extensor with the common digital extensors.

Brand has suggested removing the insertion of the extensor carpi radialis longus and transferring it to a point between the extensor carpi radialis brevis and extensor carpi ulnaris to avoid radial deviation on extension of the wrist.

Boyes devised an operation for high radial nerve palsy in which two sublimis tendons are included in the transfer. This operation satisfactorily restores function and preserves wrist control.

TECHNIQUE (BOYES). Make a long longitudinal incision on the volar side of the radial aspect of the forearm and free the insertion of the pronator teres. Perforate the extensor carpi radialis longus and brevis tendons, pass the insertion of the pronator teres through these tendons, and suture it under proper tension. Use stainless steel wire for all tendon attachments. Expose the sublimis tendons of the middle and ring fingers through a single incision in the distal palm over the metacarpal heads or through separate incision at their insertions on the middle phalanges; divide each so that the free end of its distal segment lies within its sheath. Withdraw the proximal segments of the tendons through the forearm incision. Now make a dorsal transverse incision on the wrist extending from the radial styloid toward the ulnar styloid and then curving proximally. Expose the common digital extensors proximal to the dorsal carpal ligament and incise the deep fascia. Next make a 2 cm ^{2} opening in the interosseous membrane at the proximal edge of the pronator quadratus muscle. Pass the sublimis of the middle finger to the radial side of the profundus muscle mass and between it and the flexor pollicis longus muscle mass and then through the interosseous membrane. Then attach the donor tendon to the common digital extensors. Make another opening in the interosseous membrane; pass the sublimis of the ring finger to the ulnar side of the profundus muscle mass and through the opening and attach it to the extensor pollicis longus and extensor indicis proprius. Divide the flexor carpi radialis at the wrist and suture it to the extensor pollicis brevis and abductor pollicis longus at this level.

Before the transferred tendons are sutured in place, removing the tourniquet and checking the interosseous artery for bleeding is wise. If this artery has been lacerated but not properly ligated, then serious complications, including ischemic myositis, may develop after surgery.

AFTERTREATMENT. Immobilization of some type should be carried out for 5 weeks. Usually a cast is maintained on the arm for 4 weeks followed by a spring-loaded extension splint for the wrist and fingers for another week. During the cast immobilization, the metacarpophalangeal joints should not be completely extended but should be held in about 40 degrees of flexion. However, the wrist should be fully extended with the thumb in abduction and extension. The interphalangeal joints of the fingers should be in "comfortable" flexion.

Low ulnar nerve palsy

The functional deficits caused by low ulnar nerve palsy are weakness of pinch resulting from paralysis of the adductor pollicis and first dorsal interosseus, weakness of grip produced by paralysis of most of the finger intrinsics, and sometimes clawing of the ring and little fingers associated with paralysis of all of their intrinsics (Fig. 10-36).

Paralysis of the adductor pollicis results in a major loss of function that should be restored when possible by appropriate tendon transfer (see restoration of adduction of thumb, p. 226). Normal tightness of the metacarpophalangeal joints of the ring and little fingers may limit clawing of these fingers and enable the long extensors to extend their interphalangeal joints; in this instance, no treatment is indicated for clawing, but weakness of grip is still present. However, when clawing of these fingers is troublesome, function of their intrinsics should be restored by transferring the extensor indicis proprius tendon; it is split

into two slips, is passed volar to the deep transverse metacarpal ligament, and is attached to the radial side of the extensor aponeuroses of each finger as in the Riordan transfer (Fig. 10-22). Other dynamic transfers such as that of Bunnell (p. 235) or Brand (p. 236) may be useful. As an alternative to tendon transfer, clawing of the ring and little fingers may be corrected by Zancolli capsulodeses (p. 238).

For low ulnar nerve palsy, Omer suggests the following procedure carried out in one stage (Fig. 10-37). The metacarpophalangeal joint of the thumb is arthrodesed. The insertion of the flexor digitorum sublimis of the middle finger is freed, and the tendon is split into two slips. One slip is carried across the palm parallel to the fibers of the adductor pollicis and is anchored to the insertion of that muscle. The other slip is split into two tails; one is carried through the appropriate lumbrical canal and is anchored to the radial side of the extensor aponeurosis of the ring finger, and the other is transferred in a similar manner to the little finger. Instead of the procedure just described, he sometimes transfers the brachioradialis tendon, prolonged with a free graft, through the third interosseous space to restore adduction of the thumb (p. 227); to restore abduction of the index finger, he frees the radial half of the insertion of the extensor indicis proprius, splits the tendon, and anchors the freed half of the tendon to the insertion of the first dorsal interosseus.

Burkhalter has suggested several tendon transfers, all of which ultimately end by insertion directly into the diaphysis of the proximal phalanx of the involved fingers. He believes this is a more secure attachment and also gives the advantage of a greater lever arm beyond the metacarpophalangeal joint. For motors, he has used either the brachioradialis or the extensor carpi radialis longus extended by free grafts, both of which are brought dorsally and passed through the intermetacarpal area volar to the transverse metacarpal ligament and then attached to bone (Fig. 10-38). He has also used the same bony attachment in transferring a split sublimis of the ring finger as a modification of the Stiles-Bunnell transfer. In addition, there should be a transfer to provide adduction of the thumb.

Brown has suggested several transfers for adduction of the thumb: one using the sublimis of the ring finger brought deep to the flexors of the fingers and another using the extensor indicis proprius brought into the palm through the third space of the metacarpals and then transversely across the palm, paralleling the paralyzed adductor muscle, to attach to the metacarpophalangeal joint area of the thumb. On occasion, arthrodesis of the distal thumb joint is advised to increase the power of pinch; this is accompanied at times by advancing the pulley at the metacarpophalangeal joint by sectioning it proximally to provide a greater angle of approach of the flexor pollicis longus.

High ulnar nerve palsy

The functional deficits caused by high ulnar nerve palsy are the same as those described for low ulnar nerve palsy, except that functions of the flexor digitorum profundus of the ring and little fingers and of the flexor carpi ulnaris are also lost (Fig. 10-39). The transfers described for low ulnar nerve palsy may be used, except that the sublimis of

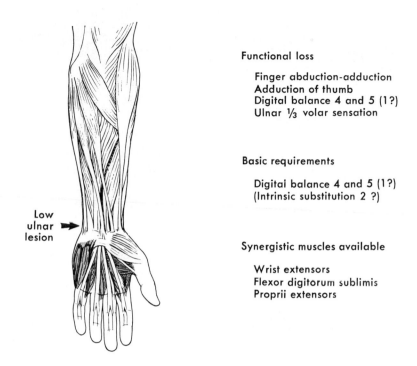

Functional loss

   Finger abduction-adduction
   Adduction of thumb
   Digital balance 4 and 5 (1?)
   Ulnar ⅓ volar sensation

Basic requirements

   Digital balance 4 and 5 (1?)
   (Intrinsic substitution 2 ?)

Synergistic muscles available

   Wrist extensors
   Flexor digitorum sublimis
   Proprii extensors

Fig. 10-36. Low ulnar nerve palsy. (From White, W.L.: Surg. Clin. North Am. **40:**427, 1960.)

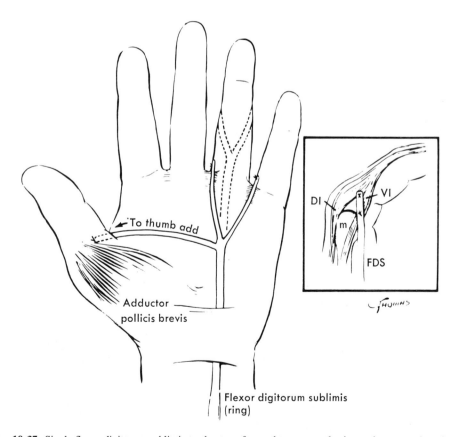

Fig. 10-37. Single flexor digitorum sublimis tendon transfer used to correct clawing and to strengthen thumb-index pinch in isolated ulnar nerve palsy. (From Omer, G.E., Jr.: Orthop. Clin. North Am. **5:**377, 1974.)

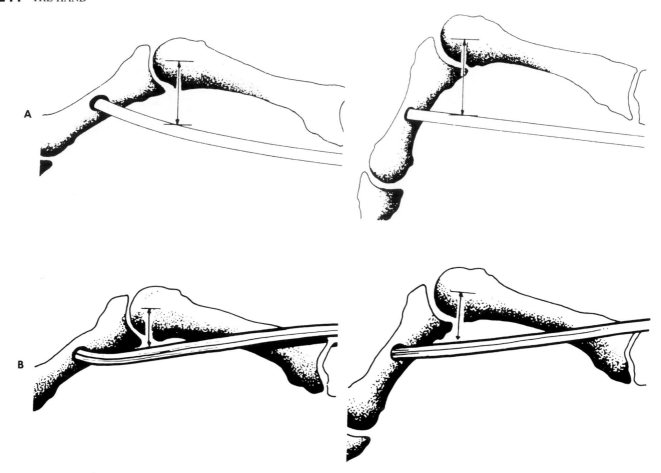

Fig. 10-38. A, Burkhalter modification of Stiles-Bunnell transfer increases distance of moment arm with increased flexion of metacarpophalangeal joint. Force applied distally varies with square of distance. **B,** With intermetacarpal route for this transfer, moment arm also increases with increasing flexion of metacarpophalangeal joint. (From Burkhalter, W.E.: Orthop. Clin. North Am. **5:**289, 1974.)

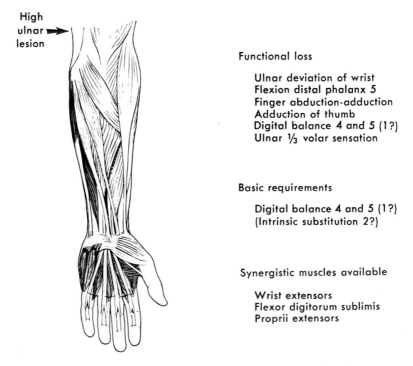

High
ulnar → lesion

Functional loss

    Ulnar deviation of wrist
    Flexion distal phalanx 5
    Finger abduction-adduction
    Adduction of thumb
    Digital balance 4 and 5 (1?)
    Ulnar ⅓ volar sensation

Basic requirements

    Digital balance 4 and 5 (1?)
    (Intrinsic substitution 2?)

Synergistic muscles available

    Wrist extensors
    Flexor digitorum sublimis
    Proprii extensors

Fig. 10-39. High ulnar nerve palsy. (From White, W.L.: Surg. Clin. North Am. **40:**427, 1960.)

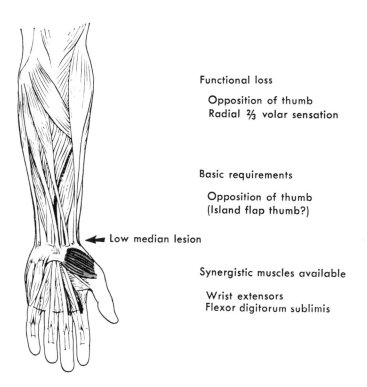

Functional loss

Opposition of thumb
Radial ⅔ volar sensation

Basic requirements

Opposition of thumb
(Island flap thumb?)

← Low median lesion

Synergistic muscles available

Wrist extensors
Flexor digitorum sublimis

Fig. 10-40. Low median nerve palsy. (From White, W.L.: Surg. Clin. North Am. **40:**427, 1960.)

the ring finger must not be transferred because the profundus of this finger is paralyzed. Flexion of the distal interphalangeal joints of the ring and little fingers may be restored by suturing the profundus tendons of these fingers to that of the middle finger. If further power is needed, transfer of the extensor carpi radialis longus into the profundus tendons of the middle, ring, and little fingers may also be done. It should be remembered that the innervation of the profundus of the middle finger may be totally ulnar at times and frequently only partially ulnar.

Low median nerve palsy

The important functional deficits caused by low median nerve palsy are loss of opposition of the thumb and loss of sensibility over the sensory distribution of the nerve; paralysis of the two radial lumbricals is of little consequence when the ulnar nerve is intact (Fig. 10-40). Restoration of thumb opposition is discussed on p. 218. Restoration of sensibility by a neurovascular island graft is discussed on p. 126.

High median nerve palsy

The important functional deficits caused by high median nerve palsy are loss of pronation of the forearm, flexion of the wrist, flexion of the index and middle fingers, flexion of the thumb, opposition of the thumb, and sensation over the median distribution (Fig. 10-41).

Function may be partly restored as follows. The flexor digitorum profundus of the index and middle fingers may be attached to the ulnar-innervated flexor digitorum profundus by side-to-side suture without sectioning of any tendons (Fig. 10-42). In addition, greater power can be

achieved by transferring the extensor carpi radialis longus into the profundus tendons of the index and middle fingers. Flexion of the thumb may be restored by transfer of the brachioradialis to the long thumb flexor at the wrist level. Opposition of the thumb may be restored by using the extensor indicis proprius as a transfer, bringing it around the ulnar side of the wrist so that construction of a pulley is not needed (see Burkhalter technique).

The restoration of sensibility by a neurovascular island graft is discussed on p. 126.

Combined low median and ulnar nerve palsy (at wrist)

Combined median and ulnar nerve lesions at the wrist (Fig. 10-43) result in complete anesthesia of the palm and loss of function of all intrinsics of both the fingers and the thumb (see the introduction to this section). When untreated, skin and joint contractures develop, and a fixed clawhand results.

Despite the palmar anesthesia, it is possible to restore some useful function after this severe paralysis. The success of treatment depends on several factors. Often the flexor tendons have been severely injured by the same trauma that caused the paralysis; in this event, the condition of the tendons is important in planning transfers. In Hansen's disease the paralysis is not accompanied by tendon injury but at times by deformity of the skin, fingernails, and bone. For tendon transfers to be successful, any contractures of the skin or joints must be corrected first because the transfers alone cannot accomplish this. Passive extension of the interphalangeal joints and flexion of the metacarpophalangeal joints of the fingers must be possible. An attempt is made to mobilize the joints by splinting; if

SEVERE PARALYSIS FROM DAMAGE TO THE CERVICAL SPINAL CORD OR OTHER CAUSES

Paralysis from spinal cord injury fortunately is not as common as paralysis caused by peripheral nerve injuries. Few surgeons have had vast experience in treatment of these patients, but excellent work has been published by Freehafer et al., Moberg, Lamb, Zancolli, and others.

A surgical procedure for quadriplegia should not be stereotyped since each patient is different, even those with cord injury at the same cervical level and even each upper extremity in the same patient. Careful analysis of the motor and sensory status is necessary to determine which surgical procedure is warranted, if any. Many patients are extremely hesitant to have any surgical procedure done for fear of losing what little function remains. The examiner not only must check for muscle function and grade the power but also should observe the patient going through his daily activities and try to determine what additional function would best accomplish a greater independence. According to Lamb, if there is no muscle power, not even a flicker, immediately after injury and again nothing in one month, then no function can be expected from this muscle. As a rule, however, surgery is begun after months of observation, usually a year or longer. In partial or incomplete quadriplegia, spasticity usually becomes a consideration since it may jeopardize the end results of surgery.

For sensibility, the two-point discrimination test with a paper clip as described by Moberg yields the accurate assessment necessary to plan treatment for improvement of grip. When sensibility is not present, then sight must substitute, and ocular substitution can be only unilateral; therefore, only one upper extremity should have surgery. Following injury, it is essential to maintain joint mobility of the fingers, wrist, elbow, and shoulder, since contractures frequently develop, especially with spasticity.

The goal of treatment by the authors mentioned usually has been to obtain key grip or key pinch. Key pinch posture provides a stronger, broader gripping surface, is cosmetically more preferable, and is more easily achieved than the "chuckjaw" or three-jawed pinch. Grasping with all fingers is desirable, but this cannot be accomplished without more available muscles. A muscle, to be useful, must be strong enough at least to move against the resistance of gravity.

There is usually no indication for arthrodesis of the wrist in this type of paralysis. The ability to project and place the hand away from the trunk is extremely important; elbow extension and shoulder abduction are involved. Methods of providing elbow extension and shoulder abduction are found elsewhere; however, attention should be called to the deltoid transfer described by Moberg.

The following are a functional classification according to the availability of muscle power and suggested treatment based on these variations.

Class 1 (brachioradialis only). When the brachioradialis muscle has exceptional grade 4 power, it can be transferred to provide wrist extension. Dissecting the muscle proximally to free its distal attachments to the forearm fascia for greater excursion is emphasized. Moberg recommends also a tenodesis of the flexor pollicis longus tendon to the radius under sufficient tension to cause metacarpophalangeal joint flexion with active wrist extension to give thumb pulp–to–radial side of index finger pinch. Thumb stability is proved by temporary "arthrodesis" of the distal joint with a transfixing Kirschner wire; this allows reversal of the procedure should the patient find that this is not helpful. In addition to this, a release of the flexor pulley system at the metacarpophalangeal joints provides a better lever arm for the passive tendon in flexing the metacarpophalangeal joint (Fig. 10-44).

Class 2 (brachioradialis and extensor carpi radialis longus and brevis). Freehafer recommends usually that no procedures be done since the patient has good wrist extensor power and therefore automatic grasp. Lamb suggests the following: transfer of the extensor carpi radialis longus to the flexor profundus of the fingers (Fig. 10-45) and the brachioradialis to the flexor pollicis longus. Both of these muscles must be dissected up high in the forearm to mo-

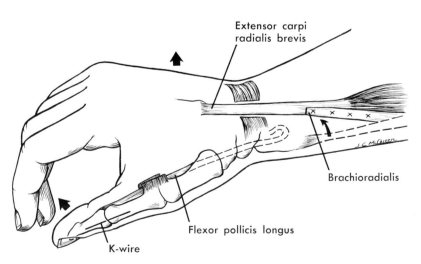

Fig. 10-44. When brachioradialis is only remaining functioning muscle unit, Moberg suggests these transfers (see text).

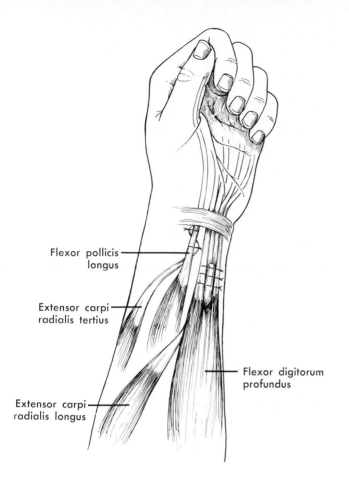

Flexor pollicis
longus

Extensor carpi
radialis tertius

Extensor carpi
radialis longus

Flexor digitorum
profundus

Fig. 10-45. When both radial wrist extensors are active, extensor carpi radialis longus may be used to help provide finger flexion; thumb flexion may be benefited by transfer of an active extensor carpi radialis tertius when present (see text).

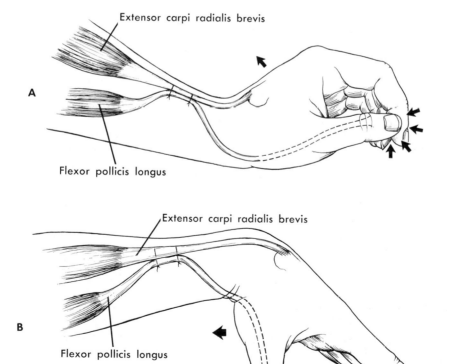

Extensor carpi radialis brevis

A

Flexor pollicis longus

Extensor carpi radialis brevis

B

Flexor pollicis longus

Fig. 10-46. A, Key pinch is obtained with wrist extension by an active extensor carpi radialis brevis with tenodesis to flexor pollicis longus. To achieve correct tension, with wrist in complete passive extension, flexor pollicis longus tendon is sutured to extensor carpi radialis when pinching is produced. **B,** With passive wrist flexion, pinch is released. (Redrawn from Zancolli, E.: Clin. Orthop. **112:**101, 1975.)

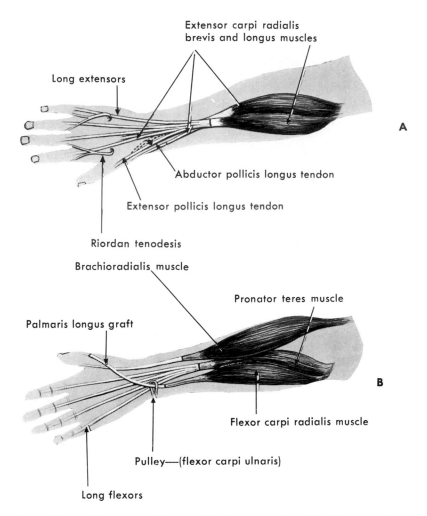

Extensor carpi radialis
brevis and longus muscles

Long extensors

A

Abductor pollicis longus tendon

Extensor pollicis longus tendon

Riordan tenodesis

Brachioradialis muscle

Pronator teres muscle

Palmaris longus graft

B

Flexor carpi radialis muscle

Pulley—(flexor carpi ulnaris)

Long flexors

Fig. 10-47. Technique of Lipscomb, Elkins, and Henderson (see text). (Modified from Lipscomb, P.R., Elkins, E.C., and Henderson, E.D.: J. Bone Joint Surg. **40-A:**1071, 1958.)

bilize them and to provide sufficient excursion to bring them around to the appropriate site for attachment. Adjustment of tension is extremely important. In addition to the transfers, it may be helpful to temporarily fix the distal joint of the thumb with a Kirschner wire to see if this improves function; also advancing the pulley at the metacarpophalangeal joint may help mechanically.

Zancolli suggests a transfer of the brachioradialis to the extensor digitorum communis and the extensor pollicis longus and of the extensor carpi radialis longus to the flexor digitorum profundus; also a tenodesis of the flexor pollicis longus by attaching it to the extensor carpi radialis brevis to produce thumb flexion with wrist extension by contraction of the muscle (Fig. 10-46). He may also, as an option, arthrodese the carpometacarpal joint.

Class 3 (brachioradialis, extensor carpi radialis longus and brevis, pronator teres, and flexor carpi radialis). Several plans for transfer again are possible. Quadriplegic patients with an active brachioradialis and extensor carpi radialis longus and brevis are much more common than those also having active pronator teres and flexor carpi radialis. Functional

pronator teres and flexor carpi radialis muscles allow multiple options for reconstruction (Fig. 10-47). In addition to the transfers described for classes 1 and 2, the flexor carpi radialis can be used for thumb opposition by construction of a pulley at the pisiform bone and prolongation of the insertion with a tendon graft.

The brachioradialis may be used as the thumb flexor when the pronator teres is used as a motor for the flexor digitorum profundus. In any of the above procedures, a Fowler tenodesis is possible for the intrinsic paralysis should there be sufficient clawing at the metacarpophalangeal joints to warrant this.

• • •

Occasionally a patient with a low cervical cord lesion will have absence only of the finger intrinsic musculature, including the thenar muscles for opposition and adduction. Several transfers are available for this; most are similar to those recommended elsewhere for low median and ulnar nerve palsy.

REFERENCES

Adams, J., and Wood, V.E.: Tendon transfers for irreparable nerve damage in the hand, Orthop. Clin. North Am. **12**:403, 1981.

Beasley, R.W.: Principles of tendon transfer, Orthop. Clin. North Am. **2**:433, 1970.

Beasley, R.W.: Tendon transfers for radial nerve palsy, Orthop. Clin. North Am. **2**:439, 1970.

Belsole, R.J., Lister, G.D., and Kleinert, H.E.: Polyarteritis: a cause of nerve palsy in the extremity, J. Hand Surg. **3**:320, 1978.

Blacker, G.J., Lister G.D., and Kleinert, H.E.: The abducted little finger in low ulnar nerve palsy, J. Hand Surg. **1**:190, 1976.

Boswick, J.A., Jr.: Tendon transfers for tendon injuries in the upper extremities, Orthop. Clin. North Am. **2**:253, 1974.

Boyes, J.H.: Tendon transfers for radial palsy, Bull. Hosp. Joint Dis. **21**:97, 1960.

Boyes, J.H.: Selection of a donor muscle for tendon transfer, Bull. Hosp. Joint Dis. **23**:1, 1962.

Boyes, J.H.: Bunnell's surgery of the hand, ed. 4, Philadelphia, 1964, J.B. Lippincott Co.

Boyes, J.H.: Problems of tendon surgery, Am. J. Surg. **109**:269, 1965.

Brand, P.W.: Paralytic claw hand: with special reference to paralysis in leprosy and treatment by the sublimis transfer of Stiles and Bunnell, J. Bone Joint Surg. **40-B**:618, 1948.

Brand, P.W.: Tendon grafting: illustrated by a new operation for intrinsic paralysis of the fingers, J. Bone Joint Surg. **43-B**:444, 1961.

Brand, P.W.: Tendon transfers for median and ulnar nerve paralysis, Orthop. Clin. North Am. **2**:447, 1970.

Brand, P.W.: Rehabilitation of the hand with motor and sensory impairment, Orthop. Clin. North Am. **4**:1135, 1973.

Braun, R.M., et al.: Preliminary experience with superficialis-to-profundus tendon transfer in the hemiplegic upper extremity, J. Bone Joint Surg. **56-A**:466, 1974.

Brooks, A.L.: Personal communication, 1969.

Brooks, D.M.: Inter-metacarpal bone graft for thenar paralysis: technique and end-results, J. Bone Joint Surg. **31-B**:511, 1949.

Brown, P.W.: Zancolli capsulorrhaphy for ulnar claw hand: appraisal of forty-four cases, J. Bone Joint Surg. **52-A**:868, 1970.

Brown, P.W.: Reconstruction for pinch in ulnar intrinsic palsy. Orthop. Clin. North Am. **2**:323, 1974.

Bunnell, S.: Tendon transfers in the hand and forearm. In American Academy of Orthopaedic Surgeons: Instructional course lectures, vol. 6, Ann Arbor, 1949, J.W. Edwards.

Burkhalter, W.E.: Early tendon transfer in upper extremity peripheral nerve injury, Clin. Orthop. **104**:68, 1974.

Burkhalter, W.E.: Restoration of power grip in ulnar nerve paralysis, Orthop. Clin. North Am. **2**:289, 1974.

Burkhalter, W.E.: Tendon transfers in median nerve palsy, Orthop. Clin. North Am. **2**:271, 1974.

Burkhalter, W.E.: Tendon transfers in brachial plexus injuries, Orthop. Clin. North Am. **2**:259, 1974.

Burkhalter, W.E. and Strait, J.L.: Metacarpophalangeal flexor replacement for intrinsic-muscle paralysis, J. Bone Joint Surg. **55-A**:1667, 1973.

Burkhalter, W.E., et al.: Extensor indicis proprius opponensplasty, J. Bone Joint Surg. **55-A**:725, 1973.

Buckwalter, J.A., Mickelson, M.R., and Emerson, R.L.: Ulnar nerve palsy in Paget's disease, Clin. Orthop. **127**:212, 1977.

Carroll, R.E., and Kleinman, W.B.: Pectoralis major transplantation to restore elbow flexion to the paralytic limb, J. Hand Surg. **4**:501, 1979.

Clippinger, F.W., Jr., and Irwin, C.E.: The opponens transfer: analysis of end results, South Med. J. **55**:33, 1962.

Cochrane, R.G.: Leprosy in theory and practice, Bristol, 1959, John Wright & Sons, Ltd.

Curtis, R.M.: Fundamental principles of tendon transfer, Orthop. Clin. North Am. **2**:231, 1974.

Curtis, R.M.: Opposition of the thumb, Orthop. Clin. North Am. **2**:305, 1974.

Curtis, R.M.: Tendon transfers in the patient with spinal cord injury, Orthop. Clin. North Am. **2**:415, 1974.

DeBenedetti, M.: Restoration of elbow extension power in the tetraplegic patient using the Moberg technique, J. Hand Surg. **4**:86, 1979.

Edgerton, M.T., and Brand, P.W.: Restoration of abduction and adduction to the unstable thumb in median and ulnar paralysis, Plast. Reconstr. Surg. **36**:150, 1965.

Enna, C.D.: Use of the extensor pollicis brevis to restore abduction in the unstable thumb, Plast. Reconstr. Surg. **46**:350, 1970.

Enna, C.D., and Riordan, D.C.: The Fowler procedure for correction of the paralytic claw hand, Plast. Reconstr. Surg. **52**:352, 1973.

Entin, M.A.: Restoration of function of paralyzed hand, Surg. Clin. North Am. **44**:1049, 1964.

Flatt, A.E.: An indication for shortening of the thumb: description of technique and brief report of five cases, J. Bone Joint Surg. **46-A**:1534, 1964.

Flynn, J.E.: Reconstruction of the hand after median-nerve palsy, N. Engl. J. Med. **256**:676, 1957.

Freehafer, A.A., and Mast, W.A.: Transfer of the brachioradialis to improve wrist extension in high spinal-cord injury, J. Bone Joint Surg. **49-A**:648, 1967.

Freehafer, A.A., et al.: Tendon transfers to improve grasp after injuries of the cervical spinal cord, J. Bone Joint Surg. **56-A**:951, 1974.

Goldner, J.L.: Tendon transfers for irreparable peripheral nerve injuries of the upper extremity, Orthop. Clin. North Am. **2**:343, 1974.

Goldner, J.L., and Irwin, C.E.: An analysis of paralytic thumb deformities, J. Bone Joint Surg. **32-A**:627, 1950.

Goldner, J.L., and Kelly, J.M.: Radial nerve injuries, South. Med. J. **51**:873, 1958.

Granberry, W.M. and Lipscomb, P.R.: Tendon transfers to the hand in brachial palsy, Am. J. Surg. **108**:840, 1964.

Groves, R.J. and Goldner, J.L.: Restoration of strong opposition after median-nerve or brachial plexus paralysis, J. Bone Joint Surg. **57-A**:112, 1975.

Hamlin, C., and Littler, J.W.: Restoration of power pinch, J. Hand Surg. **5**:396, 1980.

Henderson, E.D.: Use of sublimis tendon transfers to extend the fingers: report of case, Mayo Clin. Proc. **35**:438, 1960.

Henderson, E.D.: Transfer of wrist extensors and brachioradialis to restore opposition of the thumb, J. Bone Joint Surg. **44-A**:513, 1962.

Hentz, V.R., and Keoshian, L.A.: Changing perspectives in surgical hand rehabilitation in quadriplegic patients, Plast. Reconstr. Surg. **64**:509, 1979.

House, J.H., Gwathmey, F.W., and Lundsgaard, D.K.: Restoration of strong grasp and lateral pinch in tetraplegia due to cervical spinal cord injury, J. Hand Surg. **1**:152, 1976.

Huber, E.: Hilfsoperation bei Medianuslähmung, Deutsch. Z. Chir. **162**:271, 1921.

Irwin, C.E., and Eyler, D.L.: Surgical rehabilitation of the hand and forearm disabled by poliomyelitis, J. Bone Joint Surg. **33-A**:825, 1951.

Irwin, C.E., and Flinchum, C.E.: Piedmont Orthopaedic Society Letter, October 1957 (mimeographed).

Jacobs, B. and Thompson, T.C.: Opposition of the thumb and its restoration, J. Bone Joint Surg. **42-A**:1015, 1960.

Kaplan, I., Dinner, M., and Chait, L.: Use of extensor pollicis longus tendon as a distal extension for an opponens transfer, Plast. Reconstr. Surg. **57**:186, 1976.

Kessler, F.B.: Use of a pedicled tendon transfer with a silicone rod in complicated secondary flexor tendon repairs, Plast. Reconstr. Surg. **49**:439, 1972.

Lamphier, T.A.: Tendon transfer (or transplant) for paralysis of an interosseus muscle, Int. Surg. **27**:738, 1957.

Larsen, R.D., and Posch, J.L.: Nerve injuries in the upper extremity, Arch. Surg. **77**:469, 1958.

Leddy, J.P., et al.: Capsulodesis and pulley advancement for the correction of claw-finger deformity, J. Bone Joint Surg. **54-A**:1465, 1972.

Lipscomb, P.R., Elkins, E.C., and Henderson, E.D.: Tendon transfers to restore function of hands in tetraplegia, especially after fracture-dislocation of the sixth cervical vertebra on the seventh, J. Bone Joint Surg. **40-A**:1071, 1958.

Littler, J.W.: Tendon transfers and arthrodeses in combined median and ulnar nerve paralysis, J. Bone Joint Surg. **31-A**:225, 1949.

Littler, J.W., and Cooley, S.G.E.: Opposition of the thumb and its restoration by abductor digiti quinti transfer, J. Bone Joint Surg. **45-A**:1389, 1963.

Makin, M.: Translocation of the flexor pollicis longus tendon to restore opposition, J. Bone Joint Surg. **49-B**:458, 1967.

Mangus, D.J.: Flexor pollicis longus tendon transfer for restoration of opposition of the thumb, Plast. Reconstr. Surg. **52**:155, 1973.

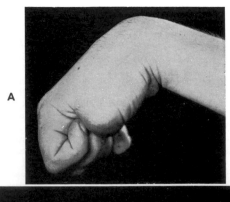

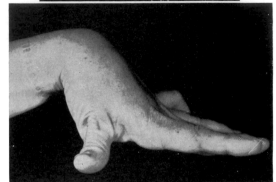

Fig. 11-1. Typical deformities of hand in severe cerebral palsy (spastic paralysis). **A,** Spastic flexor pollicis longus pulls thumb into palm. Metacarpophalangeal joint of thumb is hypermobile. Spastic wrist and finger flexors pull both wrist and fingers into fixed position. **B,** When wrist is in neutral position or in dorsiflexion, active extension of thumb and fingers is weak because extensor digitorum communis and extensor pollicis longus muscles are weak. But when wrist is in flexion, tension on extensor tendons helps these muscles to extend thumb and fingers. (From Goldner, J.L.: J. Bone Joint Surg. **37-A:**1141, 1955.)

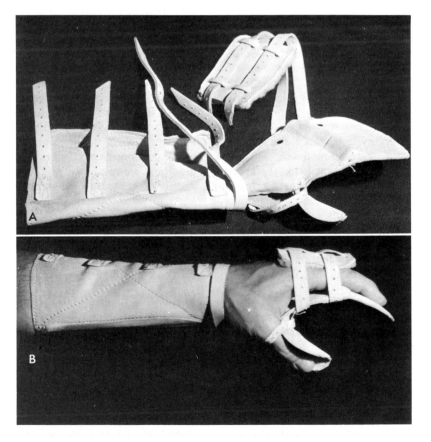

Fig. 11-2. Splint for spastic hand. **A,** Splint shaped of aluminum and padded with leather. **B,** Splint holding wrist, fingers, and thumb in proper position (see text). (From Swanson, A.B.: Surg. Clin North Am. **48:**1129, 1968.)

age of 7 years and then only after enough observation to permit careful evaluation of sensation and motor function in the extremity.

An accurate evaluation of the various deforming muscle forces is difficult because they can vary from time to time and from one examination to the next. Several examinations and possibly several years of follow-up may be required before accurate conclusions can be reached. However, a diagnostic analysis needs to be done before the possibility of performing muscle releases or tendon transfers is considered. Several examples can be cited.

When there is severe spasticity in flexion but less spasticity of the long finger flexors than of the wrist flexors, then release of the wrist flexors may be considered. In this case release or transfer of the flexor carpi ulnaris is given special consideration.

Wrist flexion is frequently accompanied by ulnar deviation and pronation. If pronation deformity is severe but the wrist and fingers can be extended with the wrist held passively in at least neutral position, then transfer of the pronator teres to provide supination may be considered.

When a thumb-in-palm deformity can be corrected by wrist flexion but is increased by wrist extension, the flexor pollicis longus tendon may be the principal deforming force and may be considered for release and transfer. However, when the adductor pollicis is the chief deforming force, wrist flexion and extension should not affect the thumb-in-palm deformity, and it may be released by myotomy.

Hypertension of the proximal interphalangeal joints of the fingers may be increased by wrist flexion and released somewhat by wrist extension. However, chronic hyperextension deformity of the proximal interphalangeal joints requires a tenodesis.

Release of flexor-pronator origin

Release of the flexor-pronator origin may improve appearance and function of a hand with severe flexion deformities of the wrist and fingers. It is not indicated in those hands that can be corrected passively but that assume a flexed position during grasp; for these, less extensive operations such as transfer of the flexor carpi ulnaris to a wrist extensor are more useful. Release of the flexor-pronator origin was described by Page in 1923 and more recently by Inglis and Cooper and by Williams and Haddad.

TECHNIQUE (INGLIS AND COOPER). Make an incision over the anterior part of the medial epicondyle of the humerus beginning 5 cm proximal to the epicondyle and continuing distally to the midpoint of the forearm over the ulna. The medial antebrachial cutaneous nerve is often seen in the distal part of the incision, and the medial brachial cutaneous nerve may be seen posterior to the medial part of the epicondyle. Next identify the ulnar nerve proximal to the epicondyle, dissect and elevate it from its groove behind the epicondyle, and carefully free it distally. Identify, free, and protect the branches of the ulnar nerve to the flexor carpi ulnaris and to the two ulnar heads of the flexor digitorum profundus (Fig. 11-3, D). Next release the origins of the flexor carpi ulnaris and flexor digitorum profundus as follows. Begin distally at about the middle of the ulna and elevate both muscles from the bone at the

subcutaneous border; the interosseous membrane is then seen around the volar surface of the bone. Then continue proximally along the ulna as far as the ulnar groove at the epicondyle. During this dissection the interosseous membrane and the fascia of the brachialis muscle are seen in the depths of the wound. Replace the ulnar nerve in its groove and divide the entire flexor-pronator muscle mass at its origin from the medial part of the epicondyle. At this point the median nerve can be seen as it passes through the pronator teres (Fig. 11-3, E). Now continue the dissection anteriorly over the flexor aspect of the elbow, dividing the lacertus fibrosus and any remaining parts of the flexor muscle origin. If a flexion contracture of the elbow persists, incise the fascia of the brachialis muscle. Then transplant the ulnar nerve anterior to the epicondyle (Fig. 11-3, F). Now the muscle mass will have been displaced 3 to 4 cm distal to its original location. Close the wound and apply a cast or plaster splints to hold the forearm in supination and the wrist and fingers in neutral positions.

AFTERTREATMENT. At 3 weeks the cast or splints and the sutures are removed. Then an extension hand splint is applied (Fig. 11-3, G); it is worn constantly for 3 months and then only at night for 3 more months, or in children, until growth is complete.

• • •

Williams and Haddad recommend a similar but more extensive release of the flexor-pronator origin than that just described. It frees completely the origins of the flexor mass almost to the wrist.

TECHNIQUE (WILLIAMS AND HADDAD). Make an incision over the medial aspect of the arm and forearm anterior to the medial epicondyle of the humerus beginning 5 cm proximal to the elbow and extending distally to about 5 cm proximal to the wrist (Fig. 11-4, A). Protecting the medial antebrachial cutaneous nerve and the basilic vein, dissect anteriorly a flap of skin and subcutaneous tissue to expose the lacertus fibrosus and the antecubital fossa (Fig. 11-4, B). Then expose the ulnar nerve as it passes between the two heads of origin of the flexor carpi ulnaris. Now, avoiding the ulnar collateral ligament and capsule of the elbow joint, divide the common tendon of origin of the superficial group of muscles just distal to the epicondyle (Fig. 11-4, C). Protecting the median nerve and its motor branches to the superficial group of muscles, free the ulnar origin of the pronator teres. Extend the dissection along the lateral border of the pronator teres to its insertion on the radius, but avoid injuring the radial artery. At this level divide the aponeurotic, radial origin of the flexor digitorum sublimis. Now retract anteriorly the ulnar nerve and the stump of the common flexor tendon, and free the origin of the flexor carpi ulnaris from the medial border of the olecranon. During this dissection ligate and divide the posterior ulnar recurrent artery. Avoiding the periosteum of the ulna, release the aponeurotic origin of the flexor carpi ulnaris and flexor digitorum profundus from the ulna throughout its entire length (Fig. 11-4, D). Next identify the common interosseous artery, its volar branch, and the anterior interosseous nerve and release from the volar aspect of the ulna and adjacent interosseous membrane the origin of the flexor digitorum profundus as far distally as

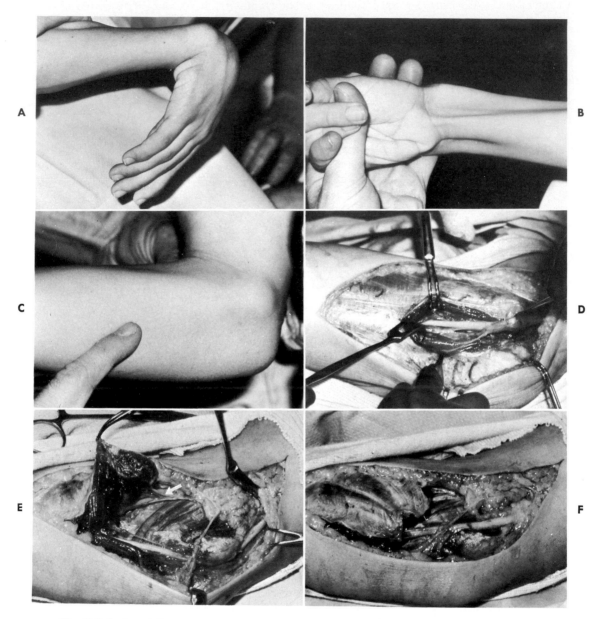

Fig. 11-3. Inglis and Cooper technique for releasing flexor-pronator origin in spastic paralysis. **A,** Posture of hand and wrist before surgery. Wrist cannot be actively extended beyond 90 degrees, and fingers cannot be actively flexed. **B,** Flexor muscles on volar surface of forearm are contracted. **C,** Contracted flexor-pronator muscle mass is attached to medial humeral epicondyle. **D,** Ulnar nerve is dissected free and elevated from behind epicondyle. Its branches to flexor carpi ulnaris and to two ulnar heads of flexor digitorum profundus are identified, freed, and protected. **E,** Entire flexor-pronator muscle mass is divided at its origin from medial part of the epidoncyle (see text). Median nerve, *arrow,* is seen as it passes through pronator teres. **F,** Ulnar nerve is transplanted anterior to epicondyle (see text).

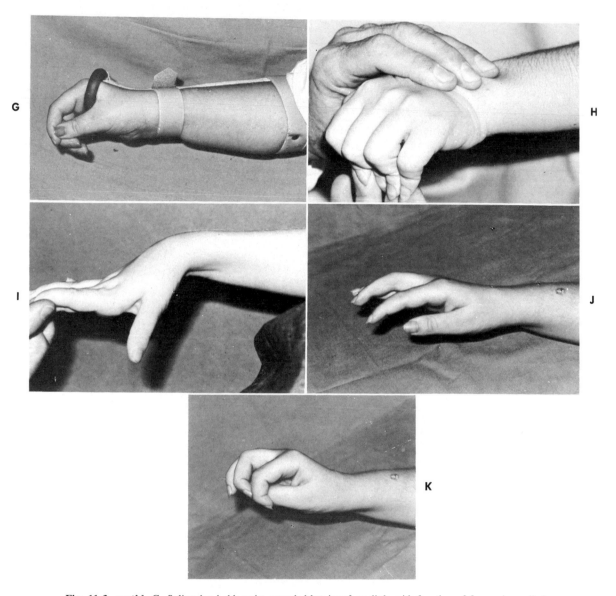

Fig. 11-3, cont'd. G, Splint that holds wrist extended but interferes little with function of fingers is applied 3 weeks after surgery. **H** and **I,** Before release of flexor-pronator origin. Active extension of wrist and fingers is absent. When wrist is passively extended, **H,** finger flexors are tight. When fingers are passively extended, **I,** wrist flexors are tight. **J** and **K,** After surgery. While wrist is extended fingers can be actively extended, **J,** and actively flexed, **K.** (From Inglis, A.E., and Cooper W.: J. Bone Joint Surg. **48-A:**847, 1966.)

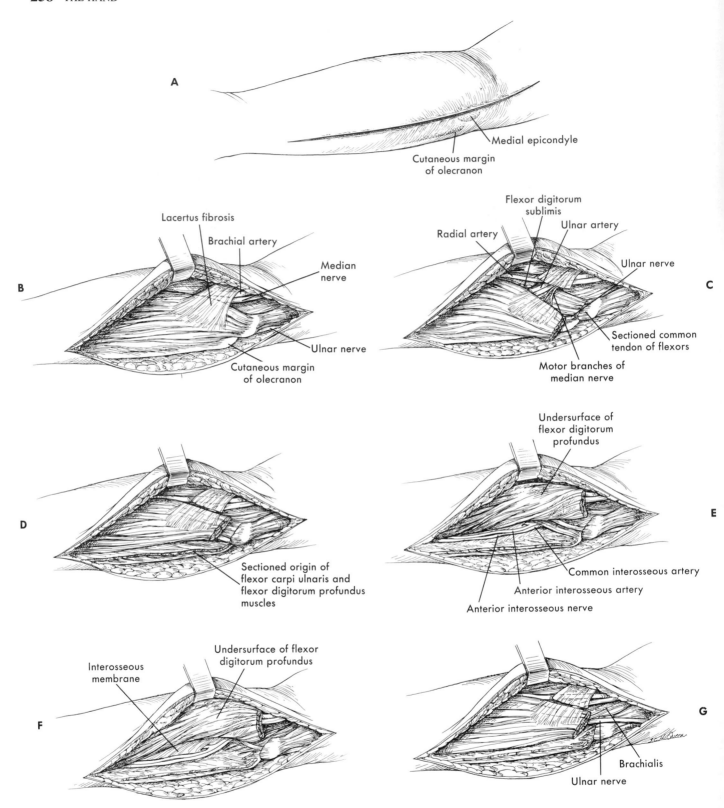

Fig. 11-4. Williams and Haddad technique of releasing flexor-pronator origin. **A,** Incision. **B,** Structures anteriorly and medially at elbow have been exposed (see text). **C,** Lacertus fibrosus has been divided, origin of superficial flexors has been released from the medial epicondyle, and origin of flexor digitorum sublimis has been released from radius (see text). **D,** Origin of flexor carpi ulnaris has been released from olecranon, and common origin of flexor carpi ulnaris and flexor digitorum profundus has been released from ulna (see text). **E,** Origin of flexor digitorum profundus has been released from volar aspect of ulna and interosseous membrane (see text). **F,** Origin of flexor digitorum profundus to index finger has been released from radius, and remaining origin of flexor digitorum sublimis has been released from coronoid process (see text). **G,** Ulnar nerve has been transplanted anteriorly into brachialis muscle (see text). (Modified from Williams, R., and Haddad, R.J.: South. Med. J. **60:**1033, 1967.)

the pronator quadratus (Fig. 11-4, *E*). Release from the radius the origin of the flexor digitorum profundus to the index finger. Then release from the medial side of the coronoid process the remaining origin of the flexor digitorum sublimis proximal to the common interosseous artery (Fig. 11-4, *F*). Extend the wrist and fingers and identify and release any remaining tight bands. If there is any tension on the ulnar nerve, transplant it anteriorly into the brachialis (Fig. 11-4, *G*), and if any elbow contracture persists, divide the brachialis tendon. If necessary, divide or lengthen the tendon of the flexor pollicis longus through a separate incision proximal to the wrist. Splint the extremity with the wrist and fingers extended and the elbow flexed.

AFTERTREATMENT. At 3 weeks the splint and sutures are removed, and another splint is applied that keeps the wrist and fingers extended and the thumb abducted. This splint is worn for 3 months, except when removed for exercises of the wrist and fingers. It is then worn only at night for 6 weeks. Occupational and physical therapy are continued as necessary.

Transfer of flexor carpi ulnaris

Transfer of the flexor carpi ulnaris dorsally removes a deforming force that pulls the hand into ulnar deviation and flexion and provides one that promotes supination of the forearm and extension of the wrist. For this operation to be effective, reasonable finger control, passive flexibility of the hand, wrist, and forearm, and a favorable diagnostic profile are all necessary. Any fixed deformity should be corrected before surgery either by successive casts or by any operations as indicated. If active supination is possible before surgery, the muscle may be carried through the interosseous membrane instead of around the ulnar side of the forearm, thus preventing it from acting as a supinator.

TECHNIQUE (GREEN AND BANKS). Make an anterior longitudinal incision extending from the flexor crease of the wrist proximally for about 3 cm to expose the insertion of the flexor carpi ulnaris on the pisiform bone. Detach the tendon from the bone, and dissect it proximally. The attachment of the muscle to the ulna often extends almost the full length of the tendon; free it by sharp dissection from the ulna, leaving the periosteum in place. The ulnar nerve now may be seen in a sheath posterior to the tendon. Next introduce a silk suture into the distal end of the tendon and by pulling on it gently, outline the course of the muscle proximally. Then beginning about 5 cm distal to the medial epicondyle of the humerus, make a second incision 7 to 10 cm long over the belly of the muscle. Define the lateral margin of the muscle and make an incision here through the deep fascia to expose this margin and the deep surface of the muscle. Once the muscle belly has been defined, dissect it from its origin on the deep surface of the deep fascia and from the ulna distally. Then pull the tendon into the proximal incision. Free the muscle further until it will pass straight from its origin across the border of the ulna to the dorsal aspect of the wrist. Take care to locate and preserve branches of the ulnar nerve to the muscle; they limit the dissection proximally. Now at a suitable level at the medial margin of the ulna, excise the intermuscular septum separating the volar and dorsal compartments of the forearm for 4 to 5 cm and expose the dorsal compartment. Then, starting just proximal to the transverse skin crease on the dorsum of the wrist and extending proximally, make a third incision about 3 cm long over the extensor carpi radialis brevis and longus tendons. Expose these tendons and choose either for insertion of the transferred tendon; that of the brevis gives a more central action in extension of the wrist, whereas that of the longus gives a better pull for supination of the forearm and radial deviation of the wrist. Using a tendon passer, direct the free end of the flexor carpi ulnaris from the proximal incision into the dorsal compartment along the path of the extensor tendons to the chosen extensor radialis tendon. Make a buttonhole in the chosen tendon and pass through it the flexor carpi ulnaris tendon; suture it there under tension with the forearm in full supination and the wrist in at least 45 degrees extension. Close the wounds. Now apply a cast extending from near the axilla to the tips of the fingers, holding the wrist in extension, the forearm in supination, the fingers in almost complete extension, and the thumb in abduction and opposition.

AFTERTREATMENT. The cast is bivalved soon after surgery and exercises are started out of the cast 4 or 5 days later. These are continued daily for at least 6 weeks, the cast remaining in place between exercise periods. Then the cast is worn at night and intermittently for several more weeks or months as necessary to keep the hand in its corrected position.

Fusion of wrist

Fusion of the wrist may improve function in the cerebral palsied hand, but it is used less frequently now than previously. It is indicated only if active extension of the fingers is sufficient while the wrist is held rigidly in a position satisfactory for fusion; this position eliminates the tenodesing effect on the extensor tendons present during active wrist flexion and often results in complete loss of extension of the fingers. Whether the hand may be improved by wrist fusion can be determined by immobilizing the wrist in a cast in various positions and observing function of the fingers. Because the epiphysis of the distal radius will be damaged, fusion must be delayed until the patient is at least 12 years old. Useful techniques are those of Haddad and Riordan and Gill-Stein (p. 114). The wrist should be fused in slight flexion and ulnar deviation, and the radius and carpus are transfixed by Kirschner wires until fusion is solid.

Carpectomy

Omer and Capen reported proximal row carpectomies to improve appearance in eight patients with cerebral palsy. At the same time, transfers of the flexor carpi ulnaris tendon around the ulna to the extensor carpi radialis brevis were carried out to strengthen wrist extension and increase supination. They warn that this does not necessarily improve function. All of their patients were older than 11 years of age. They emphasize prolonged postoperative splinting because the procedure increases the relative length of all flexor muscle–tendon units that cross the wrist and increases wrist extension and forearm supination. They further emphasize that only the proximal half of the carpal scaphoid is taken.

TECHNIQUE (OMER AND CAPEN). Make a longitudinal incision over the dorsum of the wrist. Identify the distal edge of the dorsal carpal ligament and retract the common digital extensor tendons ulnarward. Make a T-shaped incision in the dorsal capsule to expose the carpal bones. Excise the lunate and the proximal half of the scaphoid. Leave the distal half of the scaphoid with its capsular attachments. Excise the triquetrum but leave the pisiform bone. Make a longitudinal incision over the volar aspect of the wrist, beginning at the pisiform and extending proximally over the flexor carpi ulnaris tendon. Protect the neurovascular bundle and free the flexor carpi ulnaris from the intermuscular septum. Divide the muscle near its insertion and pass its tendon through a window in the interosseous membrane. (To give more supination, pass the transfer around the ulna.) Insert the flexor carpi ulnaris into the extensor carpi radialis brevis and anchor it with nonabsorbable monofilament sutures. Place the wrist in maximum passive dorsiflexion and imbricate the dorsal capsule of the wrist.

AFTERTREATMENT. The arm is placed in a bulky dressing and a volar plaster splint holding the fingers and wrist in extension. On or about the fifth day, a circular long arm cast is applied, holding the elbow flexed, the forearm supinated, and the wrist and fingers extended. This position is maintained for 6 weeks. Then a circular short arm cast, incorporating outriggers for extension of the fingers, is applied. Splinting is continued for 4 months and then used only at night for an indefinite time.

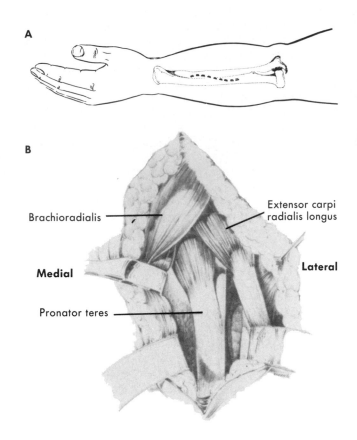

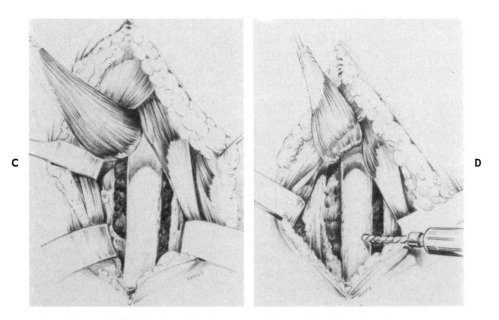

Fig. 11-5. Operative technique on left forearm. **A,** Skin incision on radial aspect, centered over insertion of pronator teres. **B,** Insertion of pronator teres is identified under brachioradialis, which has been retracted medially while extensor carpi radialis longus is retracted laterally. **C,** Tendon of pronator teres, prolonged by strip of periosteum, is detached from radius, and muscle is freed proximally. **D,** Anchoring hole is drilled in anterolateral part of radial cortex, with smaller hole through posteromedial part. **E,** Pronator teres tendon is passed posteriorly between cut edge of interosseous membrane and radius, and its end is introduced into hole drilled in radius. **F,** Drill hole in radius and technique of anchoring tendon. **G,** Transferred tendon is secured firmly in position. (From Sakellarides, H.T., Mital, M.A., and Lenzi, W.D.: J. Bone Joint Surg. **63-A:**645, 1981.)

PRONATION CONTRACTURE OF FOREARM

Sakellarides and associates have devised an operation principally to correct pronation contracture of the forearm. According to them, transferring the pronator teres tendon will produce better correction than any other transfer. This method corrects one deforming force and at the same time provides a force for supination. The tendon is released, wrapped around the radius, and inserted into the bone.

TECHNIQUE. Make a zigzag or curvilinear incision over the anterior aspect of the midforearm (Fig. 11-5, *A*). Protect the lateral cutaneous nerve of the forearm and the superficial radial nerve. Next identify the borders of the brachioradialis muscle and mobilize and retract it medially. Identify the oblique fibers that insert into bone at the musculotendinous insertion of the pronator teres (Fig. 11-5, *B*). Use sharp dissection to detach the insertion of the pronator teres along with an attached strip of periosteum (Fig. 11-5, *C*). Mobilize the muscle extraperiosteally, well proximal in the forearm. Now free the interosseous membrane from the radius as far as necessary to gain maximum passive supination. Pass the pronator teres and attached periosteum posteriorly and laterally around the radius. At the same level as the previous muscle insertion drill an anchoring hole on the anterolateral aspect of the radial cortex. Drill a smaller hole through the posteromedial part of the radius using a 1.6 mm Kirschner wire (Fig. 11-5, *D*). Enlarge the hole in the anterolateral cortex to 2.8 mm (⁷⁄₆₄ inch). Pass a suture with the tendon attached through the two holes from anterolateral to posteromedial (Fig. 11-5,

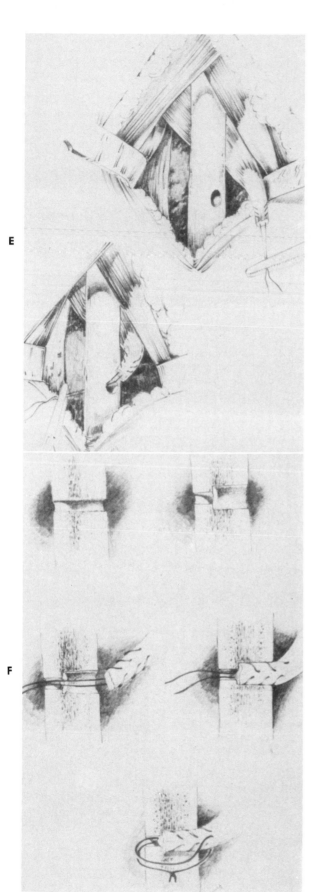

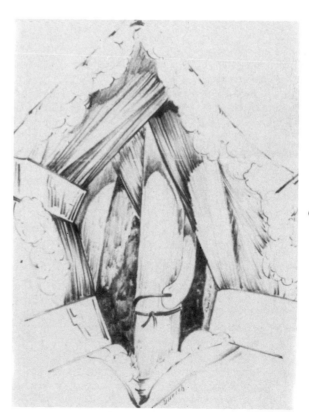

Fig. 11-5, cont'd. For legend see opposite page.

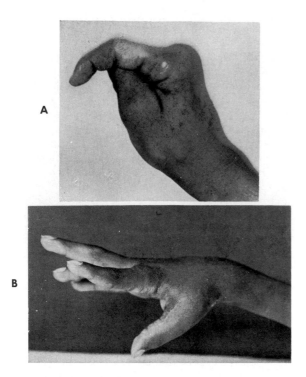

Fig. 11-6. A, Before surgery, showing thumb-in-palm deformity. Adductor pollicis and first dorsal interosseus muscles, as well as flexor pollicis longus, are spastic. **B,** After surgery. Metacarpophalangeal joint has been arthrodesed. First dorsal interosseous muscle has been stripped from first metacarpal, and adductor pollicis tendon has been divided. Flexor carpi radialis and flexor carpi ulnaris have been transferred to weak extensor pollicis longus. Now all digits can be completely extended, and pinch and grasp have been improved. (From Goldner, J.L.: J. Bone Joint Surg. **37-A:**1141, 1955.)

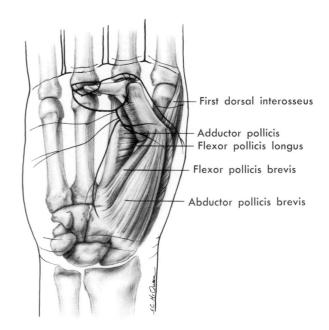

Fig. 11-7. Adducted thumb position in cerebral palsy is result of forces exerted by powerful muscles. (Redrawn from Inglis, A.E., Cooper, W., and Bruton, W.: J. Bone Joint Surg. **52-A:**253, 1970.)

E). In this manner the tendon is introduced into the larger hole and is secured (Fig. 11-5, *F* and *G*). Apply further stay sutures through the tendon as indicated. Hold the forearm in approximately 45 degrees of supination and snug the tendon up to hold this position. Allow the brachioradialis to fall in place and close the incision. Apply an axilla-to-palm cast, maintaining the elbow in 45 degrees flexion and the forearm in 60 degrees supination. Elevate the arm immediately postoperatively.

AFTERTREATMENT. After 3 weeks a new cast is applied, which may be bivalved at any time for observation. The cast is removed during the day, but is reapplied at night for at least 6 months.

THUMB-IN-PALM DEFORMITY

The second frequent and important deformity of the hand in cerebral palsy is the thumb-in-palm, adducted thumb, or clutched thumb deformity (Figs. 11-6, *A,* and 11-7). It blocks entry of objects into the palm and, in addition, prevents the thumb from assisting the fingers in grasp or pinch.

Although thumb-in-palm deformity may be caused principally by spasticity of the flexor pollicis longus muscle, it is not caused solely by this muscle. The flexor pollicis longus flexes the interphalangeal joint, the metacarpophalangeal joint, and the carpometacarpal joint, and also acts as an adductor of the thumb (Fig. 11-8). To be certain that

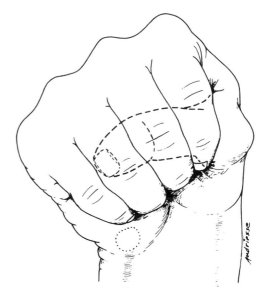

Fig. 11-8. With spastic thumb-in-palm deformity caused by tight flexor pollicis longus, thumb is flexed at interphalangeal and metacarpophalangeal joints, and carpometacarpal joint is flexed and adducted. (From Smith, R.J.: J. Hand Surg. **7:**327, 1982).

it is a principal deforming force, the patient should be able to decrease the flexion of these joints by flexing the wrist. Conversely, extending the wrist will cause an increase in deformity. The examiner should determine whether an accompanying severe adduction deformity is present, caused by contracture of muscle or other structures. A weak adductor pollicis may be overpowered by a tendon transfer; active adduction of the thumb by the adductor pollicis should be checked with the wrist palmar flexed to determine the strength of the muscle. Smith has proposed trans-

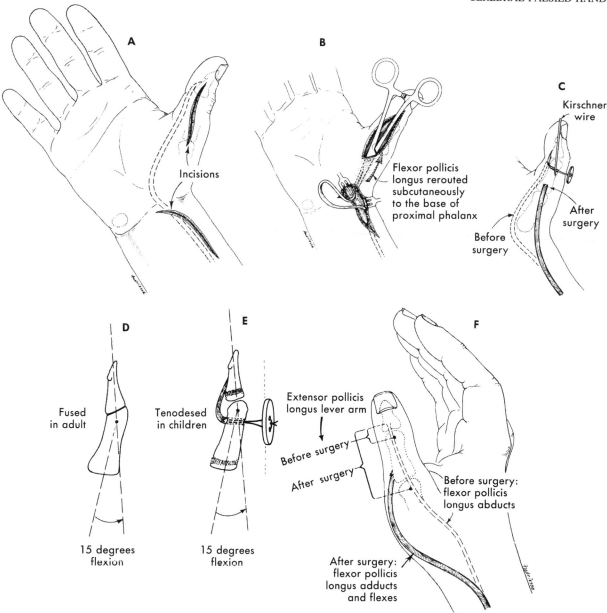

Fig. 11-9. A, Incision to radial side of thumb exposes insertion of flexor pollicis longus, interphalangeal joint, and base of proximal phalanx. Second curved incision to radial side of wrist exposes flexor pollicis longus near its musculotendinous juncture and permits tendon to be withdrawn from carpal canal. **B,** Flexor pollicis longus is transected at its insertion and withdrawn from carpal canal through wrist incision. It is then passed subcutaneously to radial side of base of proximal phalanx. Distal joint is stabilized in slight flexion. **C,** Interphalangeal joint of thumb is arthrodesed in about 15 degrees of flexion in adult. In child with open epiphysis, distal joint may be tenodesed in about 15 degrees of flexion. **D,** Transfer of flexor pollicis longus to radial side of proximal phalanx reduces adduction-flexion deformity and augments thumb abduction by transferred position of flexor pollicis longus. Interphalangeal arthrodesis improves metacarpophalangeal joint extension by increasing lever arm of extensor pollicis longus on metacarpophalangeal joint. (From Smith, R.J.: J. Hand Surg. **7:**327, 1982).

fer of the flexor pollicis tendon to the radial side of the thumb combined with tenodesis of the distal joint. He recommends the operation for patients who have spontaneous use of the affected hand, in addition to passive extension of the metacarpophalangeal joint and abduction of the carpometacarpal joint with the wrist in flexion.

TECHNIQUE (SMITH). Make a radial midlateral incision from the middle of the distal phalanx of the thumb to the neck of the first metacarpal (Fig. 11-9, *A*). Elevate a volar skin flap and transect the flexor pollicis longus tendon opposite the proximal phalanx (Fig. 11-9, *B*). Tenodese the flexor pollicus longus stump to the proximal phalanx or arthrodese the distal joint in 15 degrees flexion (Fig. 11-9, *C*). Now make a longitudinal incision in the forearm just radial to the tendon of the flexor carpi radialis, curving its distal portion ulnarward. Identify the flexor pollicis longus tendon and draw it out through this incision. Tunnel subcutaneously by blunt dissection on the radial side of the

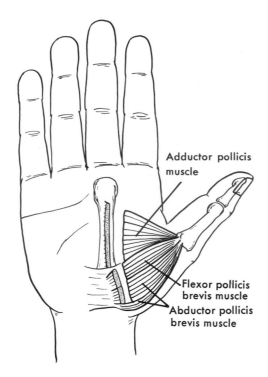

Adductor pollicis
muscle

Flexor pollicis
brevis muscle

Abductor pollicis
brevis muscle

Fig. 11-10. Myotomies of intrinsic muscles of thumb for thumb-in-palm deformity (see text). (Modified from Swanson, A.B.: Surg. Clin. North Am. **48:**1129, 1968.)

thumb to the lateral side of the metacarpophalangeal joint and pass the flexor pollicis longus tendon through this tunnel. With the wrist in neutral and the thumb at 50 degrees abduction, suture the tendon to the dorsoradial aspect of the metacarpophalangeal joint with tension (Fig. 11-9, *D*).

AFTERTREATMENT. The hand is immobilized for 6 weeks with the thumb in abduction and the wrist in 30 degrees flexion. The thumb is splinted with a C-splint in the web for an additional 6 weeks.

• • •

To release the thumb in patients in whom the deforming force is the adductor pollicis, Matev and others recommend myotomies in the palm. Myotomy of the adductor pollicis is considered better than tenotomy because it releases the first metacarpal but does not allow hyperextension of the metacarpophalangeal joint. These myotomies in themselves may be sufficient, or lengthening of the flexor pollicis longus tendon combined with fusion of the metacarpophalangeal joint or a reinforcing tendon transfer to the abductor pollicis longus and extensor pollicis brevis may be necessary later. For this transfer the brachioradialis is useful, or if it is a deforming force at the wrist, the flexor carpi radialis may be used. Release of the origin of the first dorsal interosseus muscle and occasionally also a Z-plasty of the thumb web may be useful. If the extensor pollicis longus contributes to the thumb deformity, it may be rerouted from Lister's tubercle to the radial aspect of the wrist.

TECHNIQUE OF MYOTOMY. Make an incision bordering the thenar crease in the palm, but avoid damaging the recur-

rent branch of the median nerve or the innervation of the adductor pollicis. After retracting the long flexors of the fingers, strip from the third metacarpal the origin of the adductor pollicis. Cut from the deep transverse carpal ligament about two thirds of the origin of the abductor pollicis brevis and all of the origins of the flexor pollicis brevis and opponens pollicis (Fig. 11-10). Also strip from the second metacarpal the origin of the first dorsal interosseus. If necessary, carry out a capsulorrhaphy of the metacarpophalangeal joint.

AFTERTREATMENT. A pressure dressing and a cast are applied holding the first metacarpal (not the phalanges) in wide abduction and opposition. At 3 weeks the cast and sutures are removed and a splint is applied to hold the thumb in this same position. If tendon transfers have been necessary, the cast is retained for 6 weeks. Splinting at night may be necessary for a long time if the deformity tends to recur.

SWAN-NECK DEFORMITY

When compared with other deformities of the upper extremity in cerebral palsy, swan-neck deformities of the fingers are infrequent; however, they may be quite disabling. They are caused by muscle imbalance and by secondary ligamentous and capsular relaxation at the proximal interphalangeal joints that allow these joints to hyperextend. In the involved finger the middle extensor band is relatively short as compared with the lateral bands because of tension exerted on the middle band by the long extensor and the intrinsic muscles. In this deformity the Curtis sublimis tenodesis of the proximal interphalangeal joint may improve function.

TECHNIQUE (CURTIS). The Curtis technique is described in the section on hyperextension and locking of the proximal interphalangeal joint. It employs one slip of the flexor digitorum sublimis that is left at its insertion on the bone and cut free at the bifurcation. It is then brought to the opposite side of the joint under the remaining tendons and inserted into the lateral aspect of the middle phalanx with a pull-out wire suture (Fig. 8-14). The proximal interphalangeal joint is held in flexion by a traversing Kirschner wire for 6 weeks.

REFERENCES

Chait, L.A., et al.: Early surgical correction in the cerebral palsied hand, J. Hand Surg. **5:**122, 1980.

Curtis, R.M.: Treatment of injuries of proximal interphalangeal joints of fingers. In Adams, J.P., editor: Current practice in orthopaedic surgery, vol. 2, St. Louis, 1964, The C.V. Mosby Co.

Goldner, J.L.: Reconstructive surgery of the hand in cerebral palsy and spastic paralysis resulting from injury to the spinal cord, J. Bone Joint Surg. **37-A:**1141, 1955.

Goldner, J.L.: Upper extremity reconstructive surgery in cerebral palsy or similar conditions. In American Academy of Orthopaedic Surgeons: Instructional course lectures, vol. 18, St. Louis, 1961, The C.V. Mosby Co.

Goldner, J.L.: Upper extremity tendon transfers in cerebral palsy, Orthop. Clin. North Am. **2:**389, 1974.

Goldner, J.L., and Ferlic, D.C.: Sensory status of the hand as related to reconstructive surgery of the upper extremity in cerebral palsy, Clin. Orthop. **46:**87, 1966.

Green, W.T., and Banks, H.H.: Flexor carpi ulnaris transplant and its use in cerebral palsy, J. Bone Joint Surg. **44-A:**1343, 1962.

Haddad, R.J., Jr., and Riordan, D.C.: Arthrodesis of the wrist: a surgical technique, J. Bone Joint Surg. **49-A:**950, 1967.

Hoffer, M.M., Perry, J., and Melkonian, G.J.: Dynamic electromyography and decision-making for surgery in the upper extremity of patients with cerebral palsy, J. Hand Surg. **4:**424, 1979.

Inglis, A.E., and Cooper, W.: Release of the flexor-pronator origin for flexion deformities of the hand and wrist in spastic paralysis: a study of eighteen cases, J. Bone Joint Surg. **48-A:**847, 1966.

Inglis, A.E., Cooper, W., and Bruton, W.: Surgical correction of thumb deformities in spastic paralysis, J. Bone Joint Surg. **52-A:**253, 1970.

Kaplan, E.B.: Surgical treatment of spastic hyperextension of the proximal interphalangeal joints of the fingers, accompanied by flexion of the distal phalanges: case report, Bull. Hosp. Joint Dis. **23:**35, 1962.

Keats, S.: Surgical treatment of the hand in cerebral palsy: correction of thumb-in-palm and other deformities: report of nineteen cases, J. Bone Joint Surg. **47-A:**274, 1965.

Kilgore, E.S., Jr., and Graham, W.P.: Operative treatment of swan neck deformity. III. Plast. Reconstr. Surg. **39:**468, 1967.

Lam, S.J.S.: A modified technique for stabilizing the spastic thumb, J. Bone Joint Surg. **54-B:**522, 1972.

Martz, C., and Schaffer, E.: Orthopaedic management and care of the cerebral palsied, Symposium on cerebral palsy in Indiana, 1956 (mimeographed).

Matev, I.: Surgical treatment of spastic "thumb-in-palm" deformity, J. Bone Joint Surg. **45-B:**703, 1963.

Matev, I.B.: Surgical treatment of flexion-adduction contracture of the thumb in cerebral palsy, Acta Orthop. Scand. **41:**439, 1970.

Matev, I.B.: Thumb reconstruction through metacarpal bone lengthening, J. Hand Surg. **5:**482, 1980.

McCue, F.C., et al.: Transfer of the brachioradialis for the hands deformed by cerebral palsy, J. Bone Joint Surg. **52-A:**1171, 1970.

Mital, M.A., and Sakellarides, H.T.: Surgery of the upper extremity in the retarded individual with spastic cerebral palsy, Orthop. Clin. North Am. **12:**127, 1981.

Mortens, J.: Surgery of the hand in cerebral palsy, Acta Orthop. Scand. **36:**441, 1965-1966.

Omer, G.E., and Capen, D.A.: Proximal row carpectomy with muscle transfers for spastic paralysis, J. Hand Surg. **1:**197, 1976.

Page, C.M.: An operation for the relief of flexion-contracture in the forearm, J. Bone Joint Surg. **21:**233, 1923.

Perry, J., and Hoffer, M.M.: Preoperative and postoperative dynamic electromyography as an aid in planning tendon transfers in children with cerebral palsy, J. Bone Joint Surg. **59-A:**531, 1977.

Sakellarides, H.T., Mital, M.A., and Lenzi, W.D.: Treatment of pronation contractures of the forearm in cerebral palsy, J. Bone Joint Surg. **63-A:**645, 1981.

Samilson, R.L., and Morris, J.M.: Surgical improvement of the cerebral-palsied upper limb: electromyographic studies and results of 128 operations, J. Bone Joint Surg. **46-A:**1203, 1964.

Sherk, H.H.: Treatment of severe rigid contractures of cerebral palsied upper limbs, Clin. Orthop. **125:**151, 1977.

Smith, R.J.: Flexor pollicis longus abductor-plasty for spastic thumb-in-palm deformity, J. Hand Surg. **7:**327, 1982.

Sprenger, T.R.: Pronation deformities of the forearm in cerebral palsy, Symposium on cerebral palsy in Indiana, 1956 (mimeographed).

Stein, I.: Gill turnabout radial graft for wrist arthrodesis, Surg. Gynecol. Obstet. **160:**231, 1958.

Swanson, A.B.: Surgery of the hand in cerebral palsy and the swan-neck deformity, J. Bone Joint Surg. **42-A:**951, 1960.

Swanson, A.B.: Surgery of the hand in cerebral palsy, Surg. Clin. North Am. **44:**1061, 1964.

Swanson, A.B.: Treatment of the swan-neck deformity in the cerebral palsied hand, Clin. Orthop. **48:**167, 1966.

Swanson, A.B.: Surgery of the hand in cerebral palsy and muscle origin release procedures, Surg. Clin. North Am. **48:**1129, 1968.

Tachdjian, M.O., and Minear, W.L.: Sensory disturbances in the hands of children with cerebral palsy, J. Bone Joint Surg. **40-A:**85, 1958.

White, W.F.: Flexor muscle slide in the spastic hand: the Max Page operation, J. Bone Joint Surg. **54-B:**453, 1972.

Williams, R., and Haddad, R.J.: Release of flexor origin for spastic deformities of the wrist and hand, South Med. J. **60:**1033, 1967.

Arthritic hand

Arthritic hand disorders may be caused by any of several collagen diseases that should be identified individually. Although the surgical treatment is somewhat similar for all, the surgeon should be aware of several differences. The operative indications and prognosis after surgery may differ with each disease entity. Furthermore, all have different effects on other areas of the body. In severe disorders, the diagnosis usually will have already been established by the referring physician. If not, it should be established before surgery and the necessary medical treatment should be initiated and continued during and after surgery. Operative treatment should be considered just a phase in the general management of the disease.

Arthritic patients frequently take several drugs, some of which can affect the timing of surgery. For instance, aspirin should be discontinued at least 9 days before an operation to allow the platelet count to return to normal. If the patient has been taking a steroid preparation, the dosage should be increased just before surgery and then tapered off after surgery. These and other medical problems may be handled by a medical consultant.

The goal of surgery usually is to relieve pain, restore function, improve cosmesis, and inhibit progression of the disease in that order. If pain is not the primary consideration, then the surgeon must be quite certain he can restore sufficient function to justify the surgery. On the other hand, relieving pain alone by a surgical procedure is worthwhile when adequate medical treatment has failed to do so. Cosmesis is an important consideration for some if not most patients, although they may complain principally of pain. The surgeon should be certain to discuss with the patient the expected appearance following surgery; even if pain has been relieved, the patient may be disappointed if the appearance of the hand is not improved. To inhibit the disease process locally is generally not sufficient reason for surgery. The exception may be persistent swelling on the dorsum of the wrist from synovitis that is likely to cause extensor tendon rupture.

Before surgery, the surgeon should advise the patient on what the procedure entails, including (1) the insertion of pins, (2) the location of incisions, (3) the application of splints, (4) the expected stay in the hospital, (5) the type of anesthesia, (6) the cost of the operation, (7) the aftertreatment and the rehabilitation period, and especially (8) the expected benefit from the operation. Patients with severe deformities may have developed substitution patterns that enable them to perform their daily tasks; these should not be interrupted without careful analysis of the pathologic anatomy and functional patterns. This is especially true in those who are older, who have retired, and who have no pain. The patient should be advised emphatically that surgery will not cure the disease process but only alter its course at the site of surgery.

When a general anesthetic is to be used during an operation on a rheumatoid patient, the alignment of the cervical spine should be investigated before surgery. If the disease has been generalized and prolonged, roentgenograms of the cervical spine are indicated to discover any subluxations. The degree of cervical involvement forewarns the anesthesiologist as to possible cord damage that may result from hyperextension of the neck during intubation or while maintaining a free airway.

RHEUMATOID ARTHRITIS

Rheumatoid arthritis may cause such grotesque deformities of the hands that the patient becomes almost helpless and is ashamed of his appearance. It is also one of the more painful chronic diseases. It is characterized by hypertrophic synovitis that eventually destroys the cartilage of joints, compresses or disrupts tendons, compresses adjacent nerves, and eventually dislocates and erodes the joint itself.

Almost any combination of deformities can be seen as a result of the destructive pathologic changes. Each deformity must be analyzed in detail before surgery is considered. In the hand they are usually bilateral and symmetric to the point of being mystic in their bizarre mirror imageary. Compensatory deformities may develop, such as hyperextension of one finger joint because of the inability to extend the adjacent joint; this type of deformity may develop in the interphalangeal and metacarpophalangeal joints of the thumb. Frequently one joint will dislocate and prevent dislocation of the adjoining joint by relieving the intrinsic tendon tension. For example, if the metacarpophalangeal joint becomes dislocated, the proximal interphalangeal joint frequently will not. Conversely, if the proximal interphalangeal joint becomes dislocated, the metacarpophalangeal joint may not dislocate as rapidly.

The wrist and metacarpophalangeal joints are most often affected early in rheumatoid arthritis, whereas the distal two joints are frequently affected later. The carpometacarpal joint of the thumb is not as commonly affected as the other joints, especially when compared with osteoarthritis. Tendon ruptures are common in the rheumatoid patient, especially of the extensor tendons at the distal end of the ulna and the flexor tendons at the carpal scaphoid.

OSTEOARTHRITIS

Hypertrophic synovitis is less severe in osteoarthritis than in rheumatoid arthritis. Osteoarthritis may be unilateral, but occurs as frequently in the minor hand as in the dominant one. It is not associated with tendon ruptures or triggering of fingers as is frequently seen in rheumatoid arthritis. It generally affects women more often than men and is frequently seen in the hand at the carpometacarpal joint of the thumb, sometimes as a single joint involvement. The distal interphalangeal joints of the fingers are most frequently involved often in association with Heberden nodes. Although the proximal interphalangeal joints may be affected with this form of arthritis, it is rare here when compared with rheumatoid arthritis. Spur formation, cartilage fragmentation, and limited motion but not dislocation are the frequent sequelae. During the active phase pain is severe, and the joints and overlying skin are inflamed. Direct trauma to an inflamed joint is especially painful. The most frequent operation on the osteoarthritic hand is for restoration of the carpometacarpal joint of the thumb.

LUPUS ERYTHEMATOSUS

Lupus erythematosus involves the skin especially about the nose, as well as tendons and joints (Fig. 12-1). It seems to affect the periarticular soft tissues and the tendons more than the cartilage itself; thus the soft tissues stretch, and joints dislocate without bone or cartilage damage. Eventually joints become painful and grossly deformed, especially the metacarpophalangeal joints, and less often the proximal interphalangeal joints. The disease may have a mixed pattern or exist in conjunction with rheumatoid arthritis. At surgery the tendons are found stretched yet held together in a tenacious continuity. They are functionless when stretched to a pathologic extent. Then repair is difficult because of the lack of tendon substance to hold sutures. Lupus erythematosus is not curable by surgery.

PSORIATIC ARTHRITIS

Psoriatic arthritic deformities are similar to those in rheumatoid arthritis. An estimated 7% of patients with psoriasis have some form of inflammatory arthritis. Of patients with severe skin involvement, 40% have arthritic changes. The distal interphalangeal joints are typically affected and the disease here may produce a fusiform swelling of the entire digit. Uniquely, the nails may separate from the nail bed and have a white, flaking discoloration near their distal borders; they may also be ridged.

REITER'S SYNDROME

Reiter's syndrome is described as a triad of conjunctivitis, urethritis, and synovitis. The synovitis usually involves asymmetrically four or fewer joints. Heel pain, back pain, and nail deformities may occur in this syndrome, sometimes making it difficult to distinguish from psoriatic arthritis. It affects the lower extremity more often than the upper, and 90% of the patients have remission of symptoms after several weeks; in 10% the disease may become chronic. It is typically found in young males. Surgery is rarely indicated.

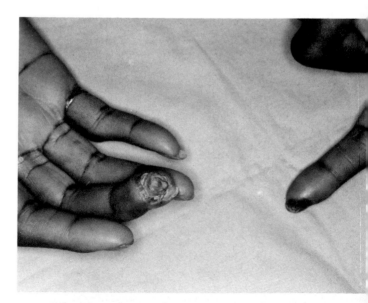

Fig. 12-1. Vasculitis associated with lupus erythematosus has resulted in necrosis of these fingertips in 33-year-old woman.

GOUT

Gout usually causes an erythematous, painful joint in adult males. The attack is often sudden with severe pain about a single joint. The joint is swollen, hot, and tender, suggesting a severe cellulitis or abscess. The area may be incised and drained by the unsuspecting surgeon. In chronic gout massive deposits of monosodium urate crystals may be found about the joints and tendon sheaths causing nerve compressions as seen in carpal tunnel syndrome. The skin may be ulcerated by pressure from within (Fig. 12-2). Amputation may be necessary because of the extreme bony disruption resulting from gout. The deposits may be visible on roentgenograms. Women do not have gouty arthritis until after menopause, and even then it is quite rare. The presence of hyperuricemia alone does not establish the diagnosis of gout, and in fact, the uric acid level may be elevated and an acute attack of gout never occur. Conversely, during an acute attack of gout the uric acid level may be normal.

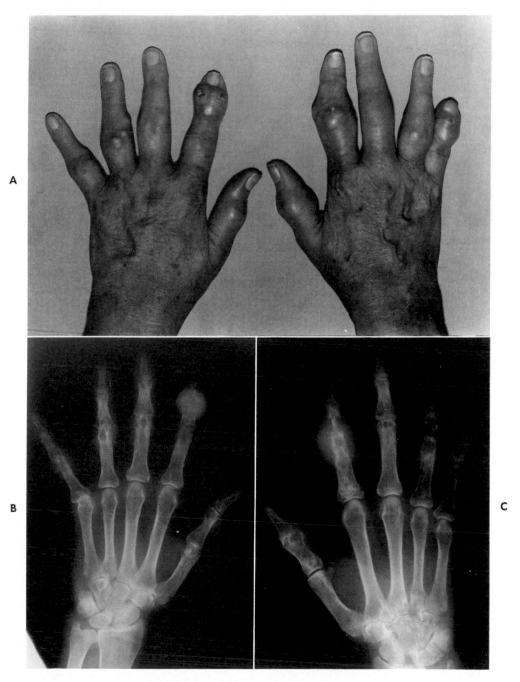

Fig. 12-2. Gout in 56-year-old woman. **A,** Heavy calcium urate deposits have caused skin erosion. **B** and **C,** Roentgenograms of hands showing destructive lesions in bones of digits.

SCLERODERMA (PROGRESSIVE SYSTEMIC SCLEROSIS)

Diffuse scleroderma, or progressive systemic sclerosis (PSS), affects both the extremities and the trunk. The disease may involve not only the skin but also the gastrointestinal tract, especially the esophagus, the heart, lungs, and kidneys. Telangiectasia may also be seen. The hand surgeon may see these patients because of calcinosis of the fingertips, ulcerations, or Raynaud's phenomenon. The age of onset is usually past 40 years.

Arthritic involvement usually causes contractures of the fingers, but synovial thickening is minimal. Involvement of tendons and tendon sheaths of the hand may cause a palpable tendon friction rub or a rather leathery crepitus as distinguished from the coarse, gritty-like crepitus palpable in osteoarthritis. These rubs may also be felt in the tendons about the foot and ankle. Ulceration of the fingertips because of vascular impairment is best treated by an extremely conservative surgical approach, including waiting for the tips to amputate spontaneously since this will retain length of the digits. Surgical sympathectomies and intraarterial injection of drugs to help dilate the vessels have been recommended. Calcification about the eroded pulps of the fingers may be excised through a lateral incision, but healing may be quite slow. Local applications of medications may be helpful. Smoking, of course, should be avoided.

NONOPERATIVE TREATMENT FOR SYNOVITIS

Persistent tenosynovitis or arthritis with obvious swelling that persists for several weeks even when treated with antiinflammatory drugs may be treated by local injections of a steroid preparation mixed with a local anesthetic. This treatment is especially applicable to trigger fingers and carpal tunnel syndrome frequently seen in rheumatoid disease or osteoarthritis of the carpometacarpal joint of the thumb. Even osteoarthritis of the distal interphalangeal joints and rheumatoid arthritis of the proximal interphalangeal joints will respond favorably to injections for a period of several weeks. However, after repeated injections the response may be less dramatic. In many instances pain may be relieved and surgery may be at least delayed by this technique.

Persistent dorsal swelling of the wrist caused by synovitis should be treated surgically, since it is frequently complicated by rupture of one or more of the extensor tendons. The swelling may or may not be accompanied by dorsal subluxation of the ulna. This subluxation of the ulna with its distal end unprotected and eroded presents an extremely potent abrasive surface, especially against the three ulnar extensor tendons (Fig. 12-3).

USE OF KIRSCHNER WIRES IN RHEUMATOID HAND

In the rheumatoid hand most Kirschner wires will eventually work loose and require removal. Fortunately, however, fusion occurs rapidly in most instances after arthrodesis. Therefore we cut off Kirschner wires under the skin at a level that makes them easily recoverable, even sometimes leaving them protruding at the proximal interphalangeal joints. The dressing is then applied over the wires.

Wires left embedded in the pulp of the fingers or near the metacarpophalangeal joint on the palmar side of the thumb are extremely painful. Wires in these areas should be inserted with the end nearest the skin on the dorsal surface. Most wires can then be removed in the office using a local anesthetic.

STAGING OF OPERATIONS

When several operations are indicated on a single hand, their order of priority must be considered. In general, when wrist arthroplasty or arthrodesis is indicated, it should be done first since the position of the wrist determines the balance of the digital flexor and extensor tendons. However, at the time of wrist surgery it may be possible to do an additional procedure such as arthrodesis of the metacarpophalangeal joint of the thumb. But extensive other surgery is contraindicated.

When multiple small-joint procedures such as metacarpophalangeal arthroplasties or proximal interphalangeal joint fusions are to be performed, plans should be made to do them all at one time to reduce the number of times operations are performed on a single hand. Frequently, a rheumatoid patient will require surgery not only on the opposite hand but also on the feet, the hips, and other joints. However, only one hand should have surgery at a given time because of the requirements for daily independent living and personal hygiene. If the lower extremities require external support, the surgeon should provide a forearm type of crutch to protect the hand.

Medications taken before surgery should usually be continued. However, aspirin and other salicylate drugs that decrease the platelet count should be withdrawn 9 days before surgery. This helps avoid abnormal and possibly unanticipated bleeding. Of course, a substitute drug may be administered.

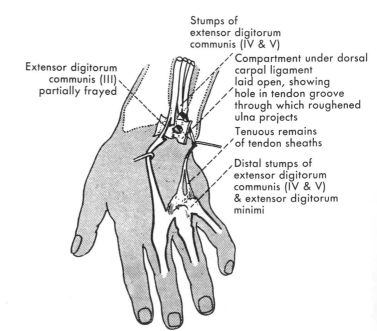

Fig. 12-3. Attrition rupture of extensor tendons caused by arthritis of inferior radioulnar joint. (From Vaughn-Jackson, D.J.: J. Bone Joint Surg. **30-B:**528, 1948.)

POSTOPERATIVE IMMOBILIZATION BY SPLINTS AND CASTS

The 3 weeks of immobilization usually recommended following hand surgery is too long for the rheumatoid patient and may result in stiffness not only in the operated digit but also in some of the other digits (Fig. 12-4). Although occasionally a tendon may rupture, protected passive movement should be started 1 or 2 days after tendon surgery. Even a few degrees of active motion should be encouraged. The metacarpophalangeal joint should not be held in complete extension to protect an extensor tendon repair. Rather the wrist is extended, but the metacarpophalangeal joint is flexed about 45 degrees and the proximal interphalangeal joint is allowed to flex.

DEFORMITIES OF FINGERS

Deformities of the finger may be caused by one or more of the following: tightness of the intrinsic muscles, displacement of the lateral bands of the extensor hood, rupture of the central slip of the hood, or rupture of the long extensor or long flexor tendons. Here abnormal forces act on joints already weakened by the disease.

In addition, flexor tenosynovitis may produce limitation of interphalangeal joint motion so that the range of active flexion of these joints is significantly less than that obtained passively.

Intrinsic plus deformity

In the intrinsic plus deformity, caused by tightness of the intrinsic muscles, the metacarpophalangeal joint cannot be fully extended while the proximal interphalangeal joint is flexed, or vice versa; often the deformity develops in combination with ulnar deviation of the fingers. If the classical Bunnell test for tightness of the intrinsics is to be accurate, it must be performed with the proximal phalanx in line with its metacarpal. Any ulnar deviation at the metacarpophalangeal joint during the test will slacken those intrinsics on the ulnar side of the finger and confuse the findings. For instance a tight first volar interosseus will

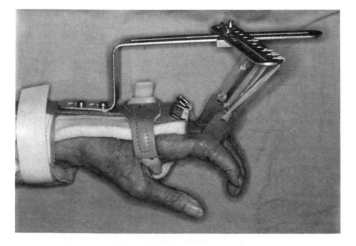

Fig. 12-4. Following removal of primary dressing at or about 10 days, splint supporting wrist and metacarpophalangeal joints of digits is useful to maintain alignment intermittently for next 6 to 8 weeks.

pull the extended index finger ulnarward, but if the finger is held in line with the second metacarpal during the test, tightness of this muscle can be demonstrated. It should be remembered that the first volar interosseus is a flexor as well as an adductor of the second metacarpophalangeal joint and that usually the first dorsal interosseus is only an abductor. Release of the volar intrinsics, especially of the abductor digiti quinti, once thought to reduce ulnar drift when performed early, is usually ineffective in itself because factors other than tight intrinsics also contribute to the deformity (p. 275).

Swan-neck deformity

Swan-neck deformity may be described as a flexion posture of the distal interphalangeal joint and hyperextension posture of the proximal interphalangeal joint with flexion at times of the metacarpophalangeal joint. It is caused by muscle imbalance and may be passively correctable depending on the fixation of the original and secondary deformities.

This deformity may begin as a mallet deformity associated with a disruption of the extensor tendon with secondary overpull of the central tendon, causing hyperextension of the proximal interphalangeal joint. The proximal interphalangeal joint may actively flex normally.

This deformity may also begin at the proximal interphalangeal joint, as hyperplastic synovitis causes herniation of the capsule, tightening of the lateral bands and central tendon, and eventual adherence of the lateral bands in a fixed dorsal position so that they can no longer flow over the condyles when the proximal interphalangeal joint is flexed. This limits proximal interphalangeal flexion. The centrally displaced lateral bands may be ineffective in extending the distal interphalangeal joint, which may secondarily assume a mallet deformity without actual rupture of the lateral tendons. This mallet deformity, however, is usually not as severe as that produced by a rupture of the lateral tendons. A swan-neck deformity may require synovectomy of the proximal interphalangeal joint, mobilization of the lateral bands, and release of the skin distal to the proximal interphalangeal joint. Wrinkles and normal laxity of the skin are lost at the level of the proximal interphalangeal joint after several weeks (see technique for release of skin and mobilization of lateral bands, p. 274).

Beckenbaugh emphasizes that flexor tenosynovitis results in ineffective support by the sublimis tendon and is an extremely important factor in initiating the development of swan-neck deformity in the rheumatoid hand. Every patient in his series had tenosynovitis with adherence of the sublimis tendon, the tendon being rendered ineffective in stabilizing the proximal interphalangeal joint against hyperextension. The overpull of the central tendon combined with synovitis of the proximal interphalangeal joint and surrounding tissue that results in stretching may cause a swan-neck or hyperextended position. Beckenbaugh treats this disorder by creating a tenodesis across the proximal interphalangeal joint with one half of the sublimis tendon. He emphasizes that postoperative immobilization of the joint is unnecessary and allows immediate movement at the joint without protective splinting. The chief complication in his technique was flexion contracture of the proximal interphalangeal joint of more than 30 degrees. Some of these were corrected by releasing the tenodesis.

When there is marked hyperextension at the proximal interphalangeal joint and a normal roentgenographic appearance with maintenance of a normal joint space, tenodesis by the flexor sublimis tendon may be combined with release of the lateral bands and the distal skin. The technique of tenodesis of the sublimis tendon is the same as for the hand in cerebral palsy (see p. 264, swan-neck deformity).

In intrinsic tightness when the metacarpophalangeal joint is held in extension, flexion of the proximal interphalangeal joint is markedly limited; however, when the metacarpophalangeal joint is flexed, the intrinsics are re-laxed and flexion of the proximal interphalangeal joint is increased. With ulnar drift of the fingers, this intrinsic tightness may be present only on the ulnar side. To test this accurately, axial alignment of the finger with the metacarpal should be maintained in checking intrinsic tightness. When indicated, intrinsic tightness may be released in conjunction with synovectomy by mobilization of the lateral band. When degeneration of the metacarpophalangeal joints requires arthroplasty, there may be sufficient resection of bone to release the intrinsic mechanism; however, it must be specifically determined at the time of the surgery when a release is necessary. A specific tendinous release of the intrinsics may be indicated (see technique [Littler], p. 305).

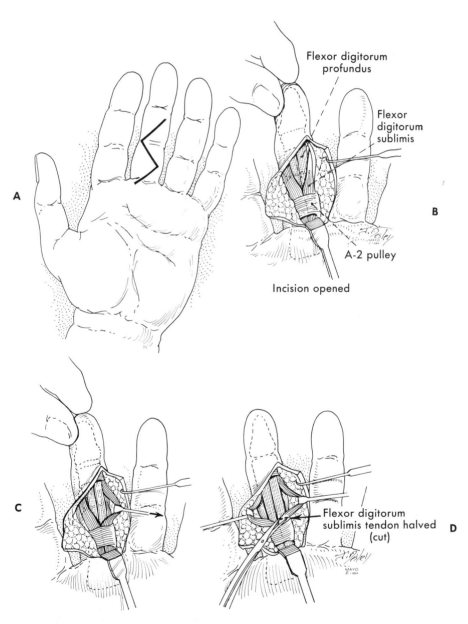

Fig. 12-5. Beckenbaugh technique for correcting hyperextension deformity of proximal interphalangeal joint (see text). (Copyright Mayo Clinic.)

When there is marked proximal interphalangeal joint extension associated with joint destruction on roentgenograms, arthrodesis may be best if there is a near normal metacarpophalangeal joint or if metacarpophalangeal joint resection arthroplasty is anticipated. Arthroplasty of the proximal interphalangeal joints of the ring and little fingers may be carried out when there are near normal metacarpophalangeal joints proximally. Arthroplasty of both the metacarpophalangeal joint and proximal interphalangeal joint of the same finger should not be done even in stages (see Swanson technique for implant at proximal interphalangeal joint, p. 275).

TECHNIQUE FOR INTRINSIC RELEASE. See p. 303.

TECHNIQUE FOR CORRECTING HYPEREXTENSION DEFORMITY OF PROXIMAL INTERPHALANGEAL JOINT (BECKENBAUGH). Make a zigzag incision over the middle and proximal phalanges (Fig. 12-5, *A*). Take care not to damage the digital nerves that may adhere to the cruciate pulley system anterior to the hyperextended proximal interphalangeal joint. Expose the cruciate pulleys by elevating medially and laterally the neurovascular bundles. Expose the A2 pulley (Fig. 12-5, *B*). Incise the central pulley centrally to expose the flexor tendons. Retract the profundus tendon and release any adhesions; then expose the sublimis tendon and its adhe-

sions and perform a synovectomy (Fig. 12-5, *C*). Pull the sublimis tendon distally and incise the decussation, splitting the tendon into two slips. If necessary, extend the incision proximally and release the adhesions at the A1 pulley level to allow distal translocation of the tendon. Pull the divided sublimis tendon distally and incise the ulnar slip, leaving a 5 cm slip of tendon attached to the ulnar side of the middle phalanx (Fig. 12-5, *D*). Pull the slip firmly to ensure that its insertion is not weakened by synovitis. In the little finger both slips are incised because a single slip is usually too small. Puncture the A2 pulley 3 to 4 mm from its distal border (Fig. 12-5, *E*). Pass a small curved hemostat through the hole distally into the sheath and clamp the tip of the sublimis tendon slip and pull it proximally through the A-2 pulley (Fig. 12-5, *F*). Now bring the slip of tendon distally and suture it to itself with nonabsorbable 4-0 sutures (Fig. 12-5, *G* and *H*). Adjust the tension so that the digit is held at only 5 degrees of flexion at the proximal interphalangeal joint. A tenodesis is accomplished by this slip of tendon fixed across the joint. Repair the cruciate pulley if feasible. Close the skin over a small drain. Several fingers may be operated on at one sitting.

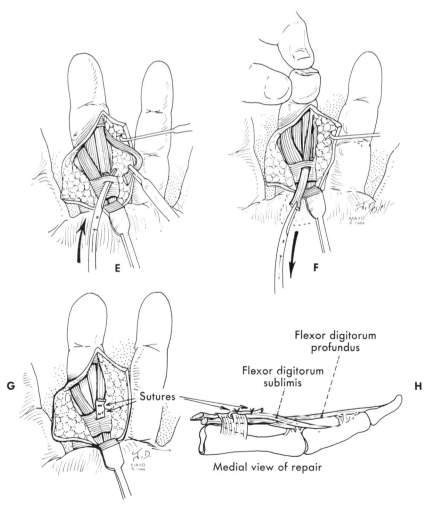

Fig. 12-5, cont'd. For legend see opposite page.

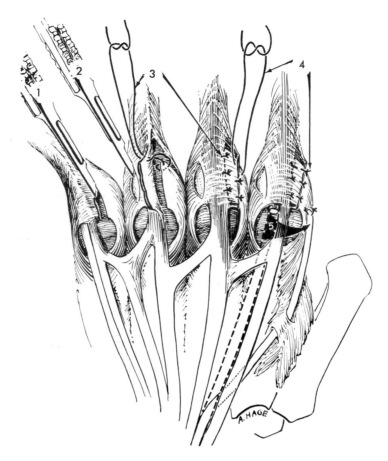

Fig. 12-11. Correction of mild to moderate ulnar drift. *1,* Joint is entered through incision in radial side of hood. *2,* Relaxing incision is made in ulnar side of hood to permit repositioning of extensor tendon. *3* and *4,* Incision in radial side of hood is closed after its edges are overlapped. *5,* Extensor indicis proprius tendon is transferred to first dorsal interosseus muscle to reinforce it. (From Flatt, A.E.: In Converse, J.M., editor: Reconstructive plastic surgery. Philadelphia, 1964, W.B. Saunders Co.)

Treatment of severe ulnar drift and metacarpophalangeal dislocation

In severe ulnar drift often one or more metacarpophalangeal joints will have dislocated (Fig. 12-14); consequently, this type of drift and dislocation of these joints will be discussed together. Here the dislocation of the metacarpophalangeal joint in effect will have released the soft structures that cross the joint, and thus by decreasing tension will have protected, at least partially, the proximal interphalangeal joint. Conversely, if it is the proximal interphalangeal joint that dislocates first, then the metacarpophalangeal joint will be partially protected. Because of the deforming forces mentioned earlier in this section, the metacarpophalangeal joints will have deviated ulnarward more and more. It should be emphasized, however, that the long flexor tendons are a major deforming force. They will have shifted ulnarward either within or without their sheaths; thus they exert a force on the finger in the ulnar direction, but in addition they exert a force on the proximal phalanx in a palmar direction that will have dislocated the metacarpophalangeal joint. For this type of ulnar drift, surgery is carried out mainly on the metacarpal head and its surrounding ligaments and tendons.

Function of a dislocated metacarpophalangeal joint may be improved by interposition arthroplasty in which bone is resected. Many different designs of interposition arthroplasty for the metacarpophalangeal joint are available, but we have had more experience with the Swanson implant than any other. From reports on hundreds of insertions, an average expected range of motion at the metacarpophalangeal joint is 55 degrees, and usually this occurs in the critical functional range. The incidence of complications is acceptable, with an infection rate of less than 1% and a breakage rate between 2% and 22%. With the new high-density silicone construction, breakage has been reduced considerably. Even though obvious fractures of the prosthesis may occur and occult fractures often may be demonstrated on tomograms, the function of the joint usually is not impaired since it is not only the prosthesis but also the encapsulating scar that provides stability and permits motion. The prostheses are easily removed when necessary. Interposition arthroplasty of the metacarpophalangeal joint can be depended on to relieve pain, maintain stability and alignment, and permit acceptable motion.

TECHNIQUE (SWANSON). Make a transverse incision on the dorsum of the hand, beginning on the radial aspect of the

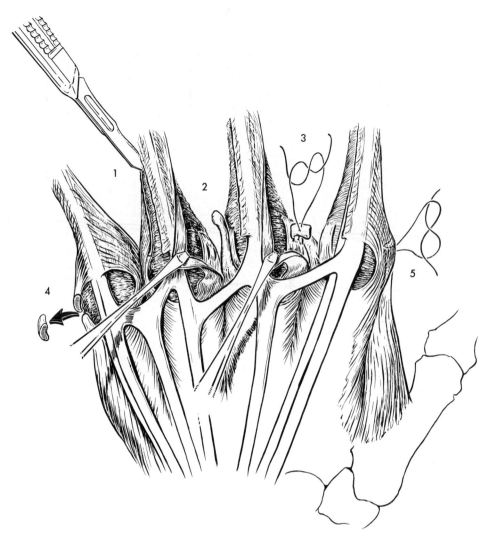

Fig. 12-12. Flatt transfer of released ulnar intrinsics to radial side of digits for ulnar drift. *1,* Incision is made on ulnar side of central tendon, releasing ulnar intrinsic insertion. *2,* Ulnar intrinsic insertion is free. *3,* Insertion is sutured to capsule on radial side of metacarpophalangeal joint of adjacent finger. *4,* Segment of abductor digiti quinti tendon is excised to relieve ulnar pull of muscle on little finger. *5,* First dorsal interosseus tendon is shortened to increase radial pull of muscle on index finger. (Courtesy Dr. A.E. Flatt.)

second metacarpophalangeal joint, and extend it ulnarward to the ulnar aspect of the fifth metacarpophalangeal joint. Carefully observe the pattern of the superficial veins, and preserve them where possible. This incision permits a slight flap that can be dissected proximally and folded back, exposing the heads of the metacarpals. Through this, incise the shroud ligament of the extensor mechanism on the radial aspect of each joint and, if necessary, on the ulnar aspect also. This permits entry into the joint capsule, which already may be ruptured dorsally, with herniation of hypertrophied synovium. Incise the capsule longitudinally, and excise it partially as well as all the synovium that presents itself, either then or after resection of the metacarpal head. With a thin osteotome or a bone-biting instrument, resect each metacarpal head to shorten the bone sufficiently to permit easy reduction of the dislocated joint. This usually requires resection proximal to the origin of

the collateral ligaments. After synovectomy, introduce into the medullary canal of the metacarpal either a square-shaped awl or, if necessary, a drilling broach to provide space for the stem of the prosthesis. The metacarpal head-neck region should be carefully cut so that it is at a 90-degree angle with the axis of the metacarpal shaft. Do not resect the concave surface of the proximal phalanx, but into this also insert the reaming device and ream the proximal phalanx to accept the distal stem of the prosthesis. Usually a No. 4 or No. 5 Swanson is used, but the largest one that can comfortably be inserted is required. Resection of bone should be sufficient to prevent buckling of the prosthesis.

To avoid pronation of the index finger, Swanson recommends that a radial slip of the volar plate be split off proximally and reattached to the radial aspect of the metacarpal to provide a mooring for the proximal phalanx (Fig.

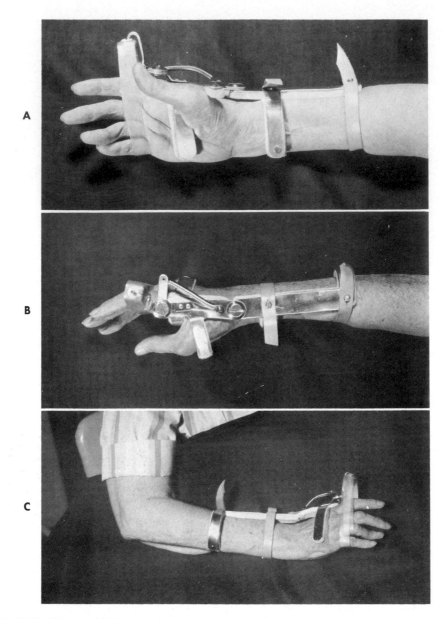

Fig. 12-13. Splint useful after synovectomy or arthroplasty of metacarpophalangeal joints in rheumatoid arthritis. Note that each finger is individually supported by bands to prevent ulnar deviation and that wrist and joints of all digits are movable. **A,** Palmar view. **B,** Radial view. **C,** Dorsal view.

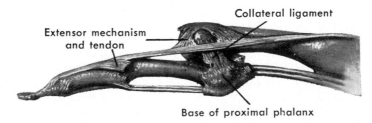

Fig. 12-14. Metacarpophalangeal dislocation in rheumatoid arthritis (see text).

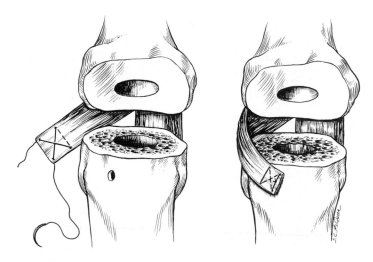

Fig. 12-15. Swanson technique for reconstruction of radial collateral ligament of index metacarpophalangeal joint by using a slip of volar plate. (Redrawn from Swanson, A.B.: Flexible implant arthroplasty in hand and extremities. St. Louis, 1973, The C.V. Mosby Co.)

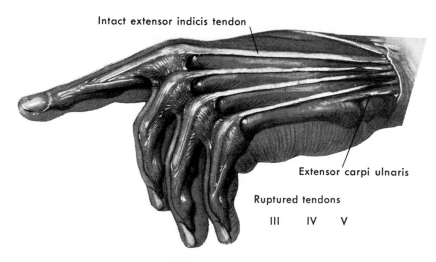

Intact extensor indicis tendon

Extensor carpi ulnaris

Ruptured tendons

III IV V

Fig. 12-16. Rupture of extensor tendons at level of dorsal carpal ligament in rheumatoid arthritis. Nearly all ruptures of common finger extensors occur at abrasive point created by dorsally dislocated distal ulna.

12-15). Remove the prosthesis from the package only after a trial prosthesis has been inserted and the exact size determined. Handle the prosthesis with instruments without sharp edges to avoid scoring or other damage. Insert the prosthesis first proximally and then distally. Accomplish reduction of the joint with a comfortable seating of the prosthesis, and demonstrate passive motion of the metacarpophalangeal joint from full extension to near 90 degrees of flexion. Check all fingers carefully for alignment and for rotary deformity. Replace and realign the extensor tendon, and be sure that an intrinsic release has been accomplished by bony resection. Use a running pull-out No. 4-0 monofilament wire or nylon suture since multiple buried sutures at this level are more likely to erode and cause inflammation with movement. Insert a drain, close the wound, and apply a supportive dressing to splint the fingers in slight radial deviation. Additional surgery may be done on the same hand at the time of the insertion of the

prostheses. Occasionally the fifth metacarpophalangeal joint may not require resection arthroplasty.

RUPTURE OF TENDONS
Extensor tendon rupture

Rupture of tendons is a major cause of deformity and disability in the rheumatoid hand. As previously stated, rheumatoid tenosynovitis is the basic cause of such ruptures.

The long extensor tendons of the middle, ring, and little fingers seem to rupture as a group, and these ruptures can be easily overlooked because of more grotesque deformities elsewhere in the hand. Dorsal subluxation of the distal ulna contributes to rupture of these three tendons because the diseased end of the bone is rough and they usually glide between it and the tight intact dorsal carpal ligament. Other extensor tendons also usually rupture at the level of this ligament (Fig. 12-16). The long extensor tendon of the thumb, because of its tortuous course, frequently ruptures at the level of Lister's tubercle where it angles through an enclosed tunnel or pulley. At surgery a white strip of connective tissue representing an effort toward regeneration of the tendon by its sheath may be seen, but it is not a true tendinous structure.

REPAIR OF RUPTURE OF EXTENSOR TENDONS

A ruptured extensor tendon may be repaired by direct suture when found within a few days. When surgery must be delayed for several days, it is well to splint the wrist in extension to relieve the constant tension on the remaining intact tendons. When the ruptured tendon is diagnosed after several weeks, a segmental tendon graft, transfer of a tendon to the distal segment of the ruptured tendon, or possibly a side-to-side suture of the proximal and distal segments of the ruptured tendon to an adjoining intact tendon are possibilities for treatment (Fig. 12-17). A synovectomy is always indicated in the region of the rupture and the repair.

When the tendon of the ring finger or little finger alone is ruptured, repair of the ring finger tendon may be possible by suturing both its distal and proximal segments to the intact middle finger extensor tendon under appropriate tension. A transfer of the extensor indicis proprius might be used as a motor to the little finger; as another alternative, a transfer of the extensor pollicis brevis as a motor is possible when it is necessary to arthrodese the metacarpophalangeal joint of the thumb for other reasons. When three extensor tendons, those of the middle, ring, and little fingers, have been ruptured for an extended period of time, the transfer of a motor is usually indicated, and an accept-

able source for this motor is the sublimis of the ring finger. This tendon has enough excursion and might be even more effective because of the tenodesing effect when the wrist is flexed. Extensor pollicis longus tendon rupture may be repaired by transfer of the extensor indicis proprius as is usual in ruptures from other causes.

Flexor tendon rupture

Flexor tendon rupture in rheumatoid patients is not as common as extensor tendon rupture but is much more difficult to treat surgically. Rupture may occur within the digit as a result of infiltrative tenosynovitis or at wrist level because of bony erosion into the tendon, especially the flexor pollicis longus tendon. Rupture of one sublimis slip may cause triggering of the finger. Rupture of a profundus tendon may be easily demonstrated, but the level of rupture may be quite difficult to determine. A ruptured profundus or ruptured sublimis may cause secondary joint stiffness. Tendon grafts for rupture of flexor tendons of the hand in rheumatoid patients almost always fail. The exception is at the wrist, where a segmental graft may occasionally be used as treatment for a ruptured flexor pollicis longus tendon. Another approach to rupture of the flexor pollicis longus is to simply fuse the distal joint of the thumb. In our experience tendon grafts other than the segmental or bridge graft mentioned above always seem to fail in rheumatoid hands.

PERSISTENT SWELLING OF PROXIMAL INTERPHALANGEAL JOINT

In persistent swelling and pain of the proximal interphalangeal joint, synovectomy is a useful operation. It can be carried out on all four fingers of one hand at the same time and in conjunction with other synovectomies.

TECHNIQUE OF SYNOVECTOMY. On one side, and occasionally on both sides, of the finger make a midlateral incision (p. 10) centered over the proximal interphalangeal joint, and on each side carry out the following procedure. Locate the transverse retinacular ligament, sever its attachment, and elevate the extensor hood. Then under the hood identify the collateral ligament. Enter the joint dorsal to this ligament and lateral to the central tendon, explore the joint, and excise as much synovium as possible. Remove the synovium from both the area behind the volar plate and the area inferior to the collateral ligament, dividing, if necessary, the accessory collateral ligament. Relocate the lateral tendon and transverse retinacular ligament. Close the incisions.

TENOSYNOVITIS OF FLEXOR TENDON SHEATHS

Often there is a progressive fusiform swelling of one or more flexor tendon sheaths extending from the middle of the palm to the distal interphalangeal joint. The swelling is painful and causes a gradual decrease in flexion of the fingers. On palpation the synovium is thickened, and nodules can be identified along the tendon sheath with tendon excursion; crepitus and grating also are not infrequent. Should passive motion at the proximal interphalangeal joint be greater than active, then synovitis within the proximal interphalangeal joint may be causing the major problem. Tenosynovectomies seem to have a lasting effect. A tenosynovectomy may increase joint motion, but while

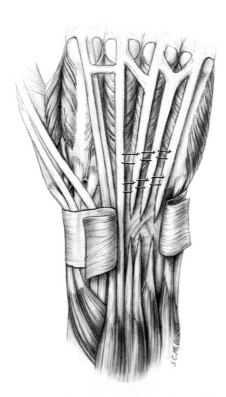

Fig. 12-17. Extensor tendon rupture under dorsal carpal ligament. Repair may be accomplished by side-to-side anastomosis with adjacent intact tendon.

synovectomy of a joint may relieve pain, increased motion cannot always be expected.

TECHNIQUE OF SYNOVECTOMY. Make a long zigzag incision (p. 10) on the palmar surface of each involved finger. Expose the flexor tendon sheath by raising flaps on each side, but take care not to damage the neurovascular bundles that lie anterolaterally (not laterally). Now excise the sheath except for pulleys 1 cm wide or wider that are left at the middle of the proximal and middle phalanges. Excise as much synovium as possible, taking care to remove it from behind the slips of the sublimis and from between the profundus and sublimis. Close the incision with interrupted sutures, apply a compression dressing, support the wrist by a volar plaster splint, and elevate the hand. Motion of the fingers is started as soon as tolerated.

DEFORMITIES OF THUMB

Frequently the first deformity of the thumb is one of hyperflexion of the metacarpophalangeal joint. Active extension of this joint is lost as the result of attenuation of the insertion of the extensor pollicis brevis caused by synovitis of the underlying joint. Furthermore, the flexed position of the joint is made worse by the intrinsics (abductor pollicis brevis and adductor pollicis), whose attachments to the extensor hood shift distally and volarward until they lie volar to the transverse axis of the joint. The extensor pollicis longus tendon may shift ulnarward and volarward until it also lies volar to the transverse axis of the joint. All of these muscles then become flexors of the joint. The extensor pollicis longus is further deforming because, with its tendon now located on the ulnar side of the metacarpophalangeal joint, it contributes to any adduction contracture of the thumb. However, this muscle can still stabilize the interphalangeal joint in extension. In carrying out the maneuver of pinch, the thumb is pushed into extension but only at the interphalangeal joint; in fact this joint eventually may become hyperextended, causing rupture of its volar plate or loss of function of the flexor pollicis longus. As the disease progresses and the deformities increase, the collateral ligaments of the metacarpophalangeal and interphalangeal joints become avulsed and effective pinch then becomes impossible.

Early in the disease when the *metacarpophalangeal* joint can be passively extended to neutral, treatment consists of synovectomy of the joint and advancement of the insertion of the extensor pollicis brevis. The extensor hood is further reinforced by advancing the insertions of the adductor pollicis and abductor pollicis brevis and attaching them dorsal to the metacarpophalangeal joint (see Inglis' technique, p. 285). If this joint is grossly unstable or its articular surfaces are destroyed, it can be arthrodesed without loss of essential function. The metacarpophalangeal joint can be arthrodesed in slight abduction to overcome a moderate adduction contracture of the first web. This increases abduction without releasing the origins of the adductor muscles from the third metacarpal shaft or their insertions from the thumb, either of which weaken the power of pinch (see technique for arthrodesis of the metacarpophalangeal joint, described below). If the interphalangeal joint is also grossly unstable, it can be arthrodesed too, but the functional loss is greater when both joints are fused instead of only one.

The *carpometacarpal* joint of the thumb is the most important single joint of the hand. When it is painful in rheumatoid arthritis and does not respond to local injection of a steroid preparation, it may be treated by arthroplasty consisting of resection of the greater multiangular or more popularly by insertion of an interposition prosthesis, especially when later fusion of the two distal joints may become necessary. Resection arthroplasty shortens the thumb and thus releases the tight thenar intrinsic muscles. While mobility is provided and pain is relieved, resection arthroplasty produces a relatively unstable joint for several months.

Interposition arthroplasty was associated early with a high incidence (18%) of subluxation or dislocation of the prosthesis. To help avoid this complication, careful capsular closure, advancement of the abductor pollicis longus, and reinforcement with a tendon slip of the flexor carpi radialis should be done as recommended by Swanson. To further improve motion after arthroplasty a release of the thumb web by Z-plasty may be necessary. Some surgeons advise arthrodesis of the carpometacarpal joint rather than arthroplasty, but usually only when the joint is affected by osteoarthritis, when it is the only joint of the thumb affected by rheumatoid arthritis, or when a strong pinch is more important than motion.

Metacarpophalangeal joint

TECHNIQUE OF ARTHRODESIS OF METACARPOPHALANGEAL JOINT OF THUMB. If the metacarpophalangeal joint of the thumb is unstable but not subluxated, bone should be resected straight across the joint. Make a straight dorsal incision and cauterize the exposed vein. Then retract the extensor pollicis longus tendon to the ulnar side and the extensor pollicis brevis to the radial side, or detach it and suture it

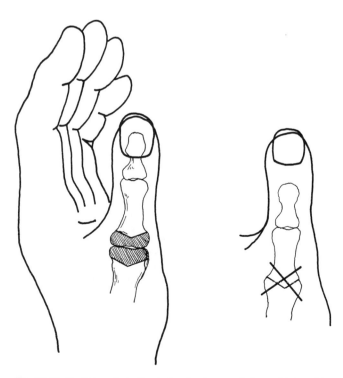

Fig. 12-18. Technique for arthrodesis of metacarpophalangeal joint of the thumb (see text).

later to the capsule. With an osteotome, cut across the articular surface of the proximal phalanx in a straight line at 90 degrees to the long axis of the bone. After the articular surface is resected, place the phalanx at an angle of 15 degrees of flexion with the metacarpal. There is a tendency to osteotomize the distal metacarpal also at 90 degrees; this is a mistake. Rather, make the osteotomy so that the metacarpophalangeal joint is flexed 15 degrees; this requires removing more bone toward the palmar aspect. The two raw surfaces should fit flush. Remove any protruding small edges of bone to smooth the site of arthrodesis. Fix the arthrodesis with three Kirschner wires inserted longitudinally. Insert them first through the metacarpal and advance them through the phalanx. Be certain that the wires do not pierce the flexor tendon or the distal joint. Cut them off under the skin and approximate the tendons with a small absorbable suture. Finally, close the wound and place the hand in a small splint to be replaced later by a cast if indicated.

When this joint is subluxated, shortening of the bone may be required. A chevron-shaped excision of bone that permits interlocking of the exposed surfaces may be used to accomplish shortening. Make a dorsal longitudinal incision over the joint and displace the extensor pollicis longus tendon to the ulnar side. Then on the proximal end of the proximal phalanx, shape a tongue of bone. On the distal end of the first metacarpal, create a V-shaped notch; these should fit like a tongue in a groove (Fig. 12-18).

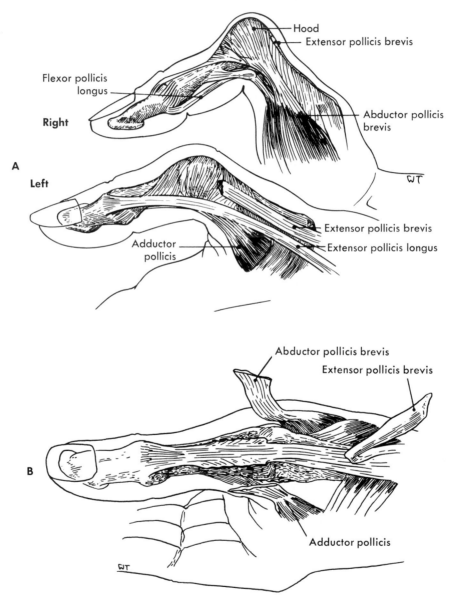

Fig. 12-19. Reconstruction of metacarpophalangeal joint of thumb in rheumatoid arthritis. **A,** Metacarpophalangeal joint of thumb with extensive tendon damage. After rupture of insertion of extensor pollicis longus tendon into base of proximal phalanx and proximal retraction, extensor hood becomes attenuated and allows abductor pollicis brevis and extensor pollicis longus to migrate volarward below center of rotation of metacarpophalangeal joint. **B,** Extensor pollicis brevis and adductor pollicis insertions are dissected free from remaining attenuated extensor tendon hood.

Thus large surfaces of bone are put in contact but the angle of fusion can be adjusted easily. Now fix the joint in proper position by two Kirschner wires that are cut off flush with the bone. Pack small fragments of bone into any spaces about the joint margins. Close the wound and apply a volar plaster splint that includes the thumb but no other digits. In both of the techniques described here be certain that the thumb is in appropriate pronation so that the pulp of the thumb can be placed against the other digits. Also, as already mentioned, if the first metacarpal is adducted some of this adduction may be overcome by fusing the joint in slight abduction. This places the thumb in proper position without releasing soft tissues in the palm.

AFTERTREATMENT. At 10 to 14 days the splint and sutures arc removed. Active use of thumb is resumed gradually despite absence of roentgenographic evidence of fusion.

TECHNIQUE OF RECONSTRUCTION OF METACARPOPHALANGEAL JOINT OF THUMB FOR RHEUMATOID ARTHRITIS (INGLIS ET AL.) Make a longitudinal incision over the dorsum of the metacarpophalangeal joint from the middle of the proximal phalanx to the midshaft of the first metacarpal. Observe the exten-

sor pollicis brevis to determine if it has become detached from the bone of the proximal phalanx and retracted proximally (Fig. 12-19, A). Split the extensor hood longitudinally between the extensor pollicis longus and extensor pollicis brevis. Detach the abductor pollicis brevis from the extensor hood on the radial side, and detach the adductor pollicis from the ulnar side (Fig. 12-19, B). Retract the remaining tendon structures laterally to expose the capsule and synovium. Preserve the collateral ligaments, but excise all the synovium within the joint (Fig. 12-19, C). This may be facilitated by flexing the joint. Drill a hole for sutures on each side of the dorsum of the base of the proximal phalanx, and make a large hole just distal to and between them for insertion of the extensor pollicis brevis tendon. Attach the extensor pollicis brevis with sufficient tension to maintain extension of the metacarpophalangeal joint, and then attach the abductor pollicis brevis and adductor pollicis dorsally to preserve the balance of this joint (Fig. 12-19, D). Maintain the metacarpophalangeal joint in extension by two transfixing Kirschner wires for 4 weeks.

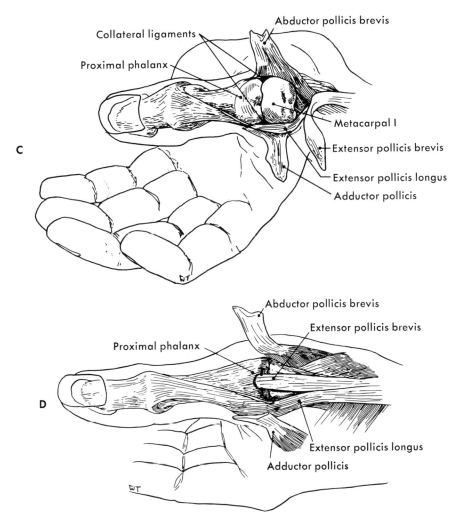

Fig. 12-19, cont'd. C, Synovectomy is facilitated by flexion of proximal phalanx. Note that collateral ligaments are preserved. **D,** Attachment of extensor pollicis brevis tendon into base of proximal phalanx. When extensor pollicis brevis cannot be advanced, tendon of extensor indicis proprius may be transferred from index finger and inserted into base of proximal phalanx (see text). (From Inglis, A.E., et al.: J. Bone Joint Surg. **54-A:**704, 1972.)

Eaton designed a prosthesis that is stabilized by passing a segment of tendon through a perforation in its body, thereby anchoring it to the adjacent bone. The prosthesis was designed in an effort to avoid subluxation or dislocation, which occurred in 10% of the patients in his series. He defines subluxation as a state in which the articular surfaces of the prosthesis and the adjoining bone are in less than 50% apposition. He has tried other tendons but now uses a slip of the abductor pollicis longus. There are two sizes of prosthesis to select from.

TECHNIQUE (EATON). Expose the carpometacarpal joint of the thumb by a palmar approach. Reflect the muscles and expose the joint capsule. Avoid injury to the superficial branch of the radial artery and nerve and the palmar cutaneous branch of the median nerve. Remove the greater multangular, preserving as much capsule and periosteum as possible. Avoid damage to the deep branch of the radial artery. Now resect a portion of the distal pole of the scaphoid along a line perpendicular to the anticipated positioning of the thumb metacarpal. Square off the thumb metacarpal perpendicular to its long axis. Remove all osteophytes. Next ream the medullary canal of the metacarpal to accept the round stem of the implant. Shave off the radial facet of the lesser multangular down to subchondral bone and gouge a channel in the bone to receive the tendon reinforcement. This channel will emerge near one of the tendons of the radial wrist extensors. See Fig. 12-21 for the direction of the channel. Pass a 28-gauge wire loop through the channel with which to pass the strip of tendon. Through a J-shaped incision over the anatomic snuff box, isolate a strip of tendon of the abductor pollicis longus, retaining its normal insertion distally on the first metacar-

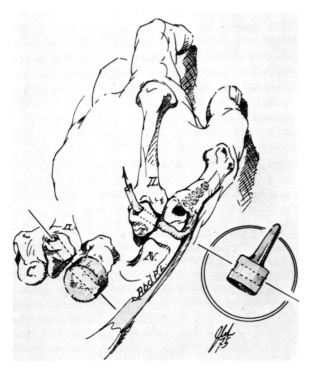

Fig. 12-21. Schema for carpometacarpal implant arthroplasty. Stabilization of implant is accomplished by passage of strip of abductor pollicis longus tendon through base of implant and into adjacent greater multangular. (From Eaton, R.G.: J. Bone Joint Surg. **61-A:**76, 1979.)

pal. Using the wire loop, pass this strip through the capsule, through the perforation in the body of the implant, and then through the channel previously created in the lesser multangular. Adjust the strip and suture it to the adjacent wrist extensor tendons after the capsule has been securely reconstructed. Keep the implant seated while holding the thumb in abduction and extension. Next divide the extensor pollicis brevis at the metacarpophalangeal joint and reattach it under moderate tension to the base of the extended thumb metacarpal. The distal free end may be used to reinforce the capsule. Plicate the intact portion of the abductor pollicis longus tendon or divide and reattach it to the base of the thumb metacarpal under moderate tension while keeping the thumb abducted. Be certain that hyperextension has not developed at the metacarpophalangeal joint. If extension is more than 20 degrees, perform a volar capsulodesis and transfix the metacarpophalangeal joint in flexion with a Kirschner wire. If the metacarpophalangeal joint is hyperextended more than 40 degrees, arthrodese it in 20 to 25 degrees of flexion. Now anchor the slip of tendon of the abductor pollicis longus emerging at the dorsum of the lesser multangular while maintaining slight tension on the adjacent wrist extensors. Avoid excessive tension because this will place a stress on the base of the implant stem. Immobilize the thumb for 5 weeks in a well-padded reinforced dressing.

TECHNIQUE FOR CONVEX CONDYLAR IMPLANT ARTHROPLASTY (SWANSON ET AL.). As in some of the previous techniques, approach the carpometacarpal joint through a palmar incision. Incise the capsule longitudinally on the radial side and resect enough of the base of the first metacarpal to allow 45 degrees of radial abduction of the bone. Ream the medullary canal of the metacarpal to receive the implant. Select the implant of appropriate size from the 11 sizes available. Align the bone, excise the ulnar distal projection of the greater multangular, and with a burr make a slightly concave surface to receive the convex implant (Fig. 12-22, A). Now reconstruct the capsule and ligamentous structures with a slip of the abductor pollicis longus as described in the previous technique. As an alternative prepare an 8 cm slip of the abductor pollicis longus, preserving its insertion on the metacarpal, loop it into the medullary canal of the metacarpal, and pull it out through a hole drilled in the radiodorsal aspect of the bone (Fig. 12-22, B). Now pass the slip through a 3 mm hole drilled in the greater multangular from inside out. Position the implant and hold the thumb in 45 degrees of abduction, tighten and interweave the slip of tendon, pass its distal end through or under the abductor pollicis longus insertion, and suture it to the radial capsular structures of the greater multangular with a nonabsorbable suture (Fig. 12-22, C). If reduction of a severely displaced metacarpal causes a hyperextension deformity of the metacarpophalangeal joint of no more than 10 to 15 degrees, the joint may be fixed in proper position by a Kirschner wire passed through the joint. If the hyperextension is more than 20 degrees, the palmar aspect of the capsule should be fixed by a capsulodesis. Severe hyperextension and instability may require arthrodesis of the metacarpophalangeal joint. The care after surgery is similar to that described for the Swanson et al. silastic implant at the carpometacarpal joint.

ARTHROPLASTY OF RHEUMATOID CARPOMETACARPAL JOINT

Most arthroplasties of the carpometacarpal joint of the thumb are performed for osteoarthritis and not for rheumatoid arthritis. Frequently only this joint is involved in osteoarthritis, whereas in rheumatoid arthritis there is almost always multiple joint involvement, often including the wrist. Resection arthroplasty (Fig. 12-23) was once a popular procedure to relieve pain, correct deformity, and provide some stability at the carpometacarpal joint. However, it did not provide a strong pinch mechanism. Milender et al. described a technique of interposition arthroplasty using a Swanson T-shaped great toe silastic prosthesis. In most instances this does not require resection of all of the greater multangular, thus preserving bone stock. In their series most patients also initially had instability and a hyperextension deformity at the metacarpophalangeal joint. This required arthrodesis of the joint as shown in Fig. 12-18. In rare instances the entire greater multangular was excised.

TECHNIQUE FOR INTERPOSITIONAL ARTHROPLASTY FOR RHEUMATOID CARPOMETACARPAL JOINT (MILLENDER ET AL.). Make a zigzag incision over the proximal one third of the first metacarpal and extend it along the first wrist extensor compartment. Identify and free the radial artery from the underlying fascia and retract it for protection. When the joint is dislocated, the artery may be displaced deep to the metacarpal. Now incise the carpometacarpal joint vertically and release the capsule from the base of the metacarpal; carefully preserve the capsule for later closure. If the joint is loose, dislocate it completely so that the end of the metacarpal is exposed.

The palmar and ulnar surfaces of the greater multangular may be eroded and may require resection or reshaping. Resect the metacarpal base perpendicular to its long axis. Shape the base to allow the insertion of a size 0 or 1 Swanson great toe prosthesis. Ream the medullary canal of the metacarpal to permit introduction of the prosthetic stem. Insert the stem and seat the base of the prosthesis on

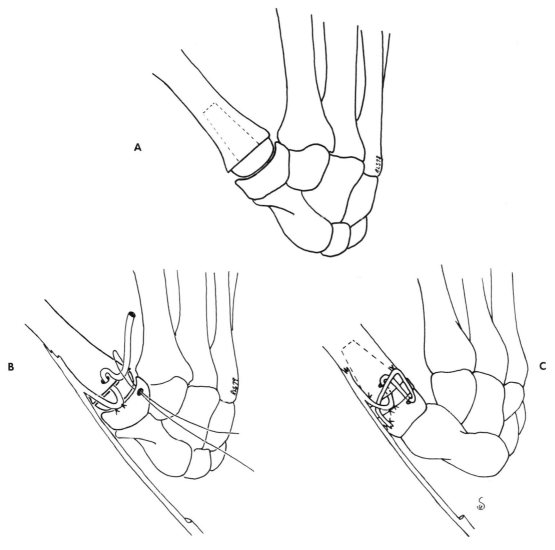

Fig. 12-22. Swanson et al. technique for convex condylar arthroplasty. (From Swanson, A., deGroot-Swanson, G., and Watermeier, J.J.: J. Hand Surg. **6:**125, 1981.)

the flat surface of the reshaped greater multangular. Insert a small Kirschner wire through the first metacarpal and the prosthesis and into the carpus to ensure proper alignment. Then close the capsule and reinforce the closure with a section of the abductor pollicis longus if indicated. A secure closure is essential. Close the wound and support the thumb on a splint for 3 weeks. Then remove the Kirschner wire but continue the splint for a total of 6 weeks.

INFECTION FOLLOWING INSERTION OF SILASTIC THUMB PROSTHESIS

If a silastic implant becomes infected, it should be removed. The wound is drained, and the thumb is then held in abduction and opposition in a cast or splint if necessary until the thumb is stable. Once the infection has cleared and the wound is healing, motion of the thumb can be started with what has now become a resection arthroplasty. In most instances the capsule will reform enough to stabilize the joint, and subluxation is unlikely if a night splint is used and motion is started gradually. We have salvaged several joints in this manner. However, several months may pass before maximum improvement is reached.

RESECTION ARTHROPLASTY OF CARPOMETACARPAL JOINT OF THUMB

See the technique of excising the greater multangular (Goldner and Clippinger) described on p. 173. Modify this technique as indicated for the given patient.

• • •

When de Quervain's disease accompanies major pathologic conditions at the carpometacarpal joint of the thumb, it can be released as can a carpal tunnel at the time of surgery on the joint. Adduction contractures of the first metacarpal shaft should be released at the time of insertion of a prosthesis or the metacarpophalangeal joint may hyperextend postoperatively because of increasing adduction of the metacarpal shaft.

SYNOVECTOMY OF INTERPHALANGEAL JOINT OF THUMB

In conjunction with other surgical procedures, a synovectomy of the interphalangeal joint of the thumb may be done to relieve pain and improve motion. Results are usually good when there is a stable joint with intact collateral ligaments.

TECHNIQUE. Make an inverted V incision over the dorsum of the interphalangeal joint with the distal point of the V just beyond the articular surface of the distal phalanx. Slope the V proximally to lie just lateral to each condyle of the proximal phalanx and create a flap over the dorsum of the joint. Do not go to the extreme lateral side of each joint so as to avoid severing the dorsal branches of the digital nerves or the branches of the superficial radial nerve. Enter the joint between the extensor pollicis longus tendon and the collateral ligament on each side. A 1 mm wide edge of the extensor tendon may be excised on each side to permit easier entry. Excise the pocket of hypertrophied synovium from under the extensor with the joint in extension. Then flex the joint to expose the articular surface of the proximal phalanx, and angulate the joint to permit exposure of the internal surface of the collateral ligament on each side for removal of synovium. Abrasive maneuvers with a small curet or pledget of gauze may be helpful in removing the synovium from under the origin of the collateral ligament (through this same incision, Heberden's nodes may be excised). Suture the skin, and place a soft wet cotton dressing over the wound.

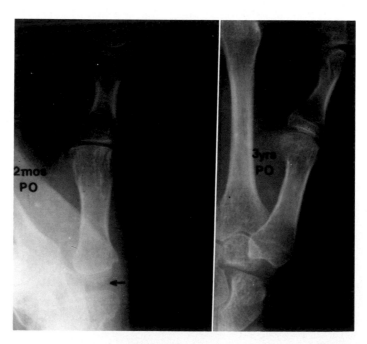

Fig. 12-23. At 2 months and at 3 years after excision of greater multangular, settling of metacarpal and narrowing of the new joint is absent, but thumb is unstable.

DEFORMITIES OF WRIST
Synovitis of wrist

Often the dorsum of the wrist is the location of the first painful swelling in rheumatoid arthritis. The swelling is, of course, caused by synovitis that often will have already caused de Quervain's disease, trigger finger, or carpal tunnel syndrome but without the diagnosis of rheumatoid arthritis having been made. The swelling may begin as a small soft mass at the distal end of the ulna; then roentgenograms may reveal a small pit at the base of the ulnar styloid as the first roentgenographic evidence of the disease. The synovitis may spread and cause a massive swelling shaped like an hourglass, its middle being constricted by the dorsal carpal ligament. Eventually destruction of joints may cause dorsal subluxation of the distal ulna, ulnar shifting of the carpal bones, radial angulation of the metacarpals, and ulnar deviation of the fingers. Finally, the wrist may even subluxate volarly. Tendons, especially those of the three ulnar finger extensors, may rupture (p. 281).

When the synovitis is only moderate and when changes in the bones are absent but pain is significant, dorsal synovectomy of the wrist seems to be of lasting benefit. Persistent swelling at the dorsum of the wrist that continues for 6 weeks or longer despite adequate medical treatment may be an indication for a dorsal synovectomy. This may be considered a prophylactic measure to avoid rupture of the extensor tendons. Their rupture is quite disabling, and function can never be restored completely. Synovectomy and resection of the distal ulnar are even more strongly indicated when the distal ulna is subluxated.

When the synovitis is severe and when the distal ulna has already dislocated dorsally, resection of the distal ulna usually is required, but Clayton warned that this operation may cause an unacceptable ulnar deviation of the wrist. To avoid this complication, resection of the distal ulna should be minimal (1 cm). Any tendons ruptured at the level of the wrist may be repaired at the time of synovectomy. Sometimes a side-to-side suture anastomosis or a free tendon graft is useful in bridging a defect in a tendon. Often when synovitis involves both the wrist and the metacarpophalangeal joints, synovectomy may be carried out at both levels through carefully planned incisions during the same operation but only on one limb at a time.

On the volar aspect of the wrist, even slight hypertrophy of the synovium undetectable clinically may cause compression of the median nerve (Fig. 12-24) and thus classical symptoms of carpal tunnel syndrome (p. 349). In fact, synovitis is considered the most frequent cause of the syndrome. Compression of the nerve in rheumatoid arthritis should be relieved surgically when conservative treatment consisting of splinting and the injection of steroids has been unsuccessful. When hypertrophy of the synovium on the volar aspect of the wrist is obvious clinically, with or without symptoms of compression of the median nerve, a volar synovectomy may be useful in relieving pain and in preventing rupture of tendons. The level of the deep transverse carpal ligament is a frequent site of rupture of flexor tendons. Furthermore, several times we have seen erosion of the distal end of the radius or carpal scaphoid into the floor of the carpal tunnel where its rough surface has caused fraying and eventual rupture of several profundus tendons. More commonly the flexor pollicis longus or index profundus is involved. Synovitis within the carpal articulations themselves as well as in the surrounding tendon sheaths is common in rheumatoid arthritis.

The various options for surgical treatment depend on the pathology involved and the severity of the disease. As already mentioned, synovectomy of the dorsal compartment is a worthwhile procedure when indicated. Repair of tendons on the extensor surface is often necessary but is best done by anastomosis to an adjoining tendon rather than by segmental grafting. Flexor tendon rupture at the wrist level is best repaired either by suture to adjoining tendons or by use of a segmental graft; however, in the thumb an arthrodesis of the distal joint is the procedure of choice. When joint destruction is severe with painful deviation and subluxation of the wrist, opinions differ as to whether an arthrodesis of the wrist or prosthetic replacement arthroplasty is preferable. However, arthrodesis is a procedure that can be depended on to relieve pain, correct deformity, and maintain stability permanently. The technique is uncomplicated, and the rate of fusion is high. Motion, however, is sacrificed.

Should bony procedures on the wrist be necessary bilaterally, then arthroplasty should be strongly considered. Some would even argue that arthroplasty is indicated initially because eventual collapse of the opposite wrist may require reconstruction. Several types of arthroplasties are available. Resection arthroplasty does not provide stability. Albright and Chase resect the distal radius to form a shelf in cases of palmar dislocation. This is done in an effort to maintain some stability, increase motion, and relieve pain without the insertion of foreign material. A silicone prostheses designed by Swanson has been used by many. This procedure does not require fixation by polymethylmethacrylate and entails less resection of bone than some procedures. Therefore it is the choice of many surgeons, despite the fact that an 8% to 10% prosthetic fracture rate has been reported.

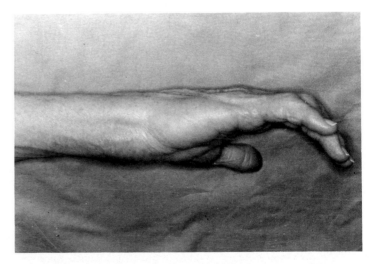

Fig. 12-24. Hypertrophic synovitis at wrist level may cause carpal tunnel syndrome.

TECHNIQUE OF DORSAL SYNOVECTOMY. Make a dorsal longitudinal incision curved only slightly ulnarward and long enough to expose both the distal ulna and the dorsal carpal ligament; avoid curving it sharply, otherwise the circulation in a flap may be impaired. Excessive scarring here has not been a problem. Preserve the larger veins, the dorsal branch of the ulnar nerve, and the superficial radial nerve. Detach from the radial side and reflect as a sheet the dorsal carpal ligament. Now carefully excise the synovium from around the finger and radial wrist extensor tendons. Excise any hypertrophied synovium from the distal ulna and the distal radioulnar joint. If the attachments of the distal ulna to the radius and carpus seem to be intact, do not disturb them. But if the distal ulna is found subluxated, excise about 1 cm of it, smooth off the remaining end, and cover the end with periosteum and surrounding soft tissues.

Incise the sheath of the extensor carpi ulnaris tendon near its attachment to the base of the fifth metacarpal. If the sheath is disintegrated and the tendon is dislocated palmarward, it then has become a flexor causing palmar flexion and ulnar deviation of the wrist. In this case, remove the tendon from the sheath as needed and return it to the dorsum of the wrist. Create a pulley with a strip of the dorsal retinaculum to keep it in position. If before surgery the patient could not actively deviate and dorsiflex the wrist from a position of radial deviation, it may be necessary to transfer the insertion of the extensor carpi radialis longus tendon to the extensor carpi ulnaris tendon to correct radial deviation. While an assistant applies traction to the hand, remove the synovium from among the carpal bones. Now pass the dorsal carpal ligament deep to the long extensor tendons, and suture its detached end in place. Elevate the hand, control bleeding by manual pressure, and release the tourniquet. Close the skin with interrupted sutures and leave a rubber drain in the wound. Apply a compression dressing and then a volar plaster splint to hold the wrist in neutral position.

AFTERTREATMENT. Active motion of the finger joints is encouraged early. The wound is periodically inspected and any hematoma beneath the skin is evacuated. At 10 to 14 days the sutures are removed and at 3 weeks the splint.

TECHNIQUE OF VOLAR SYNOVECTOMY. Make a volar longitudinal incision beginning distally at the middle of the palm and proceeding proximally to the wrist parallel to the thenar crease, then curving slightly radialward and then slightly ulnarward, and ending about 7.5 cm proximal to the wrist (Fig. 1-13, *H*). Open the deep fascia proximally and identify the median nerve. Stay on the ulnar side of this nerve and protect its recurrent branch, and if identified, its palmar branch. Now divide the deep transverse carpal ligament to expose the flexor tendons; its distal border is more distal in the palm than is usually realized. Now beginning proximally and proceeding distally and keeping constantly in mind the location of the median nerve, dissect the synovium from each flexor tendon. Do not close the deep transverse carpal ligament. Release the tourniquet, obtain hemostasis, insert a drain, and close the wound. Apply a compression dressing and a volar plaster splint from the proximal forearm to the distal palmar crease. Keep the wrist extended for a minimum of 3 weeks.

AFTERTREATMENT. The aftertreatment is the same as for dorsal synovectomy (see above).

Arthrodesis of wrist versus arthroplasty

Whether arthrodesis or arthroplasty of the wrist is best in rheumatoid arthritis is controversial. Obviously retention of wrist motion is desirable. However, wrist arthroplasty usually has a higher percentage of late complications than does arthrodesis. Arthrodesis provides a painless and stable wrist once fusion has taken place. Most consider it the procedure of choice for marked flexion deformity of the wrist and fingers, for carpal dislocation, or for a painful wrist associated with multiple ruptures of tendons. This is especially true for ruptures of the extensor carpi radialis longus and brevis since these muscles are necessary for wrist balance. Also, when wrist deformities are bilateral and require major procedures on both sides, one wrist may be arthrodesed to provide stability, especially when the use of crutches may be necessary; then an arthroplasty may be performed on the other wrist.

Furthermore, the exact position in which to fuse for maximum function is also controversial. Haddad and Riordan prefer 10 degrees of dorsiflexion, whereas Boyes prefers 30 degrees. Ferlic and Clayton prefer the neutral position with the alignment of the third metacarpal in the lateral axis of the radius, especially if bilateral wrist fusions are to be carried out. In bilateral fusions some prefer to place one wrist in dorsiflexion and the other in palmar flexion. At any rate, both wrists should not be fused in dorsiflexion because this will make it impossible for the patient to take care of personal toilet needs. Several satisfactory techniques are available for arthrodesis. Most require some type of internal fixation, usually by a medullary pin. Mannerfelt and Malmsten have described the use of a Rush pin inserted between the second and third metacarpal shafts, through the carpus, and then through the medullary canal of the radius, and a staple is inserted to prevent rotation. Clayton and associates insert a medullary Steinmann pin and bone graft the dorsum of the wrist. Millender and Nalebuff describe a new method of arthrodesing the wrist using medullary fixation with a Steinmann pin through the shaft of the third metacarpal with additional fixation by a staple or oblique pin. This permits operations to be done on any dislocated metacarpophalangeal joints at the same time. Only the last mentioned fusion procedure will be described because all the procedures are variations of one another.

TECHNIQUE FOR ARTHRODESIS (MILLENDER AND NALEBUFF). Make a dorsal straight longitudinal incision and protect the extensor tendons of the digits and wrist. Curette the cartilage and sclerotic bone from the carpus and radius down to cancellous bone. Varying amounts of bone may require resection for reduction of a dislocated wrist. Drill a Steinmann pin of appropriate size into the carpus and out distally between the second and third metacarpals. Then drill it proximally into the medullary canal of the radius and cut off its end beneath the skin. Or as an alternative method resect the head of the third metacarpal for later insertion of a joint prosthesis. Then insert the Steinmann pin through the medullary canal of the third metacarpal, then through the carpus, and finally into the radius, leaving sufficient room distally in the metacarpal in which to insert the proximal prong of a prosthesis to replace the metacarpophalangeal joint. This places the wrist in neutral position. To avoid rotational deformities, drive a staple across the radiocarpal joint or insert an oblique Kirschner wire. Insert a small plug of polymethylmethacrylate into the metacarpal shaft to prevent the Steinmann pin from shifting and protruding, or if desired, pack bone from the resected metacarpal head around the pin to accomplish the same purpose. Close the wound loosely to permit ample drainage. Now proceed with any other operations necessary on the digits. A splint rather than a solid cast, at least for the first 2 weeks, is preferred to avoid complications from swelling.

• • •

Because children with rheumatoid arthritis have such small bones, it is usually better to insert a Swanson silicone prosthesis rather than total wrist components that are too large. But silicone prostheses alone cannot be depended on to stabilize the wrist. The soft tissues must be released adequately, the bones must be aligned correctly, and the musculotendinous units must be balanced if possible to prevent recurrence of deformity.

TECHNIQUE FOR SILICONE WRIST ARTHROPLASTY (SWANSON). Make a slightly curved dorsal longitudinal incision, preserving the veins and sensory nerves. Avoid the S-shaped incision because it increases the risk of skin necrosis. Split the dorsal retinaculum over the extensor digitorum tendons and reflect it to the radial side; protect it for later use to reinforce the capsule and provide a floor for the extensor tendons. Detach the dorsal capsule from the radius and reflect it distally as a widely based flap. Detach the radial collateral ligament from the radius and carefully protect the abductor pollicis longus and extensor pollicis brevis tendons. Now hyperflex the radiocarpal joint to expose the distal radius. Remove the lunate, the proximal half of the scaphoid, and the radial side of the triquetrum. Resect the radial styloid in line with the distal articular surface of the radius at 90 degrees to the long axis of the radius. Preserve as much cortical bone as possible to provide support. When the joint is dislocated, resect more of the radius as necessary. Align the wrist and prepare the capitate and the base of the third metacarpal to receive the prosthesis. To assure proper placement of the reamer in the medullary canal of the third metacarpal, insert a Kirschner wire and check its position by roentgenograms. Then ream with an awl or if necessary with a power reamer. Do not perforate the metacarpal shaft. The size of the third metacarpal shaft will determine the size of the prosthesis. Use this shaft for size and then ream the radius to fit the opposite prong of the prosthesis. Smooth the base of the capitate and radius to eliminate any sharp bony edges that could cause a prosthetic fracture. Now seat the prosthesis against the radius and capitate so that there is no tendency for buckling. Align the hand on the wrist, avoiding ulnar deviation or flexion. See that passive flexion and extension of about 30 degrees each can be carried out without blockage. Strip the volar capsule if it is too tight. Next resect the distal ulna. Close the capsule with sutures passed through holes drilled in the dorsal cortex of the distal radius. Reattach the collateral ligament of the radius and realign the dorsal retinaculum under the finger and wrist extensors. Relocate the extensor carpi ulnaris tendon dorsally to prevent it from functioning as a wrist flexor if it has shifted palmarward; pass it through a pulley created from a segment of the

dorsal retinaculum if necessary. Repair any extensor tendons as indicated and close the wound loosely. Insert a suction drain and apply a bulky dressing and a plaster splint.

AFTERTREATMENT. The splint is worn for 5 to 6 weeks. At 3 weeks limited wrist motion is started, and at 4 weeks active motion is begun. A total motion of 60 degrees is considered satisfactory, and 95% of patients obtain relief from pain.

Total wrist prosthesis semiconstrained-type Volz design

In 1979 Volz reported on 100 total joint semiconstrained arthroplasties of the wrist. These arthroplasties were done by 15 collaborating surgeons using prostheses of his design. Good results were achieved in 86% of the wrists, poor results in 8%, and 6% failed. The most common problem after surgery was an imbalance of the musculotendinous units, especially a tendency to ulnar deviation. Since that time, Volz has redesigned the prosthesis to precisely reduplicate the instant center of motion in the normal wrist, which is located at the proximal pole of the capitate at a position vertically in line with the long axis of the third metacarpal (Fig. 12-25). He emphasized the extreme importance of a dynamic balance between the wrist flexors and extensors. Before surgery the functioning balance of these muscles should be evaluated carefully. A dorsal tenosynovitis may cause wrist extensor compromise, if not

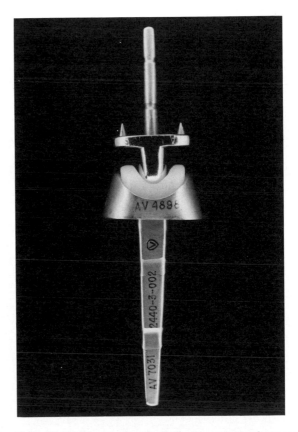

Fig. 12-25. The AMC total wrist prosthesis. The metacarpal component is press fitted into capitate–third metacarpal bed, and radial component is cemented into distal radius. (Courtesy Dr. Robert G. Volz.)

rupture. Volz has reported 25 total wrist arthroplasties using a prosthesis of the new design. Of these patients, 19 had rheumatoid arthritis, and three had osteoarthritis. The prosthesis has a potential range of motion of 90 degrees flexion and 90 degrees extension and 50 degrees radial and 50 degrees ulnar deviation. In the last six procedures the implant was inserted without the use of polymethylmethacrylate about the metacarpal component (Fig. 12-26). Long-term follow-up is not available for this prosthetic design.

TECHNIQUE (VOLZ). Make a straight dorsal incision beginning at the base of the third metacarpal and proceeding proximally to the ulnar border of the distal radius. Preserve all subcutaneous veins. If the distal radioulnar joint is diseased, resect the distal 1 cm of the ulna as follows. Expose the dorsal compartment containing the extensor carpi ulnaris tendon and incise the extensor retinaculum. Dissect the soft tissues from about the distal ulna and resect the ulna 1 cm from its distal end. Later stabilize the distal ulna by imbricating the dorsal capsule and the overlying retinaculum to create a snug sling over the ulna. Now incise the third dorsal compartment with care to protect the extensor pollicis longus tendon. Then perform a dorsal tenosynovectomy if indicated. Retract the common extensor tendons ulnarward and the extensor carpi radialis longus and brevis radially. Make a U-shaped incision in the dorsal capsule with its base directed radialward and expose the underlying carpus. Resect the distal end of the radius at the level of Lister's tubercle, or slightly more proximal, at a right angle to the long axis of the bone. Amputate the head of the capitate at the level of the lesser multangular

and the proximal carpal row. The volar capsule will be found contracted if the carpus has been subluxated. If necessary, perform a volar capsulectomy at the base of the radius to overcome the contracture and restore extensor movement at the wrist. Ream the capitate and third metacarpal shaft and rasp the distal radial canal to fit the selected prosthesis. After temporary insertion of the prosthesis, check to see that extension is not limited; if it is, resect additional bone from the distal radius and release the volar capsule as needed. Irrigate the wound copiously to remove fat, blood, and bone debris. Inject methylmethacrylate under pressure into the metacarpal, carpus, and radial canals (as noted above, Volz's later cases did not have cement injected into the third metacarpal). Make the final seating of the prosthesis as the cement hardens and check the range of motion carefully. Remove all exposed cement, close the dorsal capsule, place the extensor retinaculum under the extensor tendons, and reinforce the capsule with the retinaculum. Secure the extensor pollicis longus in the area of Lister's tubercle. Place a small drain on the repaired capsule and complete the closure. Apply a dressing and splint with hand and fingers held in neutral position or slight dorsiflexion.

AFTERTREATMENT. The wrist should be immobilized for 7 to 10 days or longer if healing of the wound is delayed. After the wound has healed, passive and active motion are started under the supervision of a therapist. For the next 6 or more weeks a splint is worn part of the time during the day and all through the night to ensure control of wrist motion.

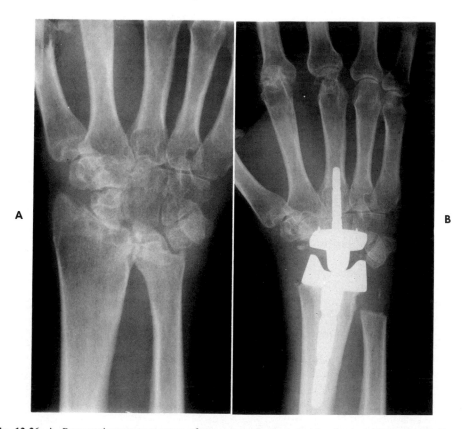

A   B

Fig. 12-26. A, Preoperative roentgenogram of a carpus badly destroyed by rheumatoid arthritis. **B,** Postoperative roentgenogram following total wrist arthroplasty. Only radial component is cemented in place. (Courtesy Dr. Robert G. Volz.)

REFERENCES

Albright, J.A., and Chase, R.A.: Palmar-shelf arthroplasty of the wrist in rheumatoid arthritis: a report of nine cases, J. Bone Joint Surg. 52-A:896, 1970.

Amadio, P.D., Millender, L.H., and Smith, R.J.: Silicone spacer or tendon spacer for trapezium resection arthroplasty: comparison of results, J. Hand Surg. 7:237, 1982.

Aptekar, R.G., Davie J.M., and Cattell, H.S.: Foreign body reaction to silicone rubber: complication of a finger joint implant, Clin. Orthop. 98:231, 1974.

Aptekar, R.G., and Duff, I.F.: Metacarpophalangeal joint surgery in rheumatoid arthritis: long-term results, Clin. Orthop. 83:123, 1972.

Aro, H., Ekfors, T., Hakkarainen, S., and Aho, A.J.: Osteolytinen vier-asesinereaktio Silastic-proteesikomplikaationa ranteesa, Suomen Kirugiyhd 3:66, 1982.

Ashworth, C.A., Blatt, G., Chuinard, R.G., and Stark, H.: Silicone rubber interposition arthroplasty, J. Hand Surg. 2:345, 1977.

Beckenbaugh, R.D.: Personal communication, November, 1984.

Bigelow, D.R.: A surgical solution to the problem of swan-neck deformity of rheumatoid arthritis, Clin. Orthop. 123:89, 1977.

Black, M.R., Boswick, J.A., Jr., and Wiedel, J.: Dislocation of the wrist in rheumatoid arthritis: the relationship to distal ulna resection, Clin. Orthop. 124:184, 1977.

Boyce, T., et al.: Clinical and experimental studies on the effect of extensor carpi radialis longus transfer in the rheumatoid hand, J. Hand Surg. 3:390, 1978.

Brannon, E.W., and Klein, G.: Experiences with a finger-joint prosthesis, J. Bone Joint Surg. 41-A:87, 1959.

Braun, R.M.: Total joint replacement at the base of the thumb: preliminary report, J. Hand Surg. 7:245, 1982.

Braun, R.M., and Chandler, J.: Quantitative results following implant arthroplasty of the proximal finger joints in the arthritic hand, Clin. Orthop. 83:135, 1972.

Brewerton, D.A.: Hand deformities in rheumatoid disease, Ann. Rheum. Dis. 16:183, 1957.

Brumfield, R.H., Jr., Conaty, J.P., and Mayes, J.D.: Surgery of the wrist in rheumatoid arthritis, Clin. Orthop. 142:159, 1979.

Buch, V.I.: Clinical and functional assessment of the hand after metacarpophalangeal capsulotomy, Plast. Reconstr. Surg. 53:452, 1974.

Bunnell, S.: Surgery of the rheumatic hand, J. Bone Joint Surg. 37-A:759, 1955.

Campbell, R.D., Jr., and Straub, L.R.: Surgical considerations for rheumatoid disease in the forearm and wrist, Am. J. Surg. 109:361, 1965.

Carroll, R.E., and Dick, H.M.: Arthrodesis of the wrist for rheumatoid arthritis, J. Bone Joint Surg. 53-A:1365, 1971.

Chamay, A., and Gabbiani, G.: Digital contracture deformity after implantation of a silicone prosthesis: light and electron microscopic study, J. Hand Surg. 3:266, 1978.

Clayton, M.L.: Surgery of the thumb in rheumatoid arthritis, J. Bone Joint Surg. 44-A:1376, 1962.

Clayton, M.L.: Surgery of the rheumatoid hand, Clin. Orthop. 36:47, 1964.

Clayton, M.L.: Surgical treatment at the wrist in rheumatoid arthritis: a review of thirty-seven patients, J. Bone Joint Surg. 47-A:741, 1965.

Clayton, M.L., and Ferlic, D.C.: Tendon transfer for radial rotation of the wrist in rheumatoid arthritis, Clin. Orthop. 100:176, 1974.

Clayton, M.L., and Ferlic, D.C.: The wrist in rheumatoid arthritis, Clin. Orthop. 106:192, 1975.

Crawford, G.P.: Ligament augmentation with replacement arthroplasty of the CMC joint, Hand 12:91, 1980.

Cregan, J.C.F.: Indications for surgical intervention in rheumatoid arthritis of the wrist and hand. Ann. Rheum. Dis. 18:29, 1959.

Crosby, E.B., Linscheid, R.L., and Dobyns, J.H.: Scaphotrapezial trapezoidal arthrosis, J. Hand Surg. 3:223, 1978.

Dell, P.C.: Compression of the ulnar nerve at the wrist secondary to a rheumatoid synovial cyst: case report and review of the literature, J. Hand Surg. 4:468, 1979.

Dell, P.C., Brushart, T.M., and Smith R.J.: Treatment of trapeziometacarpal arthritis: results of resection arthroplasty, J. Hand Surg. 3:243, 1978.

Devas, M., and Shah, V.: Link arthroplasty of the metacarpophalangeal joints: a preliminary report of a new method. J. Bone Joint Surg. 57-B:72, 1975.

Digby, J.M.: Malignant lymphoma with intranodal silicone rubber particles following metacarpophalangeal joint replacements, Hand 14:326, 1982.

Dobyns, J.H., and Linscheid, R.L.: Rheumatoid hand repairs, Orthop. Clin. North Am. 3:629, 1971.

Eaton, R.G.: Replacement of the trapezium, J. Bone Joint Surg. 61-A:76, 1979.

Ehrlich, G.E., et al.: Pathogenesis of rupture of extensor tendons at the wrist in rheumatoid arthritis, Arthritis Rheum. 2:332, 1959.

Ellison, M.R., Flatt, A.E., and Kelly, K.J.: Ulnar drift of the fingers in rheumatoid disease: treatment by crossed intrinsic tendon transfer, J. Bone Joint Surg. 53-A:1061, 1971.

Ellison, M.R., et al.: The results of surgical synovectomy of the digital joints in rheumatoid disease, J. Bone Joint Surg. 53-A:1041, 1971.

Engel, J., Tsur, H., and Farin, I.: A comparison between K-wire and compression screw fixation after arthrodesis of the distal interphalangeal joint, Plast. Reconstr. Surg. 60:611, 1977.

Ferlic, D.C., et al.: Complications of silicone implant surgery in the metacarpophalangeal joint, J. Bone Joint Surg. 57-A:991, 1975.

Ferlic, D.C., et al.: Degenerative arthritis of the carpometacarpal joint of the thumb: a clinical follow-up of eleven Niebauer prostheses, J. Hand Surg. 2:212, 1977.

Ferlic, D.C., and Clayton, M.L.: Flexor tenosynovectomy in the rheumatoid finger, J. Hand Surg. 3:364, 1978.

Flatt, A.E.: Surgical rehabilitation of the arthritic hand, Arthritis Rheum. 2:278, 1959.

Flatt, A.E.: Restoration of rheumatoid finger joint function: interim report on trial of prosthetic replacement, J. Bone Joint Surg. 43-A:753, 1961.

Flatt, A.E.: Salvage of the rheumatoid hand, Clin. Orthop. 23:207, 1962.

Flatt, A.E.: Restoration of rheumatoid finger joint function, J. Bone Joint Surg. 45-A:1101, 1963.

Flatt, A.E.: Some pathomechanics of ulnar drift, Plast. Reconstr. Surg. 37:295, 1966.

Flatt, A.E.: The care of the rheumatoid hand, ed. 3, St. Louis, 1974, The C.V. Mosby Co.

Flatt, A.E., and Ellison, M.R.: Restoration of rheumatoid finger joint function. III. A follow-up note after fourteen years of experience with a metallic hinge prosthesis, J. Bone Joint Surg. 54-A:1317, 1972.

Folmar, R.C., Nelson, C.L., and Phalen, G.S.: Ruptures of the flexor tendons in hands of non-rheumatoid patients, J. Bone Joint Surg. 54-A:579, 1972.

Freiberg, R.A., and Weinstein, A.: The scallop sign and spontaneous rupture of finger extensor tendons in rheumatoid arthritis, Clin. Orthop. 83:128, 1972.

Froimson, A.I.: Hand reconstruction in arthritis mutilans: a case report, J. Bone Joint Surg. 53-A:1377, 1971.

Girzadas, D.V., and Clayton, M.L.: Limitations of the use of metallic prosthesis in the rheumatoid hand, Clin. Orthop. 67:127, 1969.

Goldner, J.L.: Tendon transfers in rheumatoid arthritis, Orthop. Clin. North Am. 2:425, 1974.

Goldner, J.L., et al.: Metacarpophalangeal joint arthroplasty with silicone-Dacron prostheses (Niebauer type): six and a half years' experience, J. Hand Surg. 2:200, 1977.

Goodman, M.J., et al.: Arthroplasty of the rheumatoid wrist with silicone rubber: an early evaluation, J. Hand Surg. 5:114, 1980.

Gordon, M., and Bullough, P.G.: Synovial and osseous inflammation in failed silicone rubber prostheses, J. Bone Joint Surg. 64-A:574, 1982.

Granberry, W.M., and Mangum, G.L.: The hand in the child with juvenile rheumatoid arthritis, J. Hand Surg. 5:105, 1980.

Granowitz, S., and Vainio, K.: Proximal interphalangeal joint arthrodesis in rheumatoid arthritis: a follow-up study of 122 operations, Acta Orthop. Scand. 37:301, 1965–1966.

Green, D.P.: Pisotriquetral arthritis: a case report, J. Hand Surg. 4:465, 1979.

Haffajee, D.: Endoprosthetic replacement of the trapezium for arthrosis in the carpometacarpal joint of the thumb, J. Hand Surg. 2:141, 1977.

Harris, C., Jr., and Riordan, D.C.: Intrinsic contracture in the hand and its surgical treatment, J. Bone Joint Surg. 36-A:10, 1954.

Harrison, S.H.: The importance of middle or long finger realignment in ulnar drift, J. Hand Surg. 1:87, 1976.

Henderson, E.D., and Lipscomb, P.R.: Surgical treatment of rheumatoid hand, JAMA 175:431, 1961.

Howard, L.D., Jr.: Surgical treatment of rheumatic tenosynovitis, Am. J. Surg. 89:1163, 1955.

Inglis, A.E., et al.: Reconstruction of the metacarpophalangeal joint of the thumb in rheumatoid arthritis, J. Bone Joint Surg. 54-A:704, 1972.

Jackson, I.T., et al.: Ulnar head resection in rheumatoid arthritis, Hand 6:172, 1974.

Jensen, J.S.: Operative treatment of chronic subluxation of the first carpometacarpal joint, Hand **7:**269, 1975.

Kessler, I.: A new silicone implant for replacement of destroyed metacarpal heads, Hand, **6:**308, 1974.

Kessler, I., and Axter, A.: Arthroplasty of the first carpometacarpal joint with a silicone implant, Plast. Reconstr. Surg. **47:**252, 1971.

Kessler, I., and Vainio, K.: Posterior (dorsal) synovectomy for rheumatoid involvement of the hand and wrist: a follow-up study of sixty-six procedures, J. Bone Joint Surg. **48-A:**1085, 1966.

Kessler, F.B., et al.: Obliteration of traumatically induced articular surface defects using a porous implant, J. Hand Surg. **5:**328, 1980.

Kircher, T.: Silicone lymphadenopathy: a complication of silicone elastomer finger joint prostheses, Hum. Pathol. **11:**240, 1980.

Kleinert, H.E., and Frykman, G.: The wrist and thumb in rheumatoid arthritis, Orthop. Clin. North Am. **4:**1085, 1973.

Kulick, R.G., et al.: Long-term results of dorsal stabilization in the rheumatoid wrist, J. Hand Surg. **6:**272, 1981.

Laine, V.A.I., Sairanen, E., and Vainio, K.: Finger deformities caused by rheumatoid arthritis, J. Bone Joint Surg. **39-A:**527, 1957.

Lamberta, F.J., Ferlic, D.D., and Clayton, M.L.: Volz total wrist arthroplasty in rheumatoid arthritis: a preliminary report, J. Hand Surg. **5:**245, 1980.

Larmon, W.A.: Surgical management of tophaceous gout, Clin. Orthop. **71:**56, 1970.

Linscheid, R.L.: Surgery for rheumatoid arthritis: timing and techniques: the upper extremity, J. Bone Joint Surg. **50-A:**605, 1968.

Linscheid, R.L., and Dobyns, J.H.: Rheumatoid arthritis of the wrist, Orthop. Clin. North Am. **3:**649, 1971.

Lipscomb, P.R.: Surgery of the arthritic hand, Sterling Bunnell Memorial Lecture, Mayo Clin. Proc. **40:**132, 1965.

Lipscomb, P.R.: Synovectomy of the wrist for rheumatoid arthritis, JAMA **194:**655, 1965.

Lipscomb, P.R.: Synovectomy of the distal two joints of the thumb and fingers in rheumatoid arthritis, J. Bone Joint Surg. **49-A:**1135, 1967.

Lucht, U., Vang, P.S., and Munck, J.: Soft tissue interposition arthroplasty for osteoarthritis of the CMC joint of the thumb, Acta Orthop. Scand. **51:**767, 1980.

Lynn, M.D., and Lee, J.: Periarticular tenosynovial chondrometaplasia: report of a case at the wrist, J. Bone Joint Surg. **54-A:**650, 1972.

Madden, J.W., DeVore, G., and Arem, A.J.: A rational postoperative management program for metacarpophalangeal joint implant arthroplasty, J. Hand Surg. **2:**358, 1977.

Mannerfelt, L., and Malmsten, M.: Arthrodesis of the wrist in rheumatoid arthritis: a technique without external fixation, Scand. J. Plast. Reconstr. Surg. **5:**124, 1971.

Mannerfelt, L., and Andersson, K.: Silastic arthroplasty of the metacarpophalangeal joints in rheumatoid arthritis: long-term results, J. Bone Joint Surg. **57-A:**484, 1975.

Marmor, L.: The role of hand surgery in rheumatoid arthritis, Surg. Gynecol. Obstet. **116:**335, 1963.

Marmor, L.: Surgical treatment for arthritic deformities of the hands, Clin. Orthop. **39:**171, 1965.

McFarland, G.B., Jr.: Early experience with the silicone rubber prosthesis (Swanson) in the reconstructive surgery of the rheumatoid hand, South. Med. J. **65:**1113, 1972.

McMaster, M.: The natural history of the rheumatoid metacarpophalangeal joint, J. Bone Joint Surg. **54-B:**687, 1972.

Medl, W.T.: Tendonitis, tenosynovitis, "trigger finger," and Quervain's disease, Orthop. Clin. North Am. **1:**375, 1970.

Menkes, C.J., et al.: Intra-articular injection of radioisotopic beta emitters: application to the treatment of the rheumatoid hand, Orthop. Clin. North Am. **4:**1113, 1973.

Menon, J. Schoene, H.R., and Hohl, J.C.: Trapeziometacarpal arthritis: results of tendon interpositional arthroplasty, J. Hand Surg. **6:**442, 1981.

Mikkelsen, O.A.: Arthrodesis of the wrist joint in rheumatoid arthritis, Hand **12:**149, 1980.

Milford, L.W.: Some thoughts on the use of finger implant arthroplasty, Orthop. Rev. **4:**11, 1975.

Millender, L.H., and Nalebuff, E.A.: Arthrodesis of the rheumatoid wrist: functional evaluation of a modified technique. Orthop. Rev. **1:**13, 1972.

Millender, L.H., and Nalebuff, E.A.: Metacarpophalangeal joint arthroplasty utilizing the silicone rubber prosthesis, Orthop. Clin. North Am. **4:**349, 1973.

Millender, L.H., and Nalebuff, E.A.: Preventive surgery: tenosynovectomy and synovectomy, Orthop. Clin. North Am. **6:**765, 1975.

Millender, L.H., and Nalebuff, E.A.: Reconstructive surgery in the rheumatoid hand, Orthop. Clin. North Am. **6:**709, 1975.

Millender, L.H., and Nalebuff, E.A.: Evaluation and treatment of early rheumatoid hand involvement, Orthop. Clin. North Am. **6:**697, 1975.

Millender, L.H., et al.: Posterior interosseous-nerve syndrome secondary to rheumatoid synovitis, J. Bone Joint Surg. **55-A:**753, 1973.

Millender, L.H., et al,: Dorsal tenosynovectomy and tendon transfer in the rheumatoid hand, J. Bone Joint Surg. **56-A:**601, 1974.

Millender, L.H., et al.: Infection after silicone prosthetic arthroplasty in the hand, J. Bone Joint Surg. **57-A:**825, 1975.

Millender, L.H., et al.: Interpositional arthroplasty for rheumatoid carpometacarpal joint disease, J. Hand Surg. **3:**533, 1978.

Nalbandian, R.M.: Letter to editor, J. Bone Joint Surg. **65-A:**280, 1983.

Nalebuff, E.A.: Present status of rheumatoid hand surgery, Am. J. Surg. **122:**304, 1971.

Nalebuff, E.A. and Millender, L.H.: Surgical treatment of the boutonnière deformity in rheumatoid arthritis, Orthop. Clin. North Am. **6:**753, 1975.

Nalebuff, E.A, and Millender, L.H.: Surgical treatment of the swanneck deformity in rheumatoid arthritis, Orthop. Clin. North Am. **6:**733, 1975.

Nalebuff, E.A., and Patel, M.R.: Flexor digitorum sublimis transfer for multiple extensor tendon ruptures in rheumatoid arthritis, Plast. Reconstr. Surg. **52:**530, 1973.

Pahle, J.A., and Raunio, P.: The influence of wrist position on finger deviation in the rheumatoid hand: a clinical and radiological study, J. Bone Joint Surg. **51-B:**664, 1969.

Poppen, N.K., and Niebauer, J.J.: "Tie-in" trapezium prosthesis: long-term results, J. Hand Surg. **3:**445, 1978.

Resnick, D.: Arthrography in the evaluation of arthritic disorders of the wrist, Radiology **113:**331, 1974.

Riordan, D.C.: Finger deformities and tendon repairs in rheumatoid arthritis, Orthop. Rev. **2:**11, July 1973.

Schumacher, H.R., Zweiman, B., and Bora, F.W., Jr.: Corrective surgery for the deforming hand arthropathy of systemic lupus erythematosus, Clin. Orthop. **117:**292, 1976.

Smith, R.J., and Kaplan, E.B.: Rheumatoid deformities at the metacarpophalangeal joints of the fingers: a correlative study of anatomy and pathology, J. Bone Joint Surg. **49-A:**31, 1967.

Snow, J.W., Boyes, J.G., and Greider, J.L., Jr.: Implant arthroplasty of the distal interphalangeal joint of the finger for osteoarthritis, Plast. Reconstr. Surg. **60:**558, 1977.

Spar, I.: Flexor tendon ruptures in the rheumatoid hand: bilateral flexor pollicis longus rupture, Clin. Orthop. **122:**186, 1977.

Stark, H.H., et al.: Fusion of the first metacarpotrapezial joint for degenerative arthritis, J. Bone Joint Surg. **59-A:**22, 1977.

Steinberg, V.L., and Parry, C.B.: Electromyographic changes in rheumatoid arthritis. Br. Med. J. **1:**630, 1961.

Steindler, A.: Arthritic deformities of the wrist and fingers, J. Bone Joint Surg. **33-A:**849, 1951.

Straub, L.R.: The rheumatoid hand. Clin. Orthop. **15:**127, 1959.

Straub, L.R.: Surgery of the arthritic hand, West. J. Surg. **68:**5, 1960.

Straub, L.R., and Wilson, E.H., Jr.: Spontaneous rupture of extensor tendons in the hand associated with rheumatoid arthritis, J. Bone Joint Surg. **38-A:**1208, 1956.

Straub, L. R., et al.: The ulnar drift deformity in rheumatoid arthritis, J. Bone Joint Surg. **48-A:**1650, 1966.

Swanson, A.B.: Silicone rubber implants for replacement of arthritic or destroyed joints in the hand, Surg. Clin. North Amer. **48:**1113, 1968.

Swanson, A.B.: Arthroplasty in traumatic arthritis of the joints of the hand, Orthop. Clin. North Am. **1:**285, 1970.

Swanson, A.B.: Silicone rubber implants for the replacement of the carpal scaphoid and lunate bones, Orthop. Clin. North Am. **1:**299, 1970.

Swanson, A.B.: Treatment of the stiff hand and flexible implant arthroplasty in the fingers. In American Academy of Orthopaedic Surgeons: Instructional course lectures, vol. 21, St. Louis, 1972, The C.V. Mosby Co.

Swanson, A.B.:Flexible implant arthroplasty for arthritic finger joints: rationale, technique, and results of treatment, J. Bone Joint Surg. **54-A:**435, 1972.

Swanson, A.B.: Disabling arthritis at the base of the thumb: treatment by resection of the trapezium and flexible (silicone) implant arthroplasty, J. Bone Joint Surg. **54-A:**456, 1972.

Swanson, A.B.: Implant arthroplasty for disabilities of the distal radioulnar joint: use of a silicone rubber capping implant following resection of the ulnar head, Orthop. Clin. North Am. **4:**373, 1973.

Swanson, A.B.: Disabilities of the thumb joints and their surgical treatment, including flexible implant arthroplasty. In American Academy of Orthopaedic Surgeons: Instructional course lectures, vol. 22, St. Louis, 1973, The C.V. Mosby Co.

Swanson, A.B.: Implant resection arthroplasty of the proximal interphalangeal joint, Orthop. Clin. North Am. **4:** 1007, 1973.

Swanson, A.B.: Flexible implant arthroplasty for arthritic disabilities of the radiocarpal joint: a silicone rubber intramedullary stemmed flexible hinge implant for the wrist joint, Orthop. Clin. North Am. **4:**383, 1973.

Swanson, A.B., and Swanson, G. deG.: Pathogenesis and pathomechanics of rheumatoid deformities in the hand and wrist, Orthop. Clin. North Am. **4:**1039, 1973.

Swanson, A.B., deGroot-Swanson, G., Frisch E.E.: Flexible (silicone) implant arthroplasty in the small joints of the extremities: concepts, physical and biological considerations, experimental and clinical results. In Rubin, L.R., editor: Biomaterials in reconstructive surgery. St. Louis, 1983, The C.V. Mosby Co.

Swanson, A.B., Swanson-deGroot, G., and Watermeier, J.J.: Trapezium implant arthroplasty: long-term evaluation of 150 cases, J. Hand Surg. **6:**125, 1981.

Swanson, A.B., et al.: Durability of silicone implants: an in vivo study, Orthop. Clin. North Am. **4:**1097, 1973.

Taleinsnik, J.: Rheumatoid synovitis of the volar compartment of the wrist joint: its radiological signs and its contribution to wrist and hand deformity, J. Hand Surg. **4:**526, 1979.

Tubiana, R.: The hand, vol. 1, Philadelphia, 1981, W.B. Saunders Co.

Urbaniak, J.R.: Prosthetic arthroplasty of the hand, Clin. Orthop. **104:**9. 1974.

Urbaniak, J.R., McCollum, D.E., and Goldner, J.L.: Metacarpophalangeal and interphalangeal joint reconstruction: use of silicone rubber-Dacron prostheses for replacement of irreparable joints of the hand, South. Med. J. **63:**1281, 1970.

Vainio, K., Reiman, I., and Pulkki, T.: Results of arthroplasty of the metacarpophalangeal joints in rheumatoid arthritis, Reconstr. Surg. Traumatol. **9:**1, 1967.

Vaughan-Jackson, O.J.: Rupture of extensor tendons by attrition at the inferior radio-ulnar joint: report of two cases, J. Bone Joint Surg. **30-B:**528, 1948.

Vaughan-Jackson, O.J.: Rheumatoid hand deformities considered in the light of tendon imbalance. I.J. Bone Joint Surg. **44-B:**764, 1962.

Volz, R.G.: The development of a total wrist arthroplasty, Clin. Orthop. **116:**209, 1976.

Volz, R.G.: Total wrist arthroplasty, a review of 100 patients (abstract), Orthop. Trans. **3:**268, 1979.

Volz, R.G.: Personal communication, May, 1985.

Weeks, P.M.: Volar approach for metacarpophalangeal joint capsulotomy, Plast. Reconstr. Surg. **46:**473, 1970.

Wilde, A.H.: Synovectomy of the proximal interphalangeal joint of the finger in rheumatoid arthritis, J. Bone Joint Surg. **56-A:**71, 1974.

Wilkinson, M.C., and Lowry, J.H.: Synovectomy for rheumatoid arthritis, J. Bone Joint Surg. **47-B:**482, 1965.

Wilson, J.N.: Arthroplasty of the trapezio-metacarpal joint, Plast. Reconstr. Surg. **49:**143, 1972.

Wise, K.S.: The anatomy of the metacarpo-phalangeal joints, with observations of the aetiology of ulnar drift. J. Bone Joint Surg. **57-B:**485, 1975.

CHAPTER 13

Volkmann's contracture and compartment syndromes

HISTORY AND ETIOLOGY

In 1881 Volkmann stated in his classic paper that the paralytic contractures that could develop in only a few hours after injury were caused by arterial insufficiency or ischemia of the muscles. Neural injuries were initially thought to be the cause, but these were then accepted as occurring at a later stage. He suggested that tight bandages were the cause of the vascular insufficiency. This concept of extrinsic pressure as the primary cause of paralytic contracture persisted for some time in the English literature.

In 1909 Thomas studied 107 paralytic contractures. He found that some followed severe contusion of the forearm alone. In these, fractures were absent, and a splint or bandage was not applied to the limb. The idea thus was established that extrinsic pressure was not necessarily the sole cause of the ischemia. In 1914 Murphy reported that hemorrhage and effusion into the muscles could cause internal pressures to rise within the unyielding deep fascial compartments of the forearm with subsequent obstruction of the venous return. In 1928 Sir Robert Jones concluded that Volkmann's contracture could be caused by pressure from within, from without, or from both.

Compartment syndrome, already alluded to above, is the usual cause of Volkmann's contracture. It may be defined as a condition in which the circulation within a closed compartment is compromised by an increase in pressure within the compartment resulting in tissue death. This necrosis may involve not only the muscles and nerves but also eventually the skin because of excessive swelling.

Compartment syndrome may also occur as a secondary physiologic response to primary ischemia of muscle. The initial ischemia may have been incomplete, but secondary changes, including the development of massive edema within the compartment, may make it complete. The causes of compartment syndrome include crush injuries such as wringer injuries; prolonged external compression (Fig. 13-1); internal bleeding, especially after injury to a person with hemophilia; fractures; excessive exercise; burns; and intraarterial injections of drugs or sclerosing agents.

Volkmann's contracture thus is the result of an injury to the deep tissues of the limb, producing ischemia primarily of the muscles and secondarily of the nerves. Initially it is accompanied by an acute episode of severe pain that is aggravated by stretching of the muscles. *Compartment syndrome* thus is caused by an elevation in pressure of the tissue fluids within the closed fascial compartments of the limb, especially the volar compartment of the forearm in the upper extremity and the anterior compartment of the leg in the lower. The pressure interferes with circulation of the blood to the muscles and nerves in the compartment, and if it is untreated, the pressure may cause Volkmann's contracture or even gangrene.

The cycle of increasing muscle ischemia is depicted by Eaton and Greene in Fig. 13-2. In the upper extremity, the anterior compartment of the forearm is most commonly involved (Fig. 13-3) but the posterior compartment can be involved alone or in addition to the anterior. The intrinsic muscle compartments of the hand may also be involved (see Fig. 13-5).

An arterial injury usually results in an absent pulse, decreased skin temperature, and blanching of the skin. The treatment here is the repair of the artery.

DIAGNOSIS
Volkmann's ischemia

Vascular deficiency in Volkmann's ischemia may involve only a portion of an extremity, and in the upper extremity it is usually the anterior forearm. In these instances the pulse of the radial artery at the wrist may still be palpable; this may be misleading. The ischemia may be the result of a forearm fracture in an adult or child but more typically of a supracondylar fracture of the humerus in a child. Forearm fractures accounted for 22% of the contractures reported in 1967 by Eichler and Lipscomb and for 15% reported by Mubarak and Carroll in 1979. The diagnosis of ischemia in the forearm or elsewhere depends largely on the correct evaluation of the symptom of unrelenting pain. This, however, should be supported by the entire clinical picture, including swelling and any neurologic changes found by examination repeated at regular in-

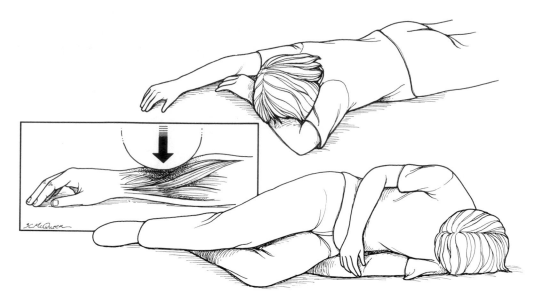

Fig. 13-1. These postures prolonged by oversedation may result in necrosis of forearm muscles, nerves, and skin.

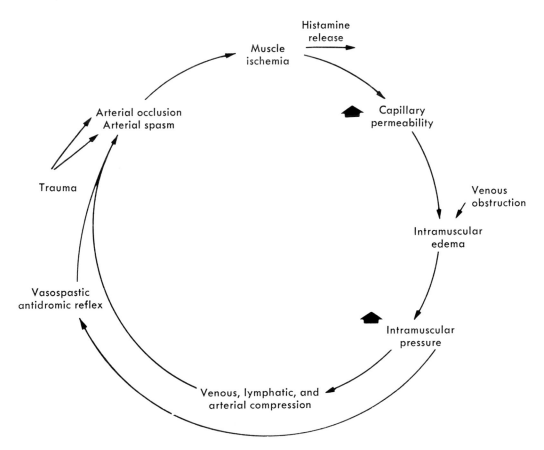

Fig. 13-2. Traumatic ischemia-edema cycle in Volkmann's contracture. (From Eaton, R.G., and Green, W.T.: Orthop. Clin. North Am. **3:**175, 1972.)

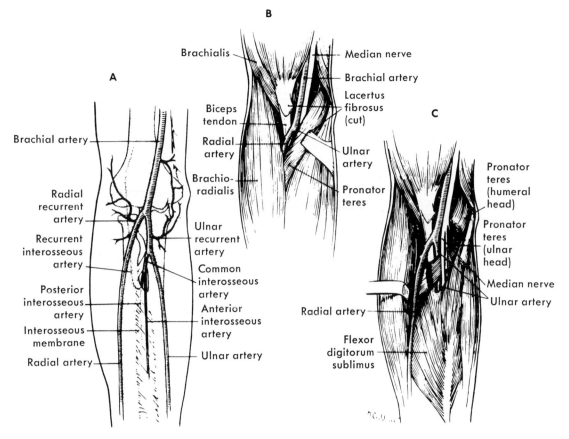

Fig. 13-3. Anatomy of Volkmann's ischemia. **A,** ''Collateral circulation'' of elbow does not communicate with vessels within flexor compartment. These elbow collaterals join radial and ulnar arteries proximal to pronator teres, proximal guardian of flexor compartment. **B,** Brachial artery and median nerve enter forearm through a tight opening formed by biceps tendon insertion laterally and pronator teres muscle medially and are tightly covered by lacertus fibrosus. Proximal angulation, hematoma, or muscle swelling within this cruciate tendon-muscle portal is capable of major compression of neurovascular bundle. **C,** Radial artery, arising from brachial artery, passes distally superficial to pronator teres and all flexor muscles. It is not crossed by any structure along this route. Ulnar artery, however, passes beneath pronator teres and lies in the deepest portions of compartment. Median nerve usually passes between humeral and ulnar heads of fleshy pronator teres, and, emerging it becomes compressed against firm arcuate band of flexor sublimis origin (see text). (From Eaton, R.G., and Green, W.T.: Orthop. Clin. North Am. **3:**175, 1972.)

tervals if necessary. Measurements of the tissue pressure as discussed below may be useful.

Acute compartment syndrome

A crush injury or fracture of the forearm or elbow, especially in the supracondylar area of the humerus, should cause suspicion that a forearm compartment syndrome may develop. The early diagnosis of impending ischemia is essential, since sometimes complete paralysis of the involved musculature may develop in only 6 to 8 hours. Pain is the most consistent symptom, although obviously it may not be discovered in a comatose patient. The forearm is tense with swelling, and sensibility of the fingertips may be diminished. Passive extension of the fingers increases the pain. The radial pulse may or may not be palpable. The dorsal compartment of the forearm may also be involved and consequently palpation with the fingers over both compartments is important. In addition, the intrinsic muscles of the hand may be involved.

The diagnosis of acute Volkmann's ischemia or acute compartment syndrome is confirmed by demonstrating an increase in compartmental pressures. In a comatose or uncooperative patient or when the diagnosis is uncertain, the tissue pressure may be measured with a wick catheter. The correlation of the clinical findings with the pressure measurements is dependable. Whitesides et al. in 1975 and in 1977 described a method of measuring tissue pressure, which has been improved. Other methods have evolved but the accepted pressure readings are basically the same.

Impending tissue ischemia may be considered when the tissue pressure reaches between 30 and 10 mm of mercury below the diastolic blood pressure. A higher pressure is a strong indication that fasciotomy should be recommended. In a hypotensive patient, of course, the acceptable pressure is lower. As a general rule, when in doubt, the compartment should be released. If it proves later to have been unnecessary only a scar is the result. However, if a fasciotomy should have been done but was not, loss of muscle tissue and worse may result.

MANAGEMENT

TECHNIQUE OF FASCIOTOMY AND ARTERIAL EXPLORATION FOR IMPENDING VOLKMANN'S CONTRACTURE AND COMPARTMENT SYNDROME. Make an anterior curvilinear incision medial to the biceps tendon. The incision may cross the crease directly or at an angle. Incise the deep fascia and possibly the lacertus fibrosa. Evacuate any hematoma. Expose the brachial artery and determine whether there is a free blood flow. If there is not a satisfactory flow, the adventitia may be removed to expose an underlying clot or spasm or intimal tear. The adventitia may have to be resected, and the artery reanastomosed. If, however, the artery appears to be normal, continue by opening the fascia of the forearm by extending the incision and then inserting a slightly opened scissors to be certain that the fascia is completely free over the forearm muscles. Should the muscles appear gray, it is of prognostic significance. However, with release within the first several hours and establishment of blood flow, this should not be a permanent condition. Observe the median nerve to be certain that it is not severed by a fracture. This may be repaired if operative conditions permit. In a supracondylar fracture, reduce the fracture and pin it in place with Kirschner wires and control the bleeding. The skin need not be closed at this time, but anticipate secondary closure later. The elbow should not be left flexed beyond 90 degrees. In the postoperative period the elbow and hand should be elevated. In some instances a dorsal incision may be required in addition to the volar incision. Usually the volar incision releases the fascia sufficiently. Dorsal compartment pressure measurement may be helpful in making the decision.

Established Volkmann's contracture of forearm

Volkmann's ischemic contracture is the result of several different degrees of tissue injury; however, the earliest changes usually involve the flexor digitorum profundus muscles in the middle third of the forearm. Tsuge has classified *established Volkmann's contracture* into three types for purposes of treatment.

MILD CONTRACTURE

During the early stages of a mild contracture, dynamic splinting to prevent wrist contracture, physical training, and active use of the muscles may be helpful. In this contracture single muscle groups, especially the wrist flexors, may be contracted but not paralyzed. The pronator teres and thumb flexors may also be included. After 3 months, the involved muscle-tendon units may be released and lengthened. However, when multiple tendon units are involved, the muscle sliding operation is better than lengthening of multiple tendons, wrist resection, or other possible procedures.

MODERATE CONTRACTURE

A moderate contracture usually involves not only the long finger flexors, but also the flexor pollicis longus and possibly the wrist flexors. In this instance the muscle sliding operation, a careful neurolysis of the median and ulnar nerves without injuring their branches, and the excision of any fibrotic muscle mass encountered may be done. When no useful amplitude of movement of the finger flexors has been retained, volar transfers of such dorsal wrist extensors as the brachioradialis and extensor carpi radialis longus and a complete release of the wrist and finger flexors may be required.

SEVERE CONTRACTURE

A severe contracture involves both the flexors and extensors of the forearm. Fractures of the forearm bones and scars about the skin also may be present. Sensory feedback is usually impaired because the nerves are strangulated by the contracted and scarred muscles surrounding them.

TECHNIQUE. Make an extensive volar forearm incision (Fig. 13-4) and excise all avascular masses of the flexor profundus and sublimus muscles leaving any muscle that might survive or appears viable. Perform a neurolysis of the median and ulnar nerves. Then correct the finger and wrist flexion deformities by releasing the musculotendinous units at their origins. At this time at least the func-

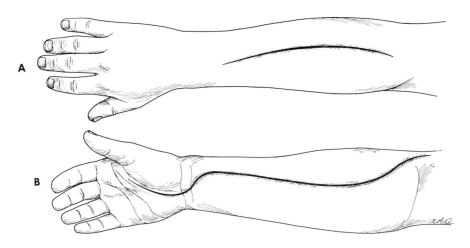

Fig. 13-4. Incisions used in forearm in severe Volkmann's contracture. **A,** Extensive opening of fascia of dorsum of forearm in dorsal compartment syndromes. **B,** Incision used for anterior forearm compartment syndromes in which skin and underlying fascia are released completely throughout. (After Gelberman, R.H., et al.: Clin. Orthop. **134:**225, 1978.)

tional position of the hand will have been restored. At a second-stage procedure, any viable extensor muscles may be transferred to the finger flexors. Remember, however, that at least one wrist extensor must be retained. Otherwise, any wrist flexor or extensor may be transferred to power the profundus and flexor pollicis longus tendons. Sometimes a free muscle transplant from the lower extremity or from the pectoralis major may be used. This, of course, is a salvage procedure that may result in only modest improvement. If the contracture is diffuse but incomplete throughout all the digital and wrist flexors, the muscle sliding technique may be considered.

MUSCLE SLIDING OPERATION OF FLEXORS FOR ESTABLISHED VOLKMANN'S CONTRACTURE

This procedure was first described by Page in 1923 and was endorsed by Scaglietti in 1957. It has been used for Volkmann's and other contractures caused by conditions such as brain damage and burns.

TECHNIQUE. Apply a pneumatic tourniquet high on the upper arm. Make an incision beginning proximal to the elbow on the medial side and extending distally along the ulnar side of the forearm and ending at the flexor crease of the wrist where it may be carried laterally over the palmaris longus tendon (see Fig. 13-4); we prefer a straight incision. Spare all subcutaneous veins if possible and any intact cutaneous nerves such as the medial cutaneous nerve of the forearm. Beginning at the cubital fossa, incise the fascia over the musculature. Locate the ulnar nerve behind the medial epicondyle and the median nerve and brachial artery and vein as the deep dissection is carried across the elbow. Free the ulnar nerve at this level and dissect distally across the forearm, carrying the proximal origin of the flexor carpi ulnaris. Protect the median nerve as dissection is carried further distally. Release the proximal origin of the pronator teres, and using a scalpel or sharp elevator, subperiosteally release the origin of the profundus finger flexors. Then release the distal origin of the pronator teres and the origins of the palmaris longus and flexor carpi radialis. Next free the origin of the flexor digitorum sublimis. This should expose the elbow joint capsule. Release now the most distal origin of the flexor carpi ulnaris, again protecting the ulnar nerve. The interosseous membrane should now be visible. Repeated passive flexion of the fingers will help determine what further parts of the muscle origins should be detached. Mature fibrous masses within the muscles may require complete excision. Avoid injuring the intact interosseous artery, vein, and nerve. Most of the blood supply enters the muscle mass at the junction of its upper and middle fourths. Individual branches of the vessels need not be dissected, only the origins of the muscles themselves. After releasing the muscle origins, including the flexor carpi ulnaris down to the wrist level, displacing the muscle mass distally 3 cm should be possible. Now transpose the ulnar nerve to the anterior side of the medial epicondyle. Although rare, it may be necessary also to lengthen one or two tendons at the wrist, usually those of the middle or ring finger profundus. Before closure, release the tourniquet and compress the muscle masses with sponges for 3 to 5 minutes. Then control any active bleeding by cautery, taking care not to damage the nerves. Reinflate the tourniquet and

close the skin. After final release of the tourniquet, be sure that the color of the fingers returns to normal. Then wrap the arm with a bulky pad, and with the elbow at 90 degrees, the fingers in full extension, and the forearm in mild supination, apply a cast.

AFTERTREATMENT. The cast is worn for 4 weeks, and a night splint is used for several additional months. Active finger motion is begun when the cast is removed.

ACUTE ISCHEMIC CONTRACTURE OF INTRINSIC MUSCULATURE OF HAND

Acute ischemia of the intrinsic muscles of the hand, like the larger muscles in the forearm, may result in contracture or necrosis of the muscle bellies. It may occur after compression injuries of the hand without fracture. The hand is swollen and tense, and the fingers are held almost rigidly in a partially flexed position with the wrist in neutral. Any passive movement of the fingers causing metacarpophalangeal joint extension usually causes considerable pain. The thenar muscles rarely are involved. The hand should be decompressed immediately.

TECHNIQUE OF RELEASE OF ACUTE ISCHEMIC CONTRACTURE OF INTRINSIC MUSCULATURE OF HAND. Make three dorsal parallel incisions through the skin beginning at the level of the metacarpophalangeal joints and extending to just distal to the wrist. Make the first radial to the second metacarpal to allow access to the first web. Make each incision down to the musculofascial area. Through each incise the fascia and release the compression of the distended muscles by allowing them to extrude into the wound if necessary. Each muscle should be identified individually to make certain that a complete release is carried out. Then passively flex the metacarpophalangeal joints and extend the proximal interphalangeal joints to stretch the muscles, being sure that all are adequately released. Do not attempt to close the wounds at this time. They may be permitted to granulate and heal, or after the swelling has decreased, they may be closed secondarily. This procedure may be done in conjunction with other more proximal releases.

Established intrinsic muscle contractures of hand

The proper surgical release of established intrinsic muscle contractures depends on the severity of the contractures. When the contractures are *mild* (Fig. 13-5)—the metacarpophalangeal joints can be passively extended completely but while they are held extended the proximal interphalangeal joints cannot be flexed (positive intrinsic tightness test)—then the distal intrinsic release of Littler may be indicated.

In contractures that are *more severe* the interosseus muscles are viable but contracted, and the intrinsic tightness test is positive. Active spreading of the fingers may be possible. In these instances the contracted muscles may be released from the metacarpal shafts by a muscle sliding operation (Fig. 13-6).

In contractures that are *most severe* the intrinsic muscles may be not only contracted but also necrotic and fibrosed so that any useful muscle excursion is absent. In these instances the tendon of each muscle must be divided to release the contractures. Other procedures such as capsulotomies and tendon transfers may also be necessary.

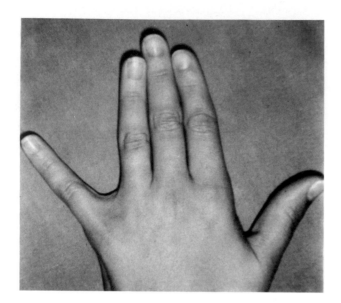

Fig. 13-5. Abduction contracture of fifth finger in patient who developed fibrosis in abductor digiti quinti, probably secondary to ischemic myositis from compressive bandage.

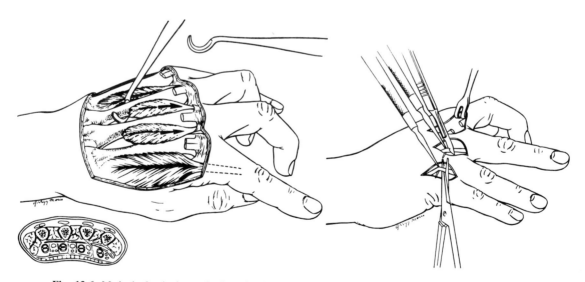

Fig. 13-6. Method of stripping and advancing interosseus muscles to slacken them, thus allowing proximal finger joints to extend and distal two to flex. Interosseus muscles of two clefts have been stripped. *Inset,* shows a cross section through middle of hand. Stripping of interossei is done only when muscles still retain considerable function. Nerve supply should be spared. (From Bunnell, S.: J. Bone Joint Surg. **35-A**:88, 1953.)

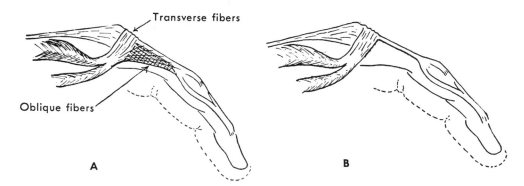

Fig. 13-7. Littler release of intrinsic contracture. **A,** Extensor aponeurosis at level of metacarpophalangeal joint consists of long extensor tendon, transverse fibers (which flex the metacarpophalangeal joint), and oblique fibers (which extend interphalangeal joints). *Crosshatched* part is resected from each side of hood. **B,** Appearance of aponeurosis after release. (From Harris, C., Jr., and Riordan, D.C.: J. Bone Joint Surg. **36-A:**10, 1954.)

TECHNIQUE (LITTLER). The same procedure is carried out on any finger as needed. Make a single midline incision on the dorsum of the proximal phalanx extending from the metacarpophalangeal joint to the proximal interphalangeal joint; thus good exposure of both sides of the extensor aponeurosis is possible. Incise the insertion of the oblique fibers of the extensor aponeurosis into the extensor tendon; make the incision parallel with the tendon (Fig. 13-7, *A*). Preserve the transverse fibers to avoid hyperextension of the metacarpophalangeal joint with its resultant clawhand deformity and limitation of extension of the interphalangeal joints. Now test the adequacy of the operation before excising any of the hood; if the dissection has not been carried far enough proximally, the interphalangeal joints cannot be fully flexed while the metacarpophalangeal joint is extended; on the other hand, if the dissection has been carried too far proximally, the metacarpophalangeal joint can be hyperextended while the interphalangeal joints are fully flexed, and a part of the aponeurosis should be sutured back to the long extensor tendon. After dissection has been extended more proximally or part of the hood has been sutured, if either is indicated, excise that part of the hood that remains as a flap. After the correct amount of hood has been resected, the interphalangeal joints can be fully flexed, and the metacarpophalangeal joint cannot be hyperextended (Fig. 13-7, *B*). Close the incision with a running suture of fine stainless steel wire. Apply a volar plaster splint from the elbow to the middle of the proximal phalanges, immobilizing the metacarpophalangeal joints in extension and permitting full motion of the interphalangeal joints.

AFTERTREATMENT. Active motion of the interphalangeal joints is begun the day after surgery, and the splint and sutures are removed at 10 to 14 days.

TECHNIQUE OF RELEASE OF SEVERE INTRINSIC CONTRACTURES WITH MUSCLE FIBROSIS (SMITH). Make a dorsal transverse incision just proximal to the metacarpophalangeal joints. Resect the lateral tendons of all the interossei and the abductor digiti quinti at the level of the metacarpophalangeal joints. If these joints remain flexed, retract the sagittal bands distally and divide each accessory collateral ligament at its insertion into the volar plate. Then free the volar plate from its attachments to the base of the proximal phalanx. With a blunt probe, separate any adhesions between the volar plate and the metacarpal head. If maintaining extension of the proximal phalanx is difficult after soft tissue release, insert a Kirschner wire obliquely through the metacarpophalangeal joint with the joint in maximum extension. When the phalanx is extended be certain that its base articulates properly with the metacarpal head before inserting the wire. If passive flexion of the proximal interphalangeal joints is incomplete with the metacarpophalangeal joints extended, resect the lateral bands at the distal half of the proximal phalanges through separate dorsal incisions.

AFTERTREATMENT. Passive and active flexion exercises of the proximal interphalangeal joints are begun within 1 day of surgery. The Kirschner wires are removed at about 3 weeks.

ADDUCTED THUMB

Only complete loss of the thumb causes more disability in the hand than a fixed severe adduction of the thumb (web contracture). The thumb is the only digit with the ability to bring its terminal sensory pad over the entire surface of any chosen finger or over the distal palmar eminence. The saddlelike first carpometacarpal joint provides the circumductive movement of the thumb necessary for pinch or grasp (Fig. 13-8). The intrinsic muscles of the thumb and the extrinsic flexors and extensors are all important in the balanced control required in performing these functions effectively: the short abductor muscle positions and stabilizes the thumb metacarpal for pinch; the adductor muscle supplies power for pinch by acting on the proximal phalanx; the long extrinsic flexor positions the distal phalanx in varying degrees of flexion and consequently controls the type of pinch, whether it be fingernail-to-fingernail opposition or pulp-to-pulp opposition with another digit. The thumb web must be supple if these important movements of the thumb are to be possible. Any contracture of the thumb web causes limited opposition of varying degrees. In severe contracture the thumb is in a position of adduction and external rotation.

The thumb web consists of skin, subcutaneous tissue,

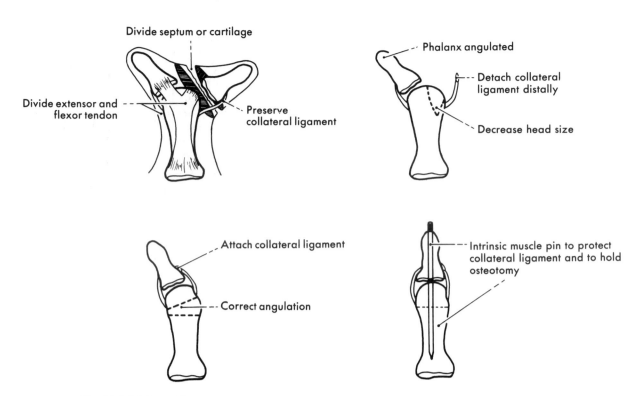

Fig. 14-4. Marks and Bayne technique for asymetric duplication of distal phalanx of thumb. Advantage of this technique is that origin of collateral ligament of distal joint is maintained while angular deformity is corrected. (From Marks, T.W., and Bayne, L.G.: J. Hand Surg. **2:**107, 1978.)

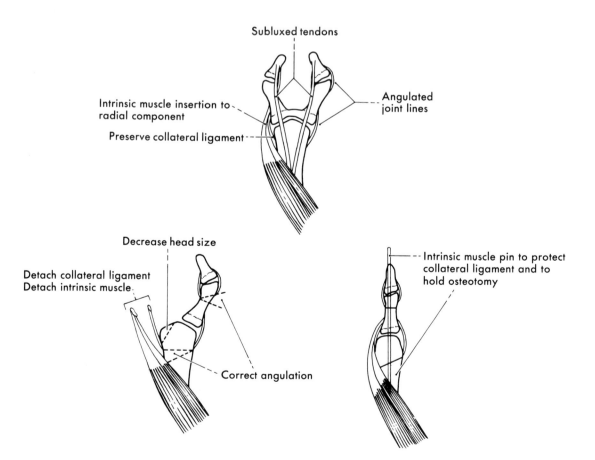

Fig. 14-5. Marks and Bayne technique for correction of duplication of thumb at proximal phalangeal level. In this instance two phalanges of thumb are separate; ulnarmost phalanges are maintained and radial phalanges are amputated. Detached collatearl ligament and intrinsic muscles are used for reattachment. Corrective osteotomies of remaining phalanges and thumb metacarpal will better align deformity. (From Marks, T.W., and Bayne, L.G.: Hand Surg. **2:**107, 1978.)

there is soft tissue separation. The collateral ligament is not involved.

When indicated, ablation of a minor-sized phalanx that includes the collateral ligament may be carried out even though it may result in an unstable distal joint; the instability may be later corrected by arthrodesis of the distal joint when epiphyseal growth is almost complete at about the age of 12 or 13 years. The ablation procedure probably should be carried out in the first year or two of life, but a functional evaluation of the part is necessary to be sure that the digit retains function. It is frequently necessary after ablation to realign and position the bifid extensor tendon and the duplicated flexor pollicis longus tendon. Surgical techniques have been devised by Marks and Bayne (Figs. 14-4 to 14-6). When removing the accessory digit, avoid a straight scar because growth of the part will continue but scars do not grow or stretch; therefore the part distal to the scar may be pulled into malalignment. Any straight scars should eventually be treated by Z-plasty. For best results a combination of operative procedures may be

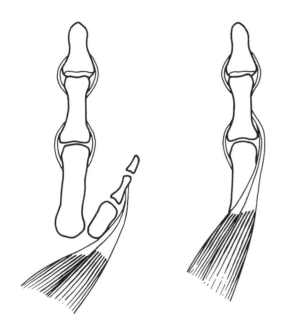

Fig. 14-6. Marks and Bayne technique for duplication of thumb at metacarpal level. Collateral ligament repair is not basic problem at this level. Power is maintained by transference of detached intrinsic when smaller digit is removed. (From Marks, T.W., and Bayne, L.G.: J. Hand Surg. **2:**107, 1978.)

necessary, such as thumb web deepening, collateral ligament reconstruction, joint arthrodesis, and tendon reconstruction.

SYNDACTYLY

Syndactyly occurs most often between the middle and ring fingers and is three times more frequent in males than females. Sometimes only the skin may be joined; when more severe, the fingernails, bones, and even nerves and tendons may be common to both fingers. One finger may be underdeveloped. If fingers are bound tightly together at their tips, extending them fully may be impossible.

When the deformity is severe, possibly involving all fingers of both hands as occurs in Apert's syndrome, surgery may be indicated as early as age 18 months; but when development of the fingers is not being retarded by the anomaly, surgery probably should be delayed until the age of 5 years. In syndactyly the skin is always deficient, and enough additional skin must always be supplied. Consequently before separating any fingers, plans must be made to cover the resulting skin defects by local flaps and full-thickness grafts because otherwise, regardless of how the incisions are made, coverage will be incomplete. An effort should be made to cover the defect on the most important finger by flaps alone because flaps often grow faster than do free grafts and a combination of grafts should be avoided. For example, when the middle and ring fingers are webbed, the middle finger is the most important one and it should be given the advantage of unhampered growth; the defect on this finger is covered by flaps from the palmar and dorsal surfaces of the ring finger, and the defect on the ring finger is covered by a free full-thickness graft. The incisions for flaps and grafts should be planned so that the resulting scars will not impair motion or cause secondary deformity during growth. The new web should be constructed so that its migration distally or the development of a transverse tension scar is unlikely; its natural slope should be established with its free edge at the palm. We prefer to construct it with a single flap in a child less than 3 or 4 years old, but a double flap as recommended by Skoog may be preferable in a child of the ideal age of 5 years. Not considering age as a factor, Flatt found no appreciable difference between using one flap or two. However, he did find that operations performed before the age of 18 months are less satisfactory than those performed later.

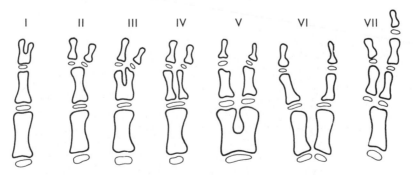

Fig. 14-7. Wassel's classification of thumb polydactyly. (From Wood, V.: J. Hand Surg. **3:**436, 1978.)

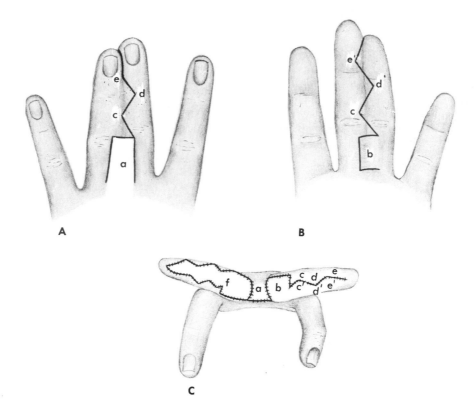

Fig. 14-8. Technique for repairing syndactyly. **A,** Incisions to be made on dorsum of fingers, *a,* Proximally based flap that will be new web between fingers, *c* to *e,* Points on medially based flap that will interdigitate with points on flaps raised from volar surface. **B,** Incisions to be made on volar surface of fingers. *b,* Medially based flap that will cover raw surface on proximal part of ring finger. *c* to *e,* Points on medially based flap that will interdigitate with those on flap raised from dorsal surface. **C,** Fingers have been separated and all flaps are sutured in place. Full-thickness free graft, *f,* has been applied to raw surface on middle finger. Flaps cover radial side of ring finger and not, as recommended in text, ulnar side of middle finger. (Redrawn from Bauer, T.B., Tondra, J.M., and Trusler, H.M.: Plast. Reconstr. Surg. **17:**385, 1956.)

TECHNIQUE (BAUER, TONDRA, AND TRUSLER). Outline all incisions carefully with methylene blue on a toothpick or sharpened applicator (Fig. 14-8). First cut a single rectangular skin flap from the dorsal surface of the two webbed fingers for the new web between these fingers. Base it proximally at the level of an adjacent normal web; make it wide enough to form a normal web and long enough to reach the palmar edge of the new web. Now separate the fingers with longitudinal dorsal and palmar incisions on the ring finger along the ulnar side of the abnormal web (toward the radial side of the ring finger); locate and curve these incisions so that they outline the flaps that will cover the ulnar side of the middle finger. Take care not to injure the digital nerve of the ring finger. Raise these flaps and remove from them any excess of subcutaneous fat to give the middle finger a more normal shape. Suture them together on the ulnar side of the middle finger; then suture the web flap in place. Cover the defect on the ring finger with a full-thickness free graft taken from the groin. Should the fingernails be confluent, they should be separated, and the matrix at their margins should be removed so that the grafts can be brought around to the edge of each nail. When a finger is webbed on both sides, it is safer to separate only one side at a time.

Place Xeroform gauze over the grafts and then carefully insert a wet contour dressing between the fingers; begin at the web space and pack distally so that the fingers are held in wide abduction and extension. Then apply a dry dressing and a plaster of Paris splint to immobilize the fingers and wrist. The hand is elevated for a week or more before redressing.

TECHNIQUE (SKOOG). With a skin pencil outline the incisions to be made on the fingers (Fig. 14-6, *A*). Design dorsal and volar flaps so that when mobilized they will cover most of the denuded side of one finger without tension (Fig. 14-9, *D* and *F*). Make the free borders of the flaps irregular by designing small triangular points at the level of the interphalangeal joints. In planning the flaps so that they fit each other, first outline the incision on one side and then establish the key points for the incision on the opposite side by pushing straight needles vertically through the web. Then elevate the tourniquet and raise the flaps consisting mainly of the skin that forms the abnormal web. In raising the flaps preserve all subcutaneous tissue and be careful not to sever digital nerves and arteries. Then release the tourniquet and control all bleeding. On one finger close the flaps as planned and cover the small

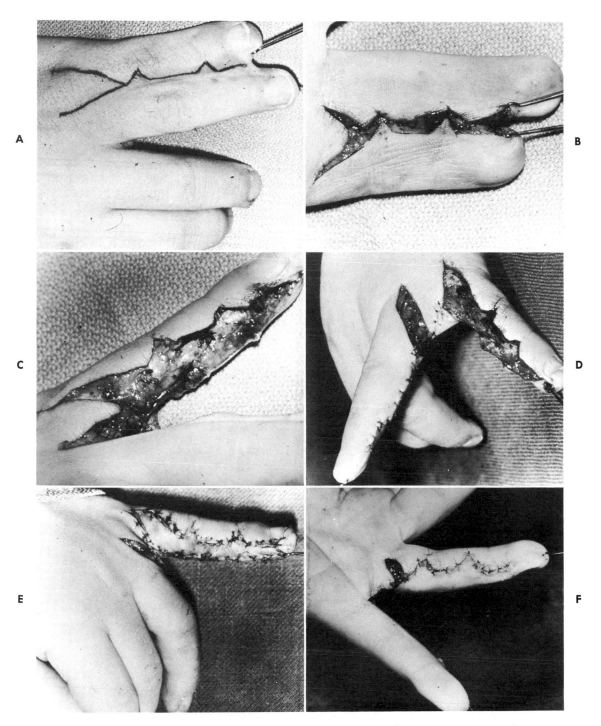

Fig. 14-9. Skoog technique for syndactyly. **A,** Proposed skin incision is outlined on dorsal surface of fingers. **B,** Incision is made on volar surface of fingers. **C,** Fingers are separated; little finger will require skin graft. **D,** Ring finger and web are covered by flaps except for small area on dorsomedial aspect of base of finger. **E,** Full-thickness skin graft is sutured in place on little finger. **F,** Ring finger is almost covered by flaps.

Continued.

APERT'S SYNDROME (ACROCEPHALOSYNDACTYLY)

The congenital deformity of the hands in Apert's syndrome is accompanied by a typical grotesque facial appearance. It is commonly associated with mental retardation, visceral anomalies, and anomalies of the lower extremities similar to those of the hand.

There is a complex syndactyly of all the fingers with a 30% incidence of webbing of the thumb with the index finger. There is bony coalition more consistently of the middle and ring fingers but sometimes of all four fingers. This produces fingers held tightly and evenly together at their distal ends, creating a funnel appearance of the hand. The soft tissue syndactyly involves all four fingers consistently. In addition, the fingers are shorter than normal, frequently fused at the distal joint, and at times fused at the proximal interphalangeal joint. The thumb is short and radially deviated at the metacarpophalangeal joint. The condition is always bilateral.

Since there is bony coalition of multiple digits, there is a perpetuation of the deformity with growth; therefore reconstructive surgery should be done early, about the age of 6 months and before the age of 1 year. First the lateral digits should be released by the usual method of flaps and skin grafting when necessary. Six months following the first procedure, the middle finger should be amputated at the metacarpophalangeal joint to leave a three-fingered hand and thumb and provide sufficient soft tissue coverage for the remaining ring finger and index finger. Attempts to salvage all four digits have usually resulted in incomplete coverage and gross deformities. The fingers usually function with hingelike metacarpophalangeal joints. Therefore usually no more than two or three operations on each hand are necessary. Both hands may be operated on at the same sitting in older children.

MACRODACTYLY

Macrodactyly is a rare congenital anomaly characterized by enlargement of all structures, especially the nerves, of one or more digits. Barsky describes two types, both present at birth: in one the enlargement does not increase disproportionately during growth and in the other it does. Often the enlargement involves not only a digit or digits but extends proximally into the palm and sometimes even into the forearm; at surgery the hypertrophic soft tissues are found to consist of neural elements and fat. According to McCarroll, hypertrophy of the soft tissues may be caused by an enlargement of a peripheral nerve, and also according to him, excessive growth of a finger ceases when the digital nerve is excised. With McCarroll's findings in mind Tsuge splits the digital nerve and excises part of its trunk and some of its branches in an effort to retard growth without eliminating all sensibility. He carries out this procedure on one side of a finger and, after about 3 months on the other side if indicated; operating on only one side at a time avoids the risk of circulatory disturbance.

When the finger is deviated as well as enlarged, osteotomy of a phalanx may be necessary (Figs. 14-11 and 14-12). The Barsky technique for shortening a finger is useful (Fig. 14-13); Tsuge suggests using this same technique at a more proximal level. Shortening is usually indicated in older children and adults, but in younger children arrest of appropriate epiphyses instead is often helpful. At the time of either shortening or epiphyseal arrest excision of fat and neural elements make the finger smaller. An enlarged fin-

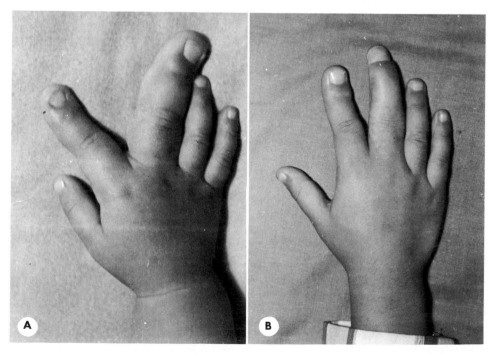

Fig. 14-11. A, Macrodactyly of index and middle fingers in 1-year-old boy. **B,** After two operations to reduce size of fingers and osteotomy of middle phalanx of middle finger to correct deviation. (From Tsuge, K.: Plast. Reconstr. Surg. **39**:590, 1967.)

gernail can be made smaller by the method of Tsuge (Fig. 14-14). Amputation is indicated only as a last resort but sometimes is the only practical treatment; unfortunately macrodactyly is often seen only after the enlargement has become so extreme that epiphyseal arrest or the other procedures just mentioned are useless.

CONSTRICTING ANNULAR GROOVES

Shallow annular grooves require no surgical treatment, but when they are deep enough to impair circulation, they should be excised down to normal tissue, and the skin should then be closed by multiple Z-plasties. Should a groove completely encircle a finger, excision of only half

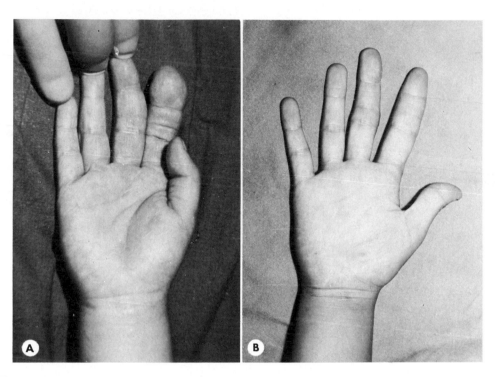

Fig. 14-12. Macrodactyly of index and middle fingers in a 3-year-old girl. **B,** Status 4 years after two operations to reduce size of fingers and osteotomy of middle finger to correct deviation. (From Tsuge, K.: Plast. Reconstr. Surg. **39**:590, 1967.)

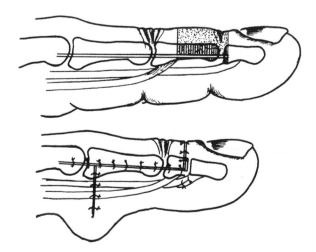

Fig. 14-13. Barsky technique for shortening finger. Appropriate amount of bone and dorsal skin *(shaded area)* is excised. Distal phalanx is carried proximally and is fixed to remaining part of middle phalanx. Profundus tendon is shortened. Excess skin on palmar surface is excised secondarily. (From Tsuge, K.: Plast. Reconstr. Surg. **39**:590, 1967.)

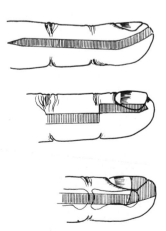

Fig. 14-14. Tsuge techniques for reducing size of finger and nail. Shaded areas are excised as indicated (From Tsuge, K.: Plast. Reconstr. Surg. **39**:590, 1967.)

of the groove at one time is safer; otherwise the circulation may be further impaired. When the finger distal to the annular groove is flail and without sensation, it may be amputated.

Constricting annular grooves may be associated with congenital shortening of the fingers and syndactylism (Fig. 14-15). An alternate method of excising the congenital grooves is a combination of a Y-V-plasty and W-plasty (Figs. 14-16 and 14-17).

KIRNER'S DEFORMITY

This entity, originally described by Kirner in 1927, consists of a spontaneous incurving of the terminal phalanx of the fifth digit (Fig. 14-18). It occurs predominantly in females from several months to 12 years of age. It is an extremely rare disorder with a 30% incidence of repeated familial occurrence. The deformity may be part of other specific syndromes too numerous to mention and must be differentiated from radial deviation or true clinodactyly. Clinodactyly frequently is familial and is characterized by

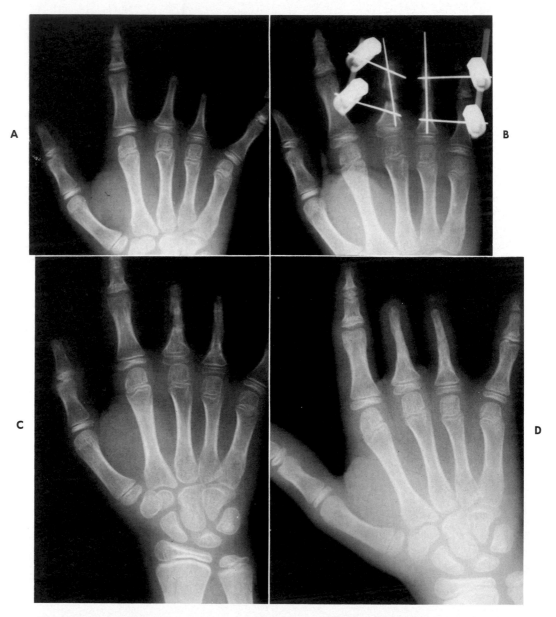

Fig. 14-15. A, Congenitally shortened central two digits in 11-year-old boy caused him considerable anxiety. **B,** Fingers were lengthened by one-state procedure using autogenous bone grafts from ilium and external fixation. **C,** Healing of bone grafts was prompt. **D,** Increase in length of 6 mm in one digit and 9 mm in other resulted in more pleasing and acceptable hand but little difference in function.

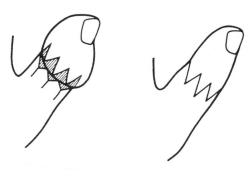

Fig. 14-16. Combination of Y-V-plasty and W-plasty. (From Miura, T.: J. Hand Surg. **9-A:**82, 1984.)

A   B   C

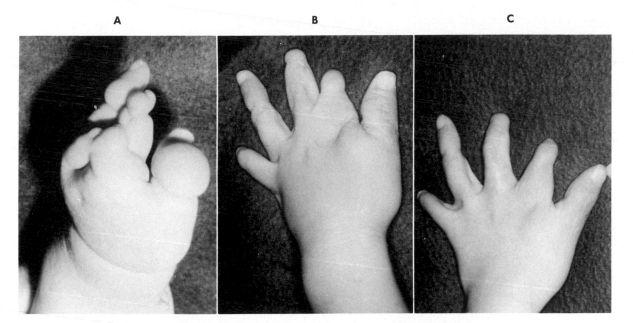

Fig. 14-17. A, Preoperative condition. **B,** Fenestrated syndactyly between index and small fingers was separated, and constricting rings were treated by combined Y-V-plasty and W-plasty. **C,** Secondary web formation between index and long fingers was separated at another operation to make a deep interdigital space. Condition after second operation. (From Miura, T.: J. Hand Surg. **9-A:**82, 1984.)

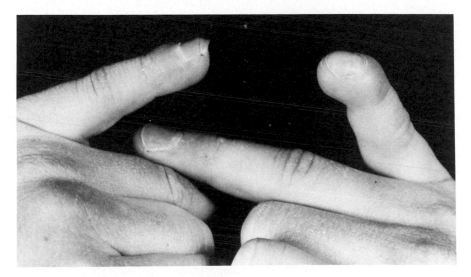

Fig. 14-18. Kirner's deformity (see text). (From Carstam, N., and Eiken, O.: J. Bone Joint Surg. **52-A:**1663, 1970.)

normal bone structure and no periarticular swelling. Therefore roentgenograms are important in differentiating the two conditions. Corrective osteotomy may yield satisfactory results cosmetically.

CAMPTODACTYLY

Camptodactyly is a flexion deformity of the proximal interphalangeal joint, usually of the little finger alone (Fig. 14-19), but sometimes of adjoining fingers too. It is apparently hereditary and is painless and gradually progressive. In children the deformity usually disappears when the wrist is flexed. It increases so gradually that functional impairment may not be noticed until suddenly the dexterity necessary for typing or playing a musical instrument is found lacking. Later in life, especially after adolescence when it increases more rapidly, the deformity becomes fixed by contracture of the skin and ligaments.

In children when the proximal interphalangeal joint can be extended while the wrist is flexed, releasing the flexor digitorum sublimis may correct the deformity or prevent its increase during growth. However, this tendon may be absent or vestigial, or may arise from the tendons of the ring finger. In older patients when the deformity is so severe that correction is indicated, extension at the joint may be increased by lengthening the appropriate flexor digitorum sublimis tendon in the palm or wrist; however, when the skin and ligaments are contracted and the extensor hood, especially the central tendon, is underdeveloped, this procedure alone is ineffective.

CONGENITAL FLEXION AND ADDUCTION CONTRACTURE OF THUMB

Congenital flexion and adduction contracture of the thumb, according to Weckesser, is a syndrome rather than an entity. It is sometimes seen in mild arthrogryposis, but usually it is an isolated anomaly apparently inherited by means of a sex-linked recessive gene. It is more frequent in males and is usually bilateral. At birth it can easily be corrected passively but later it becomes fixed by contractures of skin, joint capsules, and possibly muscles (Fig. 14-20). In most instances all power of extension at the metacarpophalangeal joint is absent because of partial or complete anatomic and functional absence of the extensor pollicis brevis; sometimes the extensor pollicis longus is also absent. The head of the proximal phalanx may be hypoplastic. When the deformity is severe, the thenar mus-

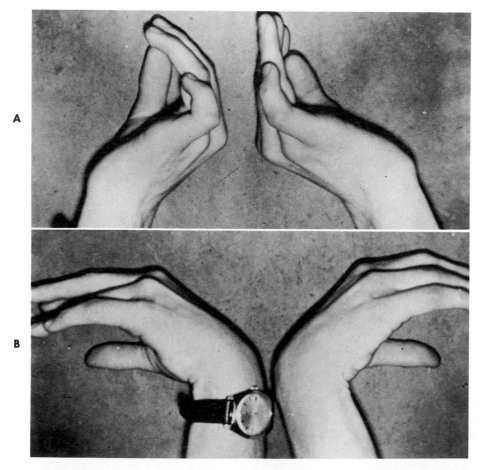

Fig. 14-19. Bilateral camptodactyly of little finger in 18-year-old man. **A,** Flexion contractures of proximal interphalangeal joints were increased when wrists were extended. **B,** Contractures almost disappeared when wrists were flexed. (From Smith, R.J., and Kaplan, E.B.: J. Bone Joint Surg. **50-A:**1187, 1968.)

cles are fibrotic and shortened and the thumb is hypoplastic; when even more severe, other digits are similarly involved. This deformity should not be confused with congenital trigger thumb (p. 309).

The treatment of congenital flexion and adduction contracture of the thumb in the newborn consists of splinting and manual stretching for several months after which some power of extension at the metacarpophalangeal joint may develop. In children after the age of 6 months a cast should be applied to prevent fixation of the contracture; after a reasonable time if power of extension fails to develop, then the extensor indicis proprius tendon may be transferred to the dorsum of the thumb and any abnormalities that might cause a fixed contracture are corrected. In older patients with fixed deformities, Z-plasty or skin grafting of the thumb web (p. 18), release of the adductor muscles (p. 264), release of the joint capsules, and fusion of the metacarpophalangeal joint in the extended position may be necessary. When the deformity is severe, several fingers may be similarly affected and in these instances the extensor indicis proprius is not available for transfer to the thumb. For this situation Littler devised an operation in which the flexor digitorum sublimis tendons of the ring and little fingers are transferred to the extensor hoods at the bases of the proximal phalanges to extend the digits.

TECHNIQUE (LITTLER). Through midlateral incisions (p. 10) on the ring and little fingers free the insertion of the flexor digitorum sublimis tendon of each finger. Then draw the freed tendons out through a short volar transverse incision proximal to the wrist (Fig. 14-21, *A*). Next split the sublimis tendon of the ring finger into three or four slips as necessary and reroute them subcutaneously deep to the dorsal carpal ligament to the bases of the affected fingers. This tendon and its slips may be rerouted around either the radial or ulnar side of the wrist. Then through short transverse incisions over the bases of the proximal phalanges of the fingers interweave each slip into the central part of the extensor hood and anchor it with interrupted sutures (Fig. 14-21, *B*). Then reroute the free sublimis tendon of the little finger subcutaneously around the radial side of the wrist deep to the abductor pollicis longus and any thumb extensor tendons and deep also to a pulley created in the aponeurotic tissue over the distal end of the first metacarpal (Fig. 14-21, *A*). While the thumb is held in full extension, anchor the tendon to the extensor hood at the base of the proximal phalanx as just described for the fingers. Close the wound and apply a plaster splint holding the wrist extended and the thumb abducted and extended.

AFTERTREATMENT. At 3 weeks the splint and sutures are removed. Then unlimited activities are allowed. Fig. 14-22 shows the results of the operation in one hand.

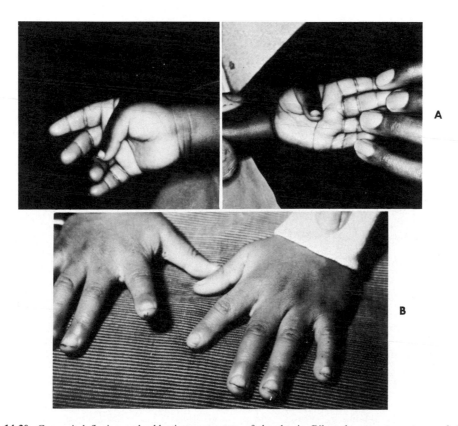

Fig. 14-20. Congenital flexion and adduction contracture of thumb. **A,** Bilateral contracture at age of 4 months. **B,** Same patient at age of 4 years. Treatment had consisted of immobilization of thumbs in extension and abduction for 3 months; extension of metacarpophalangeal joints is limited by 20 to 30 degrees. (From Weckesser, E.C., Reed, J.R., and Heiple, K.G.: J. Bone Joint Surg. **50-A:**1417, 1968.)

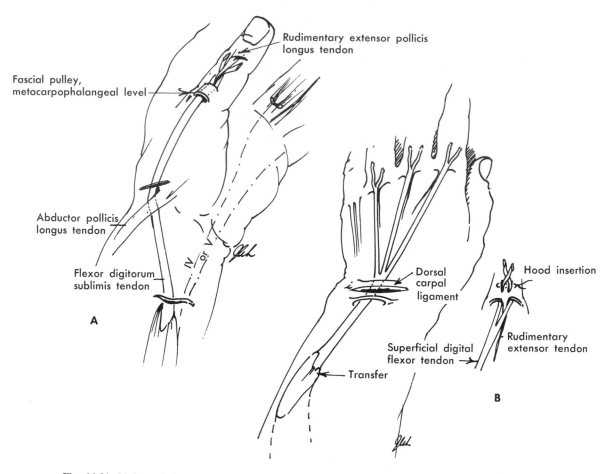

Fig. 14-21. Littler technique for correction of congenital flexion and adduction contracture of thumb (see text). (Modified from Crawford, H.H., Horton, C.E., and Adamson, J.E.: J. Bone Joint Surg. **48-A:**82, 1966.)

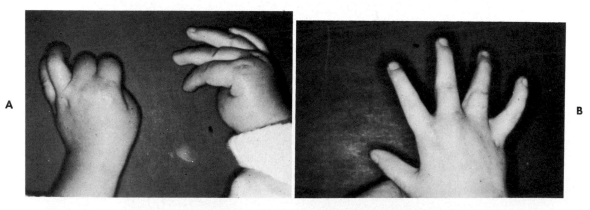

Fig. 14-22. Correction of congenital flexion and adduction contracture of thumb. **A,** Before surgery. Flexion contractures and ulnar deviation of fingers and inability to extend thumb. **B,** Right hand 9 months later after Littler technique. Fingers and thumb can be actively extended. (From Crawford, H.H., Horton, C.E., and Adamson, J.E.: J. Bone Joint Surg. **48-A:**82, 1966.)

CLEFT HAND

The cleft hand or lobster-claw hand is of two general types.

First type

In the first type a deep palmar cleft separates the two central metacarpals. One or more rays are usually absent, and the existing digits tend to be confluent and of unequal length (Fig. 14-23). The palmar cleft should be closed only if the procedure improves the grasp as well as the appearance of the hand.

TECHNIQUE. Fashion a rectangular skin flap on the medial or lateral side within the cleft, with its base distal and level with the web space of the existing fingers. This flap is to form the new web after closure of the cleft. Excise the remaining skin along the cleft and appose and suture the deep tissues. Then place the flap so that it bridges the gap near the bases of the adjoining fingers and suture it in place. Then suture the remaining dorsal and palmar skin edges. This operation forms a longitudinal scar that may hypertrophy and later require Z-plasty excision to reduce tension when the fingers are extended.

Fig. 14-24 shows the Barksy technique for this type of cleft hand. The skin on the dorsal and palmar surfaces is closed in a zigzag fashion.

Second type

In the second type of cleft hand the central rays are absent and only short radial and ulnar digits remain (Fig. 14-25). Appositional pinch between these two digits may be impossible; the phalanges may be short or absent in one or both, one or more of their joints may be stiff, or the digits may be improperly aligned. An attempt should be made to deepen the web and if necessary to shorten a digit or rotate one or both toward each other.

TECHNIQUE. It is usually safer to undertake correction in two stages. First the web is deepened by Z-plasty (p. 18), and any redundant bone segments or rudimentary digits are removed. Later a metacarpal may be shortened and one or both may be rotated to provide appositional pinch between the pulps of the digits.

REDUPLICATION OF ULNA (MIRROR HAND)

In this rare anomaly the ulna and the carpus are reduplicated, and there are seven or eight fingers but no thumb. Function may be improved by pollicizing a finger, creating a thumb web, and correcting the radial deviation of the hand (Fig. 14-26).

TECHNIQUE. Select one of the more radial digits for pollicization (Fig. 14-27). Amputate the next ulnarward digit through the base of the metacarpal, filleting it to obtain

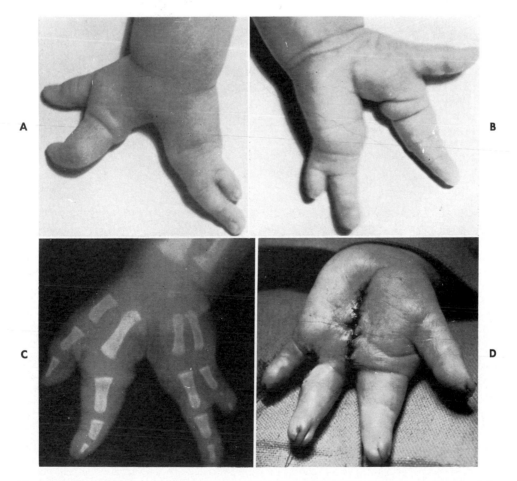

Fig. 14-23. A and **B,** Dorsal and volar views of cleft hand from which one ray is absent. **C,** Same hand. **D,** Volar surface of hand after cleft had been closed. (From Kelikian, H., and Doumanian, A.: J. Bone Joint Surg. **39-A:**1002, 1957.)

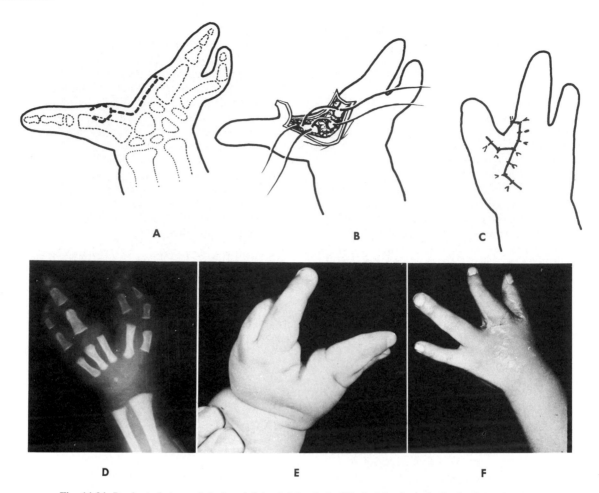

Fig. 14-24. Barsky technique of closing cleft in cleft hand. **A,** Skin incision is made *(broken line),* in which diamond-shaped flap is elevated from one side. **B,** Metacarpals are approximated by two heavy sutures passed through holes drilled in bones just proximal to their heads. **C,** Flap used to create new web and skin on dorsal and palmar surfaces is closed in zigzag fashion. **D** and **E,** Typical cleft hand. **F,** Same hand after resection of third metacarpal, closure of cleft, and separation of thumb and index finger and application of skin graft. (From Barsky, A.J.: J. Bone Joint Surg. **46-A:**1707, 1964.)

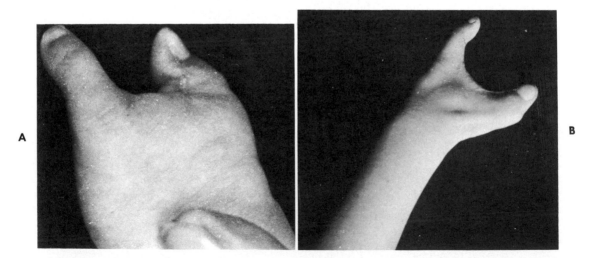

Fig. 14-25. Second or atypical type of cleft hand. **A,** Web between thumb and little finger impairs function. **B,** After web has been released by Z-plasty, function is better. (From Barsky, A.J.: J. Bone Joint Surg. **46-A:**1707, 1964.)

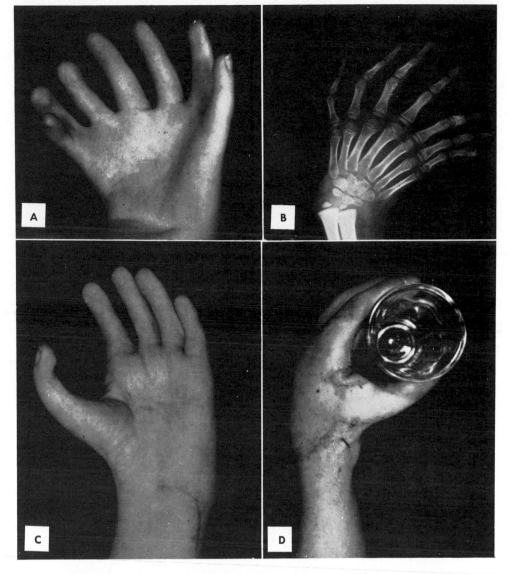

Fig. 14-26. Duplication of ulna (mirror hand). **A** and **B,** Before surgery. Middle digit corresponds to normal index finger; on each side of this digit is set of ulnar digits. **C** and **D,** Same hand 6 months after reconstruction by procedures illustrated in Fig. 14-27. (From Entin, M.A.: Surg. Clin. North Am. **40:**497, 1960.)

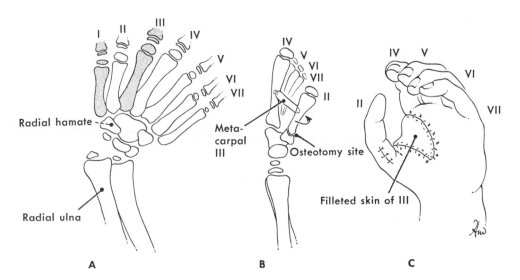

Fig. 14-27. Reconstructive procedures for duplication of ulna (mirror hand). **A,** Sketch of roentgenogram shown in Fig. 14-26. In reconstruction, digits I and III were discarded, Digit II was retained for pollicization. **B,** Sketch of pollicization of digit II. This digit was properly positioned by osteotomy through base of its metacarpal and by intermetacarpal graft cut from metacarpal of discarded digit III. **C,** Filleted skin of digit III was used as flap to cover space created by pollicization. (From Entin, M.A.: Surg. Clin. North Am. **40:**497, 1960.)

ample local skin for coverage of the pollicized ray. Amputation of one or two other digits at the periphery may also be necessary to arrive at one thumb and four fingers. Rotate the pollicized finger toward the others by osteotomy of its metacarpal; if necessary, stabilize it with a graft from the amputated metacarpal.

An arthrodesis of the wrist may be necessary at 12 years of age or later should radial deviation of the hand become severe.

CONGENITAL RADIOULNAR SYNOSTOSIS

Congenital radioulnar synostosis usually involves the proximal ends of the bones, fixing the forearm in pronation (Fig. 14-28). It is more often bilateral than unilateral. Often there is a familial predisposition, and the deformity seems to be transmitted on the paternal side of the family. Wilkie has noted two types. In the first the medullary ca-

nals of the radius and ulna are joined. The proximal end of the radius is malformed and is fused to the ulna for a distance of several centimeters. The radius is longer and larger than the ulna, and its shaft arches anteriorly more than normally. In the second type the radius is fairly normal, but its proximal end is dislocated either anteriorly or posteriorly and is fused to the proximal ulnar shaft; the fusion is neither as extensive nor as intimate as in the first type. Wilkie states that the second type is often unilateral and that sometimes other deformity, such as a supernumerary thumb, absence of the thumb, or syndactylism, is also present.

Congenital radioulnar synostosis is for several important reasons difficult to treat. The fascial tissues are short and their fibers are abnormally directed, the interosseous membrane is narrow, and the supinator muscles may be abnormal or absent. The anomalies in the forearm may be so

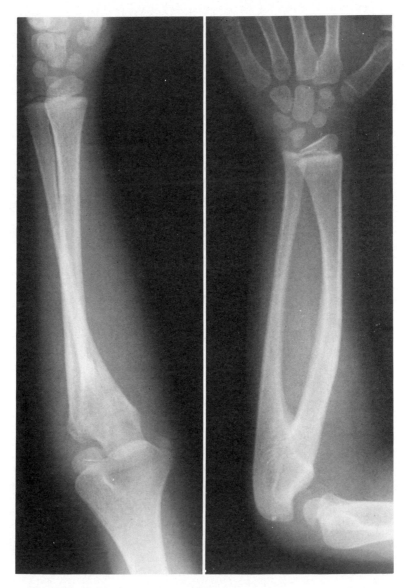

Fig. 14-28. Congenital radioulnar synostosis, involving proximal ends of bones.

widespread that sometimes no rotation is possible, even after the radius and ulna have been separated and the interosseous membrane has been split throughout its length. Simply excising the fused part of the radius never improves function. We think it inadvisable to perform any operation with the hope of obtaining pronation and supination. Fortunately most patients are not disabled enough to justify an extensive operation. Any disabling pronation deformity should be corrected by osteotomy; then motion of the shoulder, especially when the elbow is extended, compensates well for the deformity.

CONGENITAL DISLOCATION OF RADIAL HEAD

Congenital dislocation of the radial head is rare but should be suspected when the head has been dislocated for a long time but there is no evidence that the ulna has been fractured. The roentgenographic findings are fairly characteristic (Fig. 14-29). The radial shaft is abnormally long, and the ulna is usually abnormally bowed. The radial head is dislocated, usually anteriorly, is rounded, showing little if any depression for articulation with the capitulum, and is usually smaller than normal; occasionally there is an area of ossification in the tissues about it. The capitulum may also be small, and the radial notch of the ulna that should articulate with the radial head may be small or absent.

Congenital dislocation of the radial head may be familial, especially on the paternal side. It is sometimes associated with chondro-osteodystrophy.

A congenitally dislocated radial head is usually irreducible either manually or surgically because of adaptive changes in the soft tissues and the absence of normal surfaces for articulation with the ulna and humerus. Consequently open reduction of the dislocation and reconstruction of the annular ligament in childhood are inadvisable. Any disability is usually caused by restriction of rotation of the forearm, and in children physical therapy to improve this motion is the only treatment indicated. Any resection of the radial head should be postponed until growth is complete, but even then it may not improve motion because of the contractures of the soft tissues. The technique for resection of the radial head is described in Chapter 46.

MADELUNG'S DEFORMITY

Madelung's deformity is a developmental abnormality of the wrist characterized by dorsal and lateral bowing of the distal radius, shortness of the radius, and abnormal growth at the distal radial epiphysis. Sometimes the ulnar half of the epiphysis closes prematurely, resulting in a triangular appearance of the epiphysis. The distal articular surface of the radius tilts ulnarward and volarward, and commonly small excrescences are present on the ulnar bor-

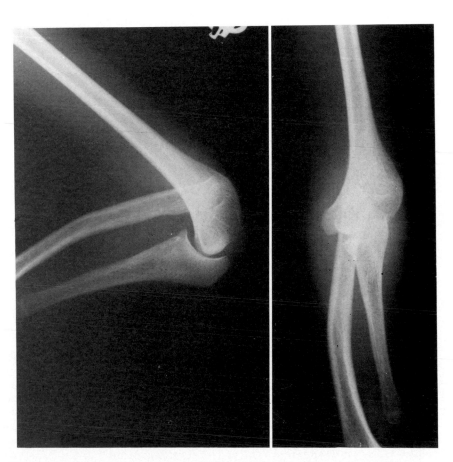

Fig. 14-29. Congenital dislocation of radial head. Head is dislocated anteriorly, and shafts of radius and ulna are abnormal (see text).

der of the distal radial metaphysis. The ulna is dorsally subluxated at its distal end, is bowed, and is relatively too long. The carpus is triangular, being wedged proximally between the abnormal radius and ulna. Thompson and Kalayjian note that sometimes in this condition the radial head is atrophic, the capitulum is hypertrophic, and the space between them is greater than normal. The deformity is more common in girls than in boys. It is not present at birth but develops during early adolescence and is probably caused by a localized form of dyschondroplasia in which the ulnar half of the distal radial epiphysis fuses prematurely.

When mild, Madelung's deformity may be asymptomatic; when moderate or severe, pain in the wrist and weakness and instability of the joint are usual. The deformity may be severe enough to justify surgery even before the distal radial epiphysis has fused. Bowing of the radius is corrected by osteotomy, and at the same time the ulna is shortened by an appropriate resection. In children the ulna should be shortened by Milch's cuff resection and in adults by Darrach's resection of the distal ulna.

TECHNIQUE. Shorten the ulna by the technique of Darrach or of Milch, depending on the patient's age as indicated. Now make a lateral longitudinal incision 2.5 cm long and expose the radius at a level 2.5 cm proximal to its distal articular surface. At this point make a transverse osteotomy and angulate the bone enough anteriorly and medially to correct the deformity. Usually the osteotomy is stable enough to make internal fixation unnecessary.

AFTERTREATMENT. With the elbow flexed to 90 degrees, the wrist extended enough to hold the distal radial fragment properly aligned, and the forearm in neutral rotation, a long arm cast is applied. The position of the fragments is then checked by roentgenograms. At 1 week the roentgenograms are repeated, and at 3 weeks the cast is removed, roentgenograms are again made, and the alignment of the wrist is inspected. A new long arm cast or occasionally a sugar-tong plaster splint is then applied and is worn for 3 weeks.

• • •

Ranawat, DeFiore, and Straub have combined Darach's resection of the distal ulna with resection of a wedge of bone from the distal radius to correct the deformity.

TECHNIQUE (RANAWAT, DEFIORE, AND STRAUB). Make a dorsal longitudinal incision over the distal forearm, detach the extensor retinaculum from the radius over the extensor digitorum communis tendons and reflect the retinaculum and the tendon of the extensor digiti minimi ulnarward. Expose the distal radioulnar joint and excise about 1 cm of the distal ulna. Next make an osteotomy parallel to the distal articular surface of the radius. Then resect an appropriate wedge of bone based radially and dorsally from the distal end of the proximal fragment of the radius and appose the raw surfaces. Stabilize the osteotomy by Kirschner wires so that the distal articular surface of the radius is facing volarward 0 to 15 degrees to the long axis of the radius and ulnarward 60 to 70 degrees. Apply a long arm cast.

AFTERTREATMENT. At 4 weeks the cast and pins are removed, and active exercises of the wrist are begun.

CONGENITAL ABSENCE OF RADIUS

Congenital absence of all or a part of the radius is not rare. The forearm is shortened, the wrist is extremely unstable, and the hand is deviated radially. Frequently the thumb ray is absent, and the ulna is bowed, shortened, and hypertrophied.

The muscles and tendons of the radial side of the limb may be absent or may be ill defined and partially confluent with their neighbors. Frequently the median nerve is the most superficial structure in the radial side of the forearm and may be extremely taut. This fact must be carefully considered when splinting is used for passive correction of the deformity. The median nerve frequently provides a sensory branch to the dorsum of the hand, in place of the absent superficial sensory branch of the radial nerve. Other gross anatomic deformities are often associated: coalition of the ribs, imperforate anus, harelip, hemivertebrae, congenital dislocation of the hip, and cardiovascular and genitourinary anomalies. Blood dyscrasias or metabolic disorders such as Fanconi's syndrome are also frequent.

Overall shortening of the forearm is obvious, accentuated by the angular deformity at the wrist. Furthermore, the articulation of the forearm with the hand is unstable, as already mentioned, because of the inability to stabilize the wrist in the neutral or dorsiflexed position. The absence of functioning radial wrist extensors reduces the power of finger flexion by at least one half. The deformity also prevents the hand positioning that is required to open doors, tie shoes, and carry plates and trays.

As in other deformities, the aim of treatment is to improve function and appearance. In this instance the ideal treatment would keep the hand on the distal end of the ulna, secure maximum but controlled wrist motion, maintain the length of the forearm as much as possible, and allow growth to proceed unhampered. A few years ago in an effort to achieve these goals the proximal fibula along with its epiphysis was transplanted to the radial side of the distal forearm. However, because the transplanted epiphysis failed to grow, radial deviation of the hand recurred and required several revisions in which the transplant was placed more distally. The trend in treatment now is to establish a stable hand properly positioned on the one-bone forearm. Most surgeons agree that treatment should begin at an early age as in congenital clubfoot. However, some surgeons recommend operative treatment much earlier than others. We believe that if by proper splinting (Fig. 14-30) the hand can be brought in line with the ulna and held there satisfactorily, surgery may be postponed until the age of 2 or 3 years.

A patient with congenital absence of the radius should be carefully evaluated medically prior to surgery since thrombocytopenia is associated with bilateral absence of the radius with thumbs present. Of those affected 30% to 40% die within the first year of life (TAR syndrome).

Stretching of the soft tissues is helpful prior to surgery, but the wearing of casts before 3 to 4 months of age frequently is impractical. Evaluation for elbow flexion is es-

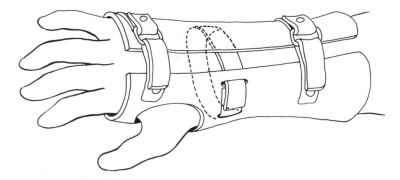

Fig. 14-30. Plastic splint for congenital absence of radius. Note especially middle strap that is placed over wrist at apex of angulation. Splint is useful for hands that can be properly aligned passively and for maintaining proper position after surgery.

sential prior to surgery, especially if the condition is bilateral. Without elbow flexion, when the wrist has been stabilized, it may be impossible to bring the hand to the mouth. Sometimes there are numerous muscle anomalies: the biceps, the brachioradialis, the extensor carpi radialis longus and brevis, and possibly the supinator or pronator quadratus may be absent. The flexor carpi radialis and the median nerve may be anomalous, with a portion of the nerve going through the carpal canal and another branch supplying the area that would ordinarily be supplied by the superficial radial nerve. The radial artery may be absent with only one large interosseous vessel.

When absence of the radius is bilateral, one hand should be surgically fixed in about 45 degrees of pronation and the other in about 45 degrees of supination. Later, after centralization, pollicization of the index finger of the right hand may be indicated for shaking hands and for pinch. Some surgeons prefer to do this on one hand only (see reconstruction of thumb, p. 195).

Bora et al. have attempted to maximize the length of the extremity and improve the function of the fingers. They suggest not only centralizing the hand on the radius (Figs. 14-31 and 14-32) but also transferring certain tendons to maintain this position. Their follow-up studies indicate that centralization alone is frequently followed by recurrence of the original deformity; thus tendon transfers as a second-stage procedure as outlined below are recommended. These include various transfers of the flexor digitorum sublimis from the central two digits, the origin of the hypothenar muscles, and the extensor carpi radialis.

Manske, McCarroll, and Swanson use the transverse ulnar incision as described by Riordan. They transfer the extensor carpi ulnaris tendon by advancement, perform an ulnar capsulodesis, transfer proximally the hypothenar muscles, and resect the redundant ulnar skin and subcutaneous tissue all at the same initial procedure. They do not however, as is suggested by Bora et al., transfer the flexor digitorum sublimis tendons for wrist extension and ulnar deviation because of possible decrease in function of the finger flexors. They emphasize the importance of stabilization of the wrist and excision of the bulbous soft tissues on the ulnar side of the wrist. Their procedure is a com-

bination of the Riordan incision, placing the distal ulna in a slot in the carpus, and selected tendon transfers as described by Bora et al. (Fig. 14-33).

TECHNIQUE FOR CENTRALIZATION OF THE HAND (RIORDAN). Make an elliptic incision transversely over the distal end of the ulna; resect the skin and the fat pad present over the distal ulna to afford later closure without redundant skin (Fig. 14-34). Dissect free the distal end of the ulna and extend the dissection deep from this point toward the radial side of the carpus, volar to any extensor tendons and dorsal to the flexor tendons. Resect bone from the central carpus to reduce the midcarpus onto the ulna. Excise a section of the lunate or a portion of the capitate to form a slot into which to place the distal ulna with minimal tension. Introduce an appropriately sized Steinmann pin or Kirschner wire through the center of the distal ulnar epiphysis and drill it proximally to come out of the ulna at the upper forearm. Now reduce the hand on the ulna so that when the pin is drilled distally, it will pass into the carpus and toward the shaft of the third metacarpal. Cut off the pin proximally under the skin.

At this time, decide whether tendon transfers should be done to help maintain the hand over the distal ulna. Bora and associates have suggested taking the sublimis flexors of the middle and ring fingers around toward the ulnar side of the wrist and dorsally to attach them to the base of the second and third metacarpals. When these muscles are not sufficiently developed, transfer the origin of the hypothenar muscles proximally toward the ulnar aspect of the distal forearm and wrist to give them a more proximal origin and this will help them to maintain the hand in the desired position. Close the wounds and apply a long arm cast.

AFTERTREATMENT. The cast is removed at 3 weeks, the sutures are removed, and a new cast is applied. The pin remains in for a total of 3 or more months unless contraindicated by irritation. The cast can be maintained for 3 months or more; then a permanent splint is made to maintain the hand in desired alignment. Usually the splint may extend below the elbow and should be worn at all times except for skin hygiene. An osteotomy may be indicated later to straighten the ulna.

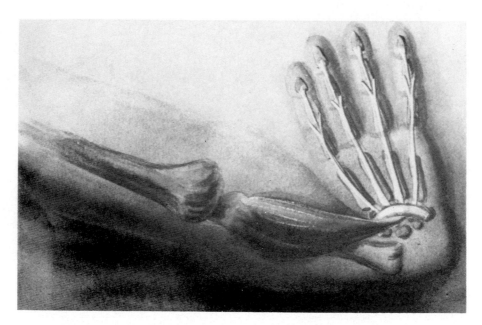

Fig. 14-31. Volar aspect of radial clubhand deformity showing right-angle relationship of hand and forearm and acute angulation of extrinsic flexor tendons. (From Bora, F.W., Jr., et al.: J. Bone Joint Surg. **52-A:**966, 1970)

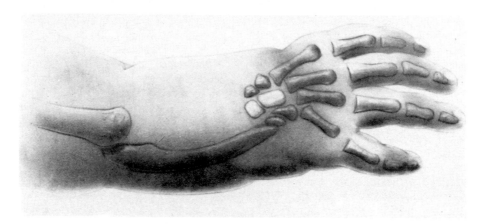

Fig. 14-32. Bones of hand, wrist, and forearm showing unstable wrist. Capitate and lunate (white) are excised when hand is centralized over ulna. (From Bora, F.W., Jr., et al.: J. Bone Joint Surg. **52-A:**966, 1970).

TECHNIQUE FOR CENTRALIZATION OF HAND AND TENDON TRANS-FERS (BORA ET AL.). Bora et al. suggest that treatment be started immediately after birth with corrective casts to stretch the radial side of the wrist. At the age of 6 to 12 months the hand is centralized surgically over the distal end of the ulna and tendon transfers are carried out 6 to 12 months later.

Stage I: Make a radial S-shaped incision and excise the radiocarpal ligament. Isolate and excise the lunate and capitate. Then make a longitudinal incision over the distal ulnar epiphysis, free it from the surrounding tissue, and preserve the tendons of the extensor carpi ulnaris and extensor digitorum quinti minimus. Transpose the distal end of the ulna through the plane between the flexor and extensor tendons and transpose it to a slot formed by the re-

moval of the lunate and capitate. With the distal end of the ulna at the base of the long finger metacarpal transfix it with a smooth Kirschner wire. Check the position of the ulna and carpus by roentgenograms in the operating room to be sure that the ulna is aligned with the long axis of the long finger metacarpal. Now suture the dorsal radiocarpal ligament over the neck of the ulna, close the skin, and apply a long-arm cast with the elbow at 90 degrees. When the deformity is unilateral, the wrist and hand should be placed in neutral, and when it is bilateral, they should be placed in 45 degrees of pronation on one side and 45 degrees of supination on the other. The cast is removed at 6 weeks, and a splint is applied for night wear.

Stage II: Three tendon transfers are performed 6 to 12 months after the centralization procedure. Before attempt-

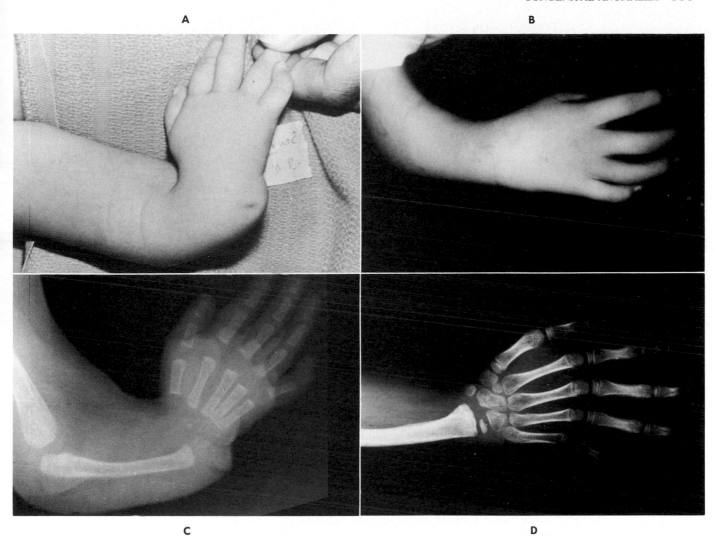

Fig. 14-33. A, Preoperative photograph of radial clubhand deformity. **B,** Status 5 years after centralization arthroplasty of radial clubhand. **C,** Preoperative roentgenogram of radial clubhand deformity. **D,** Status 5 years after centralization arthroplasty of radial clubhand. (From Manske, P.R., et al.: J. Hand Surg. **6:**423, 1981.)

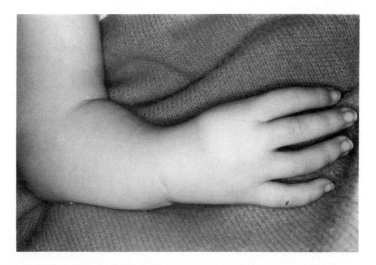

Fig. 14-34. Pleasing scar is produced by transverse incision as recommended by Riordan. Ulna was placed centrally in carpus and held with medullary wire.

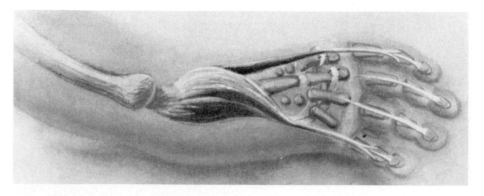

Fig. 14-35. Volar aspect of left radial clubhand after centralization and transfer of superficialis tendons of ring and long fingers to balance muscle forces about wrist (see text). (From Bora, F.W., Jr.: J. Bone Joint Surg. **52-A**:966, 1970.)

ing to transfer the flexor digitorum sublimis tendons, test for function because in some instances the sublimis tendon is nonfunctioning in one or more of the three ulnar digits. Passively maintain the metacarpophalangeal joints and the wrist joint in hyperextension and the interphalangeal joints in extension and release one finger at a time. An intact sublimis tendon will flex the proximal interphalangeal joint of the released finger.

Make a midlateral incision on the ulnar side of the long finger at the level of the proximal interphalangeal joint. Divide the sublimis tendon at the level of the middle phalanx and divide also the chiasm of the decussating fibers. Perform a similar procedure on the ring finger. Next make a short transverse incision on the volar aspect of the forearm and pull the two tendons into it. At the site of previous dorsal incision reenter the wrist and transfer the sublimis tendons subcutaneously around the ulnar side of the ulna to the dorsum of the hand. Loop the tendon from the long finger around the shaft of the index finger metacarpal and the tendon from the ring finger around the shaft of the long finger metacarpal (Fig. 14-35). Transpose the tendons extraperiostally and suture them back to themselves with the wrist in 15 degrees dorsiflexion and maximum ulnar deviation. Now transfer the extensor carpi ulnaris tendon distally along the shaft of the little finger metacarpal and transfer the origin of the hypothenar muscles proximally along the ulnar shaft. Thus an effort is made to maintain balance and prevent recurrence of the deformity. Following the procedure apply a cast that is worn for one month; after this a night splint is worn for at least 3 months. Careful follow-up should be made to observe for possible recurrence of deformity. We prefer to use a night splint for several years.

TECHNIQUE FOR CENTRALIZATION OF HAND (MANSKE ET AL.). Begin the incision just radial to the midline on the dorsum of the wrist at the level of the distal ulna and proceed ulnarward in a transverse direction to a point radial to the pisiform at the volar wrist crease. Pass the incision through the bulbous soft tissue mass on the ulnar side of the wrist, incising considerable fat and subcutaneous tissue (Fig. 14-36, *A*).

Identify and preserve the dorsal sensory branch of the ulnar nerve, which is deep to the subcutaneous tissue and lies near the extensor retinaculum. Expose the extensor retinaculum and the base of the hypothenar muscles. It is not necessary to identify the ulnar artery or nerve on the volar aspect of the wrist (Fig. 14-36, *B*).

Identify and dissect free the extensor carpi ulnaris tendon at its insertion on the base of the fifth metcarpal and detach and retract it proximally. Next identify and retract radially the extensor digitorum communis tendons. This exposes the dorsal and ulnar aspects of the wrist capsule. Incise the capsule transversely, thus exposing the distal ulna (Fig. 14-36, *C*).

The carpal bones are a cartilaginous mass deep in the wound on the radial side of the ulna. The carpoulnar junction is most easily identified by dissecting from proximal to distal along the radial side of the distal ulna. Take care not to mistake one of the intercarpal articulations for the carpoulnar junction (Fig. 14-36, *D*).

Now define the cartilaginous mass of carpal bones and excise a square segment of its midportion (measuring approximately 1 cm by 1 cm) to accommodate the distal ulna. Dissect free the distal ulnar epiphysis from the adjacent soft tissue and square it off by shaving perpendicular to the shaft. Take care not to injure the epiphyseal plate or the attached soft tissue (Fig. 14-36, *E*).

Place the distal ulna in the carpal defect and stabilize it with a smooth Kirschner wire. In practice, this is usually accomplished by passing the Kirschner wire proximally down the shaft of the distal ulna to emerge at the olecranon (or at the midshaft if the ulna is bowed.) Then pass the wire distally across the carpal notch into the third metacarpal. Cut off the proximal end of the wire beneath the skin (Fig. 14-36, *F*).

Stabilize the ulnar side of the wrist by imbricating the capsule or by suturing the distal capsule to the periosteum of the shaft of the distal ulna. (If there is insufficient distal capsule, suture the cartilaginous carpal bones to the periosteum.) Obtain additional stabilization by advancing the extensor carpi ulnaris tendon distally and reattaching it to the base of the fourth or fifth metacarpal (Fig. 14-36, *G*). Also advance the origin of the hypothenar musculature proximally and suture it to the ulnar shaft to provide additional stability to the wrist. Excise the bulbous excess of the skin and soft tissue and suture the skin. This results in a pleasing cosmetic closure and helps stabilize the hand in the ulnar position.

AFTERTREATMENT. The wrist is immobilized in a plaster cast for 6 weeks and is then placed in a removable ortho-

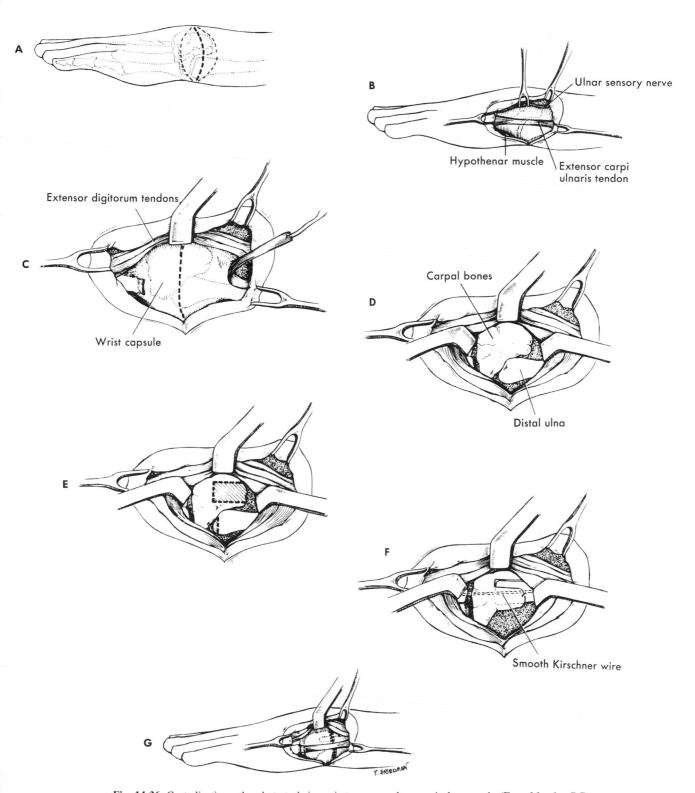

Fig. 14-36. Centralization arthroplasty technique via transverse ulnar surgical approach. (From Manske, P.R., et al.: J. Hand Surg. **6:**423, 1981.)

plast splint. Children are encouraged to wear the splint until skeletal maturity. Remove the Kirschner wire at 6 to 12 weeks.

CONGENITAL HYPOPLASIA OF THUMB WITH ABSENCE OF LONG EXTENSORS AND ABDUCTORS

Neviaser reported on 10 hands in 8 patients in which there was absence of the extrinsic thumb extensors, the abductor pollicis longus, and the thenar muscles, making it impossible to grasp large objects because of lack of thumb extension and abduction or to pinch effectively because of instability of the metacarpophalangeal joint (Fig. 14-37). His solution to this problem was to perform a sur-

gical procedure after the first year: arthrodesis of the metacarpophalangeal joint and transfer of the palmaris longus, the extensor indicis proprius tendons, and the hypothenar muscles. In assessing the thumb preoperatively one should be certain of the presence or absence of the flexor pollicis longus tendon and of its alignment. (See anomalous insertion of flexor pollicis longus.)

TECHNIQUE (NEVIASER). Because the metacarpophalangeal joint is unstable with both radial and ulnar stresses, perform an intraarticular ankylosis consisting of shaving the articular cartilage on both sides of the joint sufficiently to expose epiphyseal bone and then pinning the joint with a small Kirschner wire, which is left in for 6 weeks. Trans-

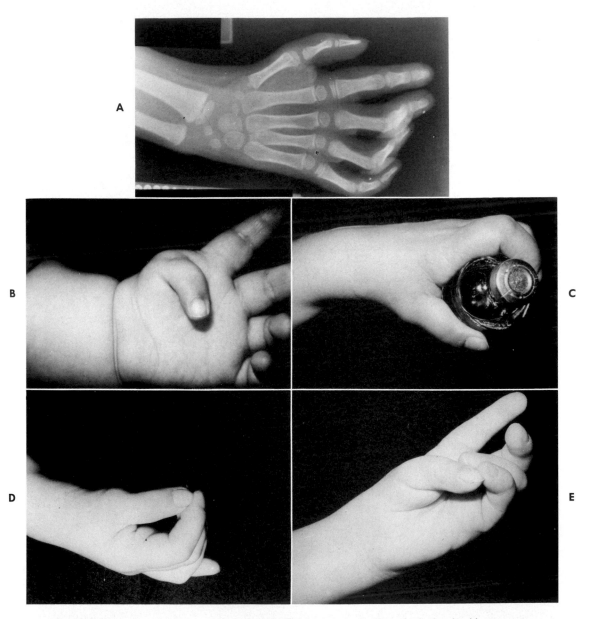

Fig. 14-37. A, Typical deformity demonstrating skeletal hypoplasia. **B,** Hypoplastic thumb with metacarpophalangeal instability, adduction contracture of web space, and absence of extrinsic extensors, long abductor, and thenar muscles. **C,** After operation, child's grasp improved. **D,** Stability also was present. **E,** Child's thumb could be opposed to tip of little finger after operation. (From Neviaser, R.J.: J. Hand Surg. **4:**301, 1979).

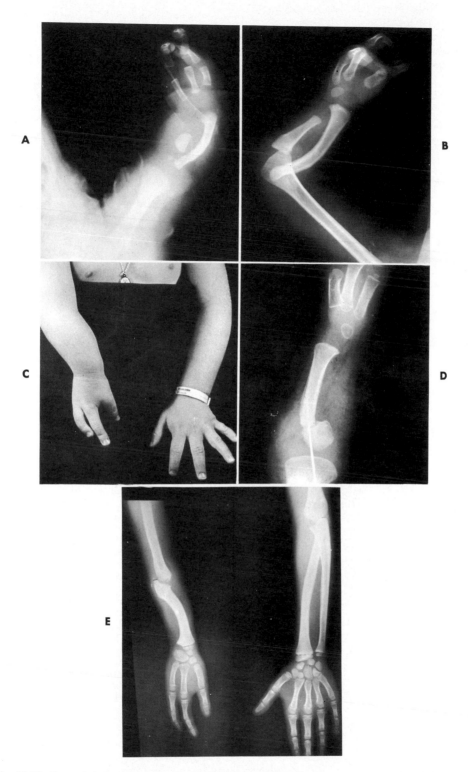

Fig. 14-38. Congenital absence of ulna. **A,** During infancy. **B** and **C,** At age of 4 years. **D,** After excision of fibrocartilaginous band and construction of one-bone forearm (see text). **E,** Both forearms 8 years after surgery. (**A** to **D** from Straub, L.R.: Am. J. Surg. **109:**300, 1965; **E** courtesy Dr. L.R. Straub.)

fer the extensor indicis proprius tendon to the base of the distal phalanx of the thumb to replace the absent extensor pollicis longus tendon. Transfer the palmaris longus to the base of the first metacarpal to act as the abductor pollicis longus. Finally transfer the abductor digiti quinti as originally described by Huber, thus providing prehension as well as bulk to the thenar eminence. Relieve any web space contracture by Z-plasty or local advancement flap. Then derotate the first metacarpal by osteotomy through its proximal third, or by careful capsulorraphy at the carpometacarpal joint. Finally pronate the thumb 90 degrees and pin it at this joint for 8 weeks. The above procedures were done in one stage and provided a useful thumb in all 10 instances.

CONGENITAL ABSENCE OF ULNA

The ulna is congenitally absent much less frequently than the radius. In congenital absence of the ulna the radius is almost always bowed, shortened, and thickened and the radial head is dislocated (Fig. 14-38, *A* and *B*). Usually the ulna is represented proximally by a vestigial bony segment that has a stable articulation with the humerus and distally by an unyielding fibrocartilaginous band that occupies the position normally held by the ulna. Frequently ulnar digits and carpal bones are also absent. Surgery is indicated to release the fibrocartilaginous band that may be deforming and to create a one-bone forearm that improves alignment of the forearm and the function of the elbow. Straub recommends such surgery at an early age.

TECHNIQUE (STRAUB). Make a curved longitudinal dorsoradial incision beginning just proximal to the elbow and ending on the middle or distal third of the forearm. Expose and excise the fibrocartilaginous band that extends distally from the ulnar fragment; in excising this band, free its proximal end by osteotomizing the distal end of the fragment. Next expose the radial nerve at the elbow and trace it distally to its interosseous branch; this branch and its enclosing supinator muscle may be grossly displaced by the dislocation of the proximal radius. While protecting the branch, strip the muscles and other soft structures extraperiosteally from the proximal radius. Develop the cleavage between the dorsal and volar muscles of the forearm while carefully protecting the important neurovascular structures in the antecubital area. Then at the level of the distal end of the ulnar fragment divide the radial shaft and excise its proximal part, including the radial head. Place the proximal end of the distal radial fragment against the distal end of the ulnar fragment and fix them together with a Kirschner wire passed distally through the olecranon (Fig. 14-38, *D*). Apply a long arm cast with the elbow flexed about 90 degrees.

REFERENCES

Almquist, E.E., Gordon, L.H., and Blue, A.I.: Congenital dislocation of the head of the radius, J. Bone Joint Surg. **51-A:**1118, 1969.

Andren, L., Carstam, N., and Linden, B.: Osteochondritis dissecans and brachymesophalangia: a hereditary syndrome, J. Hand Surg. **3:**117, 1978.

Arminio, J.A.: Congenital anomaly of the thumb: absent flexor pollicis longus tendon, J. Hand Surg. **4:**487, 1979.

Barsky, A.J.: Congenital anomalies of the hand, J. Bone Joint Surg. **33-A:**35, 1951.

Barsky, A.J.: Congenital anomalies of the hand and their surgical treatment, Springfield, Ill., 1958, Charles C. Thomas, Publisher.

Barsky, A.J.: Congenital anomalies of the thumb, Clin. Ortho. **15:**96, 1959.

Barsky, A.J.: Reconstructive surgery in congenital anomalies of the hand, Surg. Clin. North Am. **39:**449, 1959.

Barsky, A.J.: Cleft hand: classification, incidence, and treatment: review of the literature and report of nineteen cases. J. Bone Joint Surg. **46-A:**1707, 1964.

Barsky, A.J.: Macrodactyly, J. Bone Joint Surg. **49-A:**1255, 1967.

Bauer, T.B., Tondra, J.M. and Trusler, H.M.: Technical modification in repair of syndactylism, Plast. Reconstr. Surg. **17:**385, 1956.

Blackfield, D.H., and Hause, D.P.: Syndactylism, Plast. Reconstr. Surg. **16:**37, 1955.

Blair, W.F., and Omer, G.E., Jr.: Anomalous insertion of the flexor pollicis longus, J. Hand Surg. **6:**241, 1981.

Bora, F.W., Jr., et al.: Radial metromelia: the deformity and its treatment, J. Bone Joint Surg. **52-A:**966, 1970.

Broadbent, T.R., and Woolf, R.M.: Flexion-adduction deformity of the thumb: congenital clasped thumb, Plast. Reconstr. Surg. **34:**612, 1964.

Broudy, A.S., and Smith R.J.: Deformities of the hand and wrist with ulnar deficiency, J. Hand Surg. **4:**304, 1980.

Brown, J.B., and McDowell, F.: Syndactylism with absence of the pectoralis major, Surgery **7:**599, 1940.

Burrows, H.J.: An operation for the correction of Madelung's deformity and similar conditions, Proc. R. Soc. Med. **30:**31, 1937.

Call, W.H., and Strickland, J.W.: Functional hand reconstruction in the whistling-face syndrome, J. Hand Surg. **6:**148, 1981.

Carroll, R.E., and Bowers, W.H.: Congenital deficiency of the ulna, J. Hand Surg. **2:**169, 1977.

Carstam, N., and Eiken, O.: Kirner's deformity of the little finger: case reports and proposed treatment. J. Bone Joint Surg. **52-A:**1663, 1970.

Clarkson, P.: Poland's syndactyly, Guy's Hosp. Rep. **111:**335, 1962.

Clifford, R.H.: The treatment of macrodactylism: a case report, Plast. Reconstr. Surg. **23:**245, 1959.

Cohn, B.N.E.: Congenital bilateral radio-ulnar synostosis, J. Bone Joint Surg. **14:**404, 1932.

Crawford, H.H., Horton, C.E., and Adamson, J.E.: Congenital aplasia or hypoplasia of the thumb and finger extensor tendons: report of six cases, J. Bone Joint Surg. **48-A:**82, 1966.

Cronin, T.D.: Syndactylism, Tri-State Med. J. **15:**2869, 1943.

Cronin, T.D.: Syndactylism: results of zigzag incision to prevent postoperative contracture, Plast. Reconstr. Surg. **18:**460, 1956.

Dannenberg, M., Anton, J.I., and Spiegel, M.B.: Madelung's deformity: consideration of its roentgenological diagnostic criteria, Am. J. Roentgen. **42:**671, 1939.

Davis. J.S.: Plastic Surgery, London, 1919, H. Kimpton.

Davis, J.S., and German, W.J.: Syndactylism, Arch. Surg. **21:**32, 1930.

Dawson, H.G.W.: A congenital deformity of the forearm and its operative treatment, Br. Med. J. **2:**833, 1912.

Define, D.: Treatment of congenital radial club hand, Clin. Orthop. **73:**153, 1970.

Dellon, A.L., and Hansen, F.C.: Bilateral inability to grasp due to multiple (ten) congenital trigger fingers, J. Hand Surg. **5:**470, 1980.

Dick, H.M., et al.: Lengthening of the ulna in radial agenesis: a preliminary report, J. Hand Surg. **2:**175, 1977.

Dinham, J.M., and Meggitt, B.F.: Trigger thumbs in children: a review of the natural history and indications for treatment in 105 patients, J. Bone Joint Surg. **56-B:**153, 1974.

Doyle, J.R., et al.: Restoraton of the elbow flexion in arthrogryposis multiplex congenita, J. Hand Surg. **5:**149, 1980.

Ebskov, B., and Zachariae, L.: Surgical methods in syndactylism: evaluation of 208 operations, Acta Chir. Scand. **131:**258, 1966.

Edgerton, M.T., Snyder, G.B., and Webb, W.L.: Surgical treatment of congenital thumb deformities (including psychological impact of correction), J. Bone Joint Surg. **47-A:**1453, 1965.

Engber, W.D.: Syndactyly with Larsen's syndrome. J. Hand Surg. **4:**187, 1979.

Engber, W.D., and Flatt, A.E.: Camptodactyly: an analysis of sixty-six patients with twenty-four operations, J. Hand Surg. **2:**216, 1977.

Entin, M.A.: Reconstruction of congenital abnormalities of the upper extremities, J. Bone Joint Surg. **41-A:**681, 1959.

Exarhou, E.I., and Antoniou, N.K.: Congenital dislocation of the head of the radius, Acta Orthop. Scand. **41:**551, 1970.

Fahlstrom, S.: Radio-ulnar synostosis: historical review and case report, J. Bone Joint Surg. **14:**395, 1932.

Field, J.H., and Krag, D.O.: Congenital constricting bands and congenital amputations of the fingers: placental studies, J. Bone Joint Surg. **55-A:**1035, 1973.

Flatt, A.E., and Wood, V.E.: Rigid digits or symphalangism, Hand **7:**197, 1975.

Fleegler, E.J., Culver, J.E., Jr., and Jaffe, S.: Modified treatment of thumb ray dysplasia: a case report. J. Hand Surg. **5:**505, 1980.

Frykman, G.K., and Wood, V.E.: Peripheral nerve hamartoma with macrodactyly in the hand: report of three cases and review of the literature, J. Hand Surg. **3:**307, 1978.

Gelberman, R.H., and Bauman, T.: Madelung's deformity and dyschondrosteosis, J. Hand Surg. **5:**338, 1980.

Gelberman, R.H., and Goldner, J.L.: Congenital arteriovenous fistulas of the hand, J. Hand Surg. **3:**451, 1978.

Gibson, A.: A critical consideration of congenital radio-ulnar synostosis, with special reference to treatment, J. Bone Joint Surg. **5:**299, 1923.

Gold, A.H., et al.: Digital clubbing: a unique case and a new hypothesis, J. Hand Surg. **4:**60, 1979.

Goldberg, M.J., and Meyn, M.: The radial clubhand, Orthop. Clin. North Am. **7:**341, 1976.

Goldenberg, R.R.: Congenital bilateral complete absence of the radius in identical twins, J. Bone Joint Surg. **30-A:**1001, 1948.

Griffin, J.M., Vasconez, L.O., and Schatten, W.E.: Congenital arteriovenous malformations of the upper extremity, Plast. Reconstr. Surg. **62:**49, 1978.

Halpern, A.A., Wheeler, R.D., and Schurman, D.J.: Distal symphalangism: symbrachydactylism arising in a family with distal symphalangism, Clin. Orthop. **141:**251, 1979.

Hansen, O.H., and Andersen, N.O.: Congenital radio-ulnar synostosis: report of 37 cases, Acta Orthop. Scand. **41:**225, 1970.

Hartrampf, C.R., et al.: Construction of one good thumb from both parts of a congenitally bifid thumb, Plast. Reconstr. Surg. **54:**148, 1974.

Heikel, H.V.A.: Aplasia and hypoplasia of the radius: studies on 64 cases and on epiphyseal transplantation in rabbits with the imitated defect, Acta Orthop. Scand. (Suppl. 39), 1959.

Hentz, V.R., and Littler, J.W.: Abduction-pronation and recession of second (index) metacarpal in thumb agenesis, J. Hand Surg. **2:**113, 1977.

Herndon, J.H., and Swanson, A.B.: Duplication of palmar skin in a patient with multiple congenital deformities, J. Hand Surg. **3:**370, 1978.

Hoover, G.H., et al.: The hand and Apert's syndrome, J. Bone Joint Surg. **52-A:**878, 1970.

Ingram, A.J.: Pollex varus, or the thumb-clutched hand. Thesis submitted to the American Orthopaedic Association, February 1957 (unpublished).

Keats, S.: Congenital bilateral dislocation of head of the radius in a seven-year-old child, Orthop. Rev. **3:**33, August 1974.

Kelikian, H., and Doumanian, A.: Congenital anomalies of the hand: part I, J. Bone Joint Surg. **39-A:**1002, 1957.

Kelikian, H., and Doumanian, A.: Congenital anomalies of the hand: part II, J. Bone Joint Surg. **39-A:**1249, 1957.

Kessler, I., Baruch, A., and Hecht, O.: Experience with distraction lengthening of digital rays in congenital anomalies, J. Hand Surg. **2:**394, 1977.

Kettelkamp, D.B., and Flatt, A.E.: An evaluation of syndactylia repair, Surg. Gynecol. Obstet. **113:**471, 1961.

Khanna, N., et al.: Macrodactyly, Hand, **7:**215, 1975.

Kirner, J.: Doppelseitge Verkrümmung de Kleinfingergliedes als selbständiges Krankenheitsbild, Fortschr. a.d. Geb. d. Röntgenstr. **36:**804, 1927.

Kite, J.H.: Congenital syndactylism of fingers, South. Med. J. **51:**160, 1958.

Lamb, D.W.: Radial club hand: a continuing study of sixty-eight patients with one hundred and seventeen club hands, J. Bone Joint Surg. **59-A:**1, 1977.

Lewin, M.T.: Facial and hand deformity in acrocephalosyndactyly, Plast. Reconstr. Surg. **12:**138, 1953.

Lundholm, G.: Congenital manus vara, Acta Orthop. Scand. **30:**207, 1960-1961.

MacCollum, D.W.: Webbed fingers, Surg. Gynecol. Obstet. **71:**782, 1940.

Manske, P.R., and McCarroll, H.R., Jr.: Abductor digiti minimi opponensplasty in congenital radial dysplasia, J. Hand Surg. **3:**552, 1978.

Manske, P.R., McCarroll, H.R., Jr., and Swanson, K.: Centralization of radial club hand, J. Hand Surg. **6:**423, 1981.

Mardam-Bey, T., and Ger, E.: Congenital radial head dislocation, J. Hand Surg. **4:**316, 1979.

Marks, T.W., and Bayne, L.G.: Palydactyly of the thumb: abnormal anatomy and treatment, J. Hand Surg. **2:**107, 1978.

Matev, I.B.: Thumb reconstruction in children through metacarpal lengthening, Plast. Reconstr. Surg. **64:**665, 1979.

McCarroll, H.R.: Soft-tissue neoplasms associated with congenital neurofibromatosis, J. Bone Joint Surg. **38-A:**714, 1956.

McFarland, B.: Congenital dislocation of the head of the radius, Br. J. Surg. **24:**41, 1936.

Milch, H.: So-called dislocation of the lower end of the ulna, Ann. Surg. **116:**282, 1942.

Milch, H.: Triphalangeal thumb, J. Bone Joint Surg. **33-A:**692, 1951.

Minnaar, A.B. de V.: Congenital fusion of the lunate and triquetral bones in the South African Bantu, J. Bone Joint Surg. **34-B:**45, 1952.

Mital, M.A.: Congenital radioulnar synostosis and congenital dislocation of the radial head, Orthop. Clin. North Am. **7:**375, 1976.

Miura, T.: Two families with congenital nail anomalies: nail formation in ectopic areas, J. Hand Surg. **3:**348, 1978.

Miura, T.: Congenital constriction band syndrome, J. Hand Surg. **9-A:**82, 1984.

Miura, T., and Komada, T.: Simple method for reconstruction of the cleft hand with an adducted thumb, Plast. Reconstr. Surg. **64:**65, 1979.

Murray, J.F., Shore, B., and Trefler, E.: Prostheses for children with unilateral congenital absence of the hand. J. Bone Joint surg. **54-A:**1658, 1972.

Neviaser, R.J.: Congenital hypoplasia of the thumb with absence of the extrinsic extensors, abductor pollicis longus, and thenar muscles, J. Hand Surg. **4:**301, 1979.

Nogami, H., and Oohira, A.: Experimental study on pathogenesis of polydactyly of the thumb, J. Hand Surg. **5:**443, 1980.

Nutt, J. N., and Flatt, A.E.: Congenital central hand deficit, J. Hand Surg. **6:**48, 1981.

Nylen, B.: A report on the repair of congenital finger syndactylism, Acta Chir. Scand. **113:**310, 1957.

Palmieri, T.J.: The use of silicone rubber implant arthroplasty in treatment of true symphalangism, J. Hand surg. **5:**242, 1980.

Pardini, A.G., Jr.: Radial dysplasia, Clin. Orthop. **57:**153, 1968.

Pashayan, H., et al.: Bilateral aplasia of the tibia, polydactyly and absent thumb in father and daughter, J. Bone Joint Surg. **53-B:**495, 1971.

Patterson, T.J.S.: Congenital deformities of the hand, Ann. R. Coll. Surg. Engl. **25:**306, 1959.

Ranawat, C.S., DeFiore, J., and Straub, L.R.: Madelung's deformity: an end-result study of surgical treatment, J. Bone Joint Surg. **57-A:**772, 1975.

Rasmussen, L.: Kirner's deformity, Acta Orthop. Scand. **52:**36, 1981.

Rechnagel, K.: Megalodactylism: report of 7 cases, Acta Orthop. Scand. **38:**57, 1967.

Reis, N.D.: Anomalous triceps tendon as a cause for snapping elbow and ulnar neuritis: a case report, J. Hand Surg. **5:**361, 1980.

Riordan, D.C.: Congenital absence of the radius, J. Bone Joint Surg. **37-A:**1129, 1955.

Rowntree, T.: Anomalous innervation of the hand muscles, J. Bone Joint Surg. **31-B:**505, 1949.

Ruby, L., and Goldberg, M.J.: Syndactyly and polydactyly, Orthop. Clin. North Am. **7:**361, 1976.

Rushforth, A.F.: A congenital abnormality of the trapezium and first metacarpal bone, J. Bone Joint Surg. **31-B:**543, 1949.

Silver, L.: Hand abnormalities in the fetal hydantoin syndrome, J. Hand Surg. **6:**262, 1981.

Sherik, S.K., and Flatt, A.E.: The anatomy of congenital radial dysplasia: its surgical and function implications, Clin. Orthop. **66:**125, 1969.

Skoog, T.: Syndactyly: a clinical report on repair, Acta Chir. Scand. **130:**537, 1965.

Smith, R.J.: The radial club hand, Bull. Hosp. Joint Dis. **25:**85, 1964.

Smith, R.J.: Osteotomy for "delta-phalanx" deformity, Clin. Orthop. **123:**91, 1977.

Smith, R.J., and Kaplan, E.B.: Camptodactyly and similar atraumatic flexion deformities of the proximal interphalangeal joints of the fingers: a study of thirty-one cases, J. Bone Joint Surg. **50-A:**1187, 1968.

Straub, L.R.: Congenital absence of the ulna, Am. J. Surg. **109:**300, 1965.

Strauch, B., and Spinner, M.: Congenital anomaly of the thumb: absent intrinsics and flexor pollicis longus, J. Bone Joint Surg. **58-A:**115, 1976.

Su, C.T., Hoopes, J.E., and Daniel, R.: Congenital absence of the thenar muscles innervated by the median nerve: report of a case, J. Bone Joint Surg. **54-A:**1087, 1972.

Swanson, A.B.: A classification for congenital limb malformations, J. Hand Surg. **1:**8, 1976.

Swanson, A.B., and Swanson, G. deG.: The Krukenberg procedure in the juvenile amputee, Clin. Orthop. **148:**55, 1980.

Thompson, C.F., and Kalayjian, B.: Madelung's deformity and associated deformity at the elbow, Surg. Gynecol. Obstet. **69:**221, 1939.

Thompson, H.G., Martin, S.R., and Murray, J.F.: Skin grafted juvenile amputation stumps: are they durable? Plast. Reconstr. Surg. **65:**195, 1980.

Thorne, F.L., Posch, J.L. and Mladick, R.A.: Megalodctyly, Plast. Reconstr, Surg. **41:**232, 1968.

Toledo, L.C., and Ger, E.: Evaluation of the operative treatment of syndactyly, J. Hand Surg. **4:**556, 1979.

Tsuge, K.: Treatment of macrodactyly, Plast. Reconstr. Surg. **39:**590, 1967.

Tuch, B.A., et al.: A review of supernumerary thumb and its surgical management, Clin. Orthop. **125:**159, 1977.

Walsh, R.J.: Acrosyndactyly: a study of twenty-seven patients, Clin. Orthop. **71:**99, 1970.

Wassel, H.D.: The results of surgery for polydactyly of the thumb: a review, Clin. Orthop. **64:**175, 1969.

Watari, S., and Tsuge, K.: A classification of cleft hands, based on clinical findings: theory of developmental mechanism. Plast. Reconstr. Surg. **64:**381, 1979.

Watson, H.K., and Boyes, J.H.: Congenital angular deformity of the digits: delta phalanx, J. Bone Joint Surg. **49-A:**333, 1967.

Waugh, R.L., and Sullivan, R.F.: Anomalies of the carpus: with particular reference to the bipartite scaphoid (navicular), J. Bone Joint Surg. **32-A:**682, 1950.

Weckesser, E.C.: Congenital flexion-adduction deformity of the thumb (congenital "clasped thumb"), J. Bone Joint Surg. **37-A:**977, 1955.

Weckesser, E.C., Reed, J.R., and Heiple, K.G.: Congenital clasped thumb (congenital flexion-adduction deformity of the thumb): a syndrome, not a specific entity, J. Bone Joint Surg. **50-A:**1417, 1968.

Weeks, P.M.: Surgical correction of upper extremity deformities in arthrogrypotics, Plast. Reconstr. Surg. **36:**459, 1965.

White, J.R.A.: Congenital dislocation of the head of the radius, Br. J. Surg. **30:**377, 1943.

White, J.W., and Jensen, W.E.: The infant's persistent thumb-clutched hand, J. Bone Joint Surg. **34-A:**680, 1952.

Wilkie, D.P.D.: Congenital radio-ulnar synostosis, Br. J. Surg. **1:**366, 1913-1914.

Wood, V.E.: Duplication of the index finger, J. Bone Joint Surg. **52-A:**569, 1970.

Wood, V.E.: Treatment of central polydactyly, Clin. Orthop. **74:**196, 1971.

Wood, V.E.: Congenital triangular bones in the hand, J. Hand Surg. **2:**179, 1977.

Wood,V.E.: Hyperphalangism: report of a case, J. Hand Surg. **2:**79, 1977.

Wood, V.E.: Polydactyly and the triphalangeal thumb, J. Hand Surg. **3:**436, 1978.

Zielinski, C.J., and Gunther, S.F.: Congenital fusion of the scaphoid and trapezium: case report, J. Hand Surg. **6:**220, 1981.

CHAPTER 15

Dupuytren's contracture

Dupuytren's contracture is caused by a proliferative fibroplasia of the subcutaneous palmar tissue, occurring in the form of nodules and cords and resulting in secondary contractures of the finger joints. Other secondary changes include thinning of the overlying subcutaneous fat, adhesion of the skin to the lesion, and later pitting or dimpling of the skin. "Knuckle pads" are common on the dorsum of the proximal interphalangeal joints.

The activity of the lesion, and thus the rate of development of the deformity, is variable. Occasionally a finger may become markedly flexed within a few weeks or months, but the development of a severe deformity usually requires several years. In some patients, the lesion progresses steadily; in others, exacerbations and remissions occur. However, regression is rare.

Similar lesions in the medial plantar fascia (5%) of one or both feet and plastic induration of the penis (3%) may occur in patients with Dupuytren's contracture.

The cause of Dupuytren's contracture remains unknown. Recently, however, the exact origin of the fibroplasia so characteristic of the disease has been under study, and the course of the disease, with and without treatment, has been reevaluated. Evidence points to heredity as a factor in the cause; the lesion seems to occur earlier and more frequently in some families. The possibility that trauma may be a factor has been studied extensively. The presence in the lesion of hemosiderin suggests hemorrhage from tears, but in whites the lesion occurs as often in the minor hand as in the major one, thus making trauma an unlikely cause.

Dupuytren's contracture occurs 10 times more frequently in men than in women and is almost limited to the white race, although it has occasionally been reported in blacks and rarely in Orientals. The lesion is more frequent and severe in epileptics (42%) and alcoholics. Its onset, typically after age 40 years, is rare before the age of 25 years. The involvement, although often bilateral (45%), is rarely symmetric.

The lesion usually begins in line with the ring finger at the distal palmar crease and progresses to involve the ring and little fingers, these digits being affected more frequently than all others combined. Flexion contractures of the metacarpophalangeal and proximal interphalangeal joints gradually develop, their severity depending on the extent and maturity of the fibroplasia. Discomfort is rare and usually consists of itching or occasional pain over the nodules.

PATHOGENESIS

Most investigators now agree that in Dupuytren's contracture the subcutaneous nodules and cords are formed by fibroplasia and by hypertrophy of already existing fibers of the palmar fascia. But Millesi believes that the pathologic tissue arises only through changes in the existing fibers of the palmar fascia and not by the formation of new tissue. Luck has suggested that the subcutaneous nodules develop first and mature later to become cords. However, Gosset suggested that the nodules and cords do not represent two stages of the disease but two forms of it originating in two different tissues, the subcutaneous fat and palmar fascia. Furthermore, Hueston concluded that the nodules develop subcutaneously and only later may involve the palmar fascia and overlying skin. Skoog believes that in the palmar fascia only the longitudinal pretendinous bands are involved and that the transverse palmar ligament is always spared.

PROGNOSIS

The prognosis in Dupuytren's contracture seems dependent on the following factors that in turn may determine the extent of any indicated operation.

1. Heredity. A family history of the disease is an indication that the lesion is likely to progress more rapidly than usual, especially if the onset is early.
2. Sex. In women the lesion usually begins later and progresses more slowly, and women often accommodate better to the inconvenience of the resulting deformity.
3. Alcoholism or epilepsy. In these conditions the lesion is more severe, progresses more rapidly, and recurs more frequently.
4. Location and extent of disease. When bilateral and especially when associated with knuckle pads and with nodules in the plantar fascia, the disease progresses more rapidly and recurs more frequently; furthermore, it usually progresses more rapidly in the ulnar side of the hand than in the radial side.
5. Behavior of the disease. How the disease has behaved in the past, whether treated or not, is an indication of its probable behavior in the future.

341

When the proximal interphalangeal joint begins to contract, it usually progresses to a fixed deformity that becomes increasingly difficult to correct completely. Severe metacarpophalangeal joint contractures are more easily corrected surgically than even moderate proximal interphalangeal joint contractures.

TREATMENT

Although many medical remedies have been tried, the best treatment known is surgical. Roentgen therapy would be of value in the earliest fibroblastic phase, except that it destroys subcutaneous glands and may even burn the skin; thus the risk is too great. No treatment at all may be indicated in the absence of contracture or when a contracture is progressing slowly and is not disabling; then the patient should be observed every 3 months. Any operation is technically easier early in the disease when the skin is more normal, but it should be delayed until an actual contracture develops.

The five surgical procedures used in treating Dupuytren's contracture are (1) subcutaneous fasciotomy, (2) partial (selective) fasciectomy, (3) complete fasciectomy, (4) fasciectomy with skin grafting, and (5) amputation. In choosing the best procedure for a given patient, the degree of contracture, the patient's age, occupation, and general health, the nutritional status of the palmar skin, and the presence or absence of arthritis should all be considered. In general, the more severe the involvement and the more grave the prognosis, the more extensive any indicated fasciectomy should be; however, the fasciectomy may be carried out in stages and may be preceded by a subcutaneous fasciotomy.

Contracture of the metacarpophalangeal and proximal interphalangeal joints and displacement of a neurovascular bundle in a digit result from the pattern of contracture of the fascial cords.

The pretendinous cord (Fig. 15-1) is nearly always responsible for contracture of the metacarpophalangeal joint. It may attach to the base of the proximal phalanx or to the tendon sheath at this level, or it may extend to attach to the base of the middle phalanx or the skin. A spiral cord may project from the midline at the level of the metacarpophalangeal joint to insert either into the area distal to the proximal interphalangeal joint or just proximal to it. It may continue around the neurovascular bundle and rejoin the pretendinous cord. In its course it may displace the neurovascular bundle toward the midline of the finger, making dissection somewhat tedious, but it must be dissected to its insertion to afford complete release of the flexed position of the proximal interphalangeal joint. Neurovascular displacement is most commonly found on the ulnar aspect of the little and ring fingers.

The least extensive procedure, *subcutaneous fasciotomy,* is used for patients who are elderly or arthritic or whose general health is poor. The results of this procedure are more permanent when dense, mature cords are severed than when the lesion is more diffuse; when the lesion is in the involutional stage, recurrence is likely. Since this procedure will allow stretching of the palmar skin, it may also be useful as a preliminary operation to fasciectomy. Fasciotomy should be considered a temporary measure since

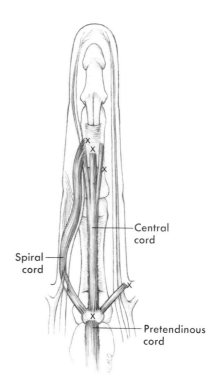

Fig. 15-1. Frequent attachments (*x*) of deforming fibrous cords are noted here. Note distorted course of digital nerve on left. (Redrawn from McFarlane, R.M.: Plast. Reconstr. Surg. **54:**31, 1974. © 1974 The Williams & Wilkins Co., Baltimore.)

72% of contractures so treated recur to a degree requiring further surgery according to Rodrigo, Niebauer, Brown, and Doyle.

Partial (selective) fasciectomy is usually indicated when only the ulnar one or two fingers are involved. It is the operation used more frequently because postsurgically morbidity is less and complications are fewer than after complete fasciectomy. Even though the rate of recurrence after partial fasciectomy is high (50%), the need for another surgical procedure is only 15%. In this operation only the involved and deforming tissue is excised. Several incisions as shown in Fig. 15-3 are useful. We prefer the zigzag incision on the fingers (Fig. 15-3, *B*) or a variant of it because it exposes the diseased tissue better. Whatever the incision used, it should be fashioned to fit the needs of the individual patient, considering the contractures of the skin and the adherence of skin to the underlying fascia. When tightness of the palmar skin limits extension of a finger, a midline incision converted to appropriate Z-plasties is indicated (Fig. 15-3, *A*); this incision allows quicker dissection under direct vision and exposes digital nerves that may have been pulled from their normal position by the fascia. Regardless of the incision, dissection is made easier by magnification, ideal lighting, and a stable surface. Pressing the knife against the taut unyielding fascial cords in a feathering motion is safer than the usual cutting movements.

Sometimes after fasciectomy, extension of the proximal interphalangeal joint is incomplete. This may be caused by a projection of the involved fascia that passes along the

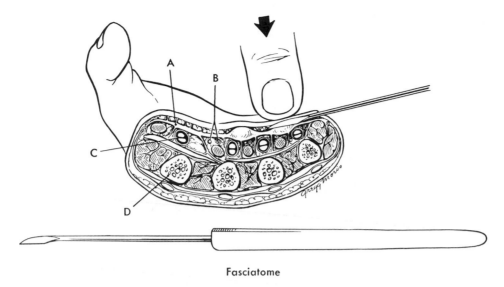

Fasciatome

Fig. 15-2. Luck subcutaneous fasciotomy. **Top,** Cross section of hand to show relations of palmar fascia and technique of subcutaneous fasciotomy. *A*, Palmar fascia; *B*, neurovascular bundle; *C*, flexor tendons; *D*, metacarpal. Fasciatome is being pressed, *arrow*, through a fascial cord. **Bottom,** Fasciatome. (From Luck, J.V.: J. Bone Joint Surg. **41-A:**635, 1959.)

proximal phalanx to the fascial structures on the dorsum of the joint; the projection can be carefully excised. When a flexion contracture of the joint is severe, a volar capsulotomy as described by Curtis may be indicated (p. 174).

The technique of Skoog is a partial or selective fasciectomy because in it only the pretendinous fibers of the palmar fascia are excised. According to Skoog, there is a definite plane between the pretendinous longitudinal fibers of the palmar fascia and the transverse palmar ligament that is limited to the midpalmar area; thus in Dupuytren's contracture the pretendinous fibers can be dissected from this ligament. He emphasizes, however, that the pretendinous fibers may seem attached to the ligament. Furthermore, according to Skoog, the interdigital or natatory ligaments do become involved in Dupuytren's contracture and prevent the fingers from spreading normally; they are indistinguishable from the transverse palmar ligament, except for their anatomic location.

Complete fasciectomy is rarely if ever indicated since it is associated frequently with complications of hematoma, joint stiffness, and delayed healing, and it does not completely prevent recurrence of the disease.

Fasciectomy with skin grafting may be indicated for young people in whom the prognosis is poor because of such factors as epilepsy, alcoholism, or the presence of the disease elsewhere in the body, and in whom the lesion has recurred after one excision. The skin and underlying abnormal fascia are excised, and a full-thickness or thick split skin graft is applied. Recurrence has not been reported in areas of the palm treated in this manner.

Amputation may be indicated if flexion contracture of the proximal interphalangeal joint, especially of the little finger, is severe and cannot be corrected enough to make the finger useful. Then the skin from the involved finger may be used to cover the defect in the palmar skin; the finger is filleted (p. 35), and the skin is folded into the

palm as a pedicle with normal neurovascular bundles.

Another alternative for the severely contracted proximal interphalangeal joint is joint resection and arthrodesis. This results in a much shortened little finger but avoids the complication of an amputation neuroma.

Subcutaneous fasciotomy

TECHNIQUE (LUCK). Using a pointed scalpel, make 3.2 mm skin puncture wounds on the ulnar side of the palmar fascia at the following levels: (1) the apex of the palmar fascia between the thenar and hypothenar eminences, (2) at or near the level of the proximal palmar crease, and (3) at the level of the distal palmar crease. (Digital nerves are more likely to be cut at the distal palm where they become more superficial and may be intertwined with the diseased collagen.) Insert a small tenotomy knife or a fasciatome (Luck) that resembles a myringotome, its blade parallel with the palm, through each of the puncture wounds in turn and pass it across the palm beneath the skin, but superficial to the fascia (Fig. 15-2). Then turn the edge of the blade toward the palmar fascia and extend the fingers to tighten the involved tissue. Carefully divide the fascial cords by pressing the fasciatome through them, using direct finger pressure over the blade or at most a gentle rocking motion; never use a sawing motion. Whenever a cord is divided, the sense of gritty firm resistance disappears, indicating that the blade has passed completely through the fascia. Now using the fasciatome blade in a plane parallel with the skin, free the latter from the underlying fascia. The corrugated skin, even though very thin at times can be safely undermined and released as necessary with little fear of necrosis.

In the fingers subcutaneous fasciotomy is safe only for a fascial cord located in the midline. Insert the blade through a puncture wound adjacent to the cord and divide it obliquely. For a laterally placed cord use a short longi-

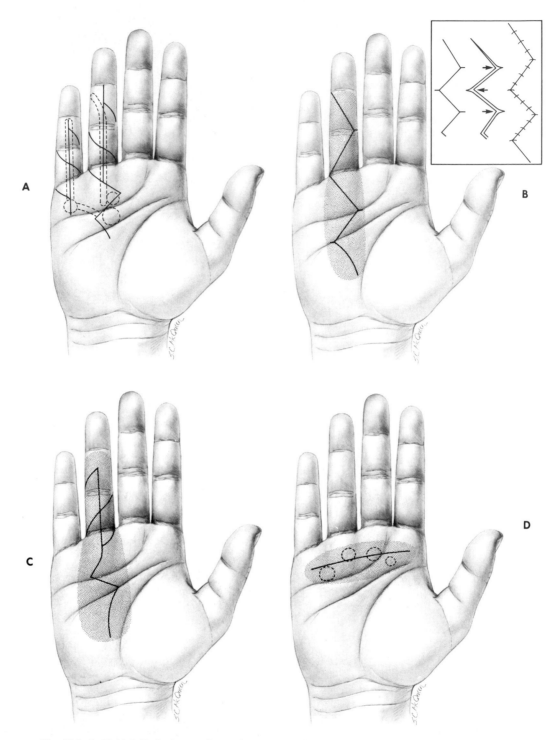

Fig. 15-3. A, Multiple Z-plasties may be employed to provide exposure and convert vertical incision to zigzag closures. Only one extension is made into palm. **B,** When skin contracture is not a major problem, a zigzag pattern may be used in making exposure, with extended corners as shown to make use of redundant skin. **C,** Extent of possible undermining of skin is shown in shaded area. **D,** When only palm is involved, a transverse incision only may be used.

tudinal incision and excise a segment under direct vision. Also enucleate larger nodules in both fingers and palm under direct vision.

AFTERTREATMENT. A pressure dressing is used for 24 hours; then a smaller dressing is applied, and active motion of the hand and fingers is encouraged.

Partial fasciectomy

TECHNIQUE. Outline the proposed incision with a skin pencil prior to inflation of the tourniquet (Fig. 15-3). Take into consideration the pits and other areas of skin with diminished vascularity by making an incision over or near these areas, thus avoiding their presence at the base of a flap. These areas may sometimes be excised when the final rotation of the skin takes place in closure.

Make a zigzag or vertical incision over the deforming pathologic structure. Zigzag incisions tend to straighten out, causing tension lines at the creases; however, the flaps created by zigzag incisions may heal more securely. Design the Z-plasty flaps to be created later for the vertical incisions so that a transverse central segment is within or near each joint crease. Continue the incision proximally into the palm, avoiding crossing the palmar creases at a right angle.

Elevate the skin and underlying normal subcutaneous tissue from the pathologic fascia from proximal to distal (Fig. 15-4). Do not create the Z-plasty flaps until the wound is ready for closure.

Excise the pathologic fascia from proximal palm to distal finger. Carefully cauterize small bleeding points, but avoid heating or burning digital nerves. Excision of all transverse palmar fascial fibers may not be necessary. Avoid entering tendon sheaths so that blood does not enter later and cause irritaton; this is more difficult to do proximal to the pulley at the metacarpophalangeal joint. Carefully excise the fascia by placing it under tension and pressing a sharp knife against it rather than using a less precise cutting motion. A frequent change of knife blades is helpful. Avoid cutting displaced digital nerves by care-

fully locating each nerve at the fatty pad at the level of the metacarpophalangeal joint and following it distally. Excise the natatory ligament if it is contracted. Be certain to follow all the bands of contracted fascia to their distal insertion. Insertions may be into tendon sheaths, bone, and skin; occasionally they are dorsolateral to the proximal interphalangeal joint. When excision of the pathologic fascia has been completed, all joints should permit full passive extension or nearly so.

Now fashion the skin flaps. If there is any extra skin, the pitted or thinned areas may be excised. Before closing, elevate the hand, compress the wound, release the tourniquet, hold for 10 minutes, and then check for and control bleeding. Using skin hooks and with minimal handling of the flaps, suture them in place with No. 5-0 or No. 6-0 monofilament nylon. Place few sutures in the palm to allow necessary drainage around a rubber drain. Apply one layer of nonadherent gauze, then a large wet cotton mass dressing that is compressed gently against the wound to conform to the contours of the palm and fingers. Apply a compression dressing over this, and use a volar plaster splint to support the wrist but leave finger motion unencumbered.

AFTERTREATMENT. All drains are removed within 48 hours. Early proximal interphalangeal motion is encouraged. The hand is kept elevated for a minimum of 48 hours. The shoulder is moved actively at intervals during this period to avoid cramping. Should there be undue pain in the hand or fever after 48 hours, the wound should be inspected for a hematoma. If a hematoma is found elevating the skin, the patient should be taken back to surgery immediately if necessary to evacuate the hematoma; the involved area of the wound should be left open. Otherwise, the first dressing is removed after about 1 week; the hand is redressed and splinted at the wrist for another week. During this second week with the wrist still splinted, the patient is encouraged to move all the finger joints.

At 2 weeks, the sutures are removed, and the hand is

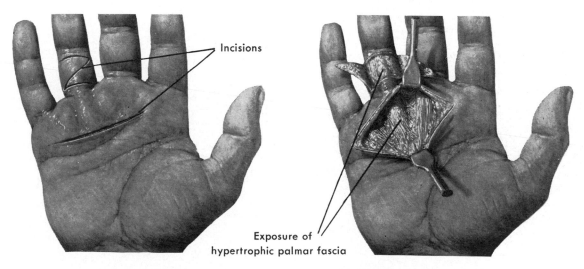

Fig. 15-4. Approaches to palmar fascia for partial fasciectomy (see text).

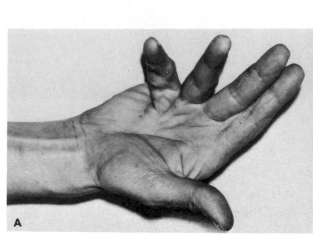
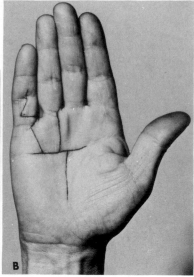

Fig. 15-5. Skoog technique of partial fasciectomy. **A,** Dupuytren's contracture that developed during period of 15 years in 55-year-old patient. **B,** Result after partial fasciectomy. **B** illustrates locations of transverse and longitudinal palmar skin incisions and of Z-plasty incision on little finger. (From Skoog, T.: Surg. Clin. North Am. **47:**433, 1967.)

left free of all dressings. The patient is warned not to place his hand in a dependent position for rest and not to soak his hand in hot water. Active exercise in warm water is permissible, but no passive stretching. Moderate use of the hand is permitted at 3 weeks; however, several months of rehabilitation may be necessary. Silicone putty is a valuable adjunct to an exercise program.

TECHNIQUE (SKOOG). Outline the proposed incision with a skin pencil. First make a transverse incision in the distal palmar crease long enough to expose the part of the palmar fascia to be excised (Fig. 15-5). Then over the cords make distal extensions of the incision and carry them to the base of the fingers. On the fingers make Z-plasty incisions. Now, from the transverse palmar incision and in a crease in the center of the palm, make a proximal extension of the incision. Raise the proximal triangular flaps thus formed no more than necessary to expose the border of the palmar fascia on either or both sides. Now excise the involved pretendinous fascia but leave intact the transverse palmar ligament and the underlying peritendinous septa. At the base of the fingers and adjacent parts of the palm free and protect the neurovascular bundles and excise any diseased fascia here and in the fingers. Now close the incision with interrupted stitches, placing a few mattress sutures along the transverse part of the incision to fix the skin edges to the transverse palmar ligament and obliterate a pocket in which a hematoma can form.

When the deformity is severe, modify the incision and use a full-thickness skin graft if necessary.

AFTERTREATMENT. The aftertreatment is as described for partial fasciectomy.

REFERENCES

Berg, E., Marino, A.A., and Becker, R.O.: Dupuytren's contracture: some associated biophysical abnormalities, Clin. Orthop. **83:**144, 1972.

Boyes, J.H.: Dupuytren's contracture: notes on the age at onset and the relationship to handedness, Am. J. Surg. **88:**147, 1954.

Boyes, J.H., and Jones, F.E.: Dupuytren's disease involving the volar aspect of the wrist, Plast. Reconstr. Surg. **41:**204, 1968.

Browne, W.E.: Dupuytren's contracture: a report of surgical correction in 83 cases (1945-1957), Clin. Orthop. **13:**255, 1959.

Carr, T.L.: Local radical fasciectomy for Dupuytren's contracture, Hand **6:**40, 1974.

Chiu, H.F., and McFarlane, R.M.: Pathogenesis of Dupuytren's contracture: a correlative clinical pathological study, J. Hand Surg. **3:**1, 1978.

Crawford, H.R.: Surgical correction of Dupuytren's contracture, Surg. Clin. North Am. **36:**793, 1956.

Davis, J.E.: One surgery of Dupuytren's contracture, Plast. Reconstr. Surg. **36:**277, 1965.

Deming, E.G.: Y-V advancement pedicles in surgery for Dupuytren's contracture, Plast. Reconstr. Surg. **29:**581, 1962.

Freehafer, A.A., and Strong, J.M.: The treatment of Dupuytren's contracture by partial fasciectomy, J. Bone Joint Surg. **45-A:**1207, 1963.

Fromison, A.I., and Zahrawi, F.: Treatment of compression neuropathy of the ulnar nerve at the elbow by epicondylectomy and neurolysis, J. Hand Surg. **5:**391, 1980.

Gelberman, R.H., et al.: Dupuytren's contracture: an electron microscopic, biochemical, and clinical correlative study, J. Bone Joint Surg. **62-A:**425, 1980.

Gosset, J.: Maladie de Dupuytren et anatomie des aponevroses palmodigitales. In Maladie de Dupuytren, Paris, 1966, L'Expansion Scientifique.

Hamlin, E., Jr.: Limited excision of Dupuytren's contracture: a follow-up study, Ann. Surg. **155:**454, 1962.

Heyse, W.E.: Dupuytren's contracture and its surgical treatment: clinical study of a local resection method, JAMA **174:**1945, 1960.

Honner, R., et al.: Dupuytren's contracture: long-term results after fasciectomy, J. Bone Joint Surg. **53-B:**240, 1971.

Hoopes, J.E., et al.: Enzymes of glucose metabolism in palmar fascia and Dupuytren's contracture, J. Hand Surg. **2**:62, 1977.

Howard, L.D., Jr.: Dupuytren's contracture: a guide for management, Clin. Orthop. **15**:118, 1959.

Hueston, J.T.: Dupuytren's contracture, Edinburgh, 1963, E. & S. Livingstone, Ltd.

Kelly, A.P., Jr., and Clifford, R.H.: Subcutaneous fasciotomy in the treatment of Dupuytren's contracture, Plast. Reconstr. Surg. **24**:505, 1959.

King, E.W., Bass, D.B., and Watson, H.K.: Treatment of Dupuytren's contracture by extensive fasciectomy through multiple Y-V-plasty incisions: short-term evaluation of 170 consecutive operations, J. Hand Surg. **4**:234, 1979.

Larsen, R.D., and Posch, J.L.: Dupuytren's contracture: with special reference to pathology, J. Bone Joint Surg. **40-A**:773, 1958.

Larsen, R.D., Takagishi, N. and Posch, J.L.: The pathogenesis of Dupuytren's contracture: experimental and further clinical observations, J. Bone Joint Surg. **42-A**:993, 1960.

Legge, J.W.H., and McFarlane, R.M.: Prediction of results of treatment of Dupuytren's disease, J. Hand Surg. **5**:608, 1980.

Luck, J.V.: Dupuytren's contracture: a new concept of the pathogenesis correlated with surgical management, J. Bone Joint Surg. **41-A**:635, 1959.

Luck, J.V.: Dupuytren's contracture: pathogenesis and surgical management: a new concept. In American Academy of Surgeons: Instructional course lectures, vol. 16, St. Louis, 1959, The C.V. Mosby Co.

Lueders, H.W., Shapiro, R.L., and Lee, H.: Cross-finger pedicle flaps for recurrent Dupuytrens, Orthop. Rev. **4**:39, July 1975.

Matev, L.B.: Asymmetric Z-plasty in the operative treatment of Dupuytren's contracture, Am. Dig. Foreign Orthop. Lit. **1st qtr**:11, 1970.

McFarlane, R.M.: Patterns of the diseased fascia in the fingers in Dupuytren's contracture: displacement of the neurovascular bundle, Plast. Reconstr. Surg. **54**:31, 1974.

McFarlane, R.M., and Jamieson, W.G.: Dupuytren's contracture: the management of one hundred patients, J. Bone Joint Surg. **48-A**:1095, 1966.

McIndoe, A., and Beare, R.L.B.: The surgical management of Dupuytren's contracture, Am. J. Surg. **95**:197, 1958.

Mennen, U., and Grabe, R.P.: Dupuytren's contracture in a Negro: a case report, J. Hand Surg. **4**:451, 1979.

Millesi, H.: The clinical and morphological course of Dupuytren's disease. In Maladie de Dupuytren, Paris, 1966, L'Expansion Scientifique.

Moberg, E.: Three useful ways to avoid amputation in advanced Dupuytren's contracture, Orthop. Clin. North Am. **4**:1001, 1973.

Reumert, T., and Zachariae, L.: Continued investigations into the effect of diuretics upon oedema of the hand following operation for Dupuytren's contracture (Bumetanide, Leo), Acta Orthop. Scand. **44**:410, 1973.

Rhode, C.M., and Jennings, W.D., Jr.: Dupuytren's contracture, Am. J. Surg. **33**:855, 1967.

Richards, H.J.: The surgical treatment of Dupuytren's contracture, J. Bone Joint Surg. **36-B**:90, 1954.

Rodrigo, J.J., et al.: Treatment of Dupuytren's contracture: long-term results after fasciotomy and fascial excision, J. Bone Joint Surg. **58-A**:380, 1976.

Skoog, T.: Dupuytren's contracture, Postgrad. Med. **21**:91, 1957.

Skoog, T.: Dupuytren's contracture: pathogenesis and surgical treatment, Surg. Clin. North Am. **47**:433, 1967.

Snyder, C.C.: The contracture of Dupuytren, Am. Surg. **23**:487, 1957.

Spiegel, D., and Chase, R.A.: The treatment of contractures of the hand using self-hypnosis, J. Hand Surg. **5**:428, 1980.

Stein, A., et al.: Dupuytren's contracture: a morphologic evaluation of the pathogenesis, Ann. Surg. **151**:577, 1960.

Tubiana, R.: Limited and extensive operations in Dupuytren's contracture, Surg. Clin. North Am. **44**:1071, 1964.

Tubiana, R., Thomine, J.M., and Brown, S.: Complications in surgery of Dupuytren's contracture, Plast. Reconstr. Surg. **39**:603, 1967.

Wakefield, A.R.: Dupuytren's contracture, Surg. Clin. North Am. **40**:483, 1960.

Wang, M.K.H., et al.: Dupuytren's contracture, an analytic and etiologic study, Plast. Reconstr. Surg. **25**:323, 1960.

Webster, G.V.: A useful incision in Dupuytren's contracture, Plast. Reconstr. Surg. **19**:514, 1957.

Weckesser, E.C.: Results of wide excision of the palmar fascia for Dupuytren's contracture: special reference to factors which adversely affect prognosis, Ann. Surg. **160**:1007, 1964.

Zachariae, L.: Operation for Dupuytren's contracture by the method of McCash, Acta Orthop. Scand. **41**:433, 1970.

Zachariae, L., et al.: The effect of a diuretic (Centyl, Leo) on the oedema of the hand following surgical treatment of Dupuytren's contracture, Acta Orthop. Scand. **41**:411, 1970.

CHAPTER 16

Carpal tunnel and ulnar tunnel syndromes and stenosing tenosynovitis

CARPAL TUNNEL SYNDROME

Carpal tunnel syndrome, also known as tardy median nerve palsy, results from compression of the median nerve within the carpal tunnel. It occurs most often in patients between 30 and 60 years old and is five times more frequent in women than in men. Any condition that crowds or reduces the capacity of the carpal tunnel may initiate the symptoms; malaligned Colles' fracture and edema from infection or trauma are among the more obvious, and tumors or tumorous conditions such as a ganglion, lipoma, or xanthoma are among the more frequent. In Colles' fracture, immobilizing the wrist in marked flexion and ulnar deviation may cause acute compression of the median nerve within the carpal tunnel immediately after reduction. (Also see arterial injury.) Systemic conditions such as obesity, diabetes mellitus, thyroid dysfunction, amyloidosis, and Raynaud's disease are sometimes associated with the syndrome.

The cause is obscure in some patients, and hence the term *spontaneous* median nerve neuropathy. The syndrome is now more frequently being found associated with nonspecific and rheumatoid tenosynovitis, as are trigger finger and de Quervain's disease. Care should be taken not to confuse this syndrome with the symptoms of a herniated cervical disc.

Occasionally a patient will have symptoms of carpal tunnel syndrome caused by an habitual sleeping posture at night in which the wrist is kept acutely flexed under the chin.

Paresthesia over the sensory distribution of the median nerve is the most frequent symptom; it occurs more often in women, and frequently causes the patient to awaken several hours after getting to sleep with burning and numbness of the hand that is relieved by exercise. Tinel's sign may also be demonstrated in most patients by percussing the median nerve at the wrist. Atrophy to some degree of the thenar muscles (which are innervated by the median nerve) has been reported in about half of the patients treated by operation. Acute flexion of the wrist for 60 seconds (Phalen's test) in some but not all patients or strenous use of the hand increases the paresthesia. Application of a blood pressure cuff on the upper arm sufficient to produce venous distention may initiate the symptoms. Gelberman and co-workers found that with the wrist in neutral the mean pressure within the carpal tunnel in 15 patients with carpal tunnel syndrome was 32 mm Hg. This pressure increased to 99 mm Hg with 90 degrees of wrist flexion and increased to 110 mm Hg with the wrist at 90 degrees of extension. The pressures in the control subjects with the wrist in neutral were 2.5 mm Hg, 31 mm Hg with the wrist in flexion, and 30 mm Hg with the wrist in extension. This syndrome may occur during pregnancy but usually disappears after delivery.

Aberrant muscles of the forearm are another cause of median nerve compression. Acute median nerve symptoms with no obvious cause may be associated with an acute thrombosis of the median artery.

Treatment

If mild symptoms have been present and there is no thenar muscle atrophy, the injection of hydrocortisone into the carpal tunnel may afford relief. Great care should be taken not to inject directly into the nerve. Incidentally injection may also be used as a diagnostic tool in patients without bony or tumorous blocking of the canal; well over 65% percent of these cases are caused by a nonspecific synovitis, and these seem to respond more favorably to injection. This helps to eliminate the possibility of other syndromes, especially cervical disc or thoracic outlet. Some patients prefer to be injected two or three times before a surgical procedure is carried out. If the response is positive and there is no muscle atrophy, conservative treatment with splinting and injection is reasonable.

When signs and symptoms are persistent and progressive, especially if they include thenar atrophy, division of the deep transverse carpal ligament is indicated. The results of surgery are good in most instances (in 85% according to Lipscomb), and benefits seem to last in most patients. Any thenar atrophy may eventually disappear. Care should be taken to avoid the palmar sensory branch of the median nerve since, when severed, it frequently causes a painful neuroma that may later require excision from the scar.

349

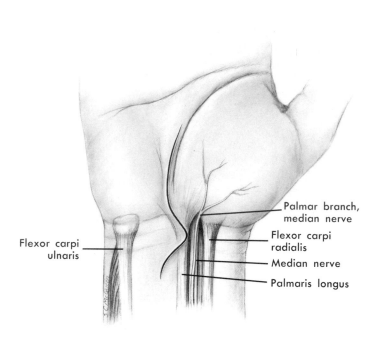

Fig. 16-1. Care should be taken in any incision about wrist to avoid cutting palmar branch of median nerve.

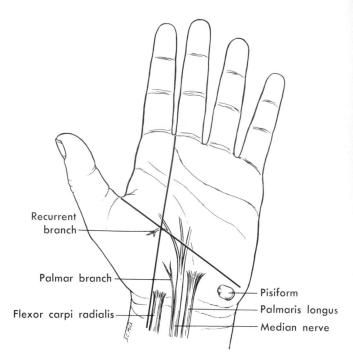

Fig. 16-2. Anatomic guidelines suggested by Kaplan to help in locating recurrent branch of median nerve.

When symptoms of median nerve compression develop during treatment of acute Colles' fracture, the constricting bandages and cast should be loosened and the wrist should be extended to neutral position. When tardy median nerve palsy develops after Colles' fracture and has been unrecognized for several weeks, then surgery is indicated without further delay.

TECHNIQUE. Make a curved incision ulnar to and paralleling the thenar crease. Extend this proximally to the flexor crease of the wrist where it may be continued farther proximally if necessary. Angle the incision toward the ulnar side of the wrist to avoid crossing the flexor creases at a right angle but especially to avoid cutting the palmar sensory branch of the median nerve that lies in the interval between the palmaris longus and the flexor carpi radialis tendons (Fig. 16-1). Should this nerve be severed, do not attempt to repair it but section it at its origin. After incising and reflecting the skin and subcutaneous tissue, divide the transverse carpal ligament along its ulnar border to avoid damage to the median nerve and its recurrent branch, which frequently perforates the distal border of the ligament and may come off the median nerve on the volar side (Fig. 16-2). The strong fibers of the transverse carpal ligament extend distally farther than is generally suspected (Fig. 16-3). Inspect the flexor tendon synovium. Tenosynovectomy may occasionally be indicated, especially in patients with rheumatoid arthritis. Close only the skin and drain the wound.

AFTERTREATMENT. A compression dressing and a volar splint are applied. The hand is actively used as soon as possible after surgery, but the dependent position is avoided. A smaller dressing may be applied after 1 week, and normal use of the hand is then encouraged. The su-

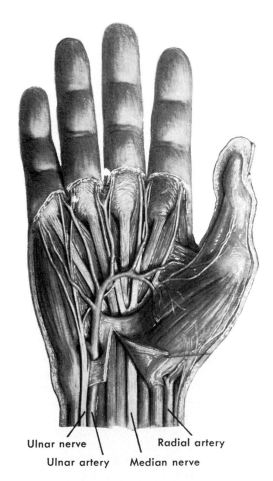

Fig. 16-3. Anatomic relations of deep transverse carpal ligament.

Fig. 16-4. Anatomic relations of structures within ulnar tunnel. **A,** Palmar view. Ulnar nerve lies medial to ulnar artery. **B,** Cross section of both carpal and ulnar tunnels. Ulnar tunnel at top right is bounded anteriorly by superficial transverse carpal ligament, posteriorly by deep transverse carpal ligament, and medially by pisiform bone and pisohamate ligament. Structures within this tunnel (*left to right*): vein, artery, vein, and nerve.

tures are removed after 10 to 14 days. The splint should be maintained for 14 to 21 days.

Unrelieved, or recurrent, carpal tunnel syndrome

In a series of explorations of previously operated patients by Langloh and Linscheid, good results were reported in one half and fair results in one third. Symptoms may recur from hyperplasia of the tenosynovium. On reexploring the canal, in most instances the deep transverse carpal ligament has reformed so completely that the site of previous division cannot be identified. Symptoms from cervical disc disease and thoracic outlet syndrome should be carefully ruled out.

ULNAR TUNNEL SYNDROME

Ulnar tunnel syndrome results from compression of the ulnar nerve within a tight triangular fibroosseous tunnel about 1.5 cm long located at the carpus. The walls of the tunnel consist of the superficial transverse carpal ligament anteriorly, the deep transverse carpal ligament posteriorly, and the pisiform bone and pisohamate ligament medially (Fig. 16-4). Like the median nerve within the carpal tunnel, the ulnar nerve is subject to compression within this tunnel. Compared with carpal tunnel syndrome, ulnar tunnel syndrome is much less common because the space occupied by the ulnar nerve at the wrist is much more yielding. The more common location of ulnar nerve constriction is at the elbow.

The exact level of compression determines whether symptoms are motor or sensory or both. Compression just distal to the tunnel affects the deep branch of the nerve that supplies most of the intrinsics (Fig. 17-31). Usually a ganglion is the cause of this compression. True or false

aneurysm of the ulnar artery (Fig. 16-5), thrombosis of the ulnar artery, or fracture of the hamate with hemorrhage may be the cause of pressure on the ulnar nerve. Other reported causes are lipoma and aberrant muscles. Occasionally in rheumatoid disease, carpal tunnel and ulnar tunnel syndromes both develop in the same hand. In the differential diagnosis, herniation of a cervical disc, thoracic outlet syndrome, and peripheral neuritis must be considered.

Treatment consists of exploration of the ulnar nerve at the wrist and removal of any ganglion or other cause of compression.

Should the ulnar artery be occluded for several millimeters, Raynaud's syndrome may be produced in the ulnar three digits because the sympathetic nerve fibers to these digits pass along the ulnar artery.

Segmental resection of the occluded section and replacement with a vein graft is the preferred procedure when it is feasible. Usually symptoms are relieved, and any weakened or atrophic intrinsic muscles recover in 3 to 12 months after surgery. For the technique of exploration, see the approach described for repair of the deep branch of the ulnar nerve, p. 123.

STENOSING TENOSYNOVITIS

Stenosing tenosynovitis occurs more often in the hand and wrist than anywhere else in the body. When the extensor pollicis brevis and the abductor pollicis longus tendons in the first dorsal compartment are affected, the condition is sometimes called de Quervain's disease. When the long flexor tendons are involved, trigger thumb, trigger finger, or snapping finger occurs. Less often, the extensor pollicis longus may be affected at the level of Lister's tubercle.

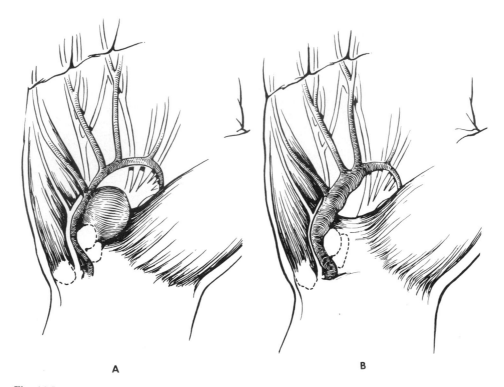

A   B

Fig. 16-5. Two types of traumatic aneurysms of ulnar artery in hand. **A,** Saccular "false" aneurysm arising from ulnar artery. **B,** "True" fusiform aneurysm of ulnar artery. (From Green, D.P.: J. Bone Joint Surg. **55-A:**120, 1973.)

Any of the other tendons that pass beneath the dorsal wrist retinaculum may also be involved. The tenosynovitis that precedes the stenosis may result from an otherwise dormant or subclinical collagen disease. Or recurrent mild trauma such as that experienced by carpenters, waitresses, or elevator operators may be the cause. Some case histories indicate that acute trauma may initiate the pathologic condition. The stenosis occurs at a point where the direction of a tendon changes, for here a sheath or fibrous theca acts as a pulley, and friction is maximum. Although the synovium lubricates the sheath, friction may cause a reaction when the repetition of a particular pattern of movement is necessary, as in winding a fine coil of wire or stacking laundry.

In our experience many cases of tenosynovitis in various locations, even stenosing tenosynovitis, respond favorably to injections of a steroid preparation mixed with a local anesthetic. Pain may increase temporarily during the initial 24 hours after the local anesthetic has been fully absorbed; therefore the patient should be warned about this possibility. It may be 3 to 7 days before the steroid becomes effective, but it avoids surgery in many instances.

The physician should determine first that the tenosynovitis is not caused by infection that could be worsened by steroid injections. Also the possibility of gout should be kept in mind.

de Quervain's disease

Stenosing tenosynovitis of the abductor pollicis longus and extensor pollicis brevis tendons occurs typically in adults between 30 and 50 years old. Women are affected 10 times more frequently than men, possibly because in women the carpometacarpal joint is more mobile. The cause is almost always occupational or is associated with rheumatoid arthritis. The presenting symptoms are usually pain and tenderness at the radial styloid. Sometimes a thickening of the fibrous sheath is palpable. Finkelstein's test is usually positive: "on grasping the patient's thumb and quickly abducting the hand ulnarward, the pain over the styloid tip is excruciating."* Although Finkelstein states that this test is "probably the most pathognomonic objective sign," it is not diagnostic; the patient's history and occupation, the roentgenograms, and the physical findings must also be considered.

Conservative treatment, consisting of rest on a splint and the injection of a steroid preparation, is most successful within the first 6 weeks after onset. But when pain persists, surgery is the treatment of choice.

Anatomic variations are numerous in this area. The point of insertion of the abductor pollicis longus tendon and the number of tendons in the sheath vary. Studies have shown that in about 75% of the wrists examined, at least one aberrant tendon was found. These tendons sometimes insert more proximally and more ulnarward than normally, into either the greater multangular (Fig. 16-6), the abductor pollicis brevis muscle (Fig. 16-7), the opponens muscle, or their fascia. Phylogenically the extensor pollicis brevis is a late tendon and is absent in about 5% of wrists. A distinct partition may be found separating the extensor pollicis brevis and the ab-

---

*From Finkelstein, H.: Stenosing tendovaginitis at the radial styloid process, J. Bone Joint Surg. **30:**509, 1930.

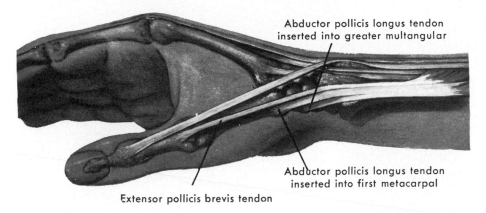

Fig. 16-6. Often abductor pollicis longus inserts on both greater multangular and base of first metacarpal through two tendons. Thus at surgery for de Quervain's disease at least one aberrant tendon is often found.

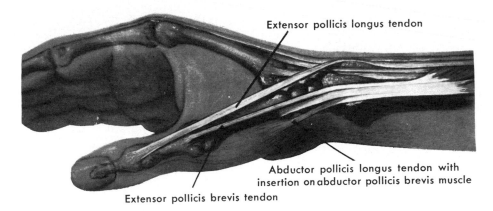

Fig. 16-7. Rarely abductor pollicis longus inserts on both fascia of abductor pollicis brevis and base of first metacarpal.

ductor pollicis longus tendons, with a common sheath enclosing both. Also, there may be another tendon and sheath separate from these two, and unless they are found at surgery, symptoms may not be relieved.

TECHNIQUE. Use a local anesthetic and a tourniquet (p. 1). Make a transverse incision parallel with the skin creases over the area of tenderness (Fig. 16-8). Avoid a longitudinal incision, since its scar will hypertrophy and perhaps limit motion. After retracting the skin edges, make a longitudinal incision in the underlying superficial fascia. Then find and protect the sensory branches of the superficial radial nerve, usually located deep to the superficial veins. Identify the tendons proximal to the stenosing dorsal ligament and sheath and open the first dorsal compartment on its ulnar side. With the thumb abducted and the wrist flexed, lift the abductor pollicis longus and the extensor pollicis brevis tendons from their groove. If they cannot be easily freed, look for additional aberrant tendons. Then close the skin incision and apply a small pressure dressing.

Failure to obtain complete relief after surgery may result from (1) formation of a neuroma in a severed branch of the superficial radial nerve, (2) volar subluxation of the tendon when too much of the sheath is removed, (3) failure to find and release a separate aberrant tendon within a separate compartment, and (4) hypertrophy of scar from a longitudinal skin incision.

AFTERTREATMENT. The small pressure dressing is removed after 48 hours, and a patch dressing is applied. Motion of the thumb and hand is immediately encouraged and is increased as tolerated.

Trigger finger and thumb

Trigger thumb may be congenital, but it also occurs in adults usually after 45 years of age. When associated with a collagen disease, several fingers may be involved—the middle and ring most often. A nodule or fusiform swelling of the flexor tendon just proximal to its theca at the distal palmar crease causes a relative stenosis of the sheath. The nodule can be palpated by the examiner's fingertip and will move with the tendon. The nodule is usually at the entry of the tendon into the proximal annulus at the level of the metacarpophalangeal joint; however, in the rheumatoid patient, a nodule distal to this point may cause triggering that will not always be relieved by sectioning the proximal annulus alone. Occasionally a partially lacerated flexor tendon at this level may heal with a nodule sufficiently large to cause triggering. Local tenderness may be present but is not a prominent complaint. Pressure accentuates the snap-

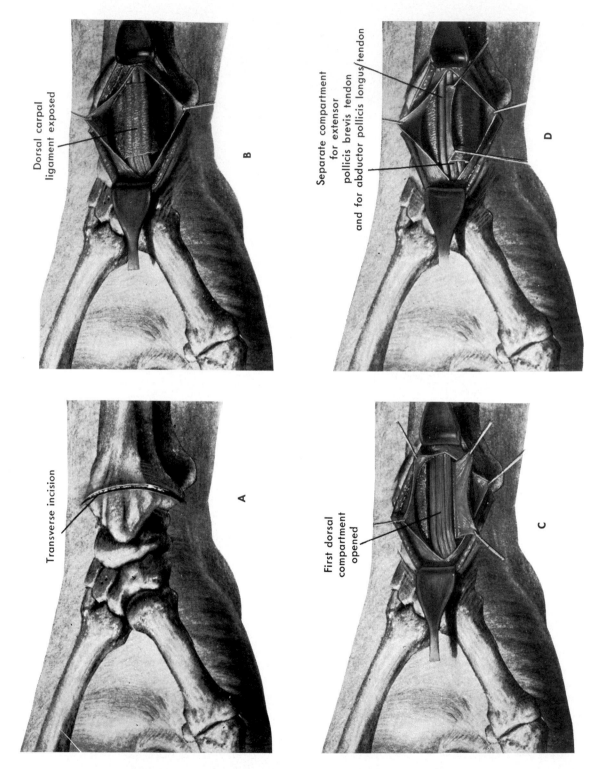

Transverse incision

A

Dorsal carpal
ligament exposed

B

First dorsal
compartment
opened

C

Separate compartment
for extensor
pollicis brevis tendon
and for abductor pollicis longus tendon

D

Fig. 16-8. Surgical treatment of de Quervain's disease. **A,** Skin incision. **B,** Dorsal carpal ligament has been exposed. **C,** First dorsal compartment has been opened on its ulnar side. **D,** Occasionally, separate compartments are found for extensor pollicis brevis and abductor pollicis longus tendons.

ping or triggering of the distal joints. It should be noted, particularly in the thumb, that the constriction is opposite the metacarpophalangeal joint, although the interphalangeal joint is the one that locks or snaps. (See also locking of joint, p. 165.) In congenital trigger digit deformities expectant observation should be the first treatment, since many will resolve within the first 6 months and nearly all within 2 years. Rarely is trigger finger seen in a child after 2 years of age. We have seen one 7-year-old child and one 11-year-old in whom the condition persisted.

TECHNIQUE. Make a transverse incision about 1.9 cm long just distal to the distal palmar crease for trigger finger (Fig. 16-9, *A*) or just distal to the flexor crease of the thumb at the metacarpophalangeal joint for trigger thumb (Fig. 1-13, *J*). Take care to avoid the digital nerves, which on the thumb are more volarward than might be anticipated. Iden-

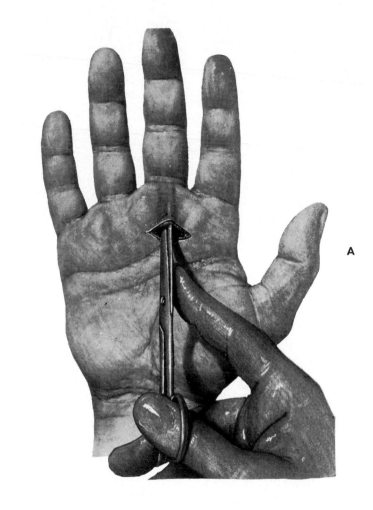

A

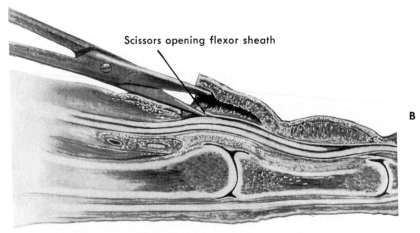

Scissors opening flexor sheath

B

Fig. 16-9. A, Surgical treatment of trigger finger (see text). **B,** One blade of scissors has been placed beneath proximal edge of tendon sheath.

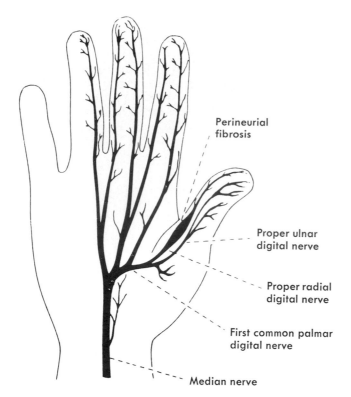

Perineurial fibrosis

Proper ulnar digital nerve

Proper radial digital nerve

First common palmar digital nerve

Median nerve

Fig. 16-10. Bowler's thumb. Distal sensory branches of median nerve in hand and location of perineural fibrosis of proper ulnar digital nerve of thumb are shown. (From Minkow, F.V., and Bassett, F.H., III: Clin. Orthop. **83:**115, 1972.)

tify with a small probe the discrete proximal edge of the flexor sheath. Now place a small knife blade or one blade of a pair of slightly opened blunt scissors just under the edge of the sheath and gently push it distally, cutting the sheath (Fig. 16-9, *B*). Thus the constriction of the tendon is released. Flex and extend the digit to be sure that the release is complete. Then close the skin and apply a small dry compression dressing.

AFTERTREATMENT. The compression dressing is removed after 48 hours, and a patch dressing is applied. Normal use of the finger or thumb is then advised.

Bowler's thumb

Bowler's thumb is a perineural fibrosis caused by repetitious compression of the ulnar digital nerve of the thumb (Fig. 16-10) while grasping a bowling ball. Bowlers with this condition are usually those who bowl 3 or 4 times a week. It is accompanied by tingling and hyperesthesia about the pulp. There is usually a palpable lump that is exceedingly tender and at times accompanied by distal skin atrophy. Early awareness of the cause can lead to protection of the thumb by a shield or splint and rest from bowling to help reduce the symptoms and to prevent the condition from becoming chronic. Occasionally neurolysis and dorsal transfer of the nerve become necessary.

REFERENCES

Adamson, J.E., et al.: The acute carpal tunnel syndrome, Plast. Reconstr. Surg. **47:**332, 1971.

Aghasi, M.K., Rzetelny, V., and Axer, A.: The flexor digitorum superficialis as a cause of bilateral carpal-tunnel syndrome and trigger wrist, J. Bone Joint Surg. **62-A:**134, 1980.

Alegado, R.B., and Meals, R.A.: An unusual complication following surgical treatment of de Quervain's disease, J. Hand Surg. **4:**185, 1979.

Ariyan, S., and Watson, S.K.: The palmar approach for the visualization and release of the carpal tunnel: an analysis of 429 cases, Plast. Reconstr. Surg. **60:**539, 1977.

Baird, D.B., and Friedenberg, Z.B.: Delayed ulnar-nerve palsy following a fracture of the hamate, J. Bone Joint Surg. **50-A:**570, 1968.

Barton, N.J.: Another cause of median nerve compression by a lumbrical muscle in the carpal tunnel, J. Hand Surg. **4:**189, 1979.

Bauman, T.D., et al.: The acute carpal tunnel syndrome, Clin. Orthop. **156:**151, 1981.

Bell, G.E., Jr., and Goldner, J.L.: Compression neuropathy of the median nerve, South. Med. J. **49:**966, 1956.

Berman, A.T., and Straub, R.R.: Importance of preoperative and postoperative electrodiagnostic studies in the treatment of the carpal tunnel syndrome, Orthop. Rev. **3:**57, June 1974.

Brian, W.R., Wright, A.D., and Wilkinson, M.: Spontaneous compression of both median nerves in the carpal tunnel: six cases treated surgically, Lancet **1:**277, 1947.

Brown, F.E., and Tanzer, R.C.: Entrapment neuropathies of the upper extremity. In Flynn, J.E., editor: Hand Surgery, ed. 3, Baltimore, 1982, Williams & Wilkins.

Brown, L.P., and Coulson, D.B.: Triggering at the carpal tunnel with incipient carpal-tunnel syndrome: report of an unusual case, J. Bone Joint Surg. **56-A:**623, 1974.

Browne, E.Z., Jr., and Snyder, C.C.: Carpal tunnel syndrome caused by hand injuries, Plast. Reconstr. Surg. **56:**41, 1975.

Bruckschwaiger, O.: An atypical form of de Quervain's disease, Can. Med. J. **71:**277, 1954.

Burman, M.: Stenosing tendovaginitis of the dorsal and volar compartments of the wrist, Arch. Surg. **65:**752, 1952.

Burnham, P.J.: Acute carpal tunnel syndrome: median artery thrombosis as cause, Arch. Surg. **87:**645, 1963.

Butler, B., Jr., and Bigley, E.C., Jr.: Aberrant index (first) lumbrical tendinous origin associated with carpal-tunnel syndrome: a case report, J. Bone Joint Surg. **53-A:**160, 1971.

Caffee, H.H.: Anomalous thenar muscle and median nerve: a case report, J. Hand Surg. **4:**446, 1979.

Cameron, B.M.: Occlusion of the ulnar artery with impending gangrene of the fingers relieved by section of the volar carpal ligament, J. Bone Joint Surg. **36-A:**406, 1954.

Cannon, B.W., and Love, J.G.: Tardy median palsy: median neuritis: median thenar neuritis amenable to surgery, Surgery **20:**210, 1946.

Carroll, R.E., and Hurst, L.C.: The relationship of the thoracic outlet syndrome and carpal tunnel syndrome, Clin. Orthop. **164:**149, 1982.

Carroll, R.E., and Green, D.P.: The significance of the palmar cutaneous nerve at the wrist, Clin. Orthop. **83:**24, 1972.

Carson, F.L., and Kingsley, W.B.: Nonamyloid green birefringence following congo red staining, Arch. Pathol. Lab. Med. **104:**333, 1980.

Conklin, J.E., and White, W.L.: Stenosing tenosynovitis and its possible relation to the carpal tunnel syndrome, Surg. Clin. North Am. **40:**531, 1960.

Craven, P.R., and Green, D.P.: Cubital tunnel syndrome: treatment by medial epicondylectomy, J. Bone Joint Surg. **62-A:**986, 1980.

Cseuz, K.A., et al.: Long-term results of operation for carpal tunnel syndrome, Mayo Clin. Proc. **41:**232, 1966.

Curtis, R.M., and Eversmann, W.W., Jr.: Internal neurolysis as an adjunct to the treatment of the carpal-tunnel syndrome, J. Bone Joint Surg. **55-A:**733, 1973.

De Abreau, L.B., and Moreira, R.G.: Median-nerve compression at the wrist, J. Bone Joint Surg. **40-A:**1426, 1958.

Dellon, A.L.: Clinical use of vibratory stimuli to evaluate peripheral nerve injury and compression neuropathy, Plast. Reconstr. Surg. **65:**466, 1980.

DeLuca, F.N., and Cowen, N.J.: Median-nerve compression complicating a tendon graft prosthesis, J. Bone Joint Surg. **57-A:**553, 1975.

Dickson, D.D., and Luckey, C.: Tenosynovitis of the extensor carpi ulnaris tendon sheath, J. Bone Joint Surg. **30-A:**903, 1948.

Doyle, J.R., and Carroll, R.E.: The carpal tunnel syndrome: a review of 100 patients treated surgically, Calif. Med. **108:**263, 1968.

Drury, B.J.: Traumatic tendovaginitis of the fifth dorsal compartment of the wrist. Arch. Surg. **80:**554, 1960.

Dupont, C., et al.: Ulnar tunnel syndrome at the wrist: a report of four cases of ulnar-nerve compression at the wrist, J. Bone Joint Surg. **47-A:**757, 1965.

Engber, W.D., Marti, L.B., and Moore, C.T.: Peripheral neuropathy: an unusual complication of meningococcemia, J. Hand Surg. **2:**404, 1977.

Enger, W.D., and Gmeiner, J.G.: Palmar cutaneous branch of the ulnar nerve, J. Hand Surg. **5:**26, 1980.

Eversmann, W.W., Jr., and Tirsick, J.A.: Intraoperative changes in motor nerve conduction latency in carpal tunnel syndrome, J. Hand Surg. **3:**77, 1978.

Fahey, J.J., and Bollinger, J.A.: Trigger-finger in adults and children, J. Bone Joint Surg. **36-A:**1200, 1954.

Fahrer, M., and Millroy, P.J.: Ulnar compression neuropathy due to an anomalous abductor digiti minimi: clinical and anatomic study, J. Hand Surg. **6:**266, 1981.

Fenton, R.: Stenosing tendovaginitis at the radial styloid involving an accessory tendon sheath, Bull. Hosp. Joint Dis. **11:**90, 1950.

Finkelstein, H.: Stenosing tendovaginitis at the radial styloid process, J. Bone Joint Surg. **12:**509, 1930.

Fissette, J., Onkelinx, A., and Fandi, N.: Carpal and Guyon tunnel syndrome in burns at the wrist, J. Hand Surg. **6:**13, 1981.

Fitton, J.M., Shea, F.W., and Goldie, W.: Lesions of the flexor carpi radialis tendon and sheath causing pain at the wrist, J. Bone Joint Surg. **40-B:**359, 1968.

Foster, R.J., and Edshage, S.: Factors related to the outcome of surgically managed compressive ulnar neuropathy at the elbow level, J. Hand Surg. **6:**181, 1981.

Freshwater, M.F., and Arons, M.S.: The effect of various adjuncts on the surgical treatment of carpal tunnel syndrome secondary to chronic synovitis, Plast. Reconstr. Surg. **61:**93, 1978.

Frymoyer, J.W., and Bland, J.: Carpal-tunnel syndrome in patients with myxedematous arthropathy, J. Bone Joint Surg. **55-A:**78, 1973.

Gama, C., and Franca, C.M.: Nerve compression by pacinian corpuscles, J. Hand Surg. **5:**207, 1980.

Garland, H., Sumner, D., and Clark, J.M.P.: Carpal-tunnel syndrome: with particular reference to surgical treatment, Br. Med. J. **1:**581, 1963.

Gelberman, R.H., Aronson, D., and Weisman, M.H.: Carpal-tunnel syndrome: results of a prospective trial of steroid injection and splinting, J. Bone Joint Surg. **62-A:**1181, 1980.

Gelberman, R.H., Hergenroeder, P.T., Hargens, A.R., Lundborg, G.N., and Akerson, W.H.: The carpal tunnel syndrome: a study of carpal tunnel pressures, J. Bone Joint Surg. **63-A:**380, 1981.

Gelberman, R.H., Szabo, R.M., Williamson, R.V., and Dimick, M.P.: Sensibility testing in peripheral nerve compression syndromes: an experimental study in humans, J. Bone Joint Surg. **65-A:**632, 1983.

Gore, D.R.: Carpometacarpal dislocation producing compression of the deep branch of the ulnar nerve, J. Bone Joint Surg. **53-A:**1387, 1971.

Green, D.P.: True and false traumatic aneurysms in the hand: report of two cases and review of the literature, J. Bone Joint Surg. **55-A:**120, 1973.

Halpern, A.A., and Nagel, D.A.: Compartment syndromes of the forearm: early recognition using tissue pressure measurements, J. Hand Surg. **4:**258, 1979.

Harris, C.M., Tanner, E., Goldstein, M.N., and Pettee, D.S.: The surgical treatment of the carpal-tunnel syndrome correlated with preoperative nerve conduction studies, J. Bone Joint Surg. **61-A:**93, 1979.

Harvey, F.J.: Locking of the metacarpophalangeal joints, J. Bone Joint Surg. **56-B:**156, 1974.

Hayes, C.W., Jr.: Ulnar tunnel syndrome from giant cell tumor of tendon sheath: a case report, J. Hand Surg. **3:**187, 1978.

Heathfield, K.W.G.: Acroparaesthesiae and the carpal-tunnel syndrome, Lancet **2:**663, 1957.

Hecht, O., and Lipsker, E.: Median and ulnar nerve entrapment caused by ectopic calcification: report of two cases, J. Hand Surg. **5:**30, 1980.

Herndon, J.H., et al.: Carpal-tunnel syndrome: an unusual presentation of osteoid-osteoma of the capitate, J. Bone Joint Surg. **56-A:**1715, 1974.

Hirasawa, Y., Sawamura, H., and Sakakida, K.: Entrapment neuropathy due to bilateral epitrochleoanconeus muscles: a case report, J. Hand Surg. **4:**181, 1979.

Howard, F.M.: Ulnar-nerve palsy in wrist fractures, J. Bone Joint Surg. **43-A:**1197, 1961.

Hybbinette, C.H., and Mannerfelt, L.: The carpal tunnel syndrome: a retrospective study of 400 operated patients, Acta Orthop. Scand. **46:**610, 1975.

Inglis, A.E.: Two unusual operative complications in the carpal-tunnel syndrome: a report of two cases, J. Bone Joint Surg. **62-A:**1208, 1980.

Jabaley, M.E.: Personal observations on the role of the lumbrical muscles in carpal tunnel syndrome, J. Hand Surg. **3:**82, 1978.

Jackson, I.T., and Campbell, J.C.: An unusual cause of carpal tunnel syndrome: a case of thrombosis of the median artery, J. Bone Joint Surg. **52-B:**330, 1970.

Jeffrey, A.K.: Compression of the deep palmar branch of the ulnar nerve by an anomalous muscle: case report and review, J. Bone Joint Surg. **53-B:**718, 1971.

Johnson, R.K., Spinner, M., and Shrewsbury, M.M.: Median nerve entrapment syndrome in the proximal forearm, J. Hand Surg. **4:**48, 1979.

Johnson, R.K., and Shrewsbury, M.M.: Anatomical course of the thenar branch of the median nerve—usually in a separate tunnel through the transverse carpal ligament, J. Bone Joint Surg. **52-A:**269, 1970.

Keon-Cohen, B.: de Quervain's disease, J. Bone Joint Surg. **33-B:**96, 1951.

Kleinert, H.E., and Hayes, J.E.: The ulnar tunnel syndrome, Plast. Reconstr. Surg. **47:**21, 1971.

Kolind-Sorensen, V.: Treatment of trigger fingers, Acta Orthop. Scand. **41:**428, 1970.

Kummel, B.M., and Zazanis, G.A.: Shoulder pain as the presenting complaint in carpal tunnel syndrome, Clin. Orthop. **92:**227, 1973.

Lacey, T., II, Goldstein, L.A., and Tobin, C.E.: Anatomical and clinical study of the variations in the insertions of the abductor pollicis longus tendon associated with stenosing tendovaginitis, J. Bone Joint Surg. **33-A:**347, 1951.

Langloh, N.D., and Linscheid, R.L.: Recurrent and unrelieved carpal-tunnel syndrome, Clin. Orthop. **83:**41, 1972.

Lanz, U.: Anatomical variations of the median nerve in the carpal tunnel, J. Hand Surg. **2:**44, 1977.

Lapidus, P.W.: Symposium on ambulant surgery, stenosing tenovaginitis, Surg. Clin. North Am. **33:**1317, 1953.

Lapidus, P.W., and Guidotti, F.P.: Stenosing tenovaginitis of the wrist and fingers, Clin. Orthop. **83:**87, 1972.

Leão, L.: de Quervain's disease: a clinical and anatomical study, J. Bone Joint Surg. **40-A:**1063, 1958.

Lichtman, D.M., Florio, R.L., and Mack, G.R.: Carpal tunnel release under local anesthesia: evaluation of the outpatient procedure, J. Hand Surg. **4:**544, 1979.

Linscheid, R.L.: Carpal tunnel syndrome secondary to ulnar bursa distention from the intercarpal joint: report of a case, J. Hand Surg. **4:**191, 1979.

Linscheid, R.L., Peterson, L.F.A., and Juergens, J.L.: Carpal-tunnel syndrome associated with vasospasm, J. Bone Joint Surg. **49-A:**1141, 1967.

Lipscomb, P.R.: Tenosynovitis of the hand and the wrist: carpal tunnel syndrome, de Quervain's disease, trigger digit, Clin. Orthop. **13:**164, 1959.

Lister, G.D., Belsole, R.B., and Kleinert, H.E.: The radial tunnel syndrome, J. Hand Surg. **4:**52, 1979.

Littler, J.W., and Li, C.S.: Primary restoration of thumb opposition with median nerve decompression, Plast. Reconstr. Surg. **39:**74, 1967.

Loomis, L.K.: Variations of stenosing tenosynovitis at the radial styloid process, J. Bone Joint Surg. **33-A:**340, 1951.

Loomis, L.K.: Flexion deformity of the infant thumb, South Med. J. **50:**1259, 1957.

Lorthioir, J., Jr.: Surgical treatment of trigger-finger by a subcutaneous method, J. Bone Joint Surg. **40-A:**793, 1958.

Lundborg, G., Gelberman, R.H., Minteer-Convery, M., Lee, Y.F., and Hargens, A.R.: Median nerve compression in the carpal tunnel: functional response to experimentally induced controlled pressure, J. Hand Surg. **7:**252, 1982.

Lutter, L.D.: A new cause of locking fingers, Clin. Orthop. **83:**131, 1972.

Lynch, A.C., and Lipscomb, P.R.: The carpal tunnel syndrome and Colles' fractures, JAMA **185:**363, 1963.

MacDonald, R.I., et al.: Complications of surgical release for carpal tunnel syndrome, J. Hand Surg. **3:**70, 1978.

MacDougal, B., Weeks, P.M., and Wray, R.C., Jr.: Median nerve compression and trigger finger in the mucopolysaccharidoses and related diseases, Plast. Reconstr. Surg. **59:**260, 1977.

Magassy, C.L., et al.: Ulnar tunnel syndrome, Orthop. Rev. **2:**21, 1973.

Mangini, U.: Some remarks on the etiology of the carpal tunnel compression of the median nerve, Bull. Hosp. Joint Dis. **22:**56, 1961.

May, J.W., and Rosen, H.: Division of the sensory ramus communicans between the ulnar and median nerves: a complication following carpal tunnel release, J. Bone Joint Surg. **63-A:**836, 1981.

McCarthy, R.E., and Nalebuff, E.A.: Anomalous volar branch of the dorsal cutaneous ulnar nerve: a case report, J. Hand Surg. **5:**19, 1980.

McCormack, R.M.: Carpal tunnel syndrome, Surg. Clin. North Am. **40:**517, 1960.

McFarland, G.B., Jr., and Hoffer, M.M.: Paralysis of the intrinsic muscles of the hand secondary to lipoma in Guyon's tunnel, J. Bone Joint Surg. **53-A:**375, 1971.

Muckart, R.D.: Stenosing tendovaginitis of abductor pollicis longus and extensor pollicis brevis at the radial styloid (de Quervain's disease), Clin. Orthop. **33:**201, 1964.

Murphy, I.D.: An unusual form of de Quervain's syndrome: report of two cases, J. Bone Joint Surg. **31-A:**858, 1949.

Nissenbaum, M., and Kleinert, H.E.: Treatment considerations in carpal tunnel syndrome with coexistent Dupuytren's disease, J. Hand Surg. **5:**544, 1980.

Ogden, J.A.: An unusual branch of the median nerve, J. Bone Joint Surg. **54-A:**1779, 1972.

Omer, G.E., and Spinner, M.: Management of peripheral nerve problems, Philadelphia, 1980, W.B. Saunders Co.

Pecket, P., et al.: Variations in the arteries of the median nerve: with special considerations on the ischemic factor in the carpal tunnel syndrome (CTS), Clin. Orthop. **97:**144, 1973.

Phalen, G.S.: The carpal tunnel syndrome. In AAOS: Instructional course lectures, vol. 14, Ann Arbor, 1957, J.W. Edwards.

Phalen, G.S.: The carpal-tunnel syndrome: seventeen years' experience in diagnosis and treatment of six hundred fifty-four hands, J. Bone Joint Surg. **48-A:**211, 1966.

Phalen, G.S.: Reflections on 21 years' experience with the carpal-tunnel syndrome, JAMA **212:**1365, 1970.

Phalen, G.S.: The carpal-tunnel syndrome: clinical evaluation of 598 hands, Clin. Orthop. **83:**29, 1972.

Phalen, G.S.: The birth of a syndrome, or carpal tunnel revisited, J. Hand Surg. **6:**109, 1981.

Phalen, G.S., Gardner W.J., and LaLonde, A.A.: Neuropathy of the median nerve due to compression beneath the transverse carpal ligament, J. Bone Joint Surg. **32-A:**109, 1950.

Phalen, G.S., and Kendrick, J.I.: Compression neuropathy of the median nerve in the carpal tunnel, JAMA **164:**524, 1957.

Pick, R.Y.: de Quervain's disease, a clinical triad, Clin. Orthop. **143:**165, 1979.

Posch, J.L., and Marcotte, D.R.: Carpal tunnel syndrome: an analysis of 1,201 cases, Orthop. Rev. **5:**25, May 1976.

Ragi, E.F.: Carpal tunnel syndrome: a statistical review, Electromyogr. Clin. Neurophysiol. **21:**373, 1981.

Richmond, D.A.: Carpal ganglion with ulnar nerve compression, J. Bone Joint Surg. **45-B:**513, 1963.

Riordan, D.C., and Stokes, H.M.: Synovitis of the extensors of the fingers associated with extensor digitorum brevis manus muscle; a case report, Clin. Orthop. **95:**273, 1973.

Ritter, M.A., and Inglis, A.E.: The extensor indicis proprius syndrome, J. Bone Joint Surg. **51-A:**1645, 1969.

Rockey, H.C.: Trigger-finger due to a tenosynovial osteochondroma. J. Bone Joint Surg. **45-A:**387, 1963.

Rowland, S.A.: A palmar incision for release of the carpal tunnel, Clin. Orthop. **103:**89, 1974.

Rubinstein, M.A.: Carpal tunnel syndrome in lymphatic leukemia, JAMA **213:**1037, 1970.

Schultz, R.J., Endler, P.M., and Huddleston, H.D.: Anomalous median nerve and an anomalous muscle belly of the first lumbrical associated with carpal-tunnel syndrome: case report, J. Bone Joint Surg. **55-A:**1744, 1973.

Shivde, A.J., Dreizin, I., and Fisher, M.A.: The carpal tunnel syndrome: a clinical electrodiagnostic analysis, Electromyogr. Clin. Neurophysiol. **21:**143, 1981.

Smith, R.J.: Anomalous muscle belly of the flexor digitorum superficialis causing carpal-tunnel syndrome: report of a case, J. Bone Joint Surg. **53-A:**1215, 1971.

Sprecher, E.E.: Trigger thumb in infants, J. Bone Joint Surg. **31-A:**672, 1949.

Stein, A.H., Jr.: The relation of median nerve compression to Sudeck's syndrome, Surg. Gynecol. Obstet. **115:**713, 1962.

Stein, A.H., Jr., and Morgan, H.C.: Compression of the ulnar nerve at the level of the wrist, Am. Pract. **13:**195, 1962.

Stern, P.J., and Kutz, J.E.: An unusual variant of the anterior interosseous nerve syndrome: a case report and review of the literature, J. Hand Surg. **5:**32, 1980.

Steuber, J.B., and Klineman, W.B.: Flexor carpi radialis tunnel syndrome, J. Hand Surg. **6:**293, 1981.

Strandell, G.: Variations of the anatomy in stenosing tenosynovitis at the radial styloid process, Acta Chir. Scand. **113:**234, 1957.

Taleisnik, J.: The palmar cutaneous branch of the median nerve and the approach to the carpal tunnel: an anatomical study, J. Bone Joint Surg. **55-A:**1212, 1973.

Tanzer, R.C.: The carpal-tunnel syndrome: a clinical and anatomical study, J. Bone Joint Surg. **41-A:**626, 1959.

Tanzer, R.C.: The carpal tunnel syndrome, Clin. Orthop. **15:**171, 1959.

Taylor, A.R.: Ulnar nerve compression at the wrist in rheumatoid arthritis: report of a case. J. Bone Joint Surg. **56-A:**142, 1974.

Tompkins, D.G.: Median neuropathy in the carpal tunnel caused by tumor-like conditions: report of two cases, J. Bone Joint Surg. **49-A:**737, 1967.

Walton, S., and Cutler, C.R.: Carpal tunnel syndrome: case report of unusual etiology, Clin. Orthop. **74:**138, 1971.

Weeks, P.M.: A cause of wrist pain: non-specific tenosynovitis involving the flexor carpi radialis, Plast. Reconstr. Surg. **62:**263, 1978.

Weilby, A.: Trigger finger: incidence in children and adults and the possibility of a predisposition in certain age groups, Acta Orthop. Scand. **41:**419, 1970.

Wissinger, H.A.: Resection of the hook of the hamate: its place in the treatment of median and ulnar nerve entrapment in the hand, Plast. Reconstr. Surg. **56:**501, 1975.

Wood, M.B., and Linscheid, R.L.: Abductor pollicis longus bursitis, Clin. Orthop. **93:**293, 1973.

Wood, M.R.: Hydrocortisone injections for carpal tunnel syndrome, Hand **12:**62, 1980.

Wood, V.E.: Nerve compression following opponensplasty as a result of wrist anomalies: report of a case, J. Hand Surg. **5:**279, 1980.

Woods, T.H.E.: de Quervain's disease: a plea for early operation: a report on 40 cases, Br. J. Surg. **51:**358, 1964.

Yamaguchi, D.M., Lipscomb, P.R., and Soule, E.H.: Carpal tunnel syndrome, Minn. Med. **48:**22, 1965.

Young, L., and Holtmann, B.: Trigger finger and thumb secondary to amyloidosis, Plast. Reconstr. Surg. **65:**68, 1980.

Tumors and tumorous conditions of the hand

Tumors of the hand are common and varied, but each has its own peculiar incidence, malignant potential, and symptoms. Any tumor that may arise in an extremity may also occur in the hand. Some tumors, such as lipoma, are often found in the proximal part of the extremities but rarely in the hand; conversely, glomus tumors seem far more frequent in the hand than elsewhere. Malignant tumors arising from tissues of the hand other than skin are so rare that even single cases warrant publication.

Since the hand is a sensitive organ, has little potential free space, and is packed with moving parts, any tumor is usually detected early because of pain, impairment of function, or obvious swelling.

The treatment of tumors of the hand is surgical, with the possible exception of warts. Rarely is biopsy needed, since complete local excision is usually indicated, and the entire tumor is then available for microscopic study. The same principles of surgical technique described in Chapter 1 are followed in treating tumors.

The following list of tumors and tumorous conditions, although not exhaustive, serves as a guide in considering those most commonly found in the hand, excluding skin blemishes, warts, and malignant skin tumors. Most of those listed are discussed here.

Benign tumors
    Lipoma
    Giant cell tumor of tendon sheath (xanthoma)
    Fibroma
    Glomus tumor
    Hemangioma
    Traumatic neuroma
    Multiple neurofibromas
    Neurilemmoma (schwannoma)
    Osteoid osteoma
    Enchondroma
    Benign osteoblastoma
    Aneurysmal bone cyst
    Giant cell tumors of bone
    Osteochondroma
Malignant tumors
    Osteogenic sarcoma
    Chondrosarcoma
    Epithelioid sarcoma
    Fibrosarcoma
    Metastatic tumors
    Rhabdomyosarcoma
    Ewing's sarcoma
Tumorous conditions
    Ganglion
    Epidermoid cyst (inclusion cyst)
    Sebaceous cyst
    Mucous cyst
    Congenital arteriovenous fistula
    Pyogenic granuloma
    Foreign body granuloma
    Gout
    Déjerine-Sottas disease
    Calcinosis
    Localized calcium deposits
    Turret exotosis
    Carpometacarpal boss
    Paget's disease
    Epidermolysis bullosa

BENIGN TUMORS

Lipoma. Lipoma, although common elsewhere in the body, is rare in the hand (Figs. 17-1 and 17-2). It may be superficial and may have the characteristic signs of a soft fluctuant bulging mass, as seen in other areas of the body, or it may occur deep in the palm. Here it may cause hypesthesia of the skin by pressing on the sensory branch of the median nerve. Unhurried, careful dissection is necessary for removal.

Giant cell tumor of tendon sheath (xanthoma). Giant cell tumor of tendon sheath is the second most common subcutaneous tumor of the hand (Figs. 17-3 and 17-4). The reported age distribution is from 8 to 80 years. It occurs in the hand more frequently than in any other part of the body and more often on the palmar side of the fingers than on the dorsal. Its growth is usually slow, and it may remain the same size for many years. Pain and tenderness are rare. If it occurs at a joint, often the proximal interphalangeal joint, its size may interfere with joint motion. It rarely erodes bone. It is always benign, but about 10% may recur even after meticulous excision of friable fragments. Excision is often difficult because the tumor may wind in and around the flexor tendons and their sheaths, the digital nerves, and even the extensor tendons and may involve as much as three fourths of the circumference of one digit.

359

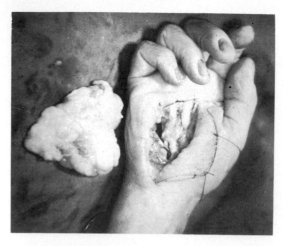

Fig. 17-1. Lipoma of palm extended through deep palmar space and between interosseus muscles to dorsum of hand.

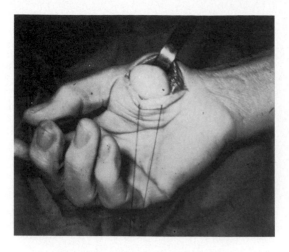

Fig. 17-2. Lipoma of thenar eminence.

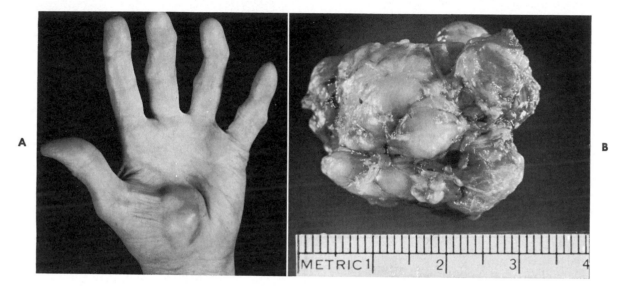

Fig. 17-3. A, Giant cell tumor of tendon sheath in palm. **B,** Gross specimen of tumor showing pseudoencapsulation. (From Phalen, G.S., McCormack, L.J., and Gazale, W.J.: Clin. Orthop. **15:**140, 1959.)

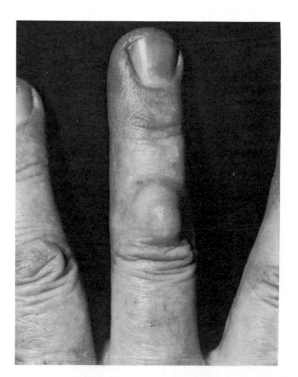

Fig. 17-4. Giant cell tumor of tendon sheath of left middle finger. (From Phalen, G.S., McCormack, L.J., and Gazale, W.J.: Clin. Orthop. **15:**150, 1959.)

Fibroma. Fibromoa is rare in the hand. It may be either deep, arising from a joint capsule, or superficial. It is encapsulated, and excision is usually curative.

Glomus tumor. Pain, cold sensitivity, and tenderness are the symptoms of a glomus tumor, a small but interesting tumor (Figs. 17-5 to 17-8). Direct pressure on the tumor by a small firm object such as a pinhead causes excruciating pain, whereas pressure slightly to one side of it is painless. It is usually less than 1 cm in diameter and often only a few millimeters. It may be deep red or purple. It occurs more often in the hand than in any other part of the body, typically under the fingernail. The tumor results from hypertrophy of a glomus, which is a normal structure of the skin, a coiled arteriovenous shunt whose function is to help regulate body temperature. It should be meticulously and completely excised but only after it has been accurately localized with the help of the conscious patient before the anesthetic is administered. The prognosis is good for relief of pain.

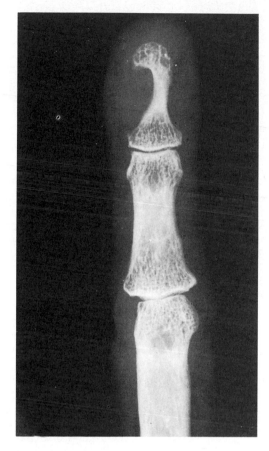

Fig. 17-6. Glomus tumor that has eroded distal phalanx. (From Posch, J.L.: J. Bone Joint Surg. **38-A:**517, 1956.)

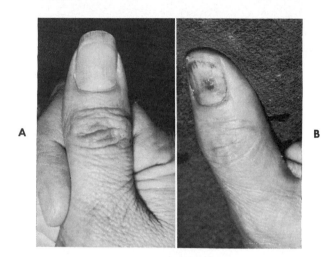

Fig. 17-5. **A,** Glomus tumor beneath thumbnail. **B,** Tumor after excision of nail (Courtesy Dr. G.S. Phalen.)

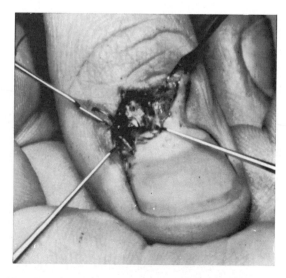

Fig. 17-7. Glomus tumor in subungual area of right thumb. (From Posch, J.L.: J. Bone Joint Surg. **38-A:**517, 1956.)

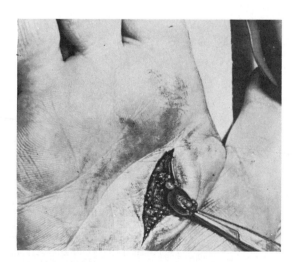

Fig. 17-8. Glomus tumor in palm. (From Posch, J.L.: J. Bone Joint Surg. **38-A:**517, 1956.)

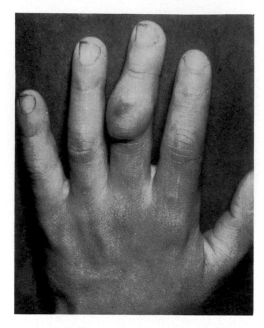

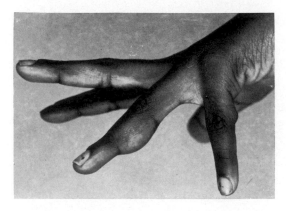

Fig. 17-10. Hemangioma of volar aspect of ring finger. (From Clifford, R.H., and Kelly, A.P., Jr.: Clin. Orthop. **13**:204, 1959.)

Fig. 17-9. Hemangioma of dorsal aspect of left middle finger. (From Posch, J.L.: J. Bone Joint Surg. **38-A:**517, 1956.)

Hemangioma. The following remarks are limited to the cavernous hemangioma (Figs. 17-9 and 17-10) and do not include the capillary superficial infantile hemangioma that tends to involute. A cavernous hemangioma may be slightly to moderately tender and may enlarge a digit with distended venous sinuses. It produces a bluish color when close to the surface and forms a soft, collapsible mass. At times the tumor is so extensive that it may have to be removed in stages. With careful tourniquet control, blood will partially fill the sinuses and will outline the extent of the tumor at the time of surgical excision. Complete excision is usually curative.

Traumatic neuroma. Traumatic neuroma is the result of an attempt by a peripheral nerve to regenerate after its fibers have been interrupted. The tumor is a bundle of all nerve elements in one tangled mass at the distal end of the proximal nerve segment. This attempted growth always occurs to some degree and is therefore not considered a true neoplasm. It may be extremely tender, particularly when involving a digital nerve and when adherent to an amputation scar unprotected by a good pad of skin and fat. The tumor is usually invisible from the exterior and is demonstrated by pressing on the suspected area with a firm object such as a pinhead or pencil point. When a painful neuroma occurs at an amputation site, enough of the nerve should be resected (usually about 6 mm in the case of a digital nerve) that its end can be well protected by a pad of subcutaneous fat and skin (see also discussions of nerve suture, p. 120, and finger amputation, p. 182). The neuroma will regenerate, of course, but will not be painful if sufficiently protected.

Multiple neurofibromas. Generalized cutaneous neurofibromas (Fig. 17-11), unlike traumatic neuromas, are not tender and are seen in clusters along the course of the nerve. They seem to be less common in the hand than in other parts of the body. The rate of malignant degeneration

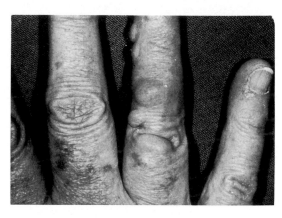

Fig. 17-11. Neurofibromas of finger. (From Gaisford, J.C.: Surg. Clin. North Am. **40**:549, 1960.)

has been reported as high as 15%, but this rate was not reported specifically of the hand. Excision is curative, but is indicated only if the tumors are being constantly irritated or are cosmetically unacceptable.

Neurilemmoma (schwannoma). Neurilemmomas arise from nerve sheaths (Fig. 17-12) and are rarely found in the hand. They are not extremely tender. They are more mobile at right angles to the course of the nerve than in line with the nerve. Excision is curative, but at times the whole nerve may have to be sacrificed if the mass cannot be dissected free from the nerve trunk. These tumors are frequently misdiagnosed as ganglions.

Osteoid osteoma. Osteoid osteoma, rare in the hand, is characterized by pain that gradually increases from mild to severe, is usually worse at night, and is dramatically relieved by aspirin. Some osteoid osteomas of the phalanges are painless presumably because of the lack of nerve fibers trapped within the tumor. Generalized swelling of the involved part and tenderness to pressure are frequent findings. The roentgenographic appearance depends on the area of bone involved. A small oval or round sclerotic nidus (Figs. 17-13 and 17-14) is surrounded first by an area of less dense bone, like a halo, and then by an area of sclerotic bone. If the lesion is in cortical bone or near the

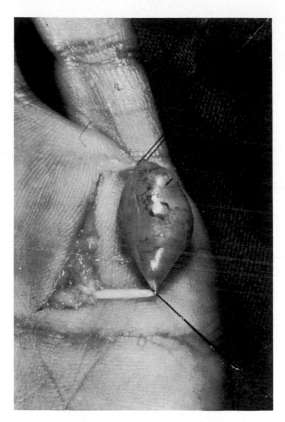

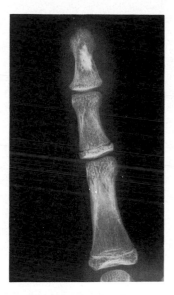

Fig. 17-12. Neurilemmoma. (From Clifford, R.H., and Kelly, A.P., Jr.: Clin. Orthop. **13**:204, 1959.)

Fig. 17-13. Osteoid osteoma of distal phalanx. Note sclerotic nidus. (From Dunitz, N.L., Lipscomb, P.R., and Ivins, J.C.: Am. J. Surg. **94**:65, 1957.)

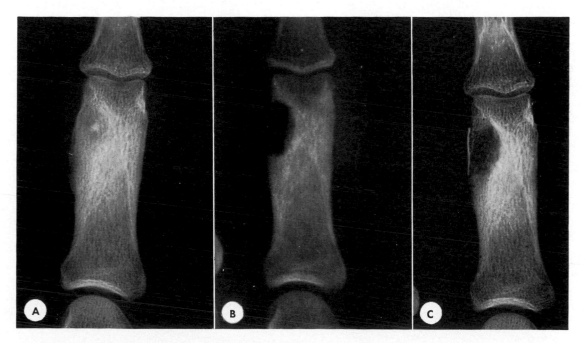

Fig. 17-14. Osteoid osteoma of proximal phalanx. **A,** Before surgery. Note sclerotic nidus surrounded by area of radiolucency. **B,** After block resection from involved part of bone. Defect was filled with graft of homogeneous iliac bone. **C,** Two months after surgery. Cortical part of graft is visible. (From Dunitz, N.L., Lipscomb, P.R., and Ivins, J.C.: Am. J. Surg. **94**:65, 1957.)

cortex, extreme sclerosis occurs and may obliterate the nidus, but at times it may be demonstrated by planograms.

Enchondroma. Enchondroma is the most destructive lesion of bone found in the hand (Figs. 17-15 and 17-16). Occasionally some enlargement of the finger is seen if the loculated medullary tumor has expanded the bony cortex. Pathologic fracture is a frequent complication, since only minimal trauma is needed to fracture the thin shell of bone. The fracture is usually allowed to heal, with thickening of the cortex by periosteal new bone, before excision of the tumor. At surgery a small window is made in the lateral aspect of the phalanx, and the soft cartilaginous material is curetted thoroughly; tiny flecks of calcium are often seen. The cavity is packed with bone chips, and the cortical window is replaced. Amputation may be necessary when all useful finger function has been destroyed. These tumors are found in multiplicity in Ollier's disease. The diagnosis can usually be made from the roentgenogram, but other destructive lesions, such as inclusion cysts, giant cell tumors, and aneurysmal bone cysts, should be considered.

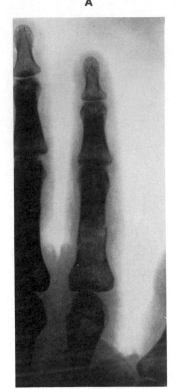

A

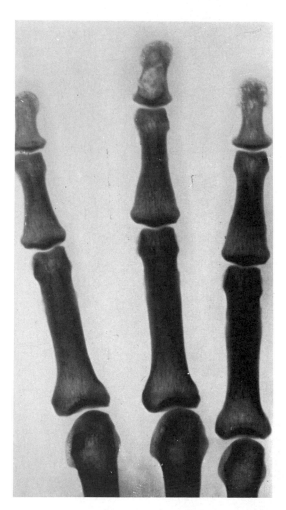

Fig. 17-15. Enchondroma of distal phalanx. In this location this tumor may be confused with epidermoid cyst (Fig. 17-33).

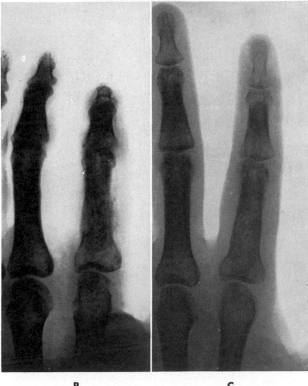

B   C

Fig. 17-16. Enchondroma of proximal phalanx. A, Before surgery. In this location this tumor may be confused with giant cell tumor (Fig. 17-17, *A*). B, After tumor has been curetted and cavity has been packed with cancellous bone. C, After bone has healed.

Benign osteoblastoma. This tumor is rare. The small bones of the hand and feet are its second most common location; it is rarely considered in the differential diagnosis of bone tumors of the hand. Fig. 17-17, *A,* shows the ground glass appearance of a benign osteoblastoma with gross deformity of the metacarpal from pressure from within, yet the cortical shell is intact. A fibular graft was interposed, and the epiphyseal plate and the distal subchondral cortex of the fifth metacarpal were preserved (Fig. 17-17, *B*). There was no evidence of recurrence of the tumor 15 months after surgery, and the metacarpal continued to grow (Fig. 17-17, *C*).

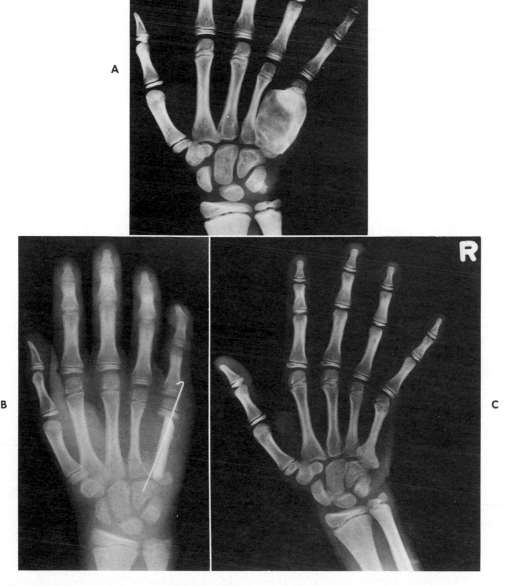

Fig. 17-17. A, Expanding intraosseous tumor of fifth metacarpal with ''ground glass'' appearance. Cortical shell is intact and has caused deformity of fourth metacarpal from pressure. **B,** Partial-thickness fibular graft has been interposed. Epiphyseal plate and subchondral cortex of proximal metacarpal have been preserved. **C,** Graft has remodeled 15 months after operation. evidence of tumor is absent, and growth plate and carpometacarpal joint have been maintained. (From Mosher, J.F., and Peckham, A.C.: J. Hand Surg. **3:**358, 1978.)

Aneurysmal bone cyst. Only about four cases of aneurysmal bone cyst of the hand have been reported, but only during the last four decades has it been recognized as a separate entity in any part of the body. When it occurs as a central lesion in a phalanx, it causes enlargement, limitation of motion, and pain (Figs. 17-18 and 17-19). In roentgenograms it is almost indistinguishable from giant cell tumor or enchondroma. The treatment is curettage and filling the cavity with bone chips.

Giant cell tumors of bone. Giant cell tumors of bone are uncommon in the hand. A literature review by Averill, Smith, and Campbell uncovered 39 reported tumors. In addition, they reported on a series of 28 tumors. In five of those hands, the lesion was eccentrically located. All lesions within the phalanges and metacarpals originated in

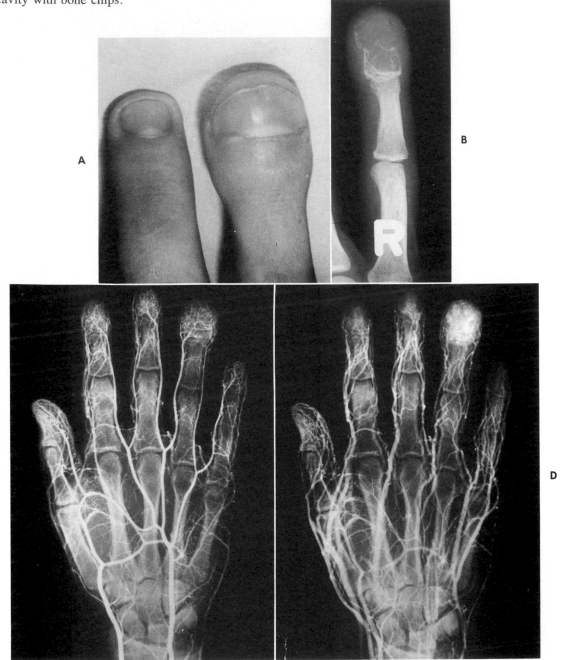

Fig. 17-18. White male, 17 years old, was first seen with swollen ring finger 1 year after tip of finger was crushed by falling railroad tie. Immediate swelling occurred with drainage from split skin at tip. Finger soon became asymptomatic as initial edema decreased; but insidious enlargement of distal phalanx, dull aching, and clubbing of nail were noted. Bulbous enlargement progressed over next 12 months. **A,** Initial examination demonstrated large bulbous distal phalanx. **B,** Roentgenogram of destroyed expanded distal phalanx. **C,** Arteriogram shows arterial filling of lesion and **D,** venous phase with large venous channels. (From Fuhs, S.E., and Herndon, J.H.: J. Hand Surg. **4:**152, 1979.)

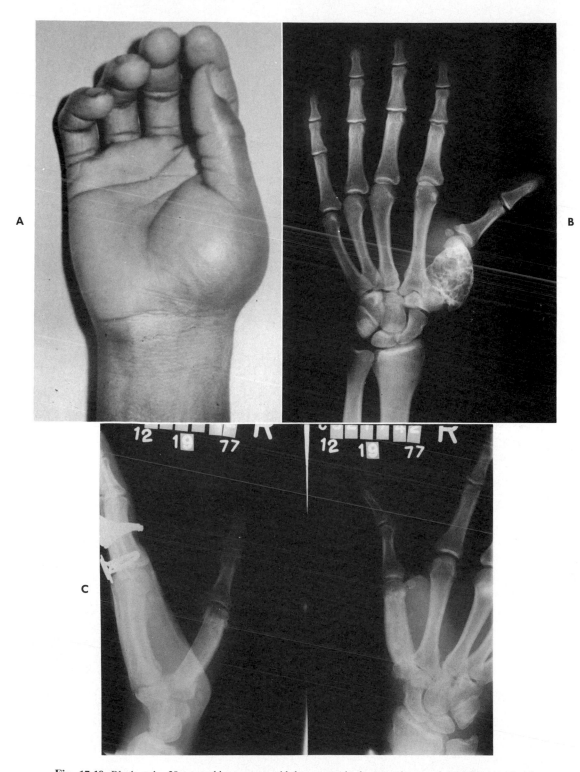

Fig. 17-19. Black male, 20 years old, was seen with bony mass in thenar eminence of right dominant hand. Two years earlier he had sustained injury to ulnar collateral ligament during high school wrestling. No medical treatment was carried out, and thumb became asymptomatic. While working on construction 5 to 6 months later, he noted dull aching about base of thumb and began having difficulty holding tools and basketball. **A,** Initial examination. Note large distorted thenar mass. **B,** Roentgenogram demonstrated cystic expanding lesion of metacarpal. Note disruption of proximal articular surface and preserved distal portion of metacarpal. **C,** Roentgenogram at 1 year after operation. Bone graft has fused to greater multangular and distal metacarpal. No tumor remains. (From Fuhs, S.E., and Herndon, J.H.: J. Hand Surg. **4:**152, 1979.)

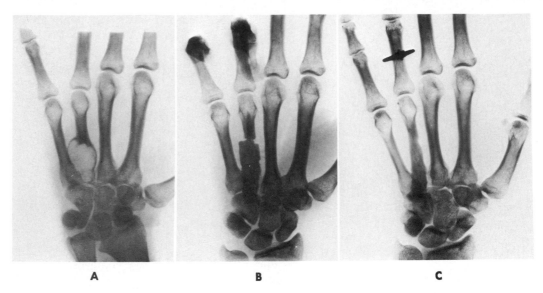

Fig. 17-20. Giant cell tumor metacarpal. **A,** Before surgery. **B,** Status 3 months after resection of proximal half of bone and insertion of tibial cortical graft. **C,** Status 9 months after surgery.

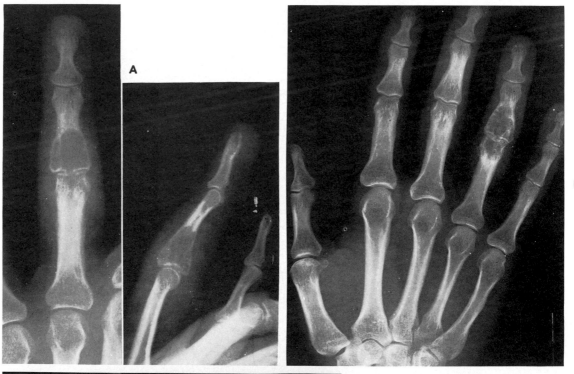

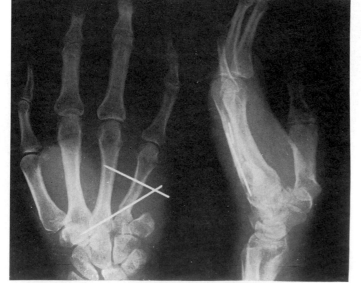

Fig. 17-21. Giant cell tumor of middle phalanx. **A,** Before excision. **B,** Status 5 years after excision. Tumor has recurred, and proximal phalanx has also become involved. **C,** After fourth ray has been amputated, and fifth ray has been transposed laterally. (From Stein, A.H., Jr.: Surg. Gynecol. Obstet. **109:**189, 1959.)

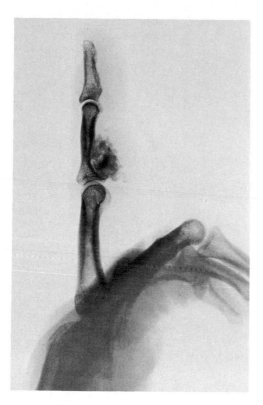

Fig. 17-22. Osteochondroma of middle phalanx.

the epiphyses; however, some later extended into the diaphysis. They noted extensive destruction of cortex and cancellous bone with marked expansion of the cortex. Most lesions had a thin covering of periosteum remaining. This tumor should not be confused with an enchondroma, and a biopsy is indicated to confirm the diagnosis.

In the 28 patients observed by Averill et al., five had cases of multicentric giant cell tumors. Tumors were found in places such as the proximal humerus, the distal radius, and the distal femur. This suggests that a full bone survey is indicated to discover remote sites of tumor when a giant cell tumor is suspected.

Generally curettage and bone grafting are not sufficient treatment for this tumor. Because giant cell tumors of the hand are just as aggressive as those found elsewhere, if the cortex is not eroded, resection of the bone and reconstructive surgery are indicated (Fig. 17-20); if cortical invasion and destruction recur ablation of the part is then indicated (Fig. 17-21). Radiation therapy for giant cell tumors of bone has been ineffective and has resulted in radiation sarcoma in as many as 20%.

Osteochondroma. Osteochondromas are rare in the hand but are occasionally seen on a phalanx (Fig. 17-22).

MALIGNANT TUMORS

In the hand, malignant tumors are rare beneath the skin. Of primary bone malignancies of the hand, chondrosarcoma is the most common. Fibrosarcoma and rhabdomyo-

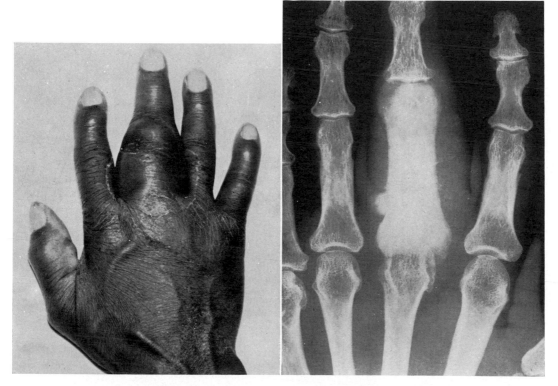

Fig. 17-23. Osteogenic sarcoma of proximal phalanx. (From Drompp, B.W.: J. Bone Joint Surg. **43-A**:199, 1961.)

sarcoma rank first and second in frequency in soft tissues, followed by synovioma. Epithelioid sarcoma in some series is reported just as frequently as fibrosarcoma and rhabdomyosarcoma.

Osteogenic sarcoma. Osteogenic sarcoma in the hand is so rare that about twelve proved cases have been reported (Fig. 17-23). Some of these are associated with irradiation from overexposure to roentgen rays or to ingestion of radium salts. Trauma seems to play no part in the cause. Careful wide local resection of the tumor offers a good prognosis; this is in contrast to the same tumor in other parts of the body.

Chondrosarcoma. Chondrosarcoma of the hand may occasionally occur in the carpus. Granberry and Bryan re-

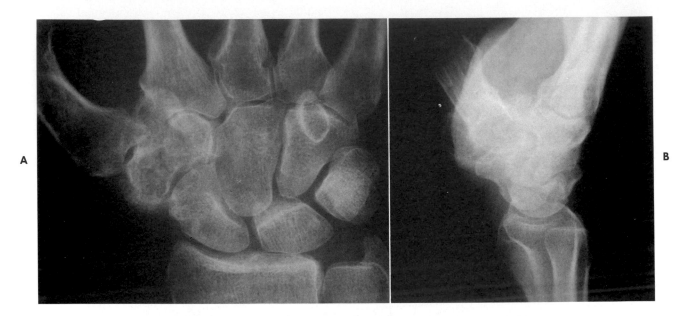

Fig. 17-24. Roentgenograms of carpus demonstrating enlargement of greater multangular with encroachment on thenar musculature. Early cystic reaction is seen in adjacent bones. Tumor at this time was thought most likely to be enchondroma. **A,** Anteroposterior view. **B,** Lateral view. (From Granberry, W.M., and Bryan, W.: J. Hand Surg. **3:**277, 1978.)

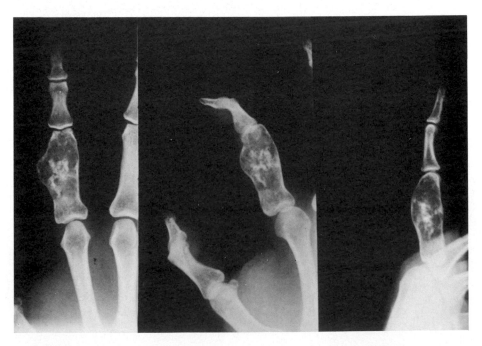

Fig. 17-25. Posteroanterior, lateral, and oblique views of right index finger, showing greatly expanded proximal phalanx in which fluffy radiopacity is clearly visible. (From Wu, K.K., Frost, H.M., and Guise, E.E.: J. Hand Surg. **8:**317, 1983.)

ported a chondrosarcoma in the greater multangular that mimicked osteoarthritis roentgenographically (Fig. 17-24). Some have been reported in preexisting enchondromas (Fig. 17-25).

Chondrosarcomas of the bones of the hands and feet are rare. At times they may be difficult to differentiate from enchondromas (Fig. 17-26). Pain is a presenting symptom, whereas it occurs rarely in other chondromas. A fracture may accompany a chondroma or enchondroma as the bone is weakened. Roentgenograms then reveal the weakened or fractured bone and the tumor. A chondrosarcoma should be suspected if a lesion recurs after routine curettage of an enchondroma. Once chondrosarcoma is diagnosed, Dahlin and Salvador report that anything short of total or en bloc

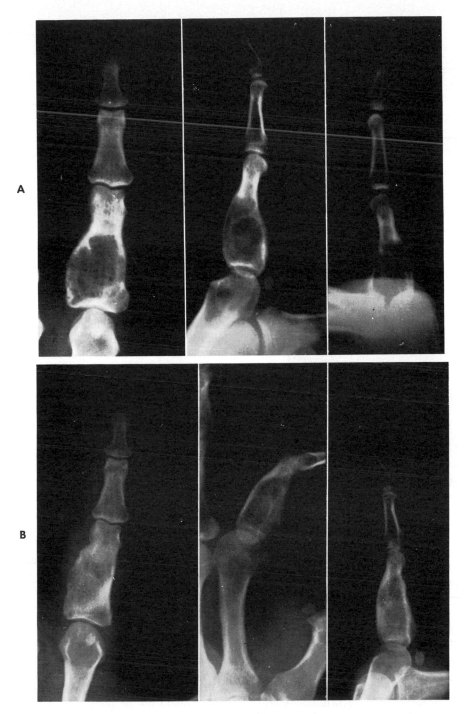

Fig. 17-26. Chondrosarcoma of proximal phalanx. **A,** Original lesion was treated by curettage and by packing cavity with bone chips. Microscopically, tumor was enchondroma. **B,** About 3 years later enlargement of bone and several areas of rarefaction but nothing suggestive of malignancy. Microscopically, recurrent tumor was chondrosarcoma. (From Sbarbaro, J.L., Jr., and Straub, L.R.: Am. J. Surg. **100:**751, 1960.)

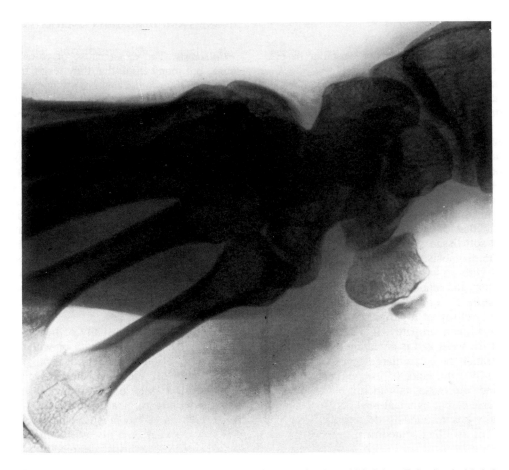

Fig. 17-40. Calcium deposit in pisiform bursa. (From Cameron, B.M., and McGehee, F.O.: South. Med. J. **51:**496, 1958.)

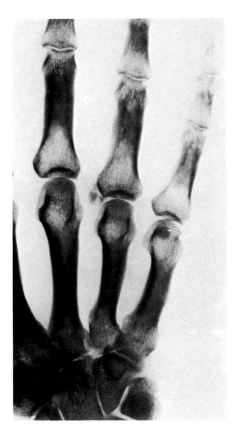

Fig. 17-41. Calcium deposit in collateral ligament of metacarpophalangeal joint of ring finger.

Fig. 17-42. Multiple calcinosis of digits, hand, and wrist.

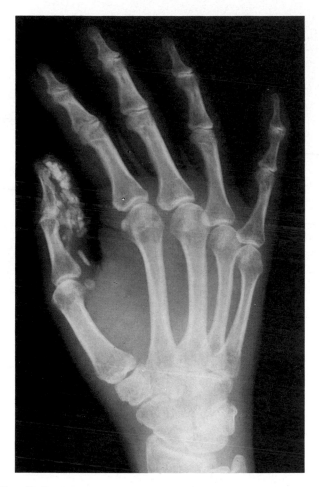

Fig. 17-43. Calcinosis circumscripta of thumb. (Courtesy Dr. E.C. Weckesser.)

nosis circumscripta is rare but frequently is preceded by Raynaud's phenomenon for many years. Deposits occur more densely over pressure areas such as fingertips and may at times erode through the skin. Partial excision may be indicated when the deposits cause pain or interfere with function, but wound breakdown and skin necrosis are frequent when dissection is extensive (Fig. 17-43).

Turret exostosis. Turret exostosis is a smooth, dome-shaped extracortical mass of bone lying beneath the extensor apparatus on the middle or proximal phalanx of a finger. It is caused by traumatic subperiosteal hemorrhage that eventually ossifies. Clinically a firm mass develops on the dorsum of the phalanx and limits excursion of the extensor apparatus, thus limiting flexion of the interphalangeal joints distal to the lesion. Roentgenograms that are negative during the first few weeks after injury later reveal subperiosteal new bone located on the dorsum of the phalanx (Fig. 17-44, *B*). Conservative treatment has not been beneficial. Any indicated surgery should be delayed until the subperiosteal bone becomes mature, usually 4 to 6 months after injury (Fig. 17-44, *C*); then recurrence is less likely.

To excise the exostosis a midlateral incision (p. 10) is made, the extensor apparatus is elevated, and the periosteum is incised laterally and carefully elevated from the underlying bone; care is taken not to tear the periosteum dorsally, thus preserving a smooth surface over which the extensor apparatus can glide. Then all of the new bone is resected, the periosteum and the wound are closed, and a wet cotton compression dressing is applied.

Carpometacarpal boss. A carpometacarpal boss is a bony, nonfluctuant, fixed lump on the dorsum of the hand at the level of the base of the second and third metacarpals and is visible on a tangential roentgenogram. It is frequently confused with a ganglion but lies distal to the common site of a ganglion. It can be confused with a fracture or tenosynovitis since occasionally an extensor tendon may sub-

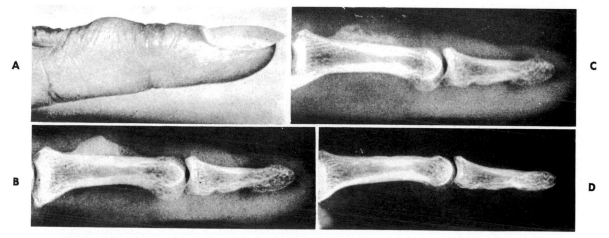

Fig. 17-44. Turret exostosis. **A,** Turret exostosis of middle phalanx. **B,** Status 4 months after injury. Subperiosteal bone is immature. **C,** Status 6 months after injury. Subperiosteal bone is mature. **D,** Status 3 months after excision of exostosis. Mass has not recurred. (From Wissinger, H.A., McClain, E.J., and Boyes, J.H.: J. Bone Joint Surg. **48-A:**105, 1966.)

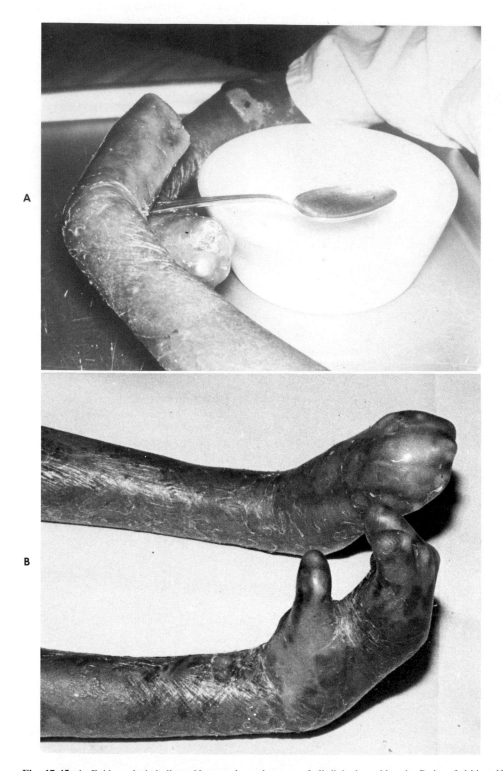

Fig. 17-45. A, Epidermolysis bullosa. Note total envelopment of all digits by epidermis. Patient fed himself by fixing a spoon between his deformed wrists. Enveloping epidermis had been removed from these hands approximately 1 year before photographs. **B,** Postoperative views of hands. Right hand was released with a pedicle flap to palm and split grafts to digits 3 months before photograph. (From Horner, R.L., et al.: J. Bone Joint Surg. **53-A:**1347, 1971.)

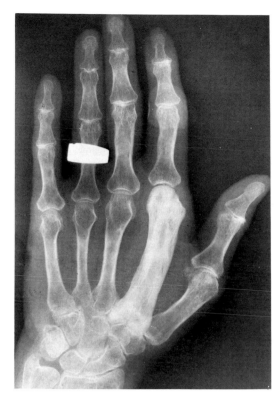

Fig. 17-46. Paget's disease of second metacarpal. (From Haverbush, T.J., et al.: J. Bone Joint Surg. **54-A:**173, 1972.)

luxate over the dome of the lesion. It may cause pain on local pressure or on forced dorsiflexion of the wrist. Some lesions are asymptomatic, whereas many lesions associated with trauma are painful; either may constitute a cosmetic problem.

There has been great hesitancy in the past to resect these lesions because of the reportedly high rate of recurrence. Recent authors do not support this but warn that even though the lesion is usually proximal to the base of the second and third metacarpals, it may involve the insertion of the extensor carpi radialis brevis.

Epidermolysis bullosa. Epidermolysis bullosa of the severe type, an hereditary disorder, occurs in 1 out of every 300,000 births. At birth or soon after, bullae are present over the extremities since the process affects the entire dermis and, at times, the mucus membranes. Its ultimate course is chronic infection of the bullae and the continuing formation of a cocoonlike epidermis over all the fingers of each hand. Surgical release of these digits is very discouraging since recurrence of the webbing and flexion contractures of the fingers is rapid. Free skin grafts and distant flaps have been used to limited advantage, but no effective treatment of the disease process is known. These patients are poor surgical risks because of the chronic infection. Some authors have reported a death rate of 25% during childhood or adolescence, apparently because of debilitation. Surgical procedures, if any are indicated, are repetitious degloving procedures that give limited hand function

over a limited period of time. The less severe types of the disease may not need surgical treatment. Splinting after degloving may be of some value (Fig. 17-45).

Paget's disease. Paget's disease may occur in the long bones of the hand, although this is very rare, especially as compared with the 3% incidence of Paget's disease of bone in the general population. Roentgenograms reveal the same sclerotic fusiform enlargement of the long bones as in the body elsewhere. It should not be confused with fibrous dysplasia (Fig. 17-46).

REFERENCES

Acharya, G., Merritt, W.H., and Theogaraj, S.D.: Hemangioendotheliomas of the hand: case reports, J. Hand Surg. **5:**181, 1980.

Aitken, A.P., and Magill, H.K.: Calcareous tendinitis of the flexor carpi ulnaris, N. Engl. J. Med. **244:**434, 1951.

Altner, P.C., and Singh, S.K.: An unusual case of ectopic ossification in a finger, J. Hand Surg. **6:**142, 1981.

Andren, L., and Eiken, O.: Arthrographic studies of wrist ganglions, J. Bone Joint Surg. **53-A:**299, 1971.

Angelides, A.C., and Wallace, P.F.: The dorsal ganglion of the wrist: its pathogenesis, gross and microscopic anatomy, and surgical treatment, J. Hand Surg. **1:**228:1976.

Arner, O., Lindholm, A., and Romanus, R.: Mucous cysts of the fingers: report of 26 cases, Acta Chir. Scand. **111:**314, 1956.

Artz, T.D., and Posch, J.L.: The carpometacarpal boss, J. Bone Joint Surg. **55-A:**747, 1973.

Averill, R.M., Smith, R.J., and Campbell, C.J.: Giant-cell tumors of the bones of the hands, J. Hand Surg. **5:**39, 1980.

Barnes, W.E., Larsen, R.D., and Posch, J.L.: Review of ganglia of the hand and wrist with analysis of surgical treatment, Plast. Reconstr. Surg. **34:**570, 1964.

Basora, J., and Fery, A.: Metastatic malignancy of the hand, Clin. Orthop. **108:**182, 1975.

Becker, H., and Chait, L.: Fibromatosis of the upper limb, J. Hand Surg. **4:**264, 1979.

Bell, J.L., and Mason, M.L.: Metastatic tumors of the hand: report of two cases, Q. Bull. Northwestern Univ. Med. School **27:**114, 1953.

Ben-Menachem, Y., and Epstein, M.J.: Post-traumatic capillary hemangioma of the hand: a case report, J. Bone Joint Surg. **56-A:**1741, 1974.

Blair, W.F.: Granular cell schwannoma of the hand, J. Hand Surg. **5:**51, 1980.

Block, R.S., and Burton, R.I.: Multiple chondrosarcomas in a hand: a case report, J. Hand Surg. **2:**310, 1977.

Bloem, J.J., Vuzevski, V.D., and Huffstadt, A.J.C.: Recurring digital fibroma of infancy, J. Bone Joint Surg. **56-B:**746, 1974.

Bogumill, G.P., Sullivan, D.J., and Baker, G.I.: Tumors of the hand, Clin. Orthop. **108:**214, 1975.

Booher, R.J.: Tumors arising from blood vessels in the hands and the feet, Clin. Orthop. **19:**71, 1961.

Booher, R.J.: Lipoblastic tumors of the hands and feet: review of the literature and report of thirty-three cases, J. Bone Joint Surg. **47-A:**727, 1965.

Booher, R.J., and McPeak, C.J.: Juvenile aponeurotic fibromas, Surgery **46:**924, 1959.

Bosch, D.T., and Bernhard, W.G.: Lipoma of the palm, Am. J. Clin. Pathol. **20:**262, 1950.

Bowers, W.H., and Hurst, L.C.: An intraarticular-intraosseous carpal ganglion, J. Hand Surg. **4:**375, 1979.

Bryan, R.S., et al.: Primary epitheloid sarcoma of the hand and forearm: a review of thirteen cases, J. Bone Joint Surg. **56-A:**458, 1974.

Bryan, R.S., et al.: Metastatic lesions of the hand and forearm, Clin. Orthop. **101:**167, 1974.

Burkhalter, W.E., Schroeder, F.C., and Eversmann, W.W., Jr.: Aneurysmal bone cysts occurring in the metacarpals: a report of three cases, J. Hand Surg. **3:**579, 1978.

Butler, E.D., et al.: Tumors of the hand: a ten-year survey and report of 437 cases, Am. J. Surg. **100:**293, 1960.

Button, M.: Epitheloid sarcoma: a case report, J. Hand Surg. **4:**368, 1979.

B

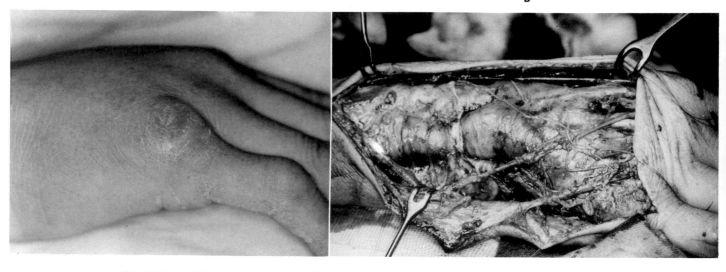

Fig. 18-14. A, Fisherman experienced swelling over dorsum of right hand and near metacarpophalangeal joint of small finger after crab bite. B, View at surgery showing extensive synovitis of extensor tendons from wrist to fingers. (From Chow, S.P., et al.: J. Hand Surg. **8:**568, 1983.)

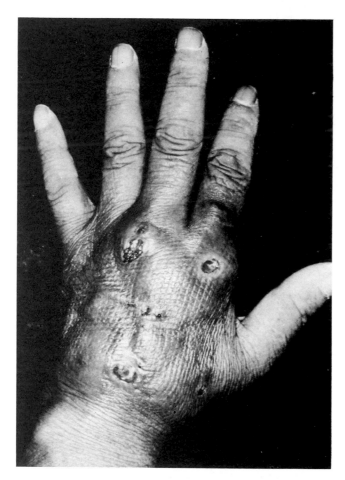

Fig. 18-15. Fisherwoman developed multiple discharging sinuses and swelling on dorsum of left hand after knocking her hand against a boat. (From Chow, S.P., et al.: J. Hand Surg. **8:**568, 1983.)

dose. This slough may involve muscles and tendons as well as skin. Accidental injections into an artery at the wrist may cause sufficient vascular impairment to result in ischemic necrosis of digits as well as skin. Venous injections about the dorsum of the hand, the dorsum of the proximal interphalangeal joint, and the lateral aspect of the digit may result in extravasation of the toxic chemical into the tissue and produce a slough with a secondary, if not primary, infection. Dorsal swelling at times may actually be chronic lymphedema and fibrosis from repeated injections instead of reaction to infection. The treatment should be expedient since many drug-addicted patients refuse adequate treatment. Debridement, antibiotics, and elevation are the essentials of treatment.

MYCOBACTERIAL INFECTIONS

Any poorly healing ulcer on the hand should have a culture for *Mycobacterium marinum* at 30° to 32° C on Lowenstein-Jensen medium. Skin testing is not as reliable for this organism as it is for tuberculosis. In the early stages, the infection is frequently confused with gout or rheumatoid arthritis, and nearly all of the reported cases have had a cortisone injection. Typically the organism is found around swimming pools or fish tanks, from whence it derives its name, "swimming pool granuloma," and may infect an open wound or abrasion. It can attack bone, joint, synovium, or skin (Figs. 18-13 to 18-15).

Mycobacterium kansasii (see Plate 4) may behave in a similar manner and should be considered when a chronic synovitis is not obviously a complication of rheumatoid arthritis, especially if only a single digit or joint is involved. A typical case is one in which a synovectomy of the wrist has been done for compression of the median nerve only to be followed by recurrence of swelling and compression of the nerve several weeks later. A slowly healing sinus may also also be present. Thus when a persistent or recurring synovitis of the wrist or finger is en-

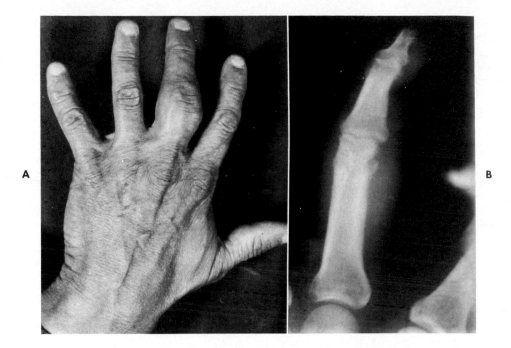

Fig. 18-16. A, Two synovectomies rid 31-year-old man of his *M. kansasii* infection of flexor sheath, but 5 years after onset he had residual infection under extensor tendon mechanism and in proximal interphalangeal joint. **B,** Bone destruction is seen under the collateral ligament origin. (From Gunther, S.F. in Flynn, J.E.: Hand Surgery, ed. 3, Baltimore, 1982, Williams & Wilkins.)

countered, a mycobacterial infection should be suspected. When a part is aspirated for routine bacterial cultures, fungus cultures and those for tuberculosis should also be ordered. The results of the cultures may not be known for several weeks, since these organisms grow slowly. (Fig. 18-16).

Treatment is by synovectomy or other excisional surgery for diagnostic and therapeutic reasons. If the diagnosis has not already been established, material is sent to bacteriology as well as pathology for identification of the organisms. When the diagnosis is established, appropriate antimicrobial therapy is started. We recommend the help of an infectious disease consultant for this therapy.

REFERENCES

Badger, S.J., Butler, T., and Kim, C.K.: Experimental *Eikenella corrodens* endocarditis in rabbits, Infect. Immun. **23:**751, 1979.

Becton, J.L., and Niebauer, J.J.: Nocardia infection of the hand, J. Bone Joint Surg. **52-A:**1443, 1970.

Belsole, R., and Fenske, N.: Cutaneous larva migrans in the upper extremity, J. Hand Surg. **5:**178, 1980.

Bielejeski, T.R.: Granular-cell tumor (myoblastoma) of the hand: report of a case, J. Bone Joint Surg. **55-A:**841, 1973.

Bilos, Z.J., Eskestrand, T., and Shivaram, M.S.: Deep fasciitis of the biceps region, J. Hand Surg. **4:**378, 1979.

Bilos, Z.J., Kucharchuk, A., and Metzger, W.: Eikenella corrodens in human bites, Clin. Orthop. **134:**320, 1978.

Bingham, D.L.C.: Acute infections of the hand, Surg. Clin. North Am. **40:**1285, 1960.

Bolton, H., Fowler, P.J., and Jepson, R.P.: Natural history and treatment of pulp space infection and osteomyelitis of the terminal phalanx, J. Bone Joint Surg. **31-B:**499, 1949.

Carter, S.J., Burman, S.O., and Mersheimer, W.L.: Treatment of digital tenosynovitis by irrigation with peroxide and oxytetracycline: review of nine cases, Ann. Surg. **163:**645, 1966.

Carter, S.J., and Mersheimer, W.L.: Infections of the hand, Orthop. Clin. North Am. **1:**455, 1970.

Chow, S.P., Stroebel, A.B., Lau, J.H.K., and Collins, R.J.: *Mycobacterium marinum* infection of deep structures of hand, J. Hand Surg. **8:**568, 1983.

Chuinard, R.G., and D'Ambrosia, R.D.: Human bite infections of the hand, J. Bone Joint Surg. **59-A:**416, 1977.

Cortez, L.M., and Pankey, G.A.: *Mycobacterium marinum* infections of the hand: report of three cases and review of the literature, J. Bone Joint Surg. **55-A:**363, 1973.

Defibre, B.K., Jr.: Bowen's disease of the nail bed: a case presentation and review of the literature, J. Hand Surg. **3:**182, 1978.

Dehaven, K.E., Wilde, A.H., and O'Duffy, J.D.: Sporotrichosis arthritis and tenosynovitis: report of a case cured by synovectomy and amphotericin B. J. Bone Joint Surg. **54-A:**874, 1972.

DeMello, F.J., and Leonard, M.S.: *Eikenella corrodens:* a new pathogen, Oral Surg. **47:**401, 1979.

Ellis, W.: Multiple bone lesions caused by Avian-Battey mycobacteria: report of a case, J. Bone Joint Surg. **56-B:**323, 1974.

Enna, C.D.: Skeletal deformities of the denervated hand in Hansen's disease, J. Hand Surg. **4:**227, 1979.

Entin, M.A.: Infections of the hand, Surg. Clin. North Am. **44:**981, 1964.

Feldman, F., Auerbach, R., and Johnston, A.: Tuberculous dactylitis in the adult, Am. J. Roentgenol. Radium Ther. Nucl. Med. **112:**460, 1971.

Fitzgerald, R.H., Jr., et al.: Bacterial colonization of mutilating hand injuries and its treatment, J. Hand Surg. **2:**85, 1977.

Flynn, J.E.: Modern considerations of major hand infections. N. Engl. J. Med. **252:**605, 1955.

Goldner, J.L.: Thumb and finger infections, Am. J. Surg. **28:**12, 1962.

Goldstein, E.J.C., Kirby, B.D., and Finegold, S.M.: Isolation of *Eikenella corrodens* from pulmonary infections, Am. Rev. Resp. Dis. **119:**55, 1979.

Goldstein, E.J., et al. Infections following clenched-fist injury: a new perspective, J. Hand Surg. **3:**455, 1978.

Gropper, P.T., Pisesky, W.A., Bowen, V., and Clement, P.B.: Flexor tenosynovitis caused by *Coccidioides immitis,* J. Hand Surg. **8:**344, 1983.

Gunther, S.: *Mycobacterium kansasii* infection in the deep structure. In Flynn, J.E.: Hand surgery, ed. 3, Baltimore, 1982, Williams & Wilkins Co.

Gunther, S.F., et al. Experience with atypical mycobacterial infection in the deep structures of the hand, J. Hand Surg. **2:**90, 1977.

Hennessy, M.J., and Mosher, T.F.: Mucormycosis infection of an upper extremity, J. Hand Surg. **6:**249, 1981.

Hooker, R.P., Eberts, T.J., and Strickland, J.A.: Primary inoculation tuberculosis, J. Hand Surg. **4:**270, 1979.

Iverson, R.E., and Vistnes, L.M.: Coccidioidomycosis tenosynovitis in the hand, J. Bone Joint Surg. **55-A:**413, 1973.

Kanavel, A.B.: Infections of the hand: a guide to the surgical treatment of acute and chronic suppurative processes in the fingers, hand, and forearm, ed. 6, Philadelphia, 1933, Lea & Febiger.

Kaplan, J.E., Zoschke, D., and Kisch, A.L.: Withdrawl of immunosuppressive agents in the treatment of disseminated coccidiodomycosis, Am. J. Med. **68:**624, 1980.

Linscheid, R.L., and Dobyns, J.H.: Common and uncommon infections of the hand, Orthop. Clin. North Am. **6:**1063, 1975.

Liseki, E.J., Curl, W.W., and Markey, K.L.: Hand and forearm infections caused by *Aeromonas hydrophilia,* J. Hand Surg. **5:**605, 1980.

Loudon, J.B., Miniero, J.D., and Scott, J.C.: Infections of the hand, J. Bone Joint Surg. **30-B:**409, 1948.

Louis, D.S., and Silva, J., Jr.: Herpetic whitlow: herpetic infections of the digits, J. Hand Surg. **4:**90, 1979.

Mandel, M.A.: Immune competence and diabetes mellitus: pyogenic human hand infections, J. Hand Surg. **3:**458, 1978.

Mann, R.J., Hoffeld, T.A., and Farmer, C.B.: Human bites of the hand: twenty years of experience, J. Hand Surg. **2:**97, 1977.

Marr, J.S., Beck, A.M., and Lugo, J.A., Jr.: An epidemiologic study of the human bite, Publ. Health Rep. **94:**514, 1979.

Mathews, R.E., Gould, J.S., and Kashlan, M.B.: Diffuse pigmented villonodular tenosynovitis of the ulnar bursa, J. Hand Surg. **6:**64, 1981.

McConnell, C.M., and Neale, H.W.: Two-year review of hand infections at a municipal hospital, Am. Surg. **45:**643, 1979.

Malinowski, R.W., Strate, R.G., Perry, J.F., Jr., and Fischer, R.P.: The management of human bite injuries of the hand, J. Trauma **19:**655, 1979.

McDonald, I.: *Eikenella corrodens* infection of the hand, Hand **11:**224, 1979.

McKay, D., et al.: Infections and sloughs in the hands of drug addicts, J. Bone Joint Surg. **55-A:**741, 1973.

Neviaser, R.J.: Closed tendon sheath irrigation for pyogenic flexor tenosynovitis, J. Hand Surg. **3:**462, 1978.

Peeples, E., Boswick, J.A., Jr., and Scott, F.A.: Wounds of the hand contaminated by human or animal saliva, J. Trauma **20:**393, 1980.

Perlman, R., Jubelirer, R.A., and Schwarz, J.: Histoplasmosis of the common palmer tendon sheath, J. Bone Joint Surg. **54-A:**676, 1972.

Rhode, C.M.: Treatment of hand infections, Am. Surg. **27:**85, 1961.

Robins, R.H.C.: Infections of the hand: a review based on 1,000 consecutive cases, J. Bone Joint Surg. **34-B:**567, 1952.

Schmidt, D.R., and Heckman, J.D.: *Eikenella corrodens* in human bite infections of the hand, J. Trauma **23:**478, 1983.

Scott, J.C., and Jones, B.V.: Results of treatment of infections of the hand, J. Bone Joint Surg. **34-B:**581, 1952.

Sehayik, R.I., and Bassett, F.H., III: Herpes simplex infection involving the hand, Clin. Orthop. **166:**138, 1982.

Simmons, E.H., Van Peteghem, K., and Trammell, T.R.: Onchocerciasis of the flexor compartment of the forearm: A case report, J. Hand Surg. **5:**502, 1980.

Sinkovics, J.G., Plager, C., and Mills, K.: *Eikenella corrodens* as pathogen, Ann. Intern. Med. **90:**991, 1979.

Smith, J., and Ruby, L.K.: Nocardia asteroides thenar space infection: a case report, J. Hand Surg. **2:**109, 1977.

Southwick, G.J., and Lister, G.D.: Actinomycosis of the hand: a case report, J. Hand Surg. **4:**360, 1979.

Williams, C.S., and Riordan, D.C.: *Mycobacterium marinum* (atypical acid-fast bacillus) infections of the hand: a report of six cases, J. Bone Joint Surg. **55-A:**1042, 1973.

CHAPTER 19

Microsurgery

Phillip E. Wright

Since the 1980 edition of this book, microsurgical techniques have been applied to an expanding range of orthopaedic problems. The surgical methods useful in the treatment of acute injuries and a variety of reconstructive problems caused by acquired and congenital disorders now include microsurgical procedures as the preferred treatment. This section includes microsurgical procedures appropriate for surgery of the hand, including the repair of small vessels and nerves, and the transfer of composite tissue grafts using microvascular techniques in the upper and lower extremities, as well as our approach to the replantation of amputated parts.

Microsurgery includes surgical procedures for structures so small that magnification by an operating microscope is required for their performance. While many procedures may be performed using magnifying loupes of up to $5\times$, magnification from $16\times$ to $40\times$ is provided by the microscope and is essential when working with structures less

eters from 50 μm to 130 μm. Nylon sutures designated as 9-0, 10-0, 11-0, and 12-0 are commercially available.

Detailed discussions of microsurgical history, microscopes, microsurgical instruments, needles, sutures, training methods, and techniques are found in many of the references at the end of this chapter.

MICROVASCULAR TECHNIQUES

TECHNIQUE FOR MICROVASCULAR ANASTOMOSIS. Expose the selected vessel by careful dissection under magnification using the operating microscope for dissection of vessels less than 2 mm in diameter. Using jeweler's forceps and microscissors, carefully remove the loose connective tissue surrounding the vessel. Mobilize each end of the vessel proximally and distally to obtain adequate length for anastomosis. Cauterize tethering side branches with bipolar electrocautery, and continue mobilization until the vessel ends can be easily approximated with minimal or no tension. Place a contrasting colored rubber or plastic sheet behind the vessel to help make it easier to see. Irrigate the operative field with heparinized Ringer's lactate solution frequently. Remove sufficient adventitia from the vessel ends to expose all the layers of the vessel wall. Adventitia may be removed either by careful circumferential trimming or by applying traction to the adventitia and transecting it in a manner similar to circumcision (Fig. 19-4, A and B).

Magnification of 6× to 10× is usually sufficient for this dissection. After the adventitia has been trimmed, continue to irrigate the field intermittently with heparinized Ringer's lactate solution. Inspect the vascular intima using magnification of 25× and 40×, and resect the vessel wall until the cut ends appear normal. Appose the vessel ends with a clamp approximator.

Use interrupted sutures to prevent vascular constriction, and place each suture through the full thickness of the vessel wall (Fig. 19-4, C to F). Place the first two sutures approximately 120 degrees apart on the vessel's circumference. Leave the ends of these sutures long for use as traction sutures. Rotate the clamp approximators to expose the posterior vessel wall and place a stitch 120 degrees from the initial two stitches. Place additional stitches in the remaining spaces to complete the anastomosis (Fig. 19-4, G). Arteries 1 mm in diameter usually require 5 to 8 stitches, and veins usually require 7 to 10 stitches.

Vessels may be gently dilated by insertion of the tips of jeweler's forceps or specially disigned dilators. The walls of the vessels may be gently grasped, but avoid rough manipulation of the intima. To overcome vascular spasm, apply topical lidocaine or papaverine. After the completion of the vascular anastomosis, remove the clamp downstream from the anastomosis first; then remove the clamp that is upstream. Minimal bleeding between stitches is of

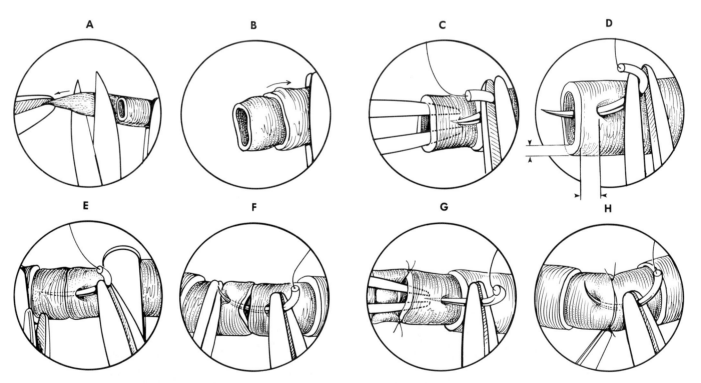

Fig. 19-4. Microvascular anastomosis, basic steps. **A,** Adventitial excision. Traction is applied to adventitia, and it is excised sufficiently to avoid intrusion into vascular lumen. **B,** Appearance of vessel end after adventitial excision. **C,** Placement of initial suture. Forceps may be used as counter pressors without internal damage. **D,** Needle is passed through full thickness of vessel wall a distance from cut edge that is slightly greater than thickness of vessel wall. **E,** Passage of needle through opposite end of vessel is accomplished a similar distance from cut edge. **F,** Forceps, used as counterpressors, assist in passage of needle through opposite end of vessel. **G,** After completion of initial sutures, vessel is stabilized, allowing completion of an even anastomosis.

no concern, but excessive bleeding should be rapidly controlled by reapplication of clamps or inflation of a pneumatic tourniquet. Place additional stitches in the areas of leakage, remove the clamps again, and deflate the tourniquet. After bleeding has stopped from the suture line, assess the patency of the anastomosis by occluding a segment of vessel with forceps distal to the anastomosis. Gently strip blood from the segment from proximal to distal. Then release the proximal clamp. Rapid filling of the emptied segment indicates a patent anastomosis. The suture line should be even, and there should be no anastomotic stenosis, dilation proximally, or stenosis distally. Small platelet clots about the anastomosis are to be expected, but avoid occlusion of the anastomosis by irrigating with heparinized solution or by gentle milking of the vessel. Following the anastomosis, close soft tissue over the vessels as soon as possible to avoid drying of the vessel wall.

TECHNIQUE FOR MICROVASCULAR END-TO-SIDE ANASTOMOSIS (FIG. 19-5). After dissecting and mobilizing the vessels as described above, carefully excise a small longitudinal elliptic portion of the recipient vessel wall using microscissors.

Cut the end of the vessel that is to be attached to the recipient vessel at an angle of about 45 degrees.

Begin the anastomosis by placing sutures at the proximal and distal ends of the ellipse. Leave the suture ends long for traction and complete the anastomosis by placing sutures evenly along the opening between the traction sutures. Release the occluding clamps or release the tourniquet and assess the patency and flow.

TECHNIQUE FOR MICROVASCULAR VEIN GRAFTING. When end-to-end vessel anastomosis cannot be performed without tension, bone shortening and vein grafting may be necessary (see Fig. 19-6, *A* to *C*). Many sizes of veins are available on the dorsum of the hand, on the dorsal and volar aspect of the forearm, and on the dorsum of the foot so that the vein graft can roughly approximate the diameter of the recipient vessel. This helps avoid thrombosis as a result of turbulence. When vein grafts are harvested, cauterize small side branches with bipolar forceps well away

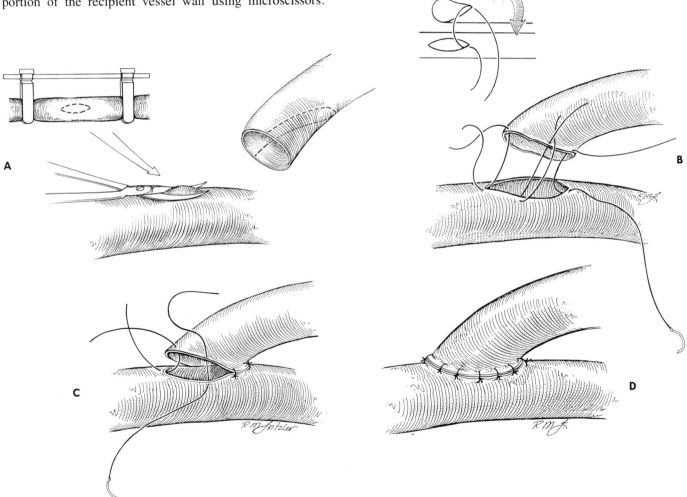

Fig. 19-5. Microvascular end-to-side anastomosis. **A,** With microvascular scissors, excise small ellipse of vessel wall between microvascular clips *(upper left)*. Suitable fit of recipient vessel achieved with oblique cut to match elliptical defect in vessel wall *(upper right)*. **B,** Suture line begins with sutures placed at each end of openings. Suture ends left long temporarily for traction. **C,** Suture line continues, placing interrupted stitches around anastomosis. **D,** Completed end-to-side microvascular anastomosis.

AGE

Replantations have been reported in patients a few weeks old and in those older than 70 years. The young patient poses particular problems, especially regarding digital replantations, because of the increased technical difficulty in microvascular anastomoses of their smaller digital vessels. Postoperative anxiety may contribute to vasospasm, and rehabilitation of children may be less predictable than of adults. Nevertheless, quite satisfactory functional results have been reported, and most authors consider for replantation amputations of almost any part including some lower extremity parts in children.

The upper age limit beyond which replantation should not be considered has not been clearly established. Poor nerve regeneration and joint stiffness are factors that limit the functional outcome. Replantation above the elbow, through the elbow, and through the proximal forearm result in little promise for hand function in the elderly; how-ever, the elbow in above-elbow amputations may be preserved in anticipation of a subsequent below-elbow reamputation to allow more satisfactory prosthetic fitting. Because the potential for return of sensibility and motion is better after replantation at and beyond the tendinous portion of the forearm, older patients may be considered as serious replantation candidates if their injury is more distal.

SEVERITY OF INJURY

The types of injuries that have the best outlook regarding survival and return of function following replantation include (1) clean, sharp "guillotine" amputations, (2) minimal local crush amputations, and (3) avulsion amputations with minimal proximal and distal vascular injury. Ideally significant additional injury to the limb should not be present, especially of the vessels, proximally and distally. Crushed and avulsed vessels require debridement and

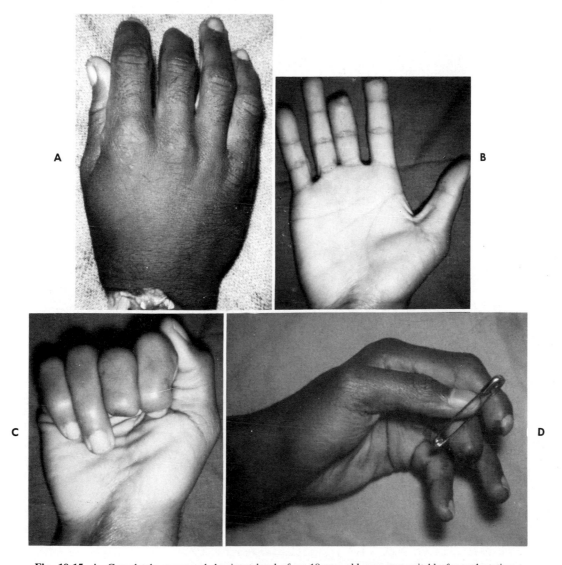

Fig. 19-15. A, Completely amputated dominant hand of an 18-year-old man was suitable for replantation because of sharp, clean wound and patient's youth. **B** and **C,** Replanted hand 18 months following initial injury. Protective sensation and two-point discrimination of 8 mm were present as well as a functional range of motion of wrist and fingers. **D,** Return of sensation allows ability to handle small objects.

the use of interpositional vein grafts as needed. Ring avulsion-degloving injuries may be revascularized and salvaged; however, if the skin has been completely degloved or if the digit has been amputated, vein grafts may be required and the outlook for useful function is extremely uncertain.

LEVEL OF INJURY AND PART AMPUTATED

Amputations through the humerus, elbow, and proximal forearm have the potential for successful replantation and useful function, especially in the young healthy patient and especially if the injury is clean and sharp (Fig. 19-14). The patient should be young enough and motivated enough to be able to await nerve regeneration sufficient for return of function. Replantation more distally, whether through the distal forearm, wrist, metacarpals, or digits should also be seriously considered because generally, the potential for sensory and motor return is good (Fig. 19-15).

Thumb amputations at almost any level should be considered for replantation in spite of nerve and tendon avulsion and joint involvement (Fig. 19-16). If the thumb can be revascularized, sensibility can be restored with nerve grafts or a neurovascular island pedicle transfer if needed, and motion can be achieved with tendon grafts or transfers.

Replantation of single and multiple digits distal to the flexor digitorum sublimus insertion should be expected to achieve satisfactory function (Fig. 19-17). Although many patients do well without replantation of single digit amputations, such a replantation may be worthwhile for some musicians, those with other special occupations, some

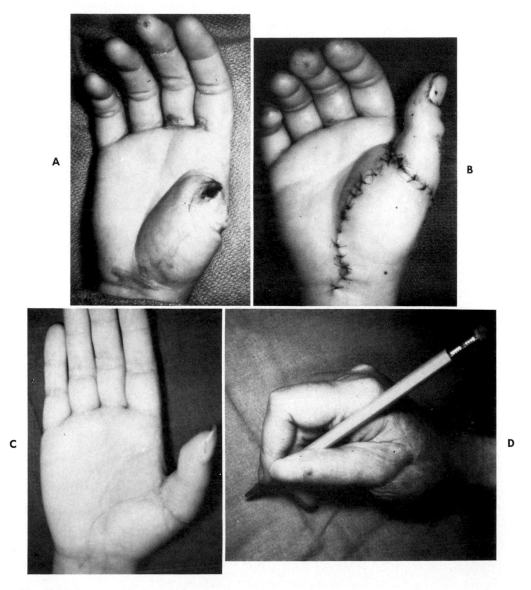

Fig. 19-16. A, Replantation of sharply amputated dominant thumb of this 32-year-old woman was a satisfactory alternative to primary closure and thumb reconstruction. **B,** Thumb immediately following replantation. **C,** Three months following injury. Appearance is satisfactory, and motion is improving. **D,** Two years after injury sensory return is sufficient to allow use of thumb.

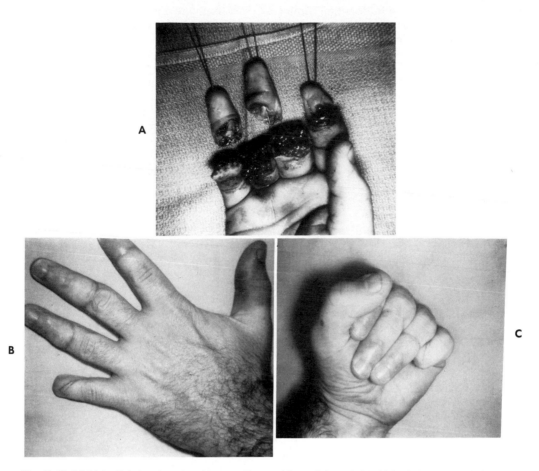

Fig. 19-17. Multiple digital replantation: 20-year-old man with saw injury. **A,** Multiple digits amputated distal to flexor sublimis insertion. **B** and **C,** Finger flexion and extension 7 months following replantation. Sensory return allows for useful finger function.

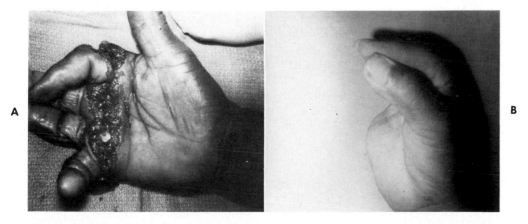

Fig. 19-18. Single digit revascularization in 54-year-old man with a crush injury on his dominant hand. **A,** See fingers devascularized. Nerves intact to index finger. **B,** Seven weeks after revascularization, useful digit is preserved.

children, and for other esthetic or social reasons. Replantation of a single digit may also be helpful if the remaining attached digits are severely damaged, especially with tendon and nerve injury over the proximal phalanx (Fig. 19-18).

If multiple digits have been amputated, replantation of at least two digits in the long and ring positions will provide a good combination of digits to use with the thumb for pinch and for power grip.

In bilateral amputations, replantation on each side should provide better function than bilateral prostheses. If replantation is not suitable or possible because of extensive injuries on one side, the best side should be selected, and at times parts from one side may be attached to the opposite, more suitable stump.

Although amputations through the joints impair the movement of those joints, a satisfactory limb can result through the use of arthrodesis, excisional or fascial arthroplasty, or in ideal circumstances silicone implant arthroplasty.

WARM ISCHEMIA (ANOXIC) TIME

Because irreversible necrotic changes begin in muscle after 6 hours of ischemia without cooling (at 20° to 25° C), it is preferable to begin the replantation of parts amputated proximal to the palm within this time. With cooling (to 4° C), this time may be extended, but the limits for parts with muscle are unknown. For parts with no muscle (digits) the allowable warm ischemia time may be 8 hours or more. With cooling this has been extended to longer than 30 hours.

Contraindications

Although contraindications may seem to be defined more easily, each patient must be evaluated individually with advantages and disadvantages considered before rejecting a given situation as hopeless.

AGE

As noted previously, age alone is not a contraindication to replantation. While older patients may obtain satisfactory function following replantation of fingers, thumbs, and hands, rarely do they achieve adequate sensibility, forearm and hand muscle strength, and coordination sufficient to warrant replantation of more proximal amputations. The patient's physiologic status, the presence of other diseases, and his general level of activity should also weigh heavily in the evaluation.

SEVERITY OF INJURY

Despite dramatic results that have been achieved using long vein grafts to salvage limbs, extensive crushing, avulsing, and segmental injuries at multiple levels damage the distal vascular tree sufficiently to frequently defeat replantation attempts. This is especially true in such digital injuries. Ring avulsion amputations through the joint are usually best treated by closure of the amputation.

Injuries contaminated extensively with soil, especially from the barnyard, carry a high risk of significant infection and should be carefully evaluated before replantation.

LEVEL OF INJURY

Because of the slowness and unpredictability of nerve regeneration and the muscle atrophy and joint stiffness that develop, replantation near the shoulder generally carries a poor prognosis regarding hand function. Replantation for salvage of the elbow for later below-elbow prosthetic fitting may be feasible in selected patients. Replantation just above the elbow, through the elbow joint, or in the proximal forearm also carries a guarded prognosis especially in older patients because of questionable nerve regeneration, limitation of elbow motion, and persistence of intrinsic muscle atrophy. But younger patients may benefit from replantation at these proximal levels, especially if the injury is sharp and clean.

Replantation of single-digit amputations proximal to the flexor sublimus insertion and especially through the proximal interphalangeal joint usually result in poor function. They are usually stiff and tend to impair the overall function of the remaining digits by getting in the way. The amputated thumb is the obvious exception to this generalization.

WARM ISCHEMIA (ANOXIC) TIME AND IMPROPER PRESERVATION OF PART

Skeletal muscle undergoes irreversible necrotic changes after about 6 hours of total warm ischemia. These changes can be slowed or minimized if the part is cooled to near 4° C. Although replantation of parts containing small amounts of muscle, such as the hand, probably is less risky, larger parts such as the forearm and arm above the elbow should probably not be replanted if they cannot be revascularized between 6 and 8 hours of amputation. The risk of renal damage resulting from myoglobinuria, acidosis, and hyperkalemia is increased following the replantation of a part with significant amounts of necrotic muscle. The risk of infection is also greater, and the long-term outlook for a functional limb is poor.

Digital viability may also be prolonged by cooling to 4° C. Because the amount of skeletal muscle in digits is not significant, the risk to the patient is minimal if digits are replanted. Cooled digits have been replanted successfully at 30 hours and more after amputation.

If amputated parts have been frozen, placed in unphysiologic solutions such as formaldehyde or alcohol, or allowed to dry excessively, the chance of survival is so low that replantation attempts are futile.

PREEXISTING DEFORMITY OR DISABILITY

If the amputated part was already deformed or disabled because of some congenital or acquired disorder, satisfactory function is unlikely to be achieved by replantation. Conditions that would fit this situation include, but are not limited to, scar deformity and contracture secondary to previous burns or mangling injury, significant residual deficits from spinal cord or peripheral nerve injuries, and deformities secondary to stroke.

OTHER CONDITIONS THAT MIGHT PRECLUDE REPLANTATION

At times in the same accident causing the amputation of a part, patients sustain significant intracranial, thoracic, cardiovascular, or major intraabdominal visceral injuries requiring lengthy lifesaving operations. In such circumstances a major limb replantation may not be possible because of excessive ischemia time. Digits may be cooled to 4° C in a refrigerator and saved for replantation later if technically feasible and if the patient's condition permits.

Patients with preexisting diseases that typically affect peripheral blood vessels are probably poor replantation candidates, especially if their vessels have an unsatisfactory appearance when inspected under the operating microscope. Diabetes mellitus, rheumatoid arthritis, lupus erythematosus, other collagen-vascular diseases, and patients with significant atherosclerosis fit into this category.

Severe chronic or uncompensated medical illnesses such as coronary artery disease, myocardial infarction, peptic ulcer disease, malignant neoplasms, and chronic renal or pulmonary disease may increase the anesthetic risk enough to preclude replantation.

Considerable judgment is required when assessing patients with psychiatric illnesses who have amputated parts. If the amputation event is an act of self-inflicted mutilation or attempted suicide during a psychiatric episode that can be treated and stabilized, replantation carries considerable risk of failure. If the amputated part is a focus in the patient's mental illness, it is likely that the part, if replanted, will be reinjured. If the amputation occurs as a true inadvertent accident, especially in a patient whose mental illness is compensated, the outlook for replantation might be better. Valid psychiatric evaluation of patients with amputated parts in an emergency room is extremely difficult. The inability of patients with profound psychiatric illness to understand their delicate postoperative condition and to cooperate with the difficult rehabilitation process further complicates their care as replantation patients.

Management and transportation of patient and part

At the scene of the injury and in the outlying hospital, the patient's condition is of utmost importance. Major injuries other than the amputated part should take precedence, and the patient's condition should be stabilized. Major stump bleeding should be controlled with pressure. No attempt should be made to clamp or ligate vessels. A pressure dressing should be applied for transporting the patient to an institution with replantation capabilities. If bleeding is persistent, the temporary use of a pneumatic tourniquet or blood pressure cuff is helpful. Elastic tourniquets should not be applied; they may later be covered with bandages and forgotten.

As noted, cooling of the amputated part to about 4° C is important to prolong the viability of the amputated part. After the amputated part has been found, it may be rinsed gently with sterile saline, Ringer's, or other physiologic

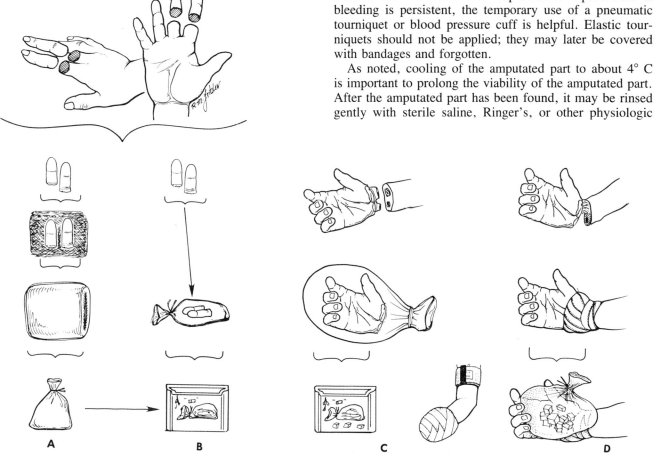

Fig. 19-19. Management of amputated parts. Digits may be managed as in **A,** by wrapping in clean or sterile gauze moistened in physiologic solution, wrapped in a cloth or towel, placed in a plastic bag, then on ice in an insulated container or **B,** by floating digits in a plastic bag with physiologic solution, refrigerated on ice in an insulated container. **C,** Larger parts, wrapped or unwrapped, are placed in plastic bag, then on ice in insulated container. Stump is wrapped and bleeding controlled with pressure dressing or pneumatic tourniquet. **D,** Incompletely severed parts are covered with sterile, nonconstricting bandages and cooled with ice pack for transportation.

solutions so that excess contamination is removed. The part should then be treated in one of two ways (Fig 19-19). (1) It may be wrapped with sterile gauze or other clean material, soaked in sterile Ringer's lactate or saline, and placed in a plastic bag, which is then sealed. (2) It may be immersed in a plastic bag containing a physiologic solution such as Ringer's lactate or physiologic saline. The bag is then placed on ice in an insulted container so that the part is not touching the ice to avoid freezing of the part. Dry ice should not be used; neither should the part be warmed. Nonphysiologic solutions such as alcohol and formaldehyde should not be used on the amputated part.

No attempt should be made to clamp, dissect, ligate, or cannulate vessels on the amputated part because this further damages vessels that may be essential to revascularization of the part.

If the part has been incompletely severed, it should be handled gently. Care should be taken to correct any kinking of the soft tissues or rotation that might compromise marginal arterial or venous flow. Sterile bandages moistened with a physiologic solution should be applied to the limb and the injured part and an ice pack is applied to the latter. The limb should then be supported with padded splints and a nonconstricting wrapping for the trip to the hospital.

When the patient is stable with an intravenous infusion is place, he along with the part may be transported. Although air transportation may be preferable for patients traveling great distances, especially in limb amputations, ground transportation is suitable if the patient can reach the replanting surgeons in 2 to 3 hours and if the amputated parts are digits that have been appropriately cooled. The receiving institution and replantation team should be contacted and alerted that the patient is being sent.

Finally, it is preferable for the patient and his family to understand that he is being referred to another hospital and other surgeons who have the capability to reattach parts and who will evaluate his particular situation and make appropriate recommendations regarding treatment. This understanding helps to minimize unrealistic expectations of patients, family, and friends who are usually quite distraught.

Preoperative preparation

Some aspects of preoperative, intraoperative, and postoperative management vary slightly among institutions; however, general agreement has been reached on many of the basic principles regarding replantation.

Having two teams to deal with replantation candidates from the time of their arrival in the emergency department is most helpful. While one team evaluates and prepares the patient, the other team assesses the amputated part.

Patient assessment and preparation should include (1) history of the injury and past medical history, including serious illnesses or previous injuries to the amputated part, (2) physical examination, especially to exclude injuries to other major organ systems, and (3) stabilization and resuscitation of the patient with the institution of an intravenous infusion, appropriate antibiotics, and tetanus prophylaxis. Blood typing and crossmatching are done, and transfusions are given if needed. An indwelling urinary catheter may

be inserted in the emergency department or in the surgical suite. Roentgenograms of the amputated part, the amputation stump, the chest, and other areas as indicated should be obtained in the emergency department.

The patient and his family are advised of the nature of the injury, the uncertainties regarding survival of the part and return of function, the possible duration of the replantation operation, the possibility of repeated operations, and the likelihood that the replanted part will never be normal.

TECHNIQUE. While the patient is being assessed and prepared, another surgeon of the replantation team takes the amputated part to the surgical suite to clean it and to evaluate the extent of injury.

Clean the part and then keep it cool by placing ice in a pan, covering the ice with a sterile plastic drape, and then placing a sterile drape sheet over the plastic and ice. Place the part on the drape sheet for dissection under loupe or microscope magnification.

Dissect the amputated part to allow exposure of the arteries, veins, nerves, tendons, joint capsule, periosteum, and other salvageable soft tissues. In digits, exposure is usually best achieved using midlateral incisions in the radial and ulnar aspects allowing reflection of dorsal and palmar flaps (Fig. 19-20). Although digital arteries and nerves are usually found with ease, locating satisfactory veins is somewhat more difficult. Careful, gentle, and meticulous dissection is required to locate these. Carefully preserve the small structures and use sutures of 8-0 or 9-0 nylon to mark them so that they can be easily located for nerve repair and vascular anastomoses.

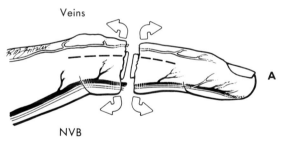

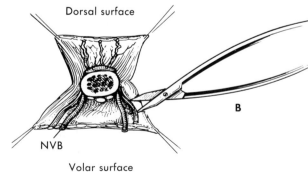

Fig. 19-20. Dissection of amputated digit. **A,** Incisions on radial and ulnar midline *(broken lines)* allow reflection of dorsal and palmar flaps. **B,** Carefully and gently dissect structures to be repaired using microsurgical instruments and meticulous technique. (Redrawn from Urbaniak, J.R.: Replantation. In Green, D.P.: Operative hand surgery, New York, 1982, Churchill Livingstone.)

Although multiple vein grafts can be used to provide tension-free anastomoses, it is our practice to shorten bone usually in the part of the digit having the most bone to spare. In digits this shortening rarely exceeds 1 cm. Next, place internal fixation in the digit. We usually insert a longitudinal Kirschner wire combined with an obliquely crossing Kirschner wire. On occasion interosseous wires are used near joints. Plates and screws are not usually needed. If the amputation has occurred through a joint, or if the extensor mechanism is irreparable, prepare for arthrodesis.

If the amputation is clean and sharp, perfusing the digital arteries before anastomosis usually is unnecessary. If the part has been crushed or avulsed, evidence of distal injury may be seen in the form of ecchymoses along the vessels or abrasions and lacerations. In these situations gently perfuse the digital artery and vascular tree using a small Silastic catheter and heparinized Ringer's solution or saline. If there is no return of the perfusate or if it extravasates from distally injured vessels, blood flow is unlikely to be maintained following anastomosis. Perfusion for brief periods may be helpful in rinsing blood and metabolites from the vascular tree of amputated hands, forearms, and arms.

The approach to the structures of the amputated hand and more proximally amputated parts is usually made through generally accepted incisions that will allow extensive exposure of the structures to be identified and repaired.

While the amputated part is being dissected, the patient is usually given an axillary brachial plexus block with the long-acting local anesthetic bupivacaine. This provides satisfactory anesthesia for digital or hand replantation in most adults and older children. For proximal amputations, younger children, anxious patients, and prolonged surgery as in multiple digital or bilateral amputations, general anesthesia is frequently preferable.

Pad the operating table well and usually apply a warming blanket to prevent body cooling during prolonged surgery. Use a pneumatic tourniquet to provide a bloodless field for initial dissection of the stump and to control any subsequent significant bleeding. Once the patient is comfortable, thoroughly cleanse the stump with an antiseptic solution, usually a povidone-iodine solution, and irrigate with normal saline.

The stump is then dissected by a hand surgeon who has microsurgery training and experience. Using gentle and meticulous technique, identify the arteries, veins, and nerves with magnifying loupes or the operating microscope and tag them with sutures of 8-0 or 9-0 nylon. Dissect tendons and hold them with 4-0 nylon sutures for later repair. Before initiating reattachment, free clots from the proximal arterial stumps and open the stumps to allow free arterial flow. If no satisfactory flow can be achieved, additional dissection, vessel resection, and possibly vein grafting may be needed.

Order of repair

After all structures have been thoroughly cleansed, debrided, and identified, repair is begun. As indicated in the discussion that follows, certain conditions or circumstances will dictate a variation in the order of repair. The following is our usual order of repair of damaged structures. Discussions of digit, hand, and arm replantations are included.

1. Shorten and internally fix bone
2. Repair extensor tendons
3. Repair flexor tendons (2 and 3 may be reversed, or flexor tendon repair may be delayed)
4. Repair arteries
5. Repair nerves
6. Repair veins
7. Close or cover wound

MANAGEMENT OF BONES AND JOINTS (FIG. 19-21)

The periosteum is stripped minimally. Bone is shortened to permit tension-free vascular anastomoses and nerve repairs. If vein grafts are used, the need for bone shortening is minimized, but survival depends on the patency of two anastomoses, rather than one. Furthermore, additional time is required to harvest the vein and to perform the anastomoses. However vein grafting may be necessary if the amputation has occurred near an undamaged joint.

Shortening of an amputated thumb should be kept to a minimum. We have found that shortening of a digit much more than 1.0 to 1.5 cm at times impairs the function of the digit. Amputations damaging digital joints are usually treated by primary arthrodesis (Fig. 19-22, *C*). But joint motion can be preserved by the insertion of a Silastic implant. This method probably is best reserved as a primary procedure for amputations that are sharp and clean, and when occupational requirements are best satisfied by having mobile joints.

Bone fixation in digits and metacarpals is usually achieved by using two parallel medullary axial Kirschner wires or a single axial Kirschner wire supplemented by an oblique Kirschner wire to control rotation (Figs. 19-21, *C*, and 19-22, *A*). Wires should be placed to allow joint motion, if possible. Occasionally when the amputation is near an undamaged joint, wire loops through drill holes are used. Care must be taken to maintain axial alignment and rotational control, especially when dealing with multiple digital amputations. We have not found it necessary to use plates and screws for digital or metacarpal fixation during replantation. This is an acceptable but often time-consuming technique. Periosteal suture with 4-0 or 5-0 absorbable suture may be done following bone fixation.

Management of the skeleton in more proximal amputations is more varied and requires more skill in the handling of medullary fixation devices, bone plates, and screws. If the amputation level is through the carpus, shortening may be achieved and motion preserved by excision of carpal bones and temporary fixation with transarticular Steinmann pins. Amputations through the forearm and arm are usually shortened 2 to 5 cm to allow tension-free vessel anastomoses and nerve repairs.

For amputations through the forearm, generally accepted principles of internal fixation are applied; however, time constraints frequently dictate modifications. Distal radial metaphyseal amputations are usually fixed with Steinmann pins; plates and screws are used less often. Amputations more proximally are fixed with plates and screws on both

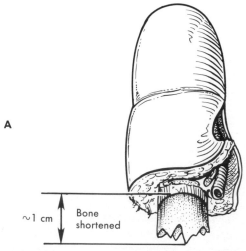

A

~1 cm — Bone shortened

Fig. 19-21. Bone management. **A,** In digits, shortening of 1 cm usually allows tension-free vessel anastomosis without excessively impairing hand function. **B,** Shortening of thumb proximal phalanx with oscillating saw in 2 to 3 mm increments until satisfactory shortening is achieved. **C,** Bone fixation with longitudinal Kirschner wires is usually sufficient.

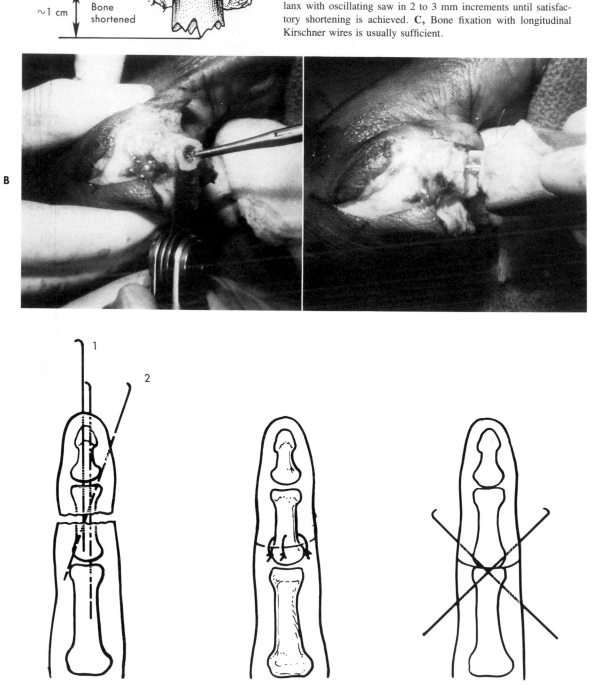

B   C

1
2

A   B   C

Fig. 19-22. Bone fixation. **A,** Fixation is usually achieved with two parallel Kirschner wires, *1,* or a single Kirschner wire supplemented with an oblique wire, *2.* **B,** Wire loop fixation suitable for amputation near undamaged joint. **C,** Primary arthrodesis with crossed wires for amputation through irreparably damaged joint.

bones, medullary fixation with Sage nails or Steinmann pins in both bones, or combinations, such as a plate and screws for the radius and medullary fixation for the ulna.

Medullary screws combined with wire loops are used for olecranon amputations. If the elbow joint is comminuted, an attempt is made to salvage sufficient bone to allow subsequent elbow arthroplasty.

Amputations through the humerus are usually fixed with plates and screws; however, fracture configuration and time considerations may require interfragmentary Steinmann pins or medullary rods.

TRANSPOSITION OF DIGITS

Because of extensive damage to amputated parts or to the amputation stump, anatomic restoration of digits is at times impossible. In these situations a functioning part may be restored by moving digits from their original anatomic position to a more suitable position (Fig. 19-23). In bilateral digital amputations, parts from one hand may be better replanted to the opposite hand. Priority should be given to restoration of the thumb position with provision for a digit in the index or long position for pinch. Consideration should also be given to providing long, ring, and little digits for cup restoration. When digital transposition is considered in bilateral amputations, the dominant hand is given priority if possible.

Tendon repair

During replantation damaged structures usually are repaired in a serial fashion from the skeletal plane to more superficial planes. This will at times delay repair of vessels in the sequence, so that deeper structures may be repaired without jeopardizing vascular anastomoses.

FLEXOR TENDONS

If the amputating injury involves crushing or avulsion of the part and if the amputation is through the digits proximal to the flexor digitorum sublimus insertion or if tendon

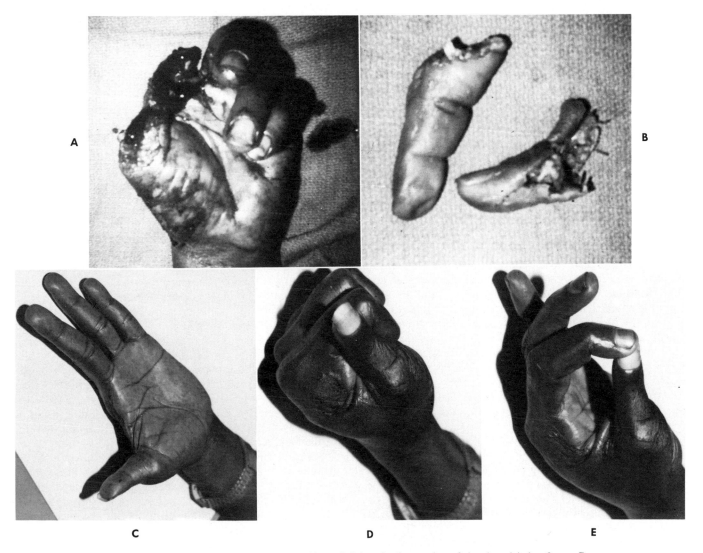

Fig. 19-23. Replantation involving transposition of digits. **A,** Amputation of thumb and index finger. **B,** Amputated thumb unsuitable for replantation. Index finger replanted to thumb position. **C, D,** and **E,** Satisfactory function following index finger–to-thumb replantation. (**C, D,** and **E** courtesy Dr. E. Hilliard.)

substance has been lost flexor tendons usually are not repaired primarily. Delayed tendon grafting is planned in these circumstances. At times Silicone rods may be inserted at the time of replantation in anticipation of two-staged tendon grafting. The condition of the wound, extent of contamination, and potential for infection should be considered before Silicone rod placement.

A flexor tendon injured distal to the flexor sublimus insertion near the distal interphalangeal joint is reattached with a pull-out wire. In injuries over the middle phalanx the distal tendon stump is tenodesed to bone or tendon sheath.

If the flexor tendons have been sharply severed, both tendons are usually repaired primarily in injuries at the proximal phalanx or more proximally. Our usual tenorrhaphy involves a rectangular (Kessler, modified, p. 43) or abbreviated crisscross configuration (Kleinert) of the suture using a single knot buried in the cut tendon end. The technique of first placing separate sutures in each end of the tendon allows nerve and vessel repair and subsequent tying of the sutures as advocated by Urbaniak. This helps prevent obstruction of the repair of vessels and nerves by the flexed finger. Nonabsorbable synthetic sutures or 4-0 wire is used for tenorrhaphy. An abbreviated, crisscross or rectangular suture configuration is used in the digit, as well as the palm. Similar configurations or mattress double-right angle sutures are used more proximally at the wrist and in the distal forearm. When technically feasible, the digital flexor sheath is repaired with 5-0 or 6-0 nonabsorbable suture, usually nylon.

If the flexor tendons have been injured at the myotendinous junction, the tendon is reattached in a fishmouth configuration with mattress sutures to the muscle belly.

EXTENSOR TENDONS

Extensor tendons are repaired using nonabsorbable sutures. For injuries between the proximal interphalangeal joint and the distal joint individual stitches or a roll stitch (p. 43) may be used. Tendons injured between the metacarpophalangeal and proximal interphalangeal joints are repaired with the roll stitch. Injuries to the extensor tendons between the metacarpophalangeal joint and the wrist extensor retinaculum are usually repaired with mattress sutures. Extensor tendons injured at the extensor retinaculum usually require excision of a portion of the retinaculum to facilitate repair and subsequent tendon gliding. A mattress stitch usually suffices at this level as well as more proximally at the myotendinous junction. This injury at the myotendinous level is repaired with insertion of the tendon into the muscle belly in a fishmouth configuration, reinforced with mattress sutures.

Vessel repair

Identifying the volar digital arteries is usually easier than finding suitable veins for anastomosis. The arteries lie just dorsal to the volar digital nerves. Although both digital arteries can usually be identified with ease, hypoplastic vessels on the radial side of the index finger and the radial side of the thumb have been frequent in our experience. In the thumb the princeps pollicis artery can provide sufficient blood flow from the dorsum if no palmar arteries are suitable for repair.

ORDER OF REPAIR

Surgeons differ regarding the order in which the vessels should be repaired. The approach may vary depending on the location of the amputation. In the digits our practice is to repair the arteries first. This allows assessment of adequacy of flow across the anastomosis and through the digit before proceeding with the replantation. If veins are repaired first, one has to await arterial anastomosis to determine whether blood will flow through the digit and across the venous anastomosis. Performing arterial repair first also allows the dorsal veins to fill, aiding in the identification of hard to find veins.

If the amputation has occurred through the palm, wrist, forearm, or more proximally and if the limb can be safely revascularized, sometimes blood loss may be minimized if 2 or 3 large veins can be repaired before the arterial repair. This should rarely be done if the ischemia time is 6 hours or more. When considerable time has passed after amputation, repairing the artery first will shorten the ischemia period and minimize the risk from revascularization of a part containing dying muscle. Carotid endarterectomy shunts and ventriculoperitoneal shunts may be used to make arterial connections if the ischemia time is 6 hours or more. Release of excessive amounts of potassium, lactic acid, and myoglobin should thus be avoided. If the artery is repaired first in such circumstances, venous repair should follow as soon as possible to avoid excessive blood loss. In such a situation, the use of the pneumatic tourniquet helps to control bleeding.

Large and small parts may benefit from perfusion of the artery, using a small, soft Silastic catheter and heparinized Ringer's lactate solution. Crushed small parts and large, muscle-containing parts may have a better chance for survival if they are perfused gently with a heparinized solution. Gentle dilatation and irrigation of the cut ends of the vessels help to clear the field of thrombogenic material.

TECHNIQUE. After the arteries have been identified and marked with a small suture, dissect the veins from the dorsal skin flap. Three or four suitable veins are usually found on the dorsum of the digit between the metacarpophalangeal joint and the midportion of the middle phalanx. Distal to this point, there may be only one or two suitable veins. Although volar veins can be seen, they are frequently less than 1 mm in diameter and may not be suitable for anastomosis. Mark the veins with small sutures and proceed to prepare the vessels for anastomosis.

Dissection of the vessels in the palm, on the dorsum of the hand, and in the forearm is somewhat less tedious than in the digits because of the larger vessel size. This frequently requires a midpalmar incision paralleling the skin creases as well as curved or zigzag incisions on the dorsum of the hand and forearm.

After all the arteries and veins have been identified and marked with sutures, mobilize them by dissecting them free from the surrounding tissues using gentle and meticulous technique. Transect small side branches and tributaries using ligatures, metal clips, or bipolar electrocautery, depending on the size of the branch being sacrificed. This mobilization will aid in a tension-free anastomosis. Once the vessels have been mobilized, the size of the gap between the vessel ends will determine whether the anastomoses can be accomplished without additional bone shortening or the use of an interpositional vein graft.

Free the vessel of any adventitia that may be causing constriction and excise the adventitia from the cut ends of the vessels. Use magnification, including the operating microscope, to determine the extent of vessel wall injury. If evidence of thrombosis in the wall is found or if the intima has been damaged, excise the damaged segment. If an avulsion appears to have caused the intima to be pulled out of the vessel ("telescoped"), excise that portion of the vessel as well. Extensive avulsing or crushing injuries may cause vessel wall injury sufficient to preclude a successful anastomosis.

After the vessel preparation has been completed, anastomose the vessels in the order noted above. Attempt to repair both digital arteries and as many veins as possible, preferably two veins per artery. Use the small vessel-approximating clips on the digital vessels. Similar clips are available for the larger vessels. Keep in mind the length of time these clips are in place, especially on small (1 mm) vessels; time elapsed should be kept to a minimum, preferably less than 30 minutes. After the vascular clips are released, bathe the vessel in lidocaine or bupivacaine to minimize spasm.

For digital vessels use 10-0 or 11-0 monofilament suture. In the hand, wrist, and distal forearm 7-0 to 9-0 suture is suitable, whereas vascular injuries near the elbow and more proximally will require larger suture, in the 6-0 and 7-0 range. Most digital arteries require six to eight sutures; digital veins require eight to 10 and sometimes more. Expect a small amount of blood leakage; this will usually stop in a few minutes. If spasm is encountered, the application of warm saline, topical lidocaine, papaverine, reserpine, and magnesium sulfate may help to relieve the spasm. Although intraoperative and postoperative systemic heparinization has been widely used, we have used low molecular weight dextran and aspirin for anticoagulation.

At times the arteries and veins may be so damaged that no satisfactory proximal vessel will be available to suture to the distal vessel, or vessel debridement will leave a gap too large to correct by simple end-to-end repair. Techniques such as interpositional grafting with arterial segments and reversed segments of vein, vein harvesting and shifting in the injured digit, and the transposition of arterial and venous pedicles from adjacent, uninjured digits may help to salvage an otherwise nonviable digit (Fig. 19-24). Vein grafts are usually harvested from unsalvageable amputated parts, the dorsum of the hand or forearm, and from the foot. In some situations a single vein graft anastomosed to a single digital artery proximally may be attached to two digital arteries distally, using side-to-end anastomoses or a Y configuration of the graft. When vein grafts are used, care should be taken to maintain the proper

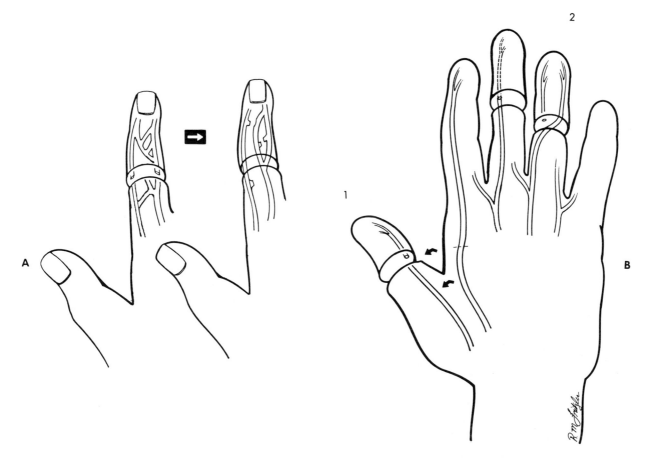

Fig. 19-24. Vessel shifting. **A,** Dorsal veins mobilized to provide additional distal veins for proximal anastomosis. **B,** Arteries mobilized to allow shifting of artery from intact digit to vascularize thumb, *1,* and within same digit to vascularize distal amputated part, *2.* (After Urbaniak, J.R.: Replantation. In Green, D.P.: Operative hand surgery, New York, 1982, Churchill Livingstone.)

(reversed) flow direction so that flow is not obstructed by venous valves. The harvested vein grafts should have approximately the same diameter as the recipient vessel.

Nerve repair

TECHNIQUE. With the dorsal and palmar skin flaps retracted locate the digital nerves in the palmar flap, superficial to the digital arteries. Usually the nerves can be repaired easily following the arterial anastomoses. Gently dissect the nerves free of the surrounding connective tissue and mobilize them so they can be repaired without excessive tension. Occasionally it may be necessary to transect small side branches for sufficient mobilization of a nerve. Once the proximal and distal ends of the nerve have been mobilized, inspect them using the operating microscope or magnifying loupes and trim 3 to 5 mm of nerve from each end. If the injury has been sharp, the nerves are usually repaired primarily. Use two to four epineurial stitches of 9-0 or 10-0 monofilament suture material to carefully align and approximate the fascicles. For more proximal injuries, dissect the respective nerve trunks using standard palmar incisions, paralleling the skin creases in the palm and extending proximally up the forearm. In a nerve trunk in the palm, at the wrist, and more proximally, use a "group fascicular" or peripheral fascicular stitch.

If the amputated part has been avulsed, or if significant crushing makes the extent of intraneural injury unclear, several techniques may be useful. Trim the nerve ends proximally and distally so that normal appearing nerve can be identified. Insert a nerve graft and secure it with microsuture techniques. Nerve grafts may be harvested from unreplantable amputated parts, the lateral antebrachial cutaneous nerve, and the sural nerve. Because of the additional operating time required for nerve grafting and the uncertainty regarding the extent of intraneural injury, we generally do not include primary nerve grafting in the replantation procedure. Instead, suture the ends of the avulsed or crushed nerve together with an 8-0 mattress suture, anticipating later nerve exploration, debridement, repair, or grafting. As an alternative, if the nerve ends cannot be brought together, secure them to the adjacent soft tissues so that they can be easily identified and mobilized later for nerve grafting.

After all structures have been repaired, close the skin primarily if the procedure has been completed promptly, if no excessive swelling is present, and if the skin edges can be approximated without tension. In both the digits and more proximal sites some areas may be left open to heal by secondary intention or to be covered with skin grafts. Nerves, vessels, bone, joints, and tendons should not be exposed if the wounds are left open. Satisfactory alternatives include closure with Z-plasties; local rotation of skin; remote, two-stage pedicle flaps; single-stage transfer of composite tissue (free flaps): and split-thickness skin grafts. In our experience primary remote pedicle flaps and free flaps have not been needed. A combination of skin flap rotation, split-thickness skin graft, and leaving the wound partially open has been satisfactory.

Apply medicated petrolatum gauze to the skin wounds, and use a bulky dressing to cover the dorsal and palmar surfaces. Fluffed cotton or synthetic material provides a soft and gently conforming dressing. Moisten the padding with physiologic saline or Ringer's lactate solution to allow blood to be absorbed into the bandage more readily and permit the bandage to conform more easily to the contours of the part. Avoid localized pressure at all times.

Aftertreatment

The part is adequately padded and a plaster splint is applied to the palmar surface to support the fingers, hand, and wrist. Excessive tightness or constriction is avoided when securing the bandage. The fingertips and small areas of skin are left exposed for evaluation of circulation (Fig. 19-25). During the first week the bandage is moistened with physiologic solutions every 8 hours to prevent dried blood from forming circumferential crusts that might have a constricting effect. Although early and frequent dressing changes may be necessary for the assessment of the circulation or to determine the source and extent of any bleeding, our policy has been to delay the initial dressing change for at least 1 week in uncomplicated replantations. This decreases the risk of disturbing the fragile vascular anastomoses and lessens the chance of stimulating vascular spasm.

The replanted part is usually positioned with the hand at heart level as long as the appearance of the part is satisfactory. If the replanted part appears congested and cyanotic because of venous obstruction, elevation on several pillows may be helpful. If the part becomes pale because of arterial insufficiency, depression of the part below the level of the heart may be required to enhance arterial flow. Depending on the extent of the injury, the patient usually is kept at bed rest for the first 3 to 7 days.

Maintaining a warm room, prohibiting smoking by the patient and visitors, and abstinence from caffeine-containing beverages are measures that help to prevent vasospasm

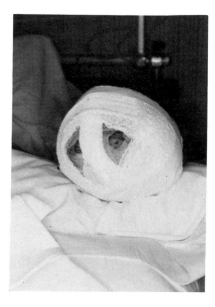

Fig. 19-25. Replantation bandage. Bulky bandage, well-padded with soft, noncompressing cotton, supported with plaster splint. Replanted digits exposed for monitoring.

in the early postoperative period. Vasospasm related to pain and emotional distress may be prevented or minimized through the use of appropriate narcotic analgesics and sedative medication such as chlorpromazine.

Nerve blocks are beneficial in the postoperative period in the prevention of vasospasm. If small Silastic catheters are left adjacent to the median and ulnar nerves, 4 to 5 ml of bupivacaine 0.25% injected every 6 to 8 hours may be sufficient. Stellate ganglion sympathetic blocks or axillary brachial plexus blocks with bupivacaine are carried out once or twice daily in situations in which controlling vasospasm appears necessary.

Various anticoagulants alone or in combination have been administered by different surgeons. Heparin, low molecular weight dextran, aspirin, dipyridamole, and coumadin have been most popular. The use of heparin has been advocated in those injuries thought to be at high risk for thrombosis, especially those with extensive crushing or avulsing injuries, those demonstrating poor flow from the cut ends of vessels before anastomosis, those with poor or equivocal flow across completed anastomoses, and those replantations done in small children. It has been our practice to use dextran, 500 ml every 24 hours for three days, combined with aspirin 600 mg twice daily for 5 to 7 days.

MANAGEMENT OF CIRCULATORY COMPROMISE FOLLOWING REPLANTATION

If the replanted part demonstrates signs of inadequate circulation prompt evaluation and management of the problem might allow salvage of a part that would otherwise be lost. Mechanical monitors of skin temperature, oxygen tension, hydrogen and fluorescein dilution, and other factors (p. 424) in many instances are sufficiently sensitive to detect significant changes in blood flow before clinically apparent ischemic changes develop. If the part is cool and has developed the pallor and loss of turgor consistent with arterial insufficiency or if it is cyanotic, congested, and turgid consistent with venous obstruction, several measures can be helpful in relieving the problem before the patient is taken to the operating room for exploration.

The room should be comfortably warm, the patient should have sufficient analgesic medication, and he should be sufficiently sedated to minimize emotional distress. As noted previously, the part should be elevated well above the level of the heart to enhance venous drainage. If arterial insufficiency is suspected, placing the part in a dependent position may be beneficial. Splints and dressings are loosened or removed to ensure that nothing is causing direct pressure on the vessels and that nothing is constricting the limb. Using gentle digital pressure, the arteries are lightly "milked" from proximally to distally and the veins from distally to proximally.

In distal injuries, if Silastic catheters have been left adjacent to the median or ulnar nerves, 4 to 5 ml of 0.25% bupivacaine are injected. Stellate ganglion sympathetic blocks as well as brachial plexus blocks have also been useful, especially in patients with troublesome vessel spasm. Although it is not part of our usual routine, many surgeons with extensive experience find it useful to administer heparin intravenously as a bolus of 3000 to 5000 units when attempting to salvage a failing replanted part.

If the replanted part does not respond to these measures, the surgeon must decide based on his knowledge of the injury and his experience, whether returning to the operating room to explore the vessels is worthwhile. This decision should be made promptly, once definite signs of impaired circulation are evident. Reoperation is more likely to be successful if done within 4 to 6 hours of the development of signs of ischemia.

REOPERATION

Although the clinical signs may indicate whether the problem is arterial or venous, once the decision to reoperate is made, all anastomoses are evaluated. First, inspect the arterial anastomoses to determine patency. If one or more arterial anastomoses are not patent, excise the anastomoses, ensure that there is adequate "spurting" flow proximally, and repair the vessels. If proximal flow is inadequate or the proximal artery appears excessively damaged, dissect more proximally, find good artery, and interpose a reversed segment of vein graft. Similar problems may be encountered in the distal arteries. If good arterial trunks cannot be found, search for other arteries to substitute. Repair any vein graft as needed. Assess the arterial flow and perfusion distally, as well as the appearance of the part. Next, inspect all venous anastomoses to assess patency. If flow cannot be restored in spite of all efforts, consider reamputation.

If on initial inspection all arterial anastomoses are patent and none appear to have spasm, torsion, pressure, or thrombosis proximally or distally, attention should next be directed to the veins. If all venous anastomoses are patent, the veins proximal and distal to the anastomoses should be inspected to exclude compression, torsion, and thrombosis. If areas of thrombosis are found, excise those segments and either repair the vessel end to end or interpose vein grafts. If the venous anastomoses are found to be obstructed by thrombi excise and repair them either end to end or with vein grafts. Next evaluate arterial and venous flow and the appearance of the part. If all available and suitable veins have been located, repaired, or grafted and satisfactory flow cannot be restored, consider reamputation. For digital injuries techniques such as pulp incisions and wedge excision of the nail to allow venous oozing may allow sufficient flow to persist long enough for a digit to survive. In such patients the hemoglobin and hematocrit should be monitored closely so that blood volume loss may be corrected promptly.

OTHER COMPLICATIONS

Although circulatory compromise related to the vessel repairs is the most pressing complication following replantation, other complications that occur in the early postreplantation period include bleeding, skin necrosis, ischemia caused by muscle compartment swelling, and infection.

Excessive bleeding may be from vessels that have not been cauterized, or it may be caused by anticoagulant therapy. Significant skin necrosis usually occurs following closure of skin that initially appears viable but later undergoes necrotic changes resulting from the magnitude of the injury sustained at the initial traumatic amputation. Additional debridement and secondary closure with local flaps or skin grafts may be required. Significant sepsis is rare following replantation and is usually satisfactorily managed with appropriate systemic antibiotics and wound debridement and drainage as needed. Although ischemia may be caused by excessive muscle compartment pressure, this usually occurs in major limb replantations, and can be treated with appropriate fasciotomies in the arm, forearm, and hand. These early complications may require wound inspection and dressing changes with the patient under anesthesia in the first week following replantation.

Later complications, such as nonunion and malunion of bones, tendon adherence, joint stiffness, and delay in return of nerve function, can usually be managed with the usual techniques appropriate to these problems. In nonunion and malunion, bone grafts and internal fixation may be required. Tendon adhesions with loss of excursion may require tendolysis, and in some situations, tendon grafting as one- or two-stage procedures. Stiff joints may require capsulotomy, or if sufficient damage has occurred, interposition arthroplasty may salvage motion in selected patients. If a primary neurorrhaphy fails to show return of function in a reasonable length of time, of if the nerves are not repaired as part of the original replantation, reexploration and repair or interpositional nerve grafting may be necessary. The nature and timing of specific reconstructive procedures depend on the individual patient's problems and needs and the judgment and experience of the surgeon.

REHABILITATION

The specific rehabilitation program for each patient depends on many factors, especially the patient's needs and motivation and the extent of injury to the part. Generally no attempt is made to begin significant movement of bone, joint, or tendon for the first 3 weeks following replantation. Then depending on the extent of injury, most replantation patients are treated in a manner similar to most patients with combined tendon, bone, and nerve injuries. After the first 3 weeks most paitents are encouraged to participate in a graduated program of active, active-assistive, and protected passive stretching and range of motion exercises supplemented by appropriate dynamic and static bracing and splinting.

Monitoring techniques following microvascular surgery

Following microvascular procedures such as replantation and free composite tissue transfer, a reliable monitoring system should be established for the replanted or transferred tissue. Although the clinical determination of the color, capillary refilling, temperature, and turgor is easily made, there is room for error because of the subjective nature of these factors, especially color and temperature. This combined with the possibility that considerable ischemic injury may occur before clear clinical signs are present has led to the development and use of a variety of mechanical monitoring devices and techniques, including

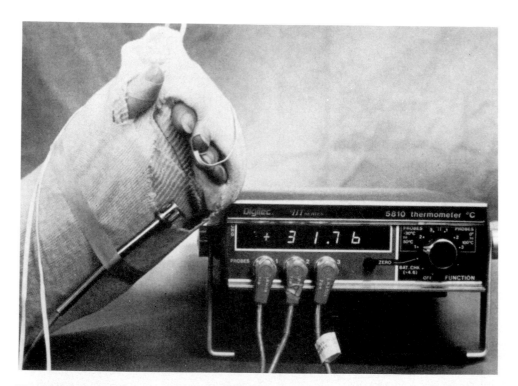

Fig. 19-26. Postreplantation temperature monitoring. Temperature probes attached to dressing for ambient temperature, on normal digit for control temperature, and on replanted digit for monitoring. Battery-operated thermistor for continual postoperative monitoring by nurses at patient's bedside. Amputated part should be within 2° to 3° C of normal part. If replanted part is more than 2° to 3° C cooler, or if there is a decrease below 30° C, circulation may be significantly impaired. (From Urbaniak, J.R.; Postoperative management in replantation. In American Academy of Orthopaedic Surgeons: Symposium on microsurgery, St. Louis, 1979, The C.V. Mosby Co.)

ultrasonic and laser Doppler probes, plethysmography, skin temperature probes, transcutaneous oxygen tension measurements, hydrogen washout techniques, and skin fluorescence measurements.

Both the Doppler probe and plethysmographic techniques are reasonably accurate indicators of arterial flow; however, they are not as accurate when venous flow is to be assessed. Although the transcutaneous oxygen tension determination, the hydrogen washout method, and the skin fluorescence measurement all have been found to be useful and sensitive assays of changes in the microcirculation, the use of skin temperature monitoring probes is presently a simple and reliable adjunct to the clinical evaluations (Fig. 19-26). With separate temperature probes attached to the revascularized tissue, adjacent normal tissue, and the dressing, relative and absolute changes in the temperature can be monitored constantly. A drop in the temperature of the replanted digit below 30° C or a fall of more than 2° to 3° C below the normal digit is considered a sign of circulatory compromise.

The work of Smith and associates, Achauer and associates, and Matsen and associates indicates that the transcutaneous oxygen measurements show changes in oxygen tension several hours before the onset of clinical signs of ischemia and before temperature changes occur. This and other techniques hold promise for the development of monitoring techniques with increasing sensitivity.

REVASCULARIZATION

Partial amputation or devitalization of tissues from serious vascular interruption may occur without complete detachment of the part. Some of these parts with impaired circulation may ultimately survive, but there may be persistant ischemia that later may cause disabling cold intol-

erance and atrophy or contracture of the intrinsic muscles of the hand. Digits with impaired circulation will demonstrate extremely slow return of the normal pink color following blanching by pressure. The management of these hand injuries is essentially the same as for replantation; however, a longer interval from the time of the accident to the anastomosis of the vessels may be tolerated, and the procedure may be done with one team. The same postoperative routine is carried out as described above. When both radial and ulnar arteries are severed at the wrist, usually at least one should be repaired.

If there is any question regarding viability of the hand, both the radial and ulnar arteries should be repaired.

SINGLE-STAGE TISSUE TRANSFER (FREE FLAPS)

Before the development of microvascular techniques, remote pedicle flaps were used to cover major soft tissue defects. In 1946 Shaw and Payne reported their extensive experience with tubed pedicle flaps based on the superficial epigastric and superficial circumflex arterial circulations. Based on that report, their analysis of the deltopectoral flap of Bakamjian, and their own experience with the groin pedicle flap, McGregor et al. explained the differences between random pattern and axial pattern flaps. The random pattern flap relies on no specific established pattern of circulation. A length to width ratio of greater than 2:1 increases the risk of failure of a random flap. An axial pattern flap relies on a definite and usually consistent arterial supply centered on one or more arteries. There are no rigid length to width ratio requirements for axial pattern flaps. These flaps are generally considered to be either cutaneous or myocutaneous, depending on the pattern of their arterial circulation. Cutaneous flaps rely on a constant circulation from a single artery passing through the underlying sub-

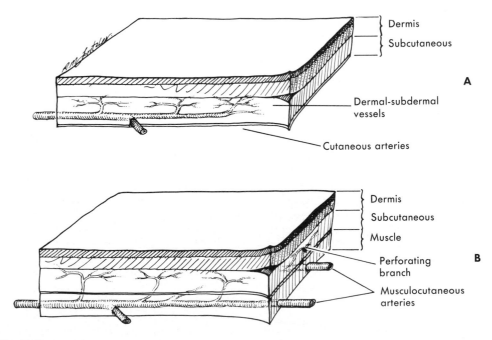

Fig. 19-27. Arterial supply to axial pattern flaps. **A,** Cutaneous flap, relying on single artery supplying dermal and subdermal vessels. **B,** Myocutaneous flap, relying on deep musculocutaneous arteries perforating muscle and fascia to overlying skin.

cutaneous tissue, supplying the overlying skin through the dermal-subdermal vessels. The myocutaneous flap receives its cutaneous arterial supply from deep vessels that perforate the muscle and fascia to reach the skin (Fig. 19-27).

Following the experimental attempts by Goldwyn, Lamb and White and the demonstration by Krizek et al. of successful microvascular free tissue transfer in dogs, McLean and Buncke in 1972 reported the first successful clinical free flap, the coverage of a scalp defect with freely vascularized omentum. In 1973 Daniel and Taylor and O'Brien et al. reported their experiences with the single-stage transfer of groin flaps to the lower extremities. Although various workers have subsequently described many free flaps from a variety of donor sites and with many different uses, this section will cover those flaps with proven application to reconstructive surgery in the extremities.

The following discussion includes the advantages and indications; the disadvantages and contraindications; a general overview regarding the selection of specific flaps; the applications of free flaps to problems as they occur in the upper and lower limbs; a general scheme of preoperative, intraoperative, and postoperative management of patients undergoing these procedures; and a discussion of the anatomic bases, surgical techniques, and more common uses for the specific free flaps in the extremities. These topics will reflect our experiences and observations at this clinic and the recommendations of Buncke, O'Brien, Zhong-Wei, Urbaniak, Kleinert, Taylor, Daniel, Steichen, Terzis, Weiland, Meyer, Ohmori, Harii, Tamai, Ikuta, Tsuge, Gilbert, Foucher, their co-workers, and others that are listed in the references at the end of this chapter.

Indications and advantages

The traditional indications for pedicle flaps are similar to those for free flaps; however, their versatility increases the number of problems that can be attacked using free flaps. Each patient must be considered individually. In most situations the use of a free flap implies that using a more traditional random or axial pattern flap would be either impossible or inappropriate. Present indications for free flaps include but are not necessarily limited to the following:

1. Secondary, and in some situations, primary coverage of extensive skin and soft tissue loss with exposure of essential structures (such as blood vessel, nerve, tendon, bone, and joint).
2. Coverage of a soft tissue bed unsatisfactory for later reconstructive procedures (such as scar, chronic draining ulcers, and chronic osteomyelitis that prevent tendon grafts, tendon transfers, nerve repairs or nerve grafts, bone stabilization, and bone grafting).
3. Replacement of unstable area scars following burns, irradiation, radical surgery for cancer, and scar contracture.
4. Coverage situations for which a suitable random or axial pattern flap is not available.
5. Coverage situations in which immobilization of the extremities for prolonged periods in awkward positions is undesirable or impossible.

6. Restoration of specific tissue to satisfy a functional need (for example, sensation in the hand or the plantar surface of foot, digital reconstruction in the hand, replacement of major skeletal muscle loss in the forearm, replacement for bone loss in the upper and lower extremities, replacement of lost or destroyed joints in the fingers, replacement of functioning epiphyses in the hand and forearm, and correction of congenital and developmental deformities, including radial clubhand and congenital pseudarthrosis of the tibia).

The advantages free flaps seem to have over more traditional techniques include the following:

1. They are usually carried out as single-stage procedures.
2. The choice of a donor site is usually not as restrictive.
3. There is usually more versatility regarding the matching of the color, texture, thickness, and hair distribution of the donor area with the recipient area.
4. In many situations the donor site can be closed primarily, without resorting to skin grafts.
5. Most donor sites are left with an acceptable appearance.
6. Well-vascularized tissue with a permanent blood supply can replace ischemic or avascular tissue.
7. When indicated, a vascularized bone graft, functioning joints, epiphyses, and skeletal muscle can be electively included in the composite graft used to reconstruct a limb.
8. Prolonged immobilization in awkward positions is not required, thus allowing the patient more freedom in his daily activities.
9. Joints adjacent to the recipient area are mobilized earlier than after conventional techniques, preventing joint stiffness and contractures.
10. Hospital stays are usually shortened.

Contraindications and disadvantages

Although the absolute contraindications to the use of free flaps are few, the surgeon should have reservations regarding their use in the following situations:

1. The surgeon has neither microsurgical training nor microsurgical experience.
2. Institutional support for a reconstructive microsurgical program is insufficient.
3. No suitable recipient vessels are available in the area requiring coverage or tissue reconstruction. Previous trauma or irradiation to the recipient area may have damaged the vessels sufficiently to preclude their use.
4. If only one major artery to the foot or the hand is present, the use of it as the recipient vessel for a free flap may jeopardize the viability of the foot or hand, even though an end-to-side anastomosis is used.
5. Age alone may not constitute a contraindication; however, if major systemic illnesses create a major anesthetic risk for the patient, an alternative method of treatment should be considered.

6. If systemic illnesses such as atherosclerosis, vasculitis, or other lesions have caused damage to the vascular system, microvascular procedures, although not certain to fail, are more likely to than are those done when the vessels are not diseased.

7. If previous operative procedures have been done in the donor area, the donor vessels may have been damaged, precluding the use of that specific donor site.

8. Obesity makes dissection of vascular pedicles difficult or impossible. Bulky, obese flaps are awkward to manipulate and difficult to inset without causing tension, torsion, or disruption of anastomoses. The fat at times causes obstruction of a clear view of the vascular pedicles, preventing the performance of satisfactory anastomoses.

The disadvantages of free tissue transfer include the following:

1. The initial operation is usually longer than are those for conventional flaps. Free flap procedures take from 4 to 10 hours, depending largely on the flap selected and the experience of the surgical team.

2. The operations may be difficult and tedious.

3. Two teams of surgeons are usually required.

4. If vascular thrombosis occurs, the risk of complete loss of the free flap is considerable.

5. Reportedly the overall risk of free flap failure compared to conventional techniques is greater. A 10% to 30% failure rate for free flaps is cited by Sharzer, Barber, and Adamson. Additionally, the reoperation rate following free flap transfers may be as high as 25%.

6. Postoperative vascular complications, which usually occur in the first 24 hours, may be seen through as late as 10 days after the procedure.

Selection of free flaps

Numerous free flaps have been described. The selection of one specific flap over another will be influenced by many factors. As Nunley has indicated, the specific tissue requirements at the recipient site are of much importance. Is full-thickness coverage needed? Will a skin graft or conventional flap suffice? Is a free flap really needed? Is the need only for simple coverage? How thick and how large should the coverage be? Is skin sensibility, bone, joint, nerve, or functioning muscle needed?

The condition and availability of donor and recipient vessels are important considerations to determine which flap will be best in a given situation. Generally the simplest procedure should be chosen that will fulfill the tissue requirements of a specific recipient area. The flap should be designed so that if the flap fails, a satisfactory salvage procedure is possible. In most situations a major factor in flap selection is likely to be the experience of the individual surgeon using specific flaps.

Single-stage transfers of composite tissue grafts (free flaps) will be discussed here as they apply to repair and reconstruction of traumatic, infectious, neoplastic, congenital, and developmental problems in the upper and lower limb. The simplest procedures, including local and remote pedicle flaps, should be considered first. In circumstances precluding more traditional techniques microsurgical procedures should be considered, and in some situations priority should be given to the use of free flaps.

UPPER EXTREMITY

In the upper extremity free tissue transfer has proven to be useful in the simple coverage of soft tissue defects, the restoration of sensibility, the reconstruction of bony defects, the replacement of nonfunctioning skeletal muscle units, and thumb and digital reconstruction by toe transfers. The transfers of vascularized toe joints to finger joints and toe and fibular epiphyses to digital and forearm epiphyses, respectively, show promise in the management of additional difficult reconstructive problems in the upper extremity.

Presently the free flaps most often used in the upper extremity include the groin cutaneous flap and the dorsalis pedis cutaneous flap for soft tissue coverage. The dorsalis pedis flap has an added advantage of having nerve supply by way of the deep and superficial peroneal nerves that can be used in restoring sensibility to the hand. As popularized by Manktelow and others, the gracilis, latissimus dorsi, and pectoralis major muscles have been used to restore skeletal muscle function to the forearm. All or portions of the great, second, and third toes have been used successfully for thumb and finger reconstruction. Vascularized bone grafts using rib, iliac crest, and fibula have been used for bone reconstruction in the upper limb and hand.

LOWER EXTREMITY

In the lower extremity, requirements for soft tissue coverage and in the management of osteomyelitis have been satisfied by using the latissimus dorsi muscle, the gracilis muscle, the tensor fasciae latae muscle, the free groin cutaneous flap, and the scapular cutaneous flap. The dorsalis pedis cutaneous flap has also been used as a neurovascular cutaneous flap to provide sensibility to the plantar surface of the foot. Although the rib and iliac crest have been used to reconstruct bony defects in the lower extremity, the curvature and relative weakness of these bones limit their usefulness. The vascularized fibula has been applied successfully to a variety of bony problems in the lower extremity, including defects caused by tumor surgery, trauma, and congenital anomalies such as congenital pseudarthrosis of the tibia. Although the vascularized fibula has been used in the treatment of avascular necrosis of the femoral head, the results are inconclusive to date because long-term results in significant numbers of patients have not been accumulated.

Preoperative requirements

Before establishing a program of reconstructive surgery involving microvascular free flaps, the principal surgeons of the operating team must develop proficiency in microvascular techniques. In addition, familiarity with the vascular anatomy of the various flaps must be acquired through cadaver dissection.

The candidate for a free flap must be evaluated before surgery. He should be healthy enough to tolerate a poten-

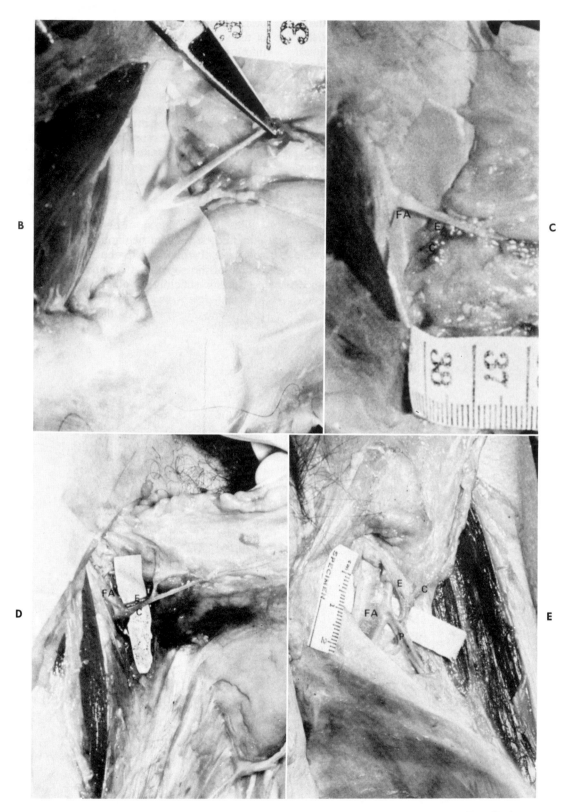

Fig. 19-30, cont'd. B, C, D, and **E,** Anatomic dissections showing variations illustrated in **A.** (From Taylor, G.I., and Daniel, R.K.: Plast. Reconstr. Surg. **56:**243, 1975.)

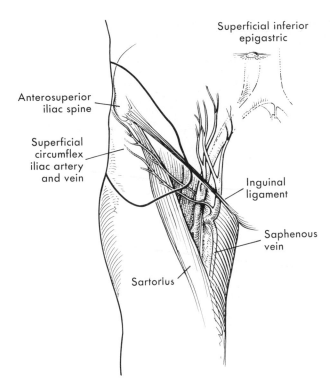

Superficial inferior epigastric

Anterosuperior iliac spine

Superficial circumflex iliac artery and vein

Inguinal ligament

Saphenous vein

Sartorius

Fig. 19-31. Anatomy of groin flap. Flap outline centered on axis of superficial circumflex iliac vessels, perforating the sartorius fascia near lateral border of sartorius muscle.

along the course of the superficial circumflex iliac artery (Fig. 19-31). Groin flaps as large as 30 cm by 20 cm may be harvested.

Begin the approach to the vessels from either the medial or the lateral end of the flap. Daniel and Taylor and Harii and Ohmori favor beginning the dissection at the lateral end of the flap, fearing damage to the artery, failure to identify the vessels, and interference with use of the flap as a pedicle flap should microvascular transfer be impossible. O'Brien et al. and Jackson recommend beginning the dissection at the medial end so that the suitability of the arterial trunk can be determined. Jackson also pointed out that four situations might make the vascular trunk unsuitable for microvascular transfer: the presence of multiple small veins unsuitable for anastomosis; a single small vein; several arteries, none large enough for anastomosis; and one extremely narrow artery.

Generally we favor starting on the medial end of the flap to assess the vessels. If care is taken and the vessels are unsuitable for anastomosis, a pedicle flap may still be fashioned if the area to be covered is in the upper extremity. Both approaches are described here, as some situations may make the use of one technique better than the other. In either case, the landmarks to keep in mind are the pubic tubercle, the anterosuperior iliac spine, the inguinal ligament, and the pulsation of the femoral artery.

When beginning the dissection medially, make a longitudinal incision over the femoral artery, centered about 5 cm inferior to the inguinal ligament. Use gentle sharp and blunt dissection and stay to the medial side of the femoral

artery, watching carefully for the superficial circumflex iliac artery to arise from the medial or anterior aspect of the femoral artery. Identify the veins and dissect them gently as well. Follow the superficial circumflex iliac artery as it passes laterally. Include the fascia overlying the sartorius until the artery can be seen to pass through the fascia into the subcutaneous fat near the lateral border of the sartorius. Before reaching that point, incise the outline of the flap on the skin as needed to permit identification of the vessels and the muscular landmarks. Once the vascular pedicle has been dissected, and suitable arteries and veins have been identified, the entire skin flap may be incised and elevated. The vessels are not transected until preparation of the recipient area is completed and suitable recipient vessels have been identified. If the vessels are not satisfactory for a microvascular transfer and if the recipient area is on the upper extremity, a pedicle flap may still be fashioned with the dissected flap.

When the dissection is begun from the lateral end of the flap, a pattern matching the recipient area is also outlined over the inguinal region. As noted, the axis of the flap is centered about 5 cm inferior to the inguinal ligament. The margins of the flap, as outlined, are incised, leaving a medial skin bridge intact. Dissect from lateral to medial, carrying the deep fascia with the flap as the lateral border of the sartorius is crossed. After the vessels are reached and identified, follow the superficial circumflex iliac artery across the femoral triangle, superficial to the iliacus and the femoral nerve to the femoral artery. Locate the superficial inferior epigastric vein on the anterior aspect of the femoral vein in the same area. Evaluate the size of the vessels. If spasm is apparent, apply topical papaverine or lidocaine to relieve it. If the vessels are satisfactory, the medial skin bridge may be transected; however, the vessels should not be sectioned until the recipient area is prepared for the skin transfer. If the vessels are not suitable for microvascular anastomosis, the flap may be used as a pedicle flap if the defect is in the upper extremity.

If the defect to be covered is in the lower extremity and the vessels are not suitable for anastomosis, another donor site must be selected or the procedure must be abandoned, irrespective of which approach is used for the vessels.

After the groin flap has been isolated on its vessels and the recipient site has been prepared, transect the artery first to allow additional venous drainage and then transect the veins. Apply suture tags to the vessels to avoid losing them, should they retract into the subcutaneous tissue. Place the free flap into the recipient defect, oriented so that the flap vessels match the recipient vessels. Place several anchoring sutures in the margins of the flap to avoid its being dislodged while the anastomoses are being performed.

Suture the artery and then the veins as promptly as possible to avoid venous congestion in the flap. While one team is working on the vessels, the other team closes the groin donor defect. This can usually be done by side-to-side direct closure of the wound. Tension on the wound is minimized by undermining the skin margins and by flexing the hip to allow closure of the wound.

AFTERTREATMENT. The general care of the patient is essentially the same as that already outlined (p. 429). The

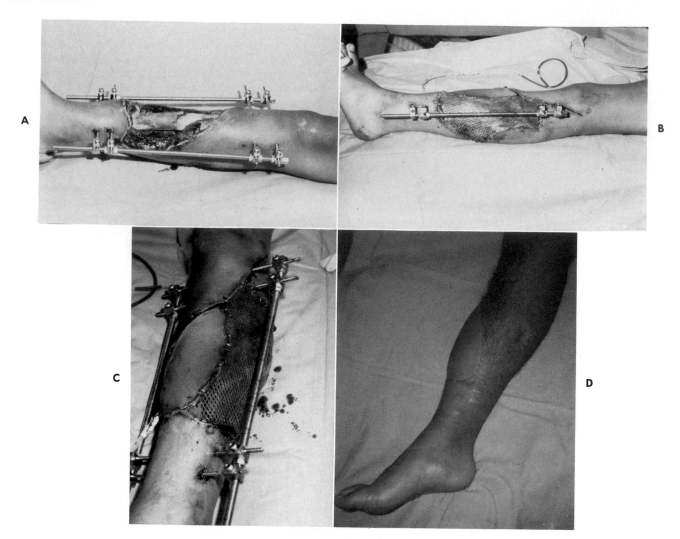

Fig. 19-32. Musculocutaneous free flap for lower extremity soft tissue coverage in 16-year-old boy. **A,** Exposed tibial fracture. **B** and **C,** Coverage achieved with latissimus dorsi myocutaneous free flap, supplemented with split-thickness skin graft. **D,** Healing of soft tissue and bone following sequestrectomy and bone grafting.

circulation to the flap is monitored, and the hip is maintained in a flexed posture for 5 to 7 days, at which time gradual extension is begun and continued for another 7 to 10 days.

Muscle and musculocutaneous free flaps

Muscle and musculocutaneous free flaps are useful in two ways. They have been widely applied for coverage of soft tissue defects in the upper and lower extremities (Fig. 19-32) and in the reconstruction of contour and soft tissue defects in the head, neck, and trunk. Their second major area of usefulness is in the transfer of functioning neuromuscular units to replace paralyzed muscular units in the face and extremities. This section will discuss these flaps as they are applied to the treatment of problems in the extremities.

Building on the 1896 reports of Tansini, who used the latissimus dorsi muscle pedicle flap for breast reconstruction, and on the extensive use of the latissimus dorsi as a muscle pedicle flap for trunk and head and neck reconstruction, Baudet et al., Harii et al., and Maxwell et al. demonstrated the successful transfer of the latissimus dorsi muscle as a free flap. This myocutaneous free flap has been used extensively for soft tissue coverage problems because of its large size and reliable vascular supply, the artery usually being 1.5 to 3 mm in diameter. Other muscle and musculocutaneous free flaps that have been safely and widely used for soft tissue coverage include the gracilis, sartorius, tensor fasciae lalae, and to a lesser extent the rectus abdominis.

The indications for the use of free muscle transfers include (1) acute (primary) wound closure when other techniques will not suffice (for example, local flaps and skin grafts), (2) improvement of soft tissue coverage to permit future procedures (for example, scar over bone, tendon, and nerve), (3) closure of chronic wounds (for example, chronic osteomyelitis, necrotic radiation ulcers), (4) restoration of contour and the filling of cavities; (5) padding of bony prominences, (6) coverage of vessels, nerves, bones, and joints, and (7) replacement of functioning neuromuscular units.

FUNCTIONING NEUROMUSCULAR TRANSFERS

Tamai et al. in 1970 first demonstrated in the canine model that a functioning neuromuscular unit could be transplanted using microvascular anastomoses and nerve repairs at the recipient site. The subsequent experimental work of Kubo, Ikuta, and Tsuge confirmed Tamai's findings. Surgeons of the Sixth People's Hospital, Shanghai, and Harii et al. were able to demonstrate the clinical success of functioning free neuromuscular transfers. Later, Manktelow et al., Buncke et al., Gordon et al., Tamai et al., Ikuta, and Schenck have shown that muscle not only survives following transfer using microvascular techniques, but that it will function following microneural repair and reinnervation. Functioning free muscle transfers may be used for replacement of the flexor and extensor compartments of the forearm, the muscles of facial expression, and the muscles in the extensor compartment of the leg.

Muscles that have been found useful for this procedure include the pectoralis major, latissimus dorsi, gracilis, rectus femoris, extensor digitorum brevis of the toes, and the serratus anterior. The semitendinosis, tensor fasciae latae, and brachioradialis also can be used for functioning muscle transfers. The selection of a given muscle depends on the strength and excursion of the donor muscle, the skin coverage requirements, the motor nerve availability, and the location of the ends of the flexor tendons. The advantage of this procedure is the transferring a viable muscle under voluntary control to restore a functional deficit with a muscle causing little functional loss following its transfer. The disadvantages include (1) the loss of a functioning muscle, (2) the long reinnervation time, (3) the requirement for microvascular skill, and (4) the length of the procedure.

Manktelow has stressed that if a simpler procedure, such as a tendon transfer, will suffice to restore the desired function, it should be used in preference to a free muscle transfer. A hgh level of patient motivation is essential for these procedures to succeed.

Manktelow, Ikuta, Egloff, Buncke, and Gordon as well as others emphasize the following factors in preoperative evaluation and planning. The needs and the neurovascular anatomy of the recipient site should be carefully assessed. A single undamaged motor nerve should be available in the recipient area to supply the muscle transplant. In the forearm, median nerve branches to the flexor digitorum sublimis or the anterior interosseous nerves have been used most often. The status of recipient nerves should be carefully evaluated by history and physical examination, electromyography, and an exploratory surgical procedure if needed. Arteriography is indicated to assess recipient vessels and in some situations the donor vessels. The muscle to be transferred should be similar in size, strength, and excursion to the muscle to be replaced.

If a large muscle is required, the latissimus dorsi or pectoralis major muscles are preferred. If small muscles are needed, the gracilis, serratus anterior, and extensor digitorum brevis may suffice. The muscle and the neurovascular pedicle should be easily accessible. The joints in the recipient extremity should be supple with a functional range of motion available in the elbow, wrist, and fingers. Stable proximal joints with balanced muscles should be present. Good skin covering the recipient site is required. If necessary, a skin island may be carried with the transferred muscle in most situations.

FREE TRANSFER OF FUNCTIONING MUSCLE

TECHNIQUE. Before this procedure is begun, preparations should be made for appropriate monitoring of vital signs and body temperature, appropriate padding of bony prominences, a heating blanket, and an indwelling urinary catheter. Two surgical teams permit a more efficient and prompt completion of the procedure and usually are essential.

Forearm preparation. Determine from previous surgical procedures on the forearm, preoperative clinical examination, electromyography, and angiography, the general location of the recipient arteries and nerves. Fashion a paper template to assist in locating the neurovascular pedicle (Fig. 19-33). This will also help to determine the area of needed skin coverage. Use a pneumatic tourniquet to allow rapid initial dissection. Inflate and deflate the tourniquet as needed after the initial dissection.

Usually, an extensive curved or a zigzag incision is required for adequate exposure. Plan to use either the radial or ulnar artery or a suitable large branch as a recipient vessel. On the flexor aspect, plan to expose the ulnar, median, and the anterior interosseous nerves as needed. On the extensor surface, plan to use branches of the radial nerve. If extensive scarring is present, carefully dissect from normal, uninjured areas into the scarred areas to avoid injury to the recipient vessels and nerves. Exposure of the anterior interosseous nerve and artery may require section of the pronator teres in a Z configuration, allowing for later repair. For venous drainage, use the venae comitantes of the arteries selected, or superficial veins in the area.

Expose the tendons of the flexor digitorum profundus and mobilize them by dissecting them free from surrounding scar to assure satisfactory gliding. Identify the flexor tendons distally for flap attachment, and plan to expose the medial epicondyle and surrounding fascia for flexor replacement attachment. Expose the lateral epicondyle and the extensor origin for extensor replacement attachment. Plan to cover the distal flexor tendon repair with a local skin flap or with skin on the transplanted muscle. Cover the proximal belly of the transplanted muscle with skin graft if needed. Before anastomosis, determine that the recipient artery demonstrates free pulsatile flow.

Transfer of functioning muscle to forearm (Manktelow). After dissecting the donor muscle, leave it attached to its major vascular pedicle and at the origin and insertion until the recipient area in the forearm has been completely prepared. Determine that the recipient vessels are suitable for anastomosis, and that all is ready for transfer to minimize the muscle ischemia time. The following method, as sug-

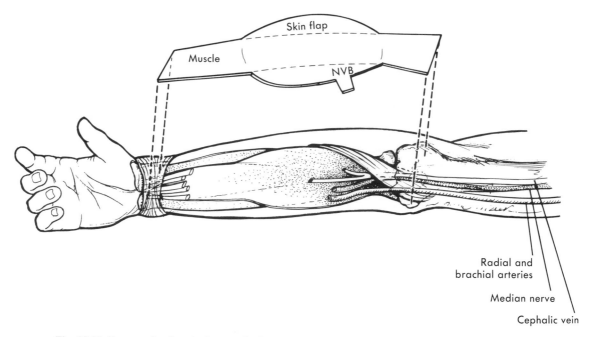

Fig. 19-33. Free transfer, functioning muscle. Paper template assists in preoperative planning of skin coverage requirements, musculotendinous attachments, and neurovascular repairs. *NVB,* neurovascular bundle. (Redrawn from Manktelow, R.T.: Free muscle flaps. In Green, D.P.: Operative hand surgery, New York, 1982, Churchill Livingstone.)

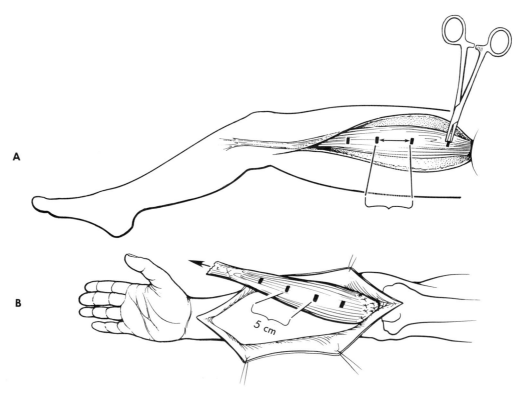

Fig. 19-34. Free transfer, functioning muscle. **A,** Gracilis transfer. Pretransfer muscle length determined by placing metal clips 5 cm apart in muscle. **B,** Transferred muscle attached proximally. Pretransfer length restored with traction on muscle sufficient to restore 5 cm interval between metal clips. (Redrawn from Manktelow, R.T.: Free muscle flaps. In Green, D.P.: Operative hand surgery, New York, 1982, Churchill Livingstone.)

gested by Manktelow, is helpful in determining the proper tension for attachment of the transferred muscle, especially if the extensor musculature is not intact. Position the extremity so that the muscle is fit to its maximum physiologic length. For the latissimus dorsi and the pectoralis major, this would be maximum humeral abduction. For the gracilis, this would be with the knee fully extended (Fig. 19-34). With the arm in the appropriate position, place suture markers on the surface of the muscle at 5 cm intervals. After transfer and revascularization, this length can be restored by stretching the muscle from its new origin to its new insertion, reestablishing the 5 cm interval between the suture markers.

Once the length determinations have been made, detach the muscle from its origin and insertion, and carefully section the arteries, veins, and nerves. Immediately transfer the muscle to the arm in the best position for ease of neurovascular repair and attachment proximally and distally. This may require reversing the ends of the muscle so that the origin becomes the insertion. On the flexor surface,

attach the origin to the medial epicondyle and surrounding fascia. On the extensor surface, attach the origin to the lateral epicondyle, fascia, and periosteum. Loosely suture the muscle in place to prevent its displacement during the neurovascular repairs. While the muscle is being attached by one team, the other closes the donor site.

Position the transferred muscle in such a way that the arterial and venous anastomoses can be easily carried out (Fig. 19-35). Also, position the vascular repairs so that the nerve repairs will be as close to the muscle as possible. This should shorten the period of muscle denervation. Manktelow reports a distance of 2 to 3 cm in most of his patients. Blood loss may be decreased by repairing one or more large veins before repairing the arteries; however, as long as the ischemia time is minimized, the order of repair is not critical. After completing the vascular repairs, connect the nerves with careful fascicular repair (p. 405) using 10-0 or 11-0 nylon suture.

Restore the predetermined 5 cm intervals on the muscle belly by pulling the transferred muscle out to length, and

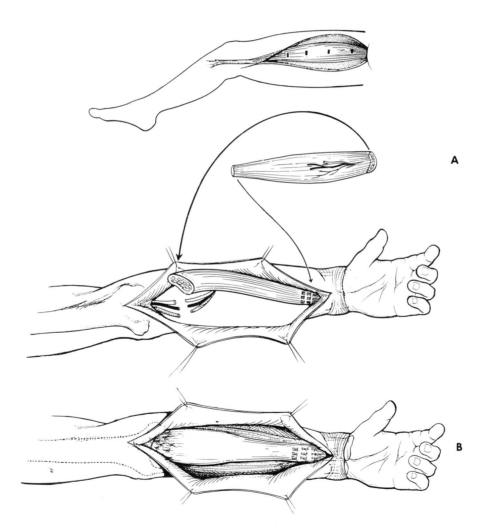

Fig. 19-35. Free transfer, functioning muscle. **A,** Scheme of transfer of functioning gracilis muscle from right thigh, reversed to match for neurovascular repairs in left forearm. If ipsilateral gracilis muscle is used, reversal is unnecessary. **B,** Transfer is completed, muscle is attached to fascia and periosteum proximally and interwoven to flexor tendons distally. (Redrawn from Manktelow, R.T.: Free muscle flaps. In Green, D.P.: Operative hand surgery, New York, 1982, Churchill Livingstone.)

mark on the recipient tendons the appropriate locations for repair. Weave the flexor digitorum profundus tendons into the transplanted tendons as marked. Before attaching the recipient tendons to the donor flap, secure the recipient tendons to each other with a side-to-side suture. If the transferred muscle has no tendon, attach the recipient tendons by sewing them into the muscle and securing them with mattress sutures.

Cover the distal musculotendinous repair with either a local skin flap or skin carried with the transferred muscle. Split-thickness skin grafts can be used to cover the muscles proximally. Close the wounds loosely to avoid constriction of vessels, and apply plaster splints with the wrist and fingers moderately flexed to relieve tension on the muscle and tendon repairs.

AFTERTREATMENT. Maintain the systemic circulation with good peripheral perfusion, ensuring adequate hydration. A postoperative anticoagulation routine may be followed, depending on the training, experience, and preference of the surgeon.

Gentle passive stretching exercises are begun 3 weeks after the operation and are continued until a full range of motion is achieved. As reinnervation occurs, usually at 2 to 4 months, active finger flexion is begun. Usually between 6 and 12 month daily exercises include active resistive exercises. Various physical therapy techniques to increase the strength and range of motion are used throughout the course. Manktelow has observed that muscle strength stabilizes between 2 and 3 years after the transplantation.

PECTORALIS MAJOR TRANSFER

The pectoralis major muscle, a humeral internal rotator and adductor, has two major components. The superior (clavicular) component arises along the medial one half of the clavicle, extending inferiorly and laterally to the lateral margin of the bicipital groove. The inferior (sternocostal) component arises along the anterior surface of the sternum, the costal cartilage of ribs 2 through 6, and the external oblique muscle aponeurosis. It passes laterally, inserting on the humerus deep to the clavicular portion of the muscle.

The pectoralis major receives its major blood supply through the pectoral branch of the thoracoacromial artery (Fig. 19-36). The pectoral artery gives off its superior branch to the clavicular portion of the pectoralis major and its inferior branch to the sternal head. These two branches are constant and are present in 100% of dissections, according to Manktelow et al. The sternal portion may also be supplied by two less constant arteries, one a branch of the pectoral artery (present in 40% of dissections), and the other a branch of the lateral thoracic artery.

The principal arteries and the accompanying veins are of sufficient size to permit anastomosis. According to Mathes and Nahai, the dominant vascular pedicle is approximately 4 cm long with a diameter of 2.5 mm. Manktelow has observed 5 to 6 nerves to the sternocostal portion. These are usually monofascicular branches and pass to the pectoralis major from above and below the pectoralis minor and through the substance of that muscle. Each nerve has a definite area of supply with little overlap.

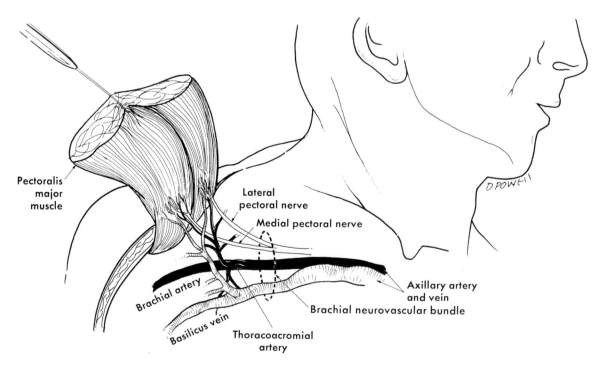

Fig. 19-36. Vascular supply to pectoralis major. Muscle receives major blood supply through pectoral branch of thoracoacromial artery. Sternocostal portion receives five to six monofascicular nerves as well. (From Ikuta, Y.: Skeletal muscle transplantation in the severely injured upper extremity. In Serafin, D., and Buncke, H.J.: Microsurgical composite tissue transplantation, St. Louis, 1979, The C.V. Mosby Co.)

TECHNIQUE (MANKTELOW-IKUTA). Prepare and drape the chest wall and arm to allow free movement of the arm. The donor muscle is usually selected from the same side as the recipient forearm. Outline an ellipse of skin on the chest wall over the muscle mass to be removed, parallel to the inferior margin of the muscle and superior to the nipple. Make an incision extending from the shoulder, curving inferiorly toward the sternum and toward the rectus abdominus insertion. Incise the outlined skin to be removed down to the fascia. Suture the dermis to the pectoral fascia to prevent shear. Remove the rectus fascia in continuity with the pectoral fascia to allow full muscle insertion attachment. Elevate and mobilize the remaining skin over the muscle.

Identify the neurovascular pedicle by one of two approaches. Elevate the lower margin of the muscle, exposing the deep surface, and identify the major branch of the inferior pectoral artery and its venae comitantes. Before encountering the vascular pedicle, identify the lateral and inferior nerves. Dissect carefully; these nerves are easily injured. Or using a second approach, separate the sternal from the clavicular parts of the pectoralis major and identify the neurovascular pedicles.

After the muscle has been dissected medially to laterally and the neurovascular pedicle has been carefully dissected, determine that the recipient area is ready. Section the insertion of the pectoralis major from the humerus. To place the muscle in the forearm, reverse it so that the origin becomes the insertion, to allow easier attachment of the neurovascular pedicle to the recipient vessels and nerves (Fig. 19-37). One team closes the donor site over suction drainage tubes while the other team attaches the muscle to the appropriate tendons and completes the microneurovascular repairs. (See general technique for transfer of functioning muscle to forearm.)

GRACILIS MUSCLE TRANSFER

Arising from the anterior body and the inferior ramus of the pubis and the ischium, the gracilis muscle passes distally in the medial thigh posterior to the adductor longus and sartorius muscles, inserting on the medial aspect of the proximal tibia posterior and deep to the sartorius tendon, and anterior to the semitendinosus muscle insertion. Its innervation comes from a branch of the obturator nerve, which has two to four fascicles entering the muscle 6 to 10 cm from the origin.

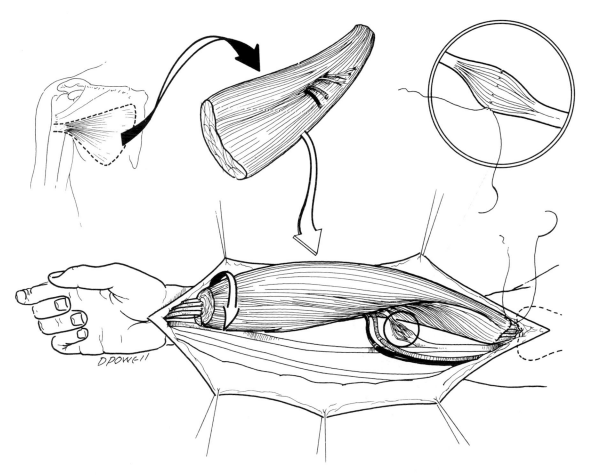

Fig. 19-37. Scheme of transfer of pectoralis major muscle. Ipsilateral muscle is rotated to allow nerve and vessel repairs on the deep surface. *Inset,* Perineurial (fascicular) neurorrhaphy. (From Ikuta, Y.: Skeletal muscle transplantation in the severely injured upper extremity. In Serafin, D., and Buncke, H.J.: Microsurgical composite tissue transplantation St. Louis, 1979, The C.V. Mosby Co.)

The obturator nerve accompanies the dominant vascular pedicle, the medial femoral circumflex artery, and its venae comitantes arising from the profunda femoris artery and vein, 8 to 12 cm from the muscle origin (Fig. 19-38). A vascular pedicle can be obtained 4 to 6 cm long with a vessel diameter of 1 to 2 mm. Two minor vascular pedicles, branches of the superficial femoral artery, are located distally and may be sacrificed. No significant functional loss can be seen following removal of the gracilis muscle.

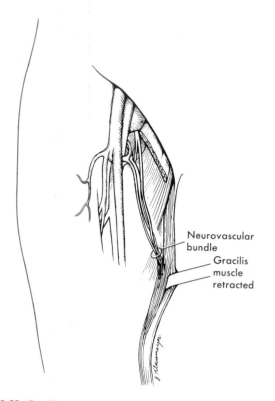

Fig. 19-38. Gracilis muscle neurovascular pedicle. Major vascular pedicle to gracilis muscle enters in proximal third of muscle and includes branch of medial femoral circumflex artery and tributary of femoral vein and anterior branch of obturator nerve.

TECHNIQUE. Prepare and drape the entire lower extremity, exposing the groin, thigh, and knee so the limb can be easily moved about. Abduct and externally rotate the hip and flex the knee to allow access to the medial side of the thigh from the groin to the knee.

Draw a straight line between the origin of the adductor longus and the tibial tuberosity. The gracilis muscle should lie posterior to such a line (Fig. 19-39).

To remove a cutaneous flap with the muscle, center the outlined flap over the proximal muscle because skin flaps in the distal portion have been found by Manktelow to be unreliable. After outlining the skin flap, make the skin incision along the line marked. Incise down to gracilis muscle and suture the dermis at the margins to the underlying muscle. Dissect anterior to the gracilis muscle, separating the adductor longus muscle from the gracilis and retracting the adductor longus anteriorly. The neurovascular structures can now be seen entering the deep surface of the gracilis muscle. Ligate or cauterize the vascular side branches to allow the development of a long pedicle. Dissect the gracilis muscle free posteriorly, and mobilize it proximally and distally by blunt digital dissection.

Ligate or cauterize the lesser vascular pedicles as they are encountered distally on the deep surface of the muscle. If the muscle is to be transferred as a functioning muscular unit, determine its physiologic length by placing marking sutures as indicated in the transfer of functioning muscle to forearm (p. 437). Make a short incision in the distal thigh, identify the gracilis tendon by blunt dissection, and section the tendon distally after determining the needed length. To avoid displacement of the muscle, suture it loosely in situ, leaving it attached to the remaining pedicle and origin until the forearm is prepared. Then section the origin and the neurovascular pedicle and deliver the muscle unit to the recipient site in the forearm.

If a cutaneous flap is not required, place the incision about 2 cm posterior to the anterior origin of the gracilis muscle in the proximal thigh. Open the fascia and dissect the adductor longus muscle, separating it anteriorly from the gracilis. The neurovascular pedicle can be seen, as

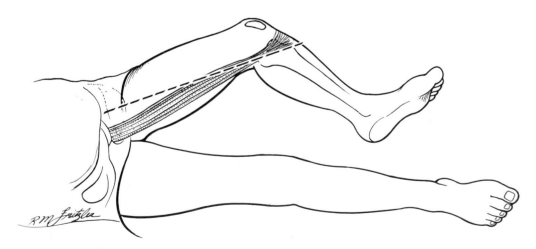

Fig. 19-39. Gracilis muscle dissection. Gracilis muscles lies posterior to line drawn between adductor origin and tibial tuberosity.

noted above, and the dissection proceeds as in the description for a muscle transfer with a skin flap. After the forearm is prepared, divide the neurovascular pedicle and place the muscle in the forearm so that the vessels and nerves can be easily repaired while a second team closes the donor site in the thigh.

LATISSIMUS DORSI TRANSFER

Arising from the thoracolumbar fascia, the iliac crest, and the lower three ribs, the latissimus dorsi muscle covers most of the lower portion of the posterior trunk and passes laterally to insert on the inferior portion of the bicipital groove of the humerus. The principle vascular supply to the latissimus dorsi is through the thoracodorsal branch of the subscapular artery with its venae comitantes. The thoracodorsal artery enters the muscle on its deep surface, 8 to 12 cm from the insertion (Fig. 19-40).

If the thoracodorsal artery is taken below the circumflex scapular branch, it has a diameter of 1.5 to 3 mm. Secondary vascular pedicles enter the muscle medially from the perforating branches of the lumbar and posterior intercostal arteries. The thoracodorsal nerve follows the artery and has two to three fascicles with a diameter of about 2 mm. All of the muscle with most of its overlying skin may be transferred with the thoracodorsal neurovascular bundle. Skin flaps of varying sizes may be oriented over the muscle as needed.

TECHNIQUE. Place the patient in the lateral decubitus position, maintaining this position with sandbags and kidney rests. Prepare and drape the patient, leaving the entire shoulder and thorax exposed anteriorly and posteriorly. Drape the entire upper extremity free so that it can be easily moved about. Draw a line along the anterior margin of the latissimus dorsi muscle from the anterior margin of the posterior axillary fold to the midportion of the iliac crest.

Although most of the skin overlying the latissimus dorsi muscle may be transplanted, it is not usually required for upper extremity reconstruction. If an island of skin is to be taken for upper extremity coverage, it may be best designed over the anteroinferior aspect of the muscle (Fig. 19-41). This will allow direct closure of the donor site. If the flap is to be removed for coverage of larger defects (as in the lower extremity), larger skin islands may be outlined, 10 to 12 cm wide, or the muscle may be removed without its overlying skin and split-thickness skin grafts may be applied to the muscle belly after the flap has been attached to the recipient site.

Make a curved incision, extending from the axilla, following the anterior margin, and include the outline of the skin flap. Identify the dorsal surface of the latissimus dorsi to avoid dissection of skin from the muscle. Separate the latissimus dorsi anteriorly from the serratus anterior muscle. Also separate the anterior margin of the muscle from the posterior iliac crest distal to the skin island. Retract the muscle posteriorly and dissect deep to the muscle medially toward the spine. Ligate or cauterize perforating vessels entering the muscle medially. With the anterior margin and the distal attachments mobilized, the muscle can be manipulated freely to allow dissection of the neurovascular pedicle. Palpate the vascular pedicle proximally near the insertion and 1 to 2 cm from the anterior margin.

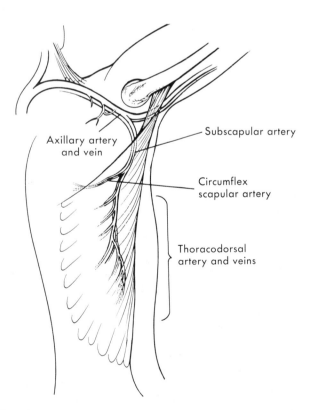

Axillary artery
and vein

Subscapular artery

Circumflex
scapular artery

Thoracodorsal
artery and veins

Fig. 19-40. Latissimus dorsi vascular supply. Thoracodorsal branch of subscapular artery with venae comitantes enters muscle on its deep surface 8 to 12 cm from humeral insertion. Thoracodorsal nerve (not shown) accompanies artery.

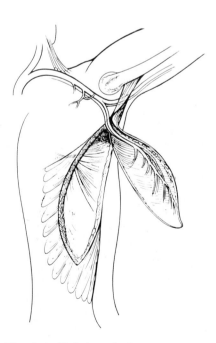

Fig. 19-41. Elevation of latissimus dorsi myocutaneous flap. Direct closure of donor site possible if overlying skin is removed from lower portion of chest wall.

Dissect carefully when mobilizing the pedicle. Using the bipolar cautery, cauterize branches perforating the chest wall musculature from the latissimus dorsi. Identify and preserve the long thoracic nerve to the serratus anterior deep and anterior to the thoracodorsal pedicle. Identify the nerve to the latissimus dorsi, accompanying the thoracodorsal pedicle. Dissect the pedicle proximal to the branch to the serratus anterior muscle to obtain a pedicle length of 8 to 12 cm.

After dissecting the neurovascular pedicle, proceed with the dissection medially and superiorly, releasing the proximal attachments of the muscle to the chest wall. Determine the amount of muscle required at the recipient site (arm or leg), and excise the medial margin of the skin flap down to the muscle. Suture the margins of the skin flap dermis to the fascia to avoid shear.

Release the tendon at its insertion and section the neurovascular pedicle only when the recipient site has been completely prepared. If a functioning muscle is required, determine the maximum functional length, as noted in transfer of functioning muscle to forearm (p. 437). Transfer the muscle to the recipient site, and close the donor site. A split-thickness skin graft may be required at the donor site, depending on the size of skin flap removed. Suction drainage is useful to avoid the development of a seroma or hematoma in the donor site.

TENSOR FASCIAE LATAE MUSCLE FLAP

The tensor fasciae latae originates from the anterior portion of the iliac crest and inserts into the fascia lata of the thigh. It has a single major arterial supply, the transverse branch of the lateral femoral circumflex artery from the profunda femoris artery. The vascular pedicle enters approximately 10 cm inferior to the iliac crest. The arterial supply lies deep to the rectus femoris muscle and gives branches to the rectus femoris, the vastus lateralis, and gluteus minimus muscles (Fig. 19-42). The venous drainage is through the two venae comitantes that accompany the arterial pedicle. A vascular pedicle 6 to 8 cm long can be dissected. The vessels have diameters of 2 to 2.5 mm.

The tensor fasciae latae receives its motor innervation through a branch of the superior gluteal nerve, which enters the muscle proximal to the vascular pedicle. The skin area of the muscle is located on the lateral thigh between lines extending from the anterosuperior iliac spine and the lateral femoral condyle anteriorly and the greater trochanter posteriorly; it extends from the iliac crest to the knee. This skin receives its sensory innervation through the cutaneous branch of the twelfth thoracic nerve and the lateral femoral cutaneous nerve. The branch of the T12 nerve enters the region in the subcutaneous tissue near the posterosuperior aspect of the flap, and the lateral femoral cutaneous nerve enters the medial portion of the flap in the subcutaneous tissue 8 to 10 cm distal to the anterosuperior iliac spine.

Although its fascial surface may not adhere to a recipient site, the muscle may be used for coverage of soft tissue defects in the upper or lower extremity, as a sensory innervated flap, as a functioning neuromuscular unit, and as an osseomusculocutaneous flap incorporating a portion of the bone of the iliac crest. Its function as an innervated

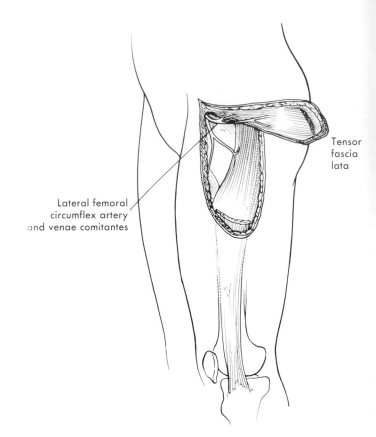

Fig. 19-42. Elevation of tensor fascia lata flap. Vascular pedicle of lateral femoral circumflex artery and venae comitantes enter the muscle 10 cm inferior to the iliac crest. Muscle with overlying skin may be taken. Donor defect usually requires skin graft coverage.

flexor substitution is limited because of the limited excursion; however, it has potential as an extensor replacement.

TECHNIQUE (HILL, NAHAI, VASCONEZ, AND MATHES) (FIG. 19-42). Outline on the proximal lateral thigh the area of skin required by the recipient area. Mark the location of the anticipated entrance of the vascular pedicle approximately 10 cm inferior to the iliac crest. If the transfer is to be a neurosensory, osseomusculocutaneous, or functional free flap, identify the areas of the lateral femoral cutaneous nerve anteromedially, the sensory branch of the T12 nerve posterolaterally, the anticipated location of the motor branch, and any required bone before making the skin incision. Plan the flap so that the required neurovascular repairs and bone placement if used can be appropriately located in the recipient site.

Outline the skin required within the skin circulation territory on the lateral aspect of the thigh. Incise the lateral, medial, and distal borders of the flap through the subcutaneous tissue to the underlying fascia lata. If sensory nerves are to be incorporated in the transfer, dissect and identify them at this point. The lateral femoral cutaneous nerve enters the flap 5 to 10 cm inferior to the anterosuperior iliac spine. Identify the nerve in the subcutaneous tissue, dissect it proximally, and divide it, providing sufficient nerve for nerve repair before identification of the vascular pedicle.

Suture the margins of the skin flap to the muscle superiorly and to the fascia lata distally to avoid shear on the flap. Dissect from distal to proximal, separating the deep surface of the fascia lata from the vastus lateralis muscle. Carry the dissection proximally, and as the junction of the tensor fasciae latae muscle and the fascia lata is encountered, take care to identify the lateral femoral circumflex artery. Retract the rectus femoris muscle to identify the transverse branch of the lateral femoral circumflex artery with its venae comitantes. Dissect the vessels medially toward the profunda femoris vessel, coagulating or ligating muscular branches.

Elevate proximally the superior margin of the flap, and incise and divide the proximal muscle superior to the entrance of the vascular pedicle. If bone is to be included, carry the dissection more proximally after incision and elevation of the superior skin to the level of the iliac crest. Using an osteotome, include the tensor fasciae latae with the underlying iliac crest. Leave the vascular pedicle intact until the recipient site is completely prepared, and then section the vessels for transfer. If a functional transplantation is to be made, extend the distal fascia lata dissection beyond the distal margins of the skin to allow use of the fascia lata for attachment to recipient distal tendon stumps.

The excessive thickness of the muscle belly, the stiffness of the fascia lata, the inability of the fascia lata to adhere to underlying structures, and the inability at times to close the donor defect are cited as disadvantages of this flap.

COMPOSITE FREE TISSUE TRANSFERS FROM FOOT

The neurovascular supply to the structures of the foot make it an unusually versatile donor site for many problems, especially in the foot and hand (Fig. 19-43). This section includes a discussion of the neurovascular anatomy of the foot as it pertains to the specific structures that are used most often as free tissue transfers. Also included are the advantages, disadvantages, various applications, and

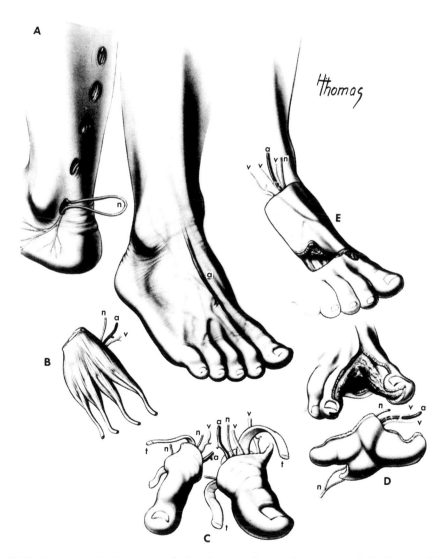

Fig. 19-43. Foot as versatile donor source for free tissue transfers. **A,** Sural nerve graft. **B,** Extensor digitorum brevis muscle flap. **C,** First and second toe free tissue transfer. **D,** First web space free neurovascular flap. **E,** Dorsalis pedis cutaneous or cutaneous neurovascular free tissue flap. All these procedures require microsurgical techniques. *a,* Artery; *v,* vein; *n,* nerve; *t,* tendon.

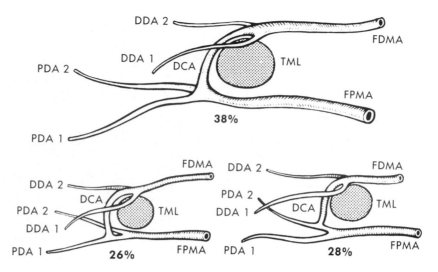

Fig. 19-52. Variations in circulation to first web space. Three patterns of communication between distal communicating artery and plantar digital arterial system have been identified. (From May, J.W., et al.: J. Hand. Surg. **2**:387, 1978.)

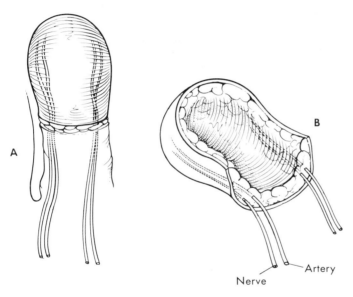

Fig. 19-53. Scheme for pulp free flap. **A,** Flap outlined, toe dissection begun for skin flap. **B,** Flap separated from toe with digital arteries, venae comitantes and nerves. (After Egloff.)

6 cm by 10 cm to 8 cm by 12 cm can be obtained from the first web. If large amounts of skin are required, the dorsalis pedis skin flap may be included with the first web skin. When planning the dissection, this should be taken into consideration.

Outline the flap in the first web with a skin marker (Fig. 19-51, *C*). Exsanguinate the limb by wrapping and inflate the pneumatic tourniquet. Begin the dissection over the artery with a curved or zigzag incision on the dorsum of the foot between the first and second metatarsals. With careful and meticulous dissection, identify the dorsalis pedis ar-

tery, follow it to the deep plantar branch, ligate and divide the deep plantar branch, and follow the first dorsal metatarsal artery distally. Elevate and divide the extensor hallucis brevis tendon to allow exposure of the artery. Include the deep peroneal nerve with the arterial pedicle. As the dissection proceeds distally, elevate the skin flaps medially and laterally to identify the venous drainage. Usually large veins can be found in the dorsum of the first web space, communicating with the large tributaries to the greater saphenous system on the medial side of the dorsum of the foot.

If the first dorsal metatarsal artery is absent or hypoplastic, make a longitudinal plantar incision between the first and second metatarsals communicating with the outline of the skin flap. Identify the plantar digital arteries, which can be found dorsal to their respective proper digital nerves. Both the plantar digital arteries and nerves can be dissected proximally so that the first plantar metatarsal artery and the common plantar digital nerve can be identified. Usually if the first dorsal metatarsal artery is hypoplastic, the first plantar metatarsal artery will have a diameter large enough to allow anastomosis. Both dorsal and plantar dissections should expose arteries, veins and nerves sufficiently long to avoid the need for interpositional grafting. After developing the neurovascular pedicles, elevate the first web skin, maintaining a plane of dissection deep to the plane of the vessels and nerves so that they are carried with the skin without devascularizing it. While the donor site is being dissected, the hand recipient site is dissected and a split-thickness skin graft for the web space closure is obtained by the recipient site team.

To harvest a pulp or hemipulp flap from the great or second toe (Fig. 19-53), outline the small amount of skin required for the recipient digit (Fig. 19-54). The vascular dissection is identical to the web dissection until it approaches the web. At the first web, if the great toe is to be the donor, dissect the digital arterial and venous branches so that those to the lateral side of the great toe are carried

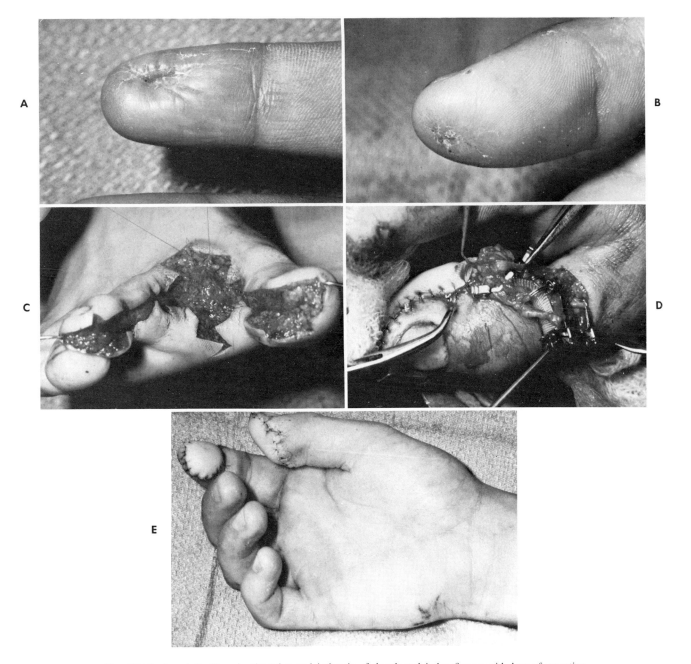

Fig. 19-54. A and **B,** Chronic ulceration and ischemia of thumb and index fingers with loss of sensation secondary to radiation injury. **C** to **E,** Small neurovascular skin flaps from the first and second toe pads to replace the thumb and index fingertip pads.

with the flap. Handle the digital nerve branches in a similar manner by dissecting the plantar digital nerves proximally and separating the nerve to the lateral side of the great toe from the common digital nerve. If the second toe is to be the donor, dissect the nerves and vessels so that they are carried with the skin and pulp on the medial side of the second toe.

If the recipient site is to be the thumb, carefully dissect and preserve the palmar digital nerves, the branches of the superficial radial nerve, the princeps pollicis artery, and dorsal digital and hand veins. If a finger is to receive the

web or pulp flap, identify and mobilize the proper digital arteries, digital nerves, and dorsal veins for anastomosis.

While one team closes the donor defect with a skin graft, the other team loosely sutures the web, pulp, or hemipulp flap into the recipient defect. In the thumb, suture of the first dorsal metatarsal or plantar metatarsal arteries are done at the princeps pollicis or radial artery at the wrist by end-to-end or end-to-side anastomoses. Perform the venous repairs on the dorsum of the hand, anastomosing the saphenous system branches to the cephalic venous system. Suture plantar digital nerves to the palmar

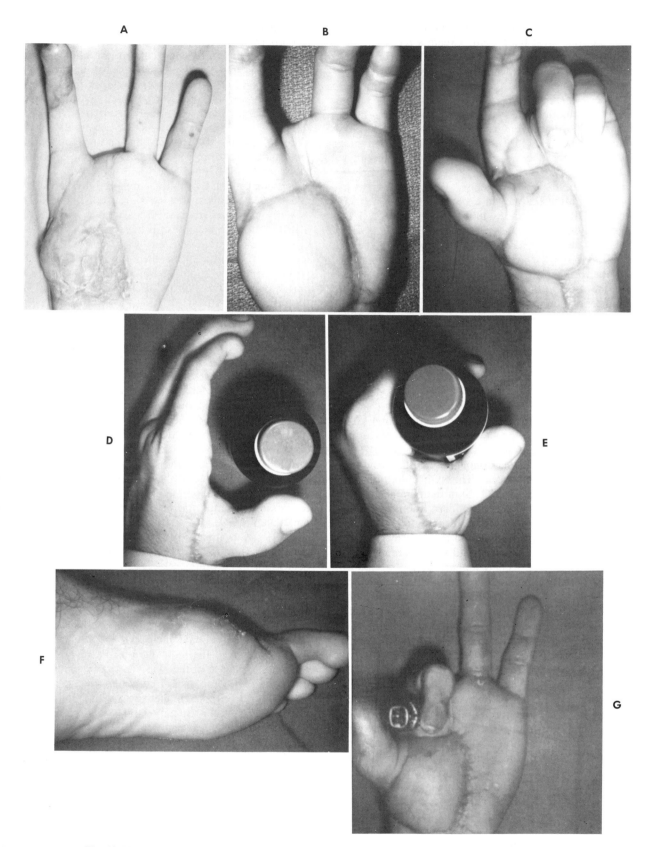

Fig. 19-60. One-stage toe-to-hand transplantation using microvascular techniques offers advantages over traditional thumb reconstructive procedures in selected patients. **A,** This 14-year-old boy lost thumb and middle finger in lawn mower accident. Appearance following split-thickness skin graft. **B,** Subsequent flap coverage of area provided suitable bed for reconstruction. **C, D,** and **E,** Six months following transplantation of great toe to hand; protective sensibility and gross grasping motion were present. **F,** Minimal morbidity follows great toe amputation when plantar skin is preserved and skin grafting of plantar surface is avoided. **G,** One year following transfer, ability to hold small objects is regained.

tion, or at the time of thumb reconstruction by incorporating a dorsal foot flap with the great toe transfer. Generally it is preferable to allow for split-thickness skin grafting to be done on the recipient hand rather than the donor foot because of the unpredictable results, especially on the dorsum of the foot.

TECHNIQUE (BUNCKE, MODIFIED). Position the patient so that the donor foot and the recipient hand are easily accessible. Provide a padded operating table with a heating and cooling blanket and esophageal or rectal temperature probes. Monitor urinary output with an indwelling urinary catheter. Two surgical teams are required, one for the hand, the other for the foot. Using skin marking pencils, outline incisions on the hand and foot, providing adequate soft tissue coverage for both areas.

Foot dissection. Based on the Doppler findings, outline the course of the dorsalis pedis artery. Allow the veins to fill by holding the foot in a dependent position over the edge of the operating table. The large superficial veins on the dorsum of the foot should be easily seen. The tributaries to the greater saphenous system can be located on the medial side of the first metatarsal. Outline the veins before exsanguinating the limb for tourniquet inflation. Exsanguinate the limb by wrapping or elevation and inflate the pneumatic tourniquet. Use straight, curved, or zigzag dorsal incisions to identify and preserve the dorsal veins and the dorsalis pedis artery and its distal continuation, the first dorsal metatarsal artery (Fig. 19-61). If preoperative evaluation reveals the first metatarsal artery to be dorsal, proceed from proximal to distal, carefully protecting the artery. Ligate or clip the side branches.

If the dominant artery has been shown to be plantar, dissect from the first web space proximally, extending a longitudinal plantar incision just lateral to the weight-bearing area of the plantar surface over the first metatarsal head. Dissect proximally to obtain sufficient length of artery. At times dividing the transverse metatarsal ligament is necessary to mobilize a plantar metatarsal artery. If the location of the dominant vessel is in doubt, begin the dissection in the first web space and dissect proximally. In the first web space ligate the artery to the second toe and mobilize the first metatarsal artery proximally until it is determined whether it can be dissected from the dorsal or plantar aspect. Do not transect the proximal vessel attachments until vascularization through the arteriovenous pedicle to the great toe is assured and until the hand dissection has been completed.

Follow the dorsal artery to the extensor hallucis brevis, divide the extensor hallucis brevis, elevate it, and expose the deep peroneal nerve lateral to the dorsalis pedis artery. Preserve the deep peroneal nerve for suture to a recipient nerve in the thumb area. Follow the first metatarsal artery to the first web, leaving all branches attached to the great toe, and ligate or cauterize the branches to the second toe. Dissect and mobilize the superficial veins so that a long venous pedicle can be developed. In the first web dissect the plantar digital nerve on the lateral side of the great toe and separate it from the digital nerve to the second toe by carefully dissecting proximally into the common digital nerve. Similarly dissect the plantar digital nerve on the medial side of the great toe and mobilize it as far proximally as possible. Attempt to preserve both digital nerves. Obtain as much length as possible, depending on the requirements of the recipient thumb area. Occasionally nerve grafts may be required.

Determine the approximate tendon length requirements in the hand. Section the extensor hallucis longus tendon near the extensor retinaculum or more proximally through

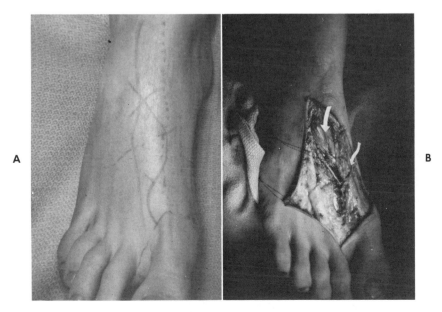

Fig. 19-61. Foot dissection for great toe transfer. **A,** Skin marking for veins *(dotted line)*, artery *(solid line)*, and skin incision *(curved, cross-hatched line.)* **B,** Vascular dissection. Markers indicate arteriovenous pedicle.

Kleinert, H.E., and Kasdan, M.L.: Anastomosis of digital vessels, J. Ky. Med. Assoc. **63:**106, 1965.

Kleinert, H.E., and Kasdan, M.L.: Salvage of devascularized upper extremities including studies on small vessel anastomosis, Clin. Orthop. **29:**29, 1963.

Kleinert, H.E., Kasdan, M.L. and Romero, J.L.: Small blood vessel anastomosis for salvage of severely injured upper extremity, J. Bone Joint Surg. **45-A:**788, 1963.

Kleinert, H.E., and Neale, H.W.: Microsurgery in hand surgery, Clin. Orthop. **104:**158, 1974.

Kleinert, H.E., Serafin, D., and Daniel, R.K.: The place of microsurgery in hand surgery, Orthop. Clin. North Am. **4:**929, 1973.

Kleinert, H.E., and Tsai, T.-M.: Microvascular repair in replantation, Clin. Orthop. **133:**205, 1978.

Komatsu, S., and Tamai, S.: Successful replantation of a completely cut-off thumb: case report, Plast. Reconstr. Surg. **42:**374, 1968.

Krizek, T.J., Tassboro, T., Desperez, Q.O., and Kiohn, C.L.: Experimental transplantation of composite grafts by microvascular anastomosis, Plast. Reconstr. Surg. **36:**358, 1956.

Kusunoki, M., Toyoshima, Y., and Okajima, M.: Successful replantation of a leg: a 7-year follow-up, Injury **16:**118, 1984.

Kutz, J.E.: Preparation of replantation. In Daniller, A.I., and Strauch, B.: Symposium on microsurgery, St. Louis, 1976, The C.V. Mosby Co.

Kutz, J.E., and Dimond, M.: Replantation in the upper extremity, Surg. Rounds, March, 1982, p. 14.

Kutz, J.E., Hanel, D., and Scheker, L., et al.: Upper extremity replantation, Orthop. Clin. North Am. **14:**873, 1983.

Lapchinsky, C.A.: Recent results of experimental transplantation of preserved limbs and kidneys and possible use of this technique in clinical practice, Ann. N.Y. Acad. Sci. **87:**539, 1960.

Lister, G.: The choice of procedure following thumb amputation, Clin. Orthop. **195:**45, 1985.

Lendvay, P.G.: Anastomosis of digital vessels, Med. J. Aust. **2:**723, 1968.

Lendvay, P.G.: Replacement of the amputated digit, Br. J. Plast. Surg. **26:**398, 1973.

Lendvay, P.G., and Owen, E.R.: Microvascular repair of completely severed digit: fate of digital vessels after six months, Med. J. Aust. **2:**818, 1970.

Leung, P.-C.: Hand replantation in an 83-year-old woman: the oldest replantation? Plast. Reconstr. Surg. **64:**416, 1979.

Lister, G.: Intraosseous wiring of the digital skeleton, J. Hand Surg. **3:**427, 1978.

Lu, S.-Y., Chiu, H.-Y., Lin, T.-W., and Chen, M.-T.: Evaluation of survival in digital replantation with thermometric monitoring, J. Hand Surg. **9-A:**805, 1984.

Lucas, G.I.: Microvascular surgery and limb replantation: editorial comment, Clin. Orthop. **133:**2, 1978.

Luck, J.V., and Boeck, W.C., Jr.: Successful replantation of upper extremity nearly amputated by "sheepfoot" roller, Clin. Orthop. **29:**50, 1963.

MacLeod, A.M., O'Brien, B.M., and Morrison, W.A.: Digital replantation: clinical experience, Clin. Orthop. **133:**26, 1978.

Malt, R.A., and McKhann, C.F.: Replantation of severed arms, JAMA **189:**716, 1964.

Malt, R.A., Remensnyder, J.P., and Harris, W.H.: Long-term utility of replanted arms, Ann. Surg. **176:**334, 1972.

Matsen, F.A., III, Bach, A.W., Wyss, C.R., and Simmons, C.W.: Transcutaneous Po_2: a potential monitor of the status of replanted limb parts, Plast. Reconstr. Surg. **65:**732, 1980.

Matsuda, M., Kato, N., and Hosoi, M.: The problems in replantation of limbs amputated through the upper arm region, J. Trauma **21:**403, 1981.

McNeill, I.F., and Wilson, J.S.P.: The problems of limb replacement, Br. J. Surg. **57:**365, 1970.

Mehl, R.L., et al.: Patency of the microcirculation in the traumatically amputated limb: a comparison of common perfusates, J. Trauma, **4:**495, 1964.

Meuli, H.C., Meyer, V., and Segmuller, G.: Stabilization of bone in replantation surgery of the upper limb, Clin. Orthop. **133:**179, 1978.

Meyer, V.E.: Microsurgery and replantation. In Evarts, M.C. (editor): Surgery of the musculoskeletal system, New York, 1983, Churchill Livingstone.

Meyer, V.E., Zhong-Wei, C., and Beasley, R.W.: Basic technical considerations in reattachment surgery, Orthop. Clin. North Am. **12:**871, 1981.

Millesi, H.: Microsurgery of peripheral nerves: neurolysis, nerve grafts, brachial plexus injuries. In Buncke, H.J., and Furnas, D.W. (editors): Symposium on clinical frontiers in reconstructive microsurgery, vol. 24, St. Louis, 1984, The C.V. Mosby Co.

Millesi, H., Meissl, G., and Berger, A.: A further experience with interfascicular grafting of the median, ulnar, and radial nerves, J. Bone Joint Surg. **58-A:**209, 1976.

Millesi, H., Miessl, G., and Berger, A.: The interfascicular nerve-grafting of the median and ulnar nerves, J. Bone Joint Surg. **54-A:**727, 1972.

Millesi, H.: Interfascicular nerve repair and secondary repair with nerve grafts. In Jewett, D.L., and McCarroll, H.R., Jr. (editors): Nerve repair and regeneration: its clinical and experimental basis, St. Louis, 1980, The C.V. Mosby Co.

Morrison, W.A., et al.: Evaluation of digital replantation: a review of 100 cases, Orthop. Clin. North Am. **8:**295, 1977.

Morrison, W.A., O'Brien, B.M., and MacLeod, A.M.: Major limb replantation, Orthop. Clin. North Am. **8**(2):343, 1977.

Morrison, W.A., O'Brien, B.M., and MacLeod, A.M.: Digital replantation and revascularization: a long term review of one hundred cases, Hand. **10:**125, 1978.

Nordzell, B.: Reimplantation of an amputated hand: long-term follow-up report, Scand. J. Plast. Reconstr. Surg. **11:**251, 1977.

Nunley, J.A., Goldner, R.D., and Urbaniak, J.R.: Thumb reconstruction by the wrap-around method, Clin. Orthop. **195:**97, 1985.

Nunley, J.A., Koman, L.A., and Urbaniak, J.R.: Arterial shunting as an adjunct to major limb revascularization, Ann. Surg. **193:**271, 1981.

O'Brien, B.M.C.: Replantation surgery, Clin. Plast. Surg. **1:**405, 1974.

O'Brien, B.M.C., et al.: Major replantation surgery in the upper limb, Hand **6:**217, 1974.

O'Brien, B.M.C., et al.: Clinical replantation of digits, Plast. Reconstr. Surg. **52:**490, 1973.

O'Brien, B.M.C., and Baxter, T.J.: Experimental digital replantation, Hand. **6:**11, 1974.

O'Brien, B.M., Franklin, J.D., Morrison, W.A., and MacLeod, A.M.: Replantation and revascularisation surgery in children, Hand **12:**12, 1980.

O'Brien, B.M.C., et al.: Clinical implantation of digits, Plast. Reconstr. Surg. **53:**490, 1973.

O'Brien, B.M., MacLeod, A.M., and Morrison, W.A.: Digital replantation. In Reid, D.A.C., and Gossett, J. (editors): Mutilating injuries of the hand, G.E.M. no. 3, New York, 1979, Churchill Livingstone.

O'Brien, B.M.C., and Miller, G.D.H.: Digital attachment and revascularization, J. Bone Joint Surg. **55-A:**714, 1973.

Paletta, F.X.: Replantation of the amputated extremity, Ann. Surg. **168:**720, 1968.

Phelps, D.B., Lilla, J.A., and Bowick, J.A., Jr.: Common problems in clinical replantation and revascularization in the upper extremity, Clin. Orthop. **133:**11, 1978.

Phelps, D.B., Rutherford, R.B. and Boswick, J.A., Jr.: Control of vasospasm following trauma and microvascular surgery, J. Hand Surg. **4:**109, 1979.

Pho, R.W.H., Chacha, P.B., and Yeo, K.Q.: Rerouting vessels and nerves from other digits in replanting an avulsed and degloved thumb, Plast. Reconstr. Surg. **64:**330, 1979.

Press, B.H.J., Sibley, R.K., and Shons, A.R.: Early histopathologic changes in experimentally replanted extremities, J. Hand Surg. **8:**549, 1983.

Ramirez, M.A., et al.: Reimplantation of limbs, Plast. Reconstr. Surg. **40:**315, 1967.

Rose, E.H., and Buncke, H.J.: Selective finger transposition and primary metacarpal ray resection in multidigit amputations of the hand, J. Hand Surg. **8:**178, 1983.

Rosenkrantz, J.G., et al.: Replantation of an infant's arm, N. Engl. J. Med. **276:**609, 1967.

Russell, R.C., O'Brien, B.M., Morrison, W.A., Pamamull, G., and MacLeod, A.: The late functional results of upper limb revascularization and replantation, J. Hand Surg. **9-A:**623, 1984.

Salibian, A.H., Anzel, S.H., Malberich, M.M., and Tesoro, V.E.: Microvascular reconstruction for close-range gunshot injuries to the distal forearm, J. Hand Surg. **9-A:**799, 1984.

Schlenker, J.D., Kleinert, H.E. and Tsai, T.: Methods and results of replantation following traumatic amputations of the thumb in sixty-four patients, J. Hand Surg. **5**:63, 1980.

Scott, F.A., Howar, J.W., and Boswick, J.A., Jr.: Recovery of function following replantation and revascularization of amputated hand parts, J. Trauma **21**:204, 1981.

Serafin, D., et al.: Replantation of a completely amputated distal thumb without venous anastomosis: case report, Plast. Reconstr. Surg. **52**:579, 1973.

Shanghai Sixth People's Hospital: Reattachment of traumatic amputation: a summing up of experience, China Med. **5**:392, 1967.

Shaw, R.S.: Treatment of the extremity suffering near or total severance with special consideration of the vascular problem, Clin. Orthop. **29**:56, 1963.

Smith, A.R., Sonneveld, G.J., Kort, W.J., and van der Meulen, J.C.: Clinical application of transcutaneous oxygen measurements in replantation surgery and free tissue transfer, J. Hand Surg. **8**:139, 1983.

Snyder, C.C. and Knowles, R.P. Amputation of extremities, Clin. Orthop. **29**:113, 1963.

Snyder, C.C., and Knowles, R.P.: Autotransplantation of extremities, Clin. Orthop. **29**: 113, 1963.

Snyder, C.C., et al.: Successful replantation of a totally severed thumb, Plast. Reconstr. Surg. **50**:553, 1972.

Stirrat, C.R., Seaber, A.V., Urbaniak, J.R. and Bright, D.S.: Temperature monitoring in digital replantation, J. Hand Surg. **3**:342, 1978.

Strauch, B., and Terzis, J.K.: Replantation of digits, Clin. Orthop. **133**:35, 1978.

Strickland, J.W.: Restoration of thumb function following partial or total amputation. In Hunter, J.M., et al. (editors): Rehabilitation of the hand, St. Louis, 1978, The C.V. Mosby Co.

Strickland, J.W., and Steichen, J.B. (editors): Difficult problems in hand surgery, St. Louis, 1982, The C.V. Mosby Co.

Tamai, S.: Digit replantation: analysis of 163 replantations in an 11-year period, Clin. Plast. Surg. **5**:195, 1978.

Tamai, S.: Twenty years' experience of limb replantation: review of 293 upper extremity replants, J. Hand Surg. **7**:549, 1982.

Tamai, S., et al.: Traumatic amputation of digits: the fate of remaining blood: an experimental and clinical study, J. Hand Surg. **2**:13, 1977.

Tamai, S., et al.: Microvascular anastomosis and its application on the replantation of amputated digits and hands, Clin. Orthop. **133**:106, 1978.

Tsai, T.-M.: Successful replantation of a forefoot, Clin. Orthop. **139**:182, 1979.

Tsai, T.-M., et al.: Primary microsurgical repair of ring avulsion amputation injuries, J. Hand Surg. **9-A**:68, 1984.

Tupper, J.W.: Techniques of bone fixation and clinical experience in replanted extremities, Clin. Orthop. **133**:165, 1978.

Urbaniak, J.R.: To replant or not to replant? That is not the question (editorial), J. Hand Surg. **8**(1):507, 1983.

Urbaniak, J.R.: Replantation. In Green, D.P. (editor): Operative hand surgery, New York, 1982, Churchill Livingstone.

Urbaniak, J.R.: Replantation of amputated parts: technique, results, and indications. In American Academy of Orthopaedic Surgeons: Symposium on microsurgery: practical use in orthopaedics, St. Louis, 1979, The C.V. Mosby Co.

Urbaniak, J.R.: Replantation of amputated hands and digits. In American Academy of Orthopaedic Surgeons: Instructional course lectures, vol. 27, St. Louis, 1979, The C.V. Mosby Co.

Urbaniak, J.R.: Results of sensibility recovery in 100 replanted digits, International Hand Surgery Congress, Melbourne, Australia, 1979.

Urbaniak, J.R., and Bright, D.S.: Replantation of amputated digits and hands in children, Interclinic Information Bull. **14**:1, 1975.

Urbaniak, J.R., Evans, J.P., and Bright, D.S.: Microvascular management of ring avulsion injuries, J. Hand Surg. **6**:25, 1981.

Urbaniak, J.R., Hayes, M.G., and Bright, D.S.: Management of bone in digital replantation: free vascularized and composite bone grafts, Clin. Orthop. **133**:184, 1978.

Van Beek, A.L., Kutz, J.E., and Zook, E.G.: Importance of the ribbon sign indicating unsuitability of the vessel in replanting a finger, Plast. Reconstr. Surg. **61**:32, 1978.

Van Giesen, P., Seaber, A., and Urbaniak, J.R.: Hypothermic preservation prior to replantation. Presented at American Academy of Orthopaedic Surgeons course-Microsurgery in reconstruction of the upper and lower extremities, Durham, N.C., May, 1979.

Vilkki, S.: Replantation, studies on clinical replantation surgery. Acta Univ. Tain., Ser. A., Vol. 156, 1983.

Wang, S.-H., Young, K.-F., and Wei, J.-N.: Replantation of severed limbs-clinical analysis of 91 cases, J. Hand Surg. **6**:311, 1981.

Weeks, P.M., and Young, V.L.: Revascularization of the skin envelope of a denuded finger, Plast. Reconstr. Surg. **69**:527, 1982.

Weiland, A.J., et al.: Replantation of digits and hands: analysis of surgical techniques and functional results in 71 patients with 86 replantations, Clin. Orthop. **133**:195, 1978.

Weiland, A.J., Robinson, H., and Futrell, J.: External stabilization of a replanted upper extremity, J. Trauma **16**:239, 1976.

Weiland, A.J., et al.: Replantation of digits and hands: analysis of surgical techniques and functional results in 71 patients with 86 replantations, J. Hand Surg. **2**:1, 1977.

Williams, G.R., et al.: Replantation of amputated extremities, Ann. Surg. **163**:788, 1966.

Yamano, Y.: Replantation of the amputated distal part of the fingers, J. Hand Surg. **10-A**:211, 1985.

Yamauchi, S., et al.: A clinical study of the order and speed of sensory recovery after digital replantation, J. Hand Surg. **8**:545, 1983.

Yoshizu, T., Katsumi, M. and Tajima, T.: Replantation of untidy amputated finger, hand and arm: experience of 99 replantations in 66 cases, J. Trauma **18**:194, 1978.

Yun ch'ing, Ç., et al.: Some problems concerning small vessel anastomosis in the reattachment of complete traumatic amputations, Chin. Med. J. **85**:79, 1966.

Zhong-Wei, C., Meyer, V.E., Kleinert, H.E., and Beasley, R.W.: Present indications and contraindications for replantation as reflected by long-term functional results, Orthop. Clin. North Am. **12**:849, 1981.

Zhong-Wei, C., Ch'ien, Y.-C., P.Y.S., and Lin, C.T.: Further experiences in the restoration of amputated limbs, Chin. Med. J. **84**:225, 1965.

Zhong-Wei, C., Ch'ien, Y.-C., Pao, Y.S., and Lin, C.T.: Some problems concerning small vessel anastomosis in the reattachment of complete traumatic amputations, Chin. Med. J. **85**:79, 1966.

Zhong-Wei, C., et al.: Salvage of the forearm following complete traumatic amputation: report of a case, Chin. Med. J. **82**:632, 1963.

Zhong-Wei, C., et al.: Further experiences in the restoration of amputated limbs: report of two cases, Chin. Med. J. **84**:225, 1965.

Revascularization

Coleman, S.S., and Anson, B.J.: Arterial patterns in the hand based upon a study of 650 specimens, Surg. Gynecol. Obstet. **113**:409, 1961.

Gelberman, R.H., et al.: The results of radial and ulnar arterial repair in the forearm: experience in three medical centers, J. Bone Joint Surg. **64-A**:383, 1982.

Kartchner, M.M., and Wilcox, W.C.: Thrombolysis of palmar and digital arterial thrombosis by intra-arterial thrombolysin, J. Hand Surg. **1**:67, 1976.

Kleinert, H.E., and Volianitis, G.J.: Thrombosis of the palmar arterial arch and its tributaries: etiology and newer concepts in treatment, J. Trauma **5**:447, 1965.

Koman, L.A.: Diagnostic study of vascular lesions, Hand Clin. **1**:217, 1985.

Koman, L.A., and Urbaniak, J.R.: Thrombosis of ulnar artery at the wrist. In American Academy of Orthopaedic Surgeons: Symposium on microsurgery: practical use in orthopaedics, St. Louis, 1979, The C.V. Mosby Co.

Koman, L.A. and Urbaniak, J.R. Ulnar atery insufficiency: a guide to treatment, J. Hand Surg. **6**:16, 1981.

Koman, L.A., and Urbaniak, J.R. Ulnar artery thrombosis, Hand Clin. **1**:311, 1985.

Koman, L.A., et al.: Isolated cold stress testing in the assessment of symptoms in the upper extremity: preliminary communication, J. Hand Surg. **9**:305, 1984.

Koman, L.A., et al.: Dynamic radionuclide imaging as a means of evaluating vascular perfusion of the upper extremity: a preliminary report, J. Hand Surg. **8**:424, 1983.

Salibian, A.H., Anzel, S.H., Mallerich, M.M., and Tesoro, V.E.: Microvascular reconstruction for close-range gunshot injuries to the distal forearm, J. Hand Surg. **9A**:799, 1984.

Tsai, Tsu-Min, et al.: Primary microsurgical repair of ring avulsion amputation injuries, J. Hand Surg. **9-A**:68, 1984.

Tupper, J.W.: Vascular defects and salvage of failed vascular repairs. In American Academy of Orthopaedic Surgeons: Symposium on microsurgery: practical use in orthopaedics, St. Louis, 1979, The C.V. Mosby Co.

Urbaniak, J. (editor): Symposium on microvascular surgery, Hand Clinics, vol. 1, Philadelphia, 1985, W.B. Saunders Co.

Urbaniak, J.R., Evans, J.P, and Bright, D.S.: Microvascular management of ring avulsion injuries, J. Hand Surg. 6:25, 1981.

Urbaniak, J.R., and Koman, L.A.: Ulnar artery thrombosis: a rationale for management. In Strickland, J.W., and Steichen, J.B. (editors): Difficult problems in hand surgery, St. Louis, 1982, The C.V. Mosby Co.

Weeks, P.M., and Young, V.L.: Revascularization of the skin envelope of a denuded finger, Plast. Reconstr. Surg. 69:527, 1982.

Wilgis, E.F.S.: Digital sympathectomy for vascular insufficiency, Hand Clin. 1:361, 1985.

Free flaps—general

Achauer, B.M., Black, K.S., and Litke, D.K.: Transcutaneous PO$_2$ in flaps: a new method of survival prediction, Plast. Reconstr. Surg. 65:738, 1980.

Acland, R.D. Outlining a free flap exactly, Plast. Reconstr. Surg. 59:113, 1977.

Acland, R.D.: The free iliac flap: a lateral modification of the free groin flap, Plast. Reconstr. Surg. 64:30, 1979.

Acland, R.D., et al.: The saphenous neurovascular free flap, Plast. Reconstr. Surg. 67:763, 1981.

Aoyagi, F., Fujino, T. and Ohshiro, T.: Detection of small vessels for microsurgery by a Doppler flowmeter, Plast. Reconstr. Surg. 55:372, 1975.

Azuma, H., Kondo, T., Mikami, M. and Harii, K.: Treatment of chronic osteomyelitis by transplantation of autogeneous omentum with microvascular anastomosis, Acta Orthop. Scand. 47:271, 1976.

Baek, S.M.: Two new cutaneous free flaps: medial and lateral thigh flaps, Plast. Reconstr. Surg. 71:354, 1983.

Bakamjian, V.Y.: A two-stage method for pharyngoesophageal reconstruction with a primary pectoral flap, Plast. Reconstr. Surg. 36:173, 1965.

Baudet, J., Guimberteau, J. and Nascimento, E.: Successful clinical transfer of two free thoraco-dorsal axillary flaps. Plast. Reconstr. Surg. 58:680, 1976.

Baudet, J., LeMaire, J., and Guimberteau, J.: Ten free groin flaps, Plast. Reconstr. Surg. 57:577, 1976.

Biemer, E., and Stock, W.: Total thumb reconstruction: a one stage reconstruction using an osteocutaneous forearm flap. Br. J. Plast. Surg. 36:52, 1983.

Boeckx, W.D., de Coninck, A., and Vanderlinden, E.: Ten free flap transfers: use of intra-arterial dye injection to outline a flap exactly, Plast. Reconstr. Surg. 57:716, 1976.

Bostwick, J., III, Nahai, F., Wallace, J.C., and Vasconez, L.O.: Sixty latissimus dorsi flaps, Plast. Reconstr. Surg. 63:31, 1979.

Brownstein, M.L., Gordon, L. and Buncke, H.J., Jr.: The use of microvascular free groin flaps for the closure of difficult lower extremity wounds, Surg. Clin. North Am. 57:977, 1977.

Buncke, H.J.: Hand reconstruction by microvascular island flap transplantation, J. Bone Joint Surg. 57-A:729, 1975.

Buncke, H.J., Alpert, B., and Shah, K.G.: Microvascular grafting, Clin. Plast. Surg. 5:185, 1978.

Caffee, H.H., and Asokan, R.: Tensor fascia lata myocutaneous free flaps, Plast. Reconstr. Surg. 68:195, 1981.

Chaikhouni, A., Dyas, C.L., Jr., Robinson, J.H., and Kellehar, J.C.: Latissimus dorsi free myocutaneous flap, J. Trauma 21:398, 1981.

Chang, N., and Mathes, S.J.: Comparison of the effect of bacterial inoculation in musculocutaneous and random-pattern flaps, Plast. Reconstr. Surg. 70:1, 1982.

Daniel, R.K.: Toward an anatomical and hemodynamic classification of skin flaps, Plast. Reconstr. Surg. 56:330, 1975.

Daniel, R.K., and Ledman, D.: Vascular complications in free flap transfers. In Buncke, H.J., and Furnas, D.W. (editors): Symposium on clinical frontiers in reconstructive microsurgery, St. Louis, 1984, The C.V. Mosby Co.

Daniel, R.K., and May, J.W., Jr.: Free flaps: an overview, Clin. Orthop. 133:122, 1978.

Daniel, R.K., and Taylor, G.I. Distant transfer of an island flap by microvascular anastomoses: a clinical technique, Plast. Reconstr. Surg. 52:111, 1973.

Daniel, R.K., and Taylor, G.I.: Anatomy and hemodynamics of free flap donor sites. In Daniller, A.I., and Strauch, B. (editors): Symposium on microsurgery, St. Louis, 1976, The C.V. Mosby Co.

Daniel, R., and Terzis, J.: Neurovascular free flaps. In Strauch, B., and Daniller, A. (editors): Symposium on microsurgery, St. Louis, 1976, The C.V. Mosby Co.

Daniel, R.K., Terzis, J., and Midgley, R.D.: Restoration to an anesthetic hand by a free neurovascular flap from the foot, Plast. Reconstr. Surg. 57:275, 1976.

Daniel, R.K., Terzis, J., and Schwarz, G.: Neurovascular free flaps: a preliminary report, Plast. Reconstr. Surg. 56:13, 1975.

Daniel, R.K., and Weiland, A.J.: Free tissue transfer for upper extremity reconstruction, J. Hand Surg. 7:66, 1982.

Daniel, R.K., and Williams, H.B.: The free transfer of skin flaps by microvascular anastomoses: an experimental study and a reappraisal, Plast. Reconstr. Surg. 52:16, 1973.

de Coninck, A., and Vanderlinden, E.: Thoradorsal skin flap: new possible donor sites in distant transfer of island flaps by microvascular anastomoses, Ann. Chir. Plast. 20:163, 1975.

Egloff, D.V.: Surgery of the hand: free tissue transfers by nerve and vascular microanastomoses, Geneva, 1984, Editions Medecine et Hygiene.

Finseth, F., May, J.W., and Smith, R.J.: Composite groin flap with iliacbone flap for primary thumb reconstruction: case report. J. Bone Joint Surg. 58-A:130, 1976.

Fogdestam, I., Hamilton, R., and Markhede, G.: Microvascular osteocutaneous groin flap in the treatment of an ununited tibial fracture with chronic osteitis: a case report, Acta Orthop. Scand. 51:175, 1980.

Fujino, T.: Contribution of the axial and perforator vasculature to circulation in flaps, Plast. Reconstr. Surg. 39:125, 1967.

Fujino, T., and Harashina, T.: Vascularized free flap transfers, Clin. Orthop. 133:154, 1978.

Gilbert, A., et al.: Transfert sur la main d'un lambeau libre sensible, Chirurgie 101:691, 1975.

Gilbert, A.: Composite tissue transfers from the foot: anatomic basis and surgical technique. In Daniller, A.I., and Strauch, B.: Symposium on Microsurgery, St. Louis, 1976, The C.V. Mosby Co.

Gilbert, A., and Teot, L.: The free scapular flap, Plast. Reconstr. Surg. 69:601, 1982.

Godina, M.: Preferential use of end-to-side arterial anastomoses in free flap transfers, Plast. Reconstr. Surg. 64:673, 1979.

Goldwyn, R.M., Lamb, D.L. and White, W.L.: An experimental study of large island flaps in dogs, Plast. Reconstr. Surg. 31:528, 1963.

Gordon, L., Buncke, H.J., and Alpert, B.S.: Free latissimus dorsi muscle flap with split-thickness skin graft cover: a report of 16 cases, Plast. Reconstr. Surg. 70:173, 1982.

Guba, A.M., Jr.: The use of free vascular tissue transfers in lower extremity injuries, Adv. Orthop. Surg. 7:60, 1983.

Hagan, K.F., Buncke, H.J., and Gonzalez, R.: Free latissimus dorsi muscle flap coverage of an electrical burn of the lower extremity, Plast. Reconstr. Surg. 69:125, 1982.

Harii, K.: Composite tissue transfer by microvascular anastomoses, Asian Med. J. 17:264, 1975.

Harii, K.: Current clinical experience in vascularized free skin flap transfers. In Strauch, B., and Daniller, A. (editors): Symposium on microsurgery, St. Louis, 1976, The C.V. Mosby Co.

Harii, K.: Free groin flaps, Plast. Reconstr. Surg. 58:120, 1976.

Harii, K.: Microvascular free flaps for skin coverage: indications and selections of donor sites, Clin. Plast. Surg. 10:37, 1983.

Harii, K.: Microvascular tissue transfer, New York, 1983, Igaku-Shoin.

Harii, K., and Ohmori, K.: Direct transfer of large free groin skin flaps to the lower extremity using microanastomoses, Chir. Plastica (Berlin) 3:1, 1975.

Harii, K., and Ohmori, K.: Free groin flaps in children, Plast. Reconstr. Surg. 55:588, 1975.

Harii, K., and Ohmori, K.: Free skin flap transfer, Clin. Plast. Surg. 3:111, 1976.

Harii, K., Ohmori, K., and Ohmori, S.: Free deltopectoral skin flaps, Br. J. Plast. Surg. 27:231, 1974.

Harii, K., Ohmori, K., and Ohmori, S.: Successful clinical transfer of ten free flaps by microvascular anastomoses, Plast. Reconstr. Surg. **53**:259, 1974.

Harii, K., et al.: Free groin skin flaps, Br. J. Plast. Surg. **28**:225, 1975.

Harrison, D.H., Girling, M., and Mott, G.: Experience in monitoring the circulation in free flap transfers, Plast. Reconstr. Surg. **68**:543, 1981.

Hentz, V. Reconstruction of individual digits, Hand Clin. **1**:335, 1985.

Ikuta, Y.: Vascularized free flap transfer in the upper limb, Hand Clin. **1**:297, 1985.

Ikuta, Y.: Skeletal muscle transplantation in the severely injured upper extremity. In Serafin, D., and Buncke, H.J., Jr. (editors): Microsurgical composite tissue transplantation, St. Louis, 1979, The C.V. Mosby Co.

Ikuta, Y.: Free flap transfer by end-to-side arterial anastomosis. In Daniel, R.K., and Terzis, J.K. (editors): Reconstructive microsurgery, Boston, 1977, Little, Brown & Co.

Ikuta, Y., et al.: Free flap transfers by end-to-side arterial anastomosis, Br. J. Plast. Surg. **28**:1, 1975.

Jackson, I.T.: Groin flaps. In Serafin, D., and Buncke, H.J. (editors): Microsurgery composite tissue transplantation, St. Louis, 1978, The C.V. Mosby Co.

Jackson, I.T., and Scheker, L.: Muscle and myocutaneous flaps on the lower limb, Injury **13**:324, 1982.

Joshi, B.B.: Neural repair for sensory restoration in a groin flap, Hand **9**:221, 1977.

Kaplan, E.N., Buncke, H.J., and Murray, D.E.: Distant transfer of cutaneous island flaps in humans by microvascular anastomoses, Plast. Reconstr. Surg. **52**:301, 1973.

Kaplan, E.N., and Pearl, R.M.: An arterial medial arm flap: vascular anatomy and clinical application, Ann. Plast. Surg. **4**:205, 1980.

Karkowski, J., and Buncke, H.J.: A simplified technique for free transfer of groin flaps by use of a Doppler probe, Plast. Reconstr. Surg. **55**:682, 1975.

Katai, K., Kido, M., and Numaguchi, Y.: Angiography of the iliofemoral arteriovenous system supplying free groin flaps and free hypogastric flaps, Plast. Reconstr. Surg. **63**:671, 1979.

Kleinert, H.E.: Bone and osteocutaneous microvascular free flaps, J. Hand Surg. **8**:735, 1983.

Kleinert, H.E., and Manstein, C.H.: Current techniques of limb reconstruction. In Buncke, H.J., and Furnas, D.W. (editors): Symposium on clinical frontiers in reconstructive microsurgery, vol. 24, St. Louis, 1984, The C.V. Mosby Co.

Koo, Boo-Chai: John Wood and his contributions to plastic surgery: the first groin flap, Br. J. Plast. Surg. **30**:9, 1977.

Krizek, T.J., et al.: Experimental transplantation of composite grafts by microsurgical vascular anastomoses, Plast. Reconstr. Surg. **36**:538, 1965.

Lanier, V.C., Jr., et al.: Microvascular procedures in reconstructive surgery, South. Med. J. **69**:1595, 1976.

LaRossa, D., Mellissinos, E., Matthews, D., and Hamilton, R.: The use of microvascular free skin-muscle flaps in management of avulsion injuries of the lower leg, J. Trauma **20**:545, 1980.

Lister, G.D., McGregor, I.A., and Jackson, I.T.: The groin flap in hand injuries, Injury **4**:229, 1973.

Maass, D.: Significance of microsurgery in the surgery of the extremities, II. Free transplantation of composite tissue with microvascular connection, Helv. Chir. Acta **43**:679, 1976.

McGregor, I.A.: Flap reconstruction in hand surgery: the evolution of presently used methods, J. Hand Surg. **4**:1, 1979.

McGregor, I.A., and Jackson, I.T.: The groin flap, Br. J. Plast. Surg. **25**:3, 1972.

Man, D., and Acland, R.D.: The microarterial anatomy of the dorsalis pedis flap and its clinical applications, Plast. Reconstr. Surg. **65**:419, 1980.

Mathes, S.J., and Nahai, F. (editors): Clinical atlas of muscle and musculotaneous flaps, St. Louis, 1979, The C.V. Mosby Co.

Mathes, S.J., and Nahai, F.: Classification of the vascular anatomy of muscles: experimental and clinical correlation, Plast. Reconstr. Surg. **67**:177, 1981.

Mathes, S.J., and Nahai, F. (editors): Clinical applications for muscle and musculocutaneous flaps, St. Louis, 1982, The C.V. Mosby Co.

Mathes, S.J., Nahai, F., and Vasconez, L.O.: Myocutaneous flap transfer, Plast. Reconstr. Surg. **62**:162, 1978.

Mathes, S.J., and Vasconez, L.O.: Free flaps (including toe transplantation). In Green, D.P. (editor): Operative hand surgery, New York, 1982, Churchill Livingstone.

Maxwell, G.P., Manson, P.N. and Hoopes, J.E.: Experience with thirteen latissimus dorsi myocutaneous free flaps, Plast. Reconstr. Surg. **64**:1, 1980.

Maxwell, G.P., Stueber, K., and Hoopes, J.K.: A free latissimus dorsi myocutaneous flap, Plast. Reconstr. Surg. **62**:462, 1978.

May, J.W., Athanasoulis, C.A., and Donelan, M.: Preoperative magnification angiography of donor and recipient sites for clinical free transfer of flaps or digits, Plast. Reconstr. Surg. **64**:483, 1979.

May, J.W., Jr., Chait, L.A., Cohen, B.E., and O'Brien, B.M.: Free neurovascular flap from the first web of the foot in hand reconstruction, J. Hand Surg. **2**:387, 1977.

May, J.W., Jr., Chait, L.A., O'Brien, B.M. and Hurley, J.V.: The no-flow phenomenon in experimental free flaps, Plast. Reconstr. Surg. **61**:256, 1976.

May, J.W., Jr., Likash, F.N., and Gallico, G.G., III: Latissimus dorsi free muscle flap in lower-extremity reconstruction, Plast. Reconstr. Surg. **68**:603, 1981.

McConnell, C.M., Hyland, W.T., and Neale, H.W.: Microvascular free groin flap for soft-tissue coverage of the extremities, J. Trauma **20**:593, 1980.

McCraw, J.B.: On the transfer of a free dorsalis pedis sensory flap to the hand, Plast. Reconstr. Surg. **59**:738, 1977.

McCraw, J.B. and Dibbell, D.C.: Experimental definition of independent myocutaneous vascular territories, Plast. Reconstr. Surg. **60**:212, 1977.

McCraw, J.B., Dibbel, D.C., and Carraway, J.H.: Clinical definition of independent myocutaneous vascular territories, Plast. Reconstr. Surg. **60**:341, 1977.

McCraw, J.B., and Furlow, L.T., Jr.: The dorsalis pedis arterialized flap: a clinical study, Plast. Reconstr. Surg. **55**:177, 1975.

McCraw, J.B., Myers, B., and Shanklin, K.D.: The value of fluorescein in predicting the viability of arterialized flaps, Plast. Reconstr. Surg. **60**:710, 1977.

McDonald, H.D., Buncke, H.J., and Goodstein, W.A.: Split-thickness skin grafts in microvascular surgery, Plast. Reconstr. Surg. **68**:731, 1981.

McGregor, I.A., and Jackson, I.T.: The groin flap, Br. J. Plast. Surg. **25**:3, 1972.

McGregor, I.A., and Morgan, G.: Axial and random pattern flaps, Br. J. Plast. Surg. **26**:202, 1973.

McLean, D.H., and Buncke, H.J., Jr.: Autotransplant of omentum to a large scalp defect with microsurgical revascularization, Plast. Reconstr. Surg. **49**:268, 1972.

Meyer, V.E.: Microsurgery and replantation. In Evarts, M.C. (editor): Surgery of the musculoskeletal system, New York, 1983, Churchill Livingstone.

Morrison, W.A., O'Brien, B.M.C., and MacLeod, A.: Clinical experiences in free flap transfer, Clin. Orthop. **133**:132, 1978.

Mühlbauer, W., Herndl, E., and Stock, W.: The forearm flap, Plast. Reconstr. Surg. **70**:336, 1982.

Nahai, F., Hill, H.L., and Hexter, T.R.: Experiences with the tensor fascia lata flap, Plast. Reconstr. Surg. **63**:788, 1979.

Nassif, T.M., Vidal, L., Bovet, J.L., and Baudet, J.: The parascapular flap: a new cutaneous microsurgical free flap, Plast. Reconstr. Surg. **69**:591, 1982.

Nunley, J.A.: Elective microsurgery for orthopaedic reconstruction, part I. Donor site selection for cutaneous and myocutaneous free flaps. In American Academy of Orthopaedic Surgeons: Instructional course lectures, vol. 33, St. Louis, 1984, The C.V. Mosby Co.

O'Brien, B.M.: Microvascular free flap and omental transfer. In O'Brien, B.M.: Microvascular reconstructive surgery, Edinburgh, 1977, Churchill Livingstone.

O'Brien, B.M., MacLeod, A.M., Hayhurst, J.W., and Morrison, W.A.: Successful transfer of a large island flap from the groin to the foot by microvascular anastomoses. Plast. Reconstr. Surg. **52**:271, 1973.

O'Brien, B.M., MacLeod, A.M., and Morrison, W.A.: Microvascular free flap transfer, Orthop. Clin. North Am. **8**(2):349, 1977.

O'Brien, B.M., Morrison, W.A., Ishida, H., MacLeod, A.M., and Gilbert, A.: Free flap transfers with microvascular anastomoses, Br. J. Plast. Surg. **27**:220, 1974.

O'Brien, B.M., and Shanmugan, N.: Experimental transfer of composite free flaps with microvascular anastomosis, Aust. N.Z. J. Surg. **43**:285, 1973.

Ohmori, K.: Versatility of composite tissue transplantation in soft tissue reconstruction of the upper extremity. In Serafin, D., and Buncke, H.J., Jr., (editors): Microsurgical composite tissue transplantation, St. Louis, 1979, The C.V. Mosby Co.

Ohmori, K.: Free flaps in children. In Daniel, R.K., and Terzis, J.K. (editors): Reconstructive microsurgery, Boston, 1977, Little, Brown & Co.

Ohmori, K., and Harii, K.: Free dorsalis pedis sensory flap to the hand with microneurovascular anastomoses, Plast. Reconstr. Surg. **58**:546, 1976.

Ohmori, K., and Harii, K.: Free groin flaps: their vascular basis, Br. J. Plast. Surg. **28**:238, 1975.

Ohtsuka, H., Fujita, K., and Shioya, N.: Replantation and free flap transfers by microvascular surgery, Plast. Reconstr. Surg. **58**:708, 1976.

Ohtsuka, H., Kamiishi, H., Saito, H., Ito, M., and Shioya, N.: Successful free flap transfers with diseased recipient vessels, Br. J. Plast. Surg. **29**:5, 1976.

Reinisch, J.F., Winters, R., and Puckett, C.L.: The use of the osteocutaneous groin flap in gunshot wounds of the hand, J. Hand Surg. **9-A**:12, 1984.

Rigg, B.M.: Transfer of a free groin flap to the heel by microvascular anastomoses, Plast. Reconstr. Surg. **55**:36, 1975.

Robinson, D.W.: Microsurgical transfer of the dorsalis pedis neurovascular island flap, Br. J. Plast. Surg. **29**:209, 1976.

Rubinstein, Z.J., Shafir, R., and Tsur, H.: The value of angiography prior to the use of the latissimus dorsi myocutaneous flap, Plast. Reconstr. Surg. **63**:374, 1979.

Schenck, R.R.: Free muscle and composite skin transplantation by microvascular anastomoses, Orthop. Clin. North Am. **8**:367, 1977.

Serafin, D., and Buncke, H.J. (editors): Microsurgical composite tissue transplantation, St. Louis, 1979, The C.V. Mosby Co.

Serafin, D., and Georgiade, N.G.: Microsurgical composite tissue transplantation, Ann. Surg. **187**:620, 1978.

Serafin, D., Georgiade, M.G, and Smith, D.H.: Comparison of free flaps with pedicled flaps for coverage of defects of the leg or foot, Plast. Reconstr. Surg. **59**:492, 1977.

Serafin, D., et al.: Transcutaneous PO$_2$ monitoring for assessing viability and predicting survival of skin flaps: experiences and clinical correlations, J. Microsurg. **2**:165, 1981.

Serafin, D., Sabatier, R.E., Morris, R.L, and Georgiade, N.G.: Reconstruction of the lower extremity with vascularized composite tissue: improved tissue survival and specific indications, Plast. Reconstr. Surg. **66**:230, 1980.

Serafin, D., Shearin, J.C., and Georgiade, N.G.: The vascularization of free flaps, Plast. Reconstr. Surg. **60**:233, 1977.

Serafin, D., Villareal-Rios, A., and Georgiade, N.: Fourteen free groin flap transfers, Plast. Reconstr. Surg. **57**:707, 1976.

Shah, K.G., Garrett, J.C., and Buncke, H.J., Jr.: Free groin flap transfer to the upper extremity, Hand **11**:315, 1979.

Sharzer, L.A., et al.: Clinical applications of free flap transfer in the burn patient, J. Trauma **15**:766, 1975.

Shaw, D.T., and Payne, R.L., Jr.: One-stage tubed abdominal flaps: single pedicle tubes, Surg. Gynecol. Obstet. **83**:205, 1946.

Sinclair, S.W., and Blake, G.B.: The groin flap in hand injuries, N.Z. Med. J. **84**:393, 1976.

Smith, P.J., Foley, B., McGregor, I.A., and Jackson, I.: The anatomical basis for the groin flap, Plast. Reconstr. Surg. **49**:41, 1972.

Song, R., Song, Y., and Yu, Y.: The upper arm free flap, Clin. Plast. Surg. **9**:27, 1982.

Steichen, J.: Microvascular free groin flaps. In Urbaniak, J.R., and Bright, D.S. (editors): American Academy of Orthopaedic Surgeons symposium on microsurgery, St. Louis, 1979, The C.V. Mosby Co.

Stern, P.J., Neale, H.W., Gregory, R.O., and McDonough, J.J.: Functional reconstruction of an extremity by free tissue transfer of the latissimus dorsi, J. Bone Joint Surg. **65-A**:729, 1983.

Strauch, B., and Greenstein, B. Neurovascular flaps to the hand, Hand Clin. **1**:327, 1985.

Strauch, B., and Murray, D.E.: Transfer of composite graft with immediate suture anastomosis of its vascular pedicle measuring less than 1 mm. in external diameter using microsurgical techniques, Plast. Reconstr. Surg. **40**:325, 1967.

Strauch, B., and Shafiroff, B.B.: The foot: a versatile source of donor tissue. In Serafin, D., and Buncke, H.J., Jr. (editors): Microsurgical composite tissue transplantation, St. Louis, 1979, The C.V. Mosby Co.

Strauch, B., Sharzer, L., and Brauman, D.: Innervated free flaps for sensibility and coverage. In Smith, J.E., and Sherman, J.E. (editors): Hand surgery: a concise guide to clinical practice, Boston, 1984, Little, Brown & Co.

Strauch, B., and Tsur, H. Restoration of sensation to the hand by a free neurovascular flap from the first web space of the foot, Plast. Reconstr. Surg. **62**:361, 1978.

Swartz, W.M.: Immediate reconstruction of the wrist and dorsum of the hand with a free osteocutaneous groin flap, J. Hand Surg. **9-A**:18, 1984.

Tamai, S.: Experimental neuromuscular transplantation. In Serafin, D., and Buncke, H.J., editors: Microsurgical composite tissue transplantation, St. Louis, 1979, The C.V. Mosby Co.

Tamai, S.: Replantation of the upper arm and forearm. In Buncke, H.J., and Furnas, D.W. (editors): Symposium on clinical frontiers in reconstructive microsurgery, vol. 24, St. Louis, 1984, The C.V. Mosby Co.

Tamai, S.: Vascularized fibular transplantation: congenital pseudarthrosis and radial club hand. In Buncke, H.J., and Furnas, D.W. (editors): Symposium on clinical frontiers in reconstructive microsurgery, vol. 24, St. Louis, 1984, The C.V. Mosby Co.

Taylor, G.I.: Tissue defects in the limbs: replacement with free vascularized tissue transfers, Aust. N.Z. J. Surg. **47**:276, 1977.

Taylor, G.I., and Daniel, R.K.: The anatomy of several free flap donor sites, Plast. Reconstr. Surg. **56**:243, 1975.

Taylor, G.I., and Watson, N.: One-stage repair of compound leg defects with free, vascularized flaps of groin skin and iliac bone, Plast. Reconstr. Surg. **46**:219, 1970.

Terzis, J.K., Sweet, R.C., Dykes, R.W., and Williams, H.B.: Recovery of function in free muscle transplants using microneurovascular anastomoses, J. Hand Surg. **3**:37, 1978.

Thorvaldsson, S.E., and Grabb, W.C.: The intravenous fluorescein test as a measure of skin flap viability, Plast. Reconstr. Surg. **53**:576, 1974.

Tobin, G.R., Schusterman, M., Peterson, G.H., Nichols, G., and Bland, K.I.: The intramuscular neurovascular anatomy of the latissimus dorsi muscle: the basis for splitting the flap, Plast. Reconstr. Surg. **67**:637, 1981.

Tsuge, K.: Special surgical techniques. In Omer, G.E., and Spinner, M. (editors): Management of peripheral nerve problems, Philadelphia, 1980, W.B. Saunders.

Urbaniak, J.R., Koman, L.A., Goldner, R.D., Armstrong, N.B., and Nunley, J.A.: The vascularized cutaneous scapular flap, Plast. Reconstr. Surg. **69**:772, 1982.

Watson, J.S., Brough, M.D., and Orton, C.: Simultaneous coverage of both heels with one free flap, Plast. Reconstr. Surg. **64**:269, 1979.

Watson, J.S., Craig, P., and Orton, C.I.: The free latissimus dorsi myocutaneous flap, Plast. Reconstr. Surg. **64**:299, 1979.

Zhong-Wei, C., Meyer, V.E., Kleinert, H.E., and Beasley, R.W.: Basic technical considerations in reattachment surgery, Orthop. Clin. North Am. **12**(4):871, 1981.

Free flaps—muscle

Batchelor, A., Kay, S., and Evans, D.: A simple and effective method of monitoring free muscle transfers: a preliminary report, Br. J. Plast. Surg. **35**:343, 1982.

Buncke, H.J.: Hand reconstruction by microvascular island flap transplantation, J. Bone Joint Surg. **57-A**:729, 1975.

Caffee, H., and Asokan, R.: Tensor fascia lata myocutaneous free flaps, Plast. Reconstr. Surg. **68**:195, 1981.

Gordon, L., and Buncke, H.J.: Heterotopic free skeletal muscle autotransplantation with utilization of a long nerve graft and microsurgical techniques: a study in the primate, J. Hand Surg. **4**:103, 1979.

Gordon, L., Buncke, H.J., and Alpert, B.S.: Free latissimus dorsi muscle flap with split-thickness skin graft cover: a report of 16 cases, Plast. Reconstr. Surg. **70**:173, 1982.

Hagan, K.F., Buncke, H.J., and Gonzalez, R.: Free latissimus dorsi muscle flap coverage of an electrical burn of the lower extremity, Plast. Reconstr. Surg. **69**:125, 1982.

Harii, K., Ohmori, K., and Sekiguchi, J.: The free musculocutaneous flap, Plast. Reconstr. Surg. **57**:295, 1976.

Harii, K., Ohmori, K., and Torii, S.: Free gracilis muscle transplantation, with microneurovascular anastomoses for the treatment of facial paralysis, Plast. Reconstr. Surg. **57:**133, 1976.

Hill, H.L., Nahai, F., and Vasconez, L.O.: The tensor fascia lata myocutaneous free flap, Plast. Reconstr. Surg. **61:**517, 1978.

Ikuta, Y.: Skeletal muscle transplantation in the severely injured upper extremity. In Serafin, D., and Buncke, H.J., Jr. (editors): Microsurgical composite tissue transplantation, St. Louis, 1979, The C.V. Mosby Co.

Ikuta, Y., Hatano, E., and Yoshioka, K.: Free muscle graft: clinical and experimental studies. In Buncke, H.J., and Furnas, D.W. (editors): Symposium on clinical frontiers in reconstructive microsurgery, St. Louis, 1984, The C.V. Mosby Co.

Ikuta, K., Kubo, T., and Tsuge, K.: Free muscle transplantation by microsurgical technique to treat severe Volkmann's contracture, Plast. Reconstr. Surg. **58:**407, 1976.

Ikuta, Y., Yoshioka, K. and Tsuge, K. Free muscle transfer. Aust. N.Z. J. Surg. **50:**401, 1980.

Kubo, T., Ikuta, Y., and Tsuge, K.: Free muscle transplantation in dogs by microneurovascular anastomoses, Plast. Reconstr. Surg. **57:**495, 1976.

Manktelow, R.T.: Free muscle flaps. In Green, D.P. (editor): Operative hand surgery, New York, 1982, Churchill Livingstone.

Manktelow, R.T.: Muscle transplantation: making it function. In Buncke, H.J., and Furnas, D.W. (editors): Symposium on clinical frontiers in reconstructive microsurgery, St. Louis, 1984, The C.V. Mosby Co.

Manktelow, R.T., and McKee, N.H.: Free muscle transplantation to provide active finger flexion, J. Hand Surg. **3:**416, 1978.

Manktelow, R.T., McKee, N.H., and Vettese, T.: An anatomical study of the pectoralis major muscle as related to functioning free muscle transplantation, Plast. Reconstr. Surg. **65:**610, 1980.

Manktelow, R.T., Zuker, R.M., and McKee, N.H.: Functioning free muscle transplantation, J. Hand Surg. **9-A:**32, 1984.

Mathes, S.J., and Nahai, F.: Classification of the vascular anatomy of muscles: experimental and clinical correlation, Plast. Reconstr. Surg. **67:**177, 1981.

Nahai, F., Hill, H.L., and Hexter, T.R.: Experiences with the tensor fascia lata flap, Plast. Reconstr. Surg. **63:**788, 1979.

Pennington, D., and Pelly, A.: The rectus abdominis myocutaneous free flap, Br. J. Plast. Surg. **33:**277, 1980.

Schenck, R.R.: Rectus femoris muscle and composite skin transplantation by microneurovascular anastomoses for avulsion of forearm muscles: a case report, J. Hand Surg. **3:**60, 1978.

Stern, P.J., Neale, H.W., Gregory, R.O., and McDonough, J.J.: Functional reconstruction of an extremity by free tissue transfer of the latissimus dorsi, J. Bone Joint Surg. **65-A:**729, 1983.

Tamai, S., et al.: Free muscle transplants in dogs with microsurgical neurovascular anastomoses, Plast. Reconstr. Surg. **46:**219, 1970.

Terzis, J.K., Sweet, R.C., and Dykes, R.W., and Williams, A.B.: Recovery of function in free muscle transplants using microneurovascular anastomoses, J. Hand Surg. **3:**37, 1978.

Thompson, N.: Autogenous free grafts of skeletal muscle, Plast. Reconstr. Surg. **48:**11, 1971.

Thompson, N.: Investigation of autogenous skeletal muscle free grafts in the dog, Transplantation **12:**353, 1971.

Watson, J.S., Craig, P., and Orton, C.I.: The free latissimus dorsi myocutaneous flap, Plast. Reconstr. Surg. **64:**299, 1979.

Free flaps—bone

Adelaar, R.S., Soucacos, P.N., and Urbaniak, J.R.: Autologous cortical grafts with microsurgical anastomosis of periosteal vessels, Surg. Forum. **25:**487, 1974.

Andersen, K.S.: Operative treatment of congenital pseudarthrosis of the tibia, Acta Orthop. Scand. **45:**935, 1974.

Berggren, A., Weiland, A.J., Ostrup, L.T., and Dorfman, H.: Microvascular free bone transfer with revascularization of the medullary and periosteal circulation and of the periosteal circulation alone: a comparative experimental study, J. Bone Joint Surg. **64-A:**73, 1982.

Brookes, M.: The vascularization of long bones in the human foetus. J. Anat. **92:**261, 1958.

Campbell, W.C.: Transference of the fibula as an adjunct free bone graft in tibial deficiency, J. Bone Joint Surg. **7:**625, 1919.

Carrel, A.: Results of the transplantation of blood vessels, organs and limbs, JAMA **51:**1662, 1908.

Clark, K.: A case of replacement of the upper end of the humerus by a fibular graft: reviewed after 29 years, J. Bone Joint Surg. **41-B:**365, 1959.

Daniel, R.K., and Weiland, A.J.: Free tissue transfer for upper extremity reconstruction, J. Hand Surg. **7:**66, 1982.

Doi, K., Tominaga, S., and Shibata, T.: Bone grafts with microvascular anastomoses of vascular pedicles: an experimental study in dogs, J. Bone Joint Surg. **59-A:**809, 1977.

Dos Santos, L.F.: The scapular flap: a new microsurgical free flap, Bol. Chir. Plast. **70:**133, 1980.

Finseth, F., May, J.W., and Smith, R.J. Composite groin flap with iliac-bone flap for primary thumb reconstruction: case report, J. Bone Joint Surg. **58-A:**130, 1976.

Fogdestam, K., Hamilton, R., and Markhede, G.: Microvascular osteocutaneous groin flap in the treatment of an ununited tibial fracture with chronic osteitis: a case report, Acta Orthop. Scand. **51:**175, 1980.

Gilbert, A., and Toot, L.: The free scapular flap, Plast. Reconstr. Surg. **69:**601, 1982.

Hagan, K.F., and Buncke, H.J.: Treatment of congenital pseudarthrosis of the tibia with free vascularized bone graft, Clin. Orthop. **166:**34, 1982.

Herndon, C.H., and Chase, S.W.: Experimental studies in the transplantation of whole joints, J. Bone Joint Surg. **34-A:**564, 1952.

Johnson, R.W., Jr.: A physiological study of the blood supply of the diaphysis, J. Bone Joint Surg. **9:**153, 1927.

Kelly, P.J.: Anatomy, physiology and pathology of the blood supply of bones, J. Bone Joint Surg. **50-A:**766, 1968.

Lexer, E.: Substitution of whole or half joints from freshly amputated extremities by free plastic operation, Surg. Gynecol. Obstet. **6:**601, 1908.

McCullough, D.W., and Fredrickson, J.M.: Neovascularized rib grafts to reconstruct mandibular defects, Can. J. Otolaryngol. **2:**96, 1973.

Medgyesi, S.: Observations on pedicle bone grafts in goats: vascular connections between soft tissues and bones, Scand. J. Plast. Reconstr. Surg. **7:**110, 1973.

Miller, R.C., and Phalen, G.S.: The repair of defects of the radius with fibular bone grafts, J. Bone Joint Surg. **29:**629, 1947.

Moore, J.R., Phillips, T.W., Weiland, A.J., and Randolph, M.A.: Allogenic transplants of bone revascularized by microvascular anastomoses: a preliminary study, J. Orthop. Res. **1:**352, 1984.

O'Brien, B.M.: Microvascular free bone and joint transfer. In Microvascular reconstructive surgery, Edinburgh, 1977, Churchill Livingstone.

O'Brien, B.M., Morrison, W.A., and Dooley, B.J.: Microvascular osteocutaneous transfer using the groin flap and iliac crest and the dorsalis pedis flap and second toe, Br. J. Plast. Surg. **32:**188, 1979.

Ostrup, L.T.: The free, living bone graft: an experimental study, Linkoping University Medical Dissertations, Linkoping, Sweden, 1975.

Ostrup, L.T., and Fredickson, J.M.: Distant transfer of a free, living bone graft by microvascular anastomoses: an experimental study, Plast. Reconstr. Surg. **54:**274, 1974.

Reichel, S.M.: Vascular system of the long bones of the rat, Surgery **22:**146, 1947.

Restrepo, J., Katz, D., and Gilbert, A.: Arterial vascularization of the proximal epiphysis and the diaphysis of the fibula, Int. J. Microsurg. **2:**48, 1980.

Serafin, D., Villarreal-Rios, A., and Georgiade, N.A.: A rib-containing free flap to reconstruct mandibular defects, Br. J. Plast. Surg. **30:**263, 1977.

Snyder, C.C., Bateman, J.M., Davis, C.W., and Warden, G.D.: Mandibulo-facial restoration with live osteocutaneous flaps, Plast. Reconstr. Surg. **45:**14, 1970.

Solonen, K.A.: Free vascularized bone graft in the treatment of pseudarthrosis, Int. Orthop. (SICOT) **6:**9, 1982.

Strauch, B., Bloomburg, A.E., and Lewin, M.L.: An experimental approach to mandibular replacement: island vascular composite rib grafts, Br. J. Plast. Surg. **24:**334, 1971.

Taylor, G.: Current status of free vascularized bone grafts, Clin. Plast. Surg. **10:**185, 1978.

Taylor, G.I.: Microvascular free bone transfer: a clinical technique, Orthop. Clin. North Am. **8:**425, 1977.

Taylor, G.I., Miller, G.D.H., and Ham, F.J.: The free vascularized bone graft: a clinical extension of microvascular techniques, Plast. Reconstr. Surg. **55**:533, 1975.

Taylor, G.I., Townsend, P., and Corlett, R.: Superiority of the deep circumflex iliac vessels as the supply for free groin flaps: clinical work, Plast. Reconstr. Surg. **64**:745, 1979.

Trias, A., and Fery, A.: Cortical circulation of long bones, J. Bone Joint Surg. **61-A**:1052, 1979.

Trueta, J., and Caladias, A.X.: A study of the blood supply of the long bones, Surg. Gynecol. Obstet. **118**:485, 1964.

Tsai, Tsu-Min, Jupiter, J.B., Kutz, J.E., and Kleinert, H.E.: Vascularized autogenous whole joint transfer in the hand: a clinical study, J. Hand Surg. **7**:335, 1982.

Urbaniak, J.R., Hayes, M.G., and Bright, D.S.: Management of bone in digital replantation: free vascularized and composite bone grafts, Clin. Orthop. **133**:184, 1978.

Watari, S., et al.: Vascular pedicle fibular transplantation as treatment for bone tumor, Clin. Orthop. **133**:158, 1978.

Weiland, A.J.: Elective microsurgery for orthopaedic reconstruction, part III. Vascularized bone transfers. In American Academy of Orthopaedic Surgeons: Instructional course lectures, vol. 33, St. Louis, 1984, The C.V. Mosby Co.

Weiland, A.J.: Current concepts review: vascularized free bone transplants, J. Bone Joint Surg. **63-A**:166, 1981.

Weiland, A.J., and Daniel, R.K.: Vascularized bone grafts. In Green, D.P. (editor): Operative hand surgery, New York, 1982, Churchill Livingstone.

Weiland, A.J., and Daniel, R.K.: Microvascular anastomoses for bone grafts in the treatment of massive defects in bone, J. Bone Joint Surg. **61-A**:98, 1979.

Weiland, A.J., Daniel, R.K., and Riley, L.H., Jr.: Application of the free vascularized bone graft in the treatment of malignant or aggressive bone tumors, Johns Hopkins Med. J. **140**:85, 1977.

Weiland, A.J., Kleinert, H.E., Kutz, J.E., and Daniel, R.K.: Free vascularized bone grafts in surgery of the upper extremity, J. Hand Surg. **4**:129, 1979.

Wood, M.B., Cooney, W.P., III, and Irons, G.B.: Posttraumatic lower extremity reconstruction by vascularized bone graft transfer, Orthopedics **7**:255, 1984.

Wray, R.C., Mathes, S.M., Young, V.L., and Weeks, P.M.: Free vascularized whole-joint transplants with ununited epiphyses, Plast. Reconstr. Surg. **67**:591, 1981.

Weiland, A.J., Kleinert, H.E., Kutz, J.E., and Daniel, R.K.: Vascularized bone grafts in the upper extremity. In Serafin, D., and Buncke, H.J., Jr. (editors): Microsurgical composite tissue transplantation, St. Louis, 1979, The C.V. Mosby Co.

Zhong-Wei, C., Zhong-Jia, Y., and Yean, W.: A new method of treatment of congenital pseudarthrosis using free vascularized fibular grafts, Ann. Acad. Med. Singapore, **8**:465, 1979.

Free flaps—scapular and periscapular

Biemer, E., and Stock, W. Total thumb reconstruction: A one-stage reconstruction using an osteocutaneous forearm flap, Br. J. Plast. Surg. **36**:52, 1983.

Daniel, R.K., Terzis, J., and Midgley, R.D.: Restoration to an anesthetic hand by a free neurovascular flap from the foot, Plast. Reconstr. Surg. **57**:275, 1976.

Krizek, T.J., Tassboro, T., Desperez, Q.O., and Kiohn, C.L.: Experimental transplantation of composite grafts by microvascular anastomosis, Plast. Reconstr. Surg. **36**:358, 1956.

Nassif, T.M., Vidal, L., Bovet, J.L., and Baudet, J.: The parascapular flap: a new cutaneous microsurgical free flap, Plast. Reconstr. Surg. **69**:591, 1982.

Nunley, J.A.: Elective microsurgery for orthopaedic reconstruction, part I. Donor site selection for cutaneous and myocutaneous free flaps. In American Academy of Orthopaedic Surgeons: Instructional course lectures, vol. 33, St. Louis, 1984, The C.V. Mosby Co.

Toe transplantation

Ben Menachem, Y., and Butler, J.E.: Arteriography of the foot in congenital deformities, J. Bone Joint Surg. **56-A**:1625, 1974.

Biemer, E.: Reconstruction of the hand with two-toe block transfers and neurovascular flaps from the foot. In Buncke, H.J., and Furnas, D.W. (editors): Symposium on clinical frontiers in reconstructive microsurgery, vol. 24, St. Louis, 1984, The C.V. Mosby Co.

Buncke, H.J.: Toe digital transfer, Clin. Plast. Surg. **3**:49, 1976.

Buncke, H.J., Jr., Buncke, C.M., and Schulz, W.P.: Immediate Nicoladoni procedure in the rhesus monkey, or hallux-to-hand transplantation, utilizing microminiature vascular anastomoses, Br. J. Plast. Surg. **19**:332, 1966.

Buncke, H.J., Jr., et al.: Thumb replacement: great toe transplantation by microvascular anastomosis, Br. J. Plast. Surg. **26**:194, 1973.

Buncke, H.J., et al.: Thumb replacement: great toe transplantation by microvascular anastomosis, Br. J. Plast. Surg. **26**:194, 1973.

Buncke, H.J., and Rose, E.H. Free toe-to-fingertip neurovascular flaps, Plast. Reconstr. Surg. **63**:609, 1979.

Buncke, H.J., and Shah, K.: Toe-digital transfers. In Serafin, D., and Buncke, H.J. (editors): Microsurgical composite tissue transplantation, St. Louis, 1978, The C.V. Mosby Co.

Clarkson, P.: Reconstruction of hand digits by toe transfers, J. Bone Joint Surg. **37-A**:270, 1955.

Clarkson, P.: On making thumbs, Plast. Reconstr. Surg. **29**:325, 1962.

Cobbett, J.R.: Free digital transfer: report of a case of transfer of a great toe to replace an amputated thumb, J. Bone Joint Surg. **51-B**:677, 1969.

Cobbett, J.R.: Free digital transfer, J. Bone Joint Surg. **51-B**:13, 1975.

Doi, K., et al.: New procedure on making a thumb: one-stage reconstruction with free neurovascular flap and iliac bone graft, J. Hand Surg. **6**:346, 1981.

Doi, K., Tominaga, S., Hono, H., Ogoshi, E., and Nakamura, S.: Reconstruction of an amputated thumb in one stage: case report—free neurovascular-flap transfer with iliac-bone graft, J. Bone Joint Surg. **61-A**:1254, 1979.

Doi, K., Kuwata, N., and Kawai, S.: Reconstruction of the thumb with a free wrap-around flap from the big toe and an iliac-bone graft, J. Bone Joint Surg. **67-A**:439, 1985.

Edwards, E.A.: Anatomy of the small arteries of the foot and toes, Acta Anat. **41**:81, 1960.

Edwards, E.A.: Organization of the small arteries of the hand and digits, Am. J. Surg. **99**:837, 1960.

Egloff, D.V.: Surgery of the hand: free tissue transfers by nerve and vascular microanastomoses, Geneva, 1984, Editions Medecine et Hygiene.

Foucher, G., Merle, M., Maneaud, M., and Michon, J.: Microsurgical free partial toe transfer in hand reconstruction: a report of 12 cases, Plast. Reconstr. Surg. **65**:616, 1980.

Furnas, D.W., and Achauer, B.M.: Microsurgical transfer of the great toe to the radius to provide prehension after partial avulsion of the hand, J. Hand Surg. **8**:453, 1983.

Gilbert, A.: Reconstruction of congenital hand defects with microvascular toe transfer, Hand Clin. **1**:351, 1985.

Gordon, L., Leitner, D.W., Buncke, H.J., and Alpert, B.S.: Hand reconstruction for multiple amputations by double microsurgical toe transplantation, J. Hand Surg. **10-A**:218, 1985.

Gordon, L., Rose, J., Alpert, B.S., and Buncke, H.J.: Free microvascular transfer of second toe ray and serratus anterior muscle for management of thumb loss at the carpometacarpal joint level, J. Hand Surg. **9-A**:642, 1984.

Gosset, J.: Reconstruction of an amputated thumb. In Reid, D.A.C, and Gosset, J. (editors): Mutilating injuries of the hand, Edinburgh, 1979, Churchill Livingstone.

Hmura, M.S., and Buncke, H.J.: Biomechanical analysis of toe-to-thumb transplants: a look at both sides, J. Hand Surg. **1**:81, 1976.

Holle, J., Freilinger, G., Mandl, H., and Frey, M.: Grip reconstruction by double-toe transplantation in cases of a fingerless hand and a handless arm, Plast. Reconstr. Surg. **69**:962, 1982.

Kartchinov, K.D.: Reconstruction and sensitization of the amputated thumb, J. Hand Surg. **9A**:478, 1984.

Leung, P.C.: Thumb reconstruction using second-toe transfer, Hand Clin. **1**:285, 1985.

Leung, P.C., and Kok, L.C.: Transplantation of the second toe to the hand: a preliminary report of sixteen cases, J. Bone Joint Surg. **62-A**:990, 1980.

Leung, P.C., and Wong, W.L.: The vessels of the first metatarsal web space: an operative and radiographic study, J. Bone Joint Surg. **65-A**:235, 1983.

Lichtman, D.M., Ahbel, D.E., Murphy, R.B., and Buncke, H.J.: Microvascular double toe transfer for opposable digits: case report and rationale for treatment, J. Hand Surg. **7**:279, 1982.

Lister, Graham: The choice of procedure following thumb amputation, Clin. Orthop. **195**:45, 1985.

Lister, G.D, Kalisman, M., and Tsai, T.M. Reconstruction of the hand with free microneurovascular toe-to-hand transfer: experience with 54 toe transfers, Plast. Reconstr. Surg. **71**:372, 1983.

May, J.W., and Bartlett, S.P.: Great toe-to-hand free tissue transfer for thumb reconstruction, Hand Clin. **1**:271, 1985.

May, J.W., Jr., and Daniel, R.K.: Great toe to hand free tissue transfer, Clin. Orthop. **133**:140, 1978.

May, J.W., Jr., Smith, R.J., and Peimer, C.A.: Toe-to-hand free tissue transfer for thumb construction with multiple digit aplasia, Plast. Reconstr. Surg. **67**:205, 1981.

Morrison, W.A., O'Brien, B.M., and Hamilton, R.B.: Neurovascular free foot flaps in reconstruction of the mutilated hand, Clin. Plast. Surg. **5**:265, 1978.

Morrison, W.A., O'Brien, B.M., and MacLeod, A.M.: Thumb reconstruction with a free neurovascular wrap-around flap from the big toe, J. Hand Surg. **5**:575, 1980.

Murray, J.F.: The missing thumb. In Littler, J.W., Cramer, L.M., and Smith, J.W. (editors): Symposium on reconstructive hand surgery, St. Louis, 1974. The C.V. Mosby Co.

Nicoladoni, C.: Daumenplastik, Wein. Klin. Wochenschr. **10**:663, 1897.

Nicoladoni, C.: Daumenplastik und organischer Ersatz der Fingerspitze, Arch. Klin. Chir. **61**:606, 1900.

Nunley, J.A., Goldner, R.D., and Urbaniak, J.R.: Thumb reconstruction by the wrap-around method, Clin. Orthop. **195**:97, 1985.

O'Brien, B.M., et al.: Microvascular great toe transfer for congenital absence of the thumb, Hand **10**:113, 1978.

O'Brien, B., Brennen, M.D., and McLeod, A.M.: Simultaneous double toe transfer for severely disabled hands, Hand **10**:232, 1978.

O'Brien, B.M., Brennen, M.D., and MacLeod, A.M.: Microvascular free toe transfer, Clin. Plast. Surg. **5**:223, 1978.

O'Brien, B., MacLeod, A.M., Sykes, P.J., Browning, F.S.C., and Threlfall, G.N.: Microvascular second toe transfer for digital reconstruction, J. Hand Surg. **3**:123, 1978.

O'Brien, B.M., and Shanmugan, N.: Experimental transfer of composite free flaps with microvascular anastomoses, Aust. N.Z. J. Surg. **43**:285, 1973.

Ohtsuka, H., Torigai, K., and Shioya, N.: Two toe-to-finger transplants in one hand, Plast. Reconstr. Surg. **60**:561, 1977.

Poppen, N.K., Norris, T.R., and Buncke, H.J., Jr.: Evaluation of sensibility and function with microsurgical free tissue transfer of the great toe to the hand for thumb reconstruction, J. Hand Surg. **8**(Part 1):516, 1983.

Rose, E.H., and Buncke, H.J.: Free transfer of a large sensory flap from the first web space and dorsum of the foot including the second toe for reconstruction of a mutilated hand, J. Hand Surg. **6**:196, 1981.

Rose, E.H., and Hendel, P.: Primary toe-to-thumb transfer in the acutely avulsed thumb, Plast. Reconstr. Surg. **67**:214, 1981.

Smith, R.J., and Dworecka, F. Treatment of the one-digit hand, J. Bone Joint Surg. **55-A**:113, 1973.

Tomita, Y., Kurota, K., and Okubo, K.: Wrap-around flap with tip of distal phalanx. Paper presented at the Seventh Symposium of the International Society of Reconstructive Microsurgeons, New York, June 20, 1983.

Wilson, C.S., Buncke, H.J., Alpert, B.S., and Gordon, L.: Composite metacarpophalangeal joint reconstruction in great toe-to-hand free tissue transfers, J. Hand Surg. **9-A**:645, 1984.

Yang, D., and Gu, Y.: Thumb reconstruction utilizing second toe transplantation by microvascular anastomoses: a report of 78 cases, Chin. Med. J. **92**:295, 1979.

Yoshimura, M.: Toe-to-hand transfer, Plast. Reconstr. Surg. **66**:74, 1980.

Yoshimura, M., et al.: Toe-to-hand transfer: experience with thirty-eight digits, Aust. N.Z. Surg. **50**:248, 1980.

Zhong-Wei, C.: Reconstruction by autogenous toe transplantations for total hand amputation, Orthop. Clin. North Am. **12**:835, 1981.

Zhong-Wei, C., Meyer, V.E., and Beasley, R.W.: The versatile second toe microvascular transfer, Orthop. Clin. North Am. **12**:827, 1981.

Vascularized free flaps and nerve grafts

Brown, K.L., et al.: Epiphyseal growth after free fibular transfer with and without microvascular anastomosis: experimental study in the dog, J. Bone Joint Surg. **65-B**:493, 1983.

Herndon, C.H., and Chase, S.W.: Experimental studies in the transplantation of whole joints, J. Bone Joint Surg. **34-A**:564, 1952.

Lexer, E.: Substitution of whole or half joints from freshly-amputated extremities by free plastic operation, Surg. Gynecol. Obstet. **6**:601, 1908.

Mathes, S.J., Buchanan, R. and Weeks, P.M.: Microvascular joint transplantation with epiphyseal growth, J. Hand Surg. **5**:586, 1980.

Rose, E.: Reconstruction of central metacarpal ray defects of the hand with a free vascularized double metatarsal-metatarsophalangeal joint transfer, J. Hand Surg. **9-A**:28, 1984.

Taylor, G.I.: Nerve grafting with simultaneous microvascular reconstruction, Clin. Orthop. **133**:56, 1978.

Tsai, Tsu-Min, Jupiter, J.B., Kutz, J.E., and Kleinert, H.E.: Vascularized autogenous whole joint transfer in the hand: a clinical study, J. Hand Surg. **7**:335, 1982.

Weiland, A.J.: Elective microsurgery for orthopaedic reconstruction, part III. Vascularized bone tranfers. In American Academy of Orthopaedic Surgeons: Instructional course lectures, vol. 33, St. Louis, 1984, The C.V. Mosby Co.

Weiland, A.J.: Current concepts review: vascularized free bone transplants, J. Bone Joint Surg. **63-A**:166, 1981.

Weiland, A.J., and Daniel, R.K.: Vascularized bone grafts. In Green, D.P. (editor): Operative hand surgery, New York, 1982, Churchill Livingstone.

Weiland, A.J., and Daniel R.K.: Microvascular anastomoses for bone grafts in the treatment of massive defects in bone, J. Bone Joint Surg. **61-A**:98, 1979.

Weiland, A.J., Daniel, R.K., and Riley, L.H., Jr.: Application of the free vascularized bone graft in the treatment of malignant or aggressive bone tumors, Johns Hopkins Med. J. **140**:85, 1977.

Weiland, A.J., Kleinert, H.E., Kutz, J.E., and Daniel, R.K.: Free vascularized bone grafts in surgery of the upper extremity, J. Hand Surg. **4**:129, 1979.

Weiland, A.J., Kleinert, H.E., Kutz, J.E., and Daniel R.K.: Vascularized bone grafts in the upper extremity. In Serafin, D., and Buncke, H.J., Jr. (editors): Microsurgical composite tissue transplantation, St. Louis, 1979, The C.V. Mosby Co.

Wray, R.C., Mathes, S.M., Young, V.L., and Weeks, P.M.: Free vascularized whole-joint transplants with ununited epiphyses, Plast. Reconstr. Surg. **67**:591, 1981.

Wilson, C.S., Buncke, H.J., Alpert, B.S., and Gordon, L.: Composite metacarpophalangeal joint reconstruction in great toe-to-hand free tissue transfers, J. Hand Surg. **9-A**:645, 1984.